Lecture Notes in Computer Science 12734

More information about this subseries at http://www.springer.com/series/7407

Didier Buchs · Josep Carmona (Eds.)

Application and Theory of Petri Nets and Concurrency

42nd International Conference, PETRI NETS 2021
Virtual Event, June 23–25, 2021
Proceedings

 Springer

Editors
Didier Buchs
Université de Genève
Carouge, Switzerland

Josep Carmona 🆔
Universitat Politècnica de Catalunya
Barcelona, Spain

ISSN 0302-9743   ISSN 1611-3349   (electronic)
Lecture Notes in Computer Science
ISBN 978-3-030-76982-6   ISBN 978-3-030-76983-3   (eBook)
https://doi.org/10.1007/978-3-030-76983-3

LNCS Sublibrary: SL1 – Theoretical Computer Science and General Issues

This Springer imprint is published by the registered company Springer Nature Switzerland AG
The registered company address is: Gewerbestrasse 11, 6330 Cham, Switzerland

Preface

This volume constitutes the proceedings of the 42st International Conference on Application and Theory of Petri Nets and Concurrency (Petri Nets 2021). This series of conferences serves as an annual meeting place to discuss progress in the field of Petri nets and related models of concurrency. These conferences provide a forum for researchers to present and discuss both applications and theoretical developments in this area. Novel tools and substantial enhancements to existing tools can also be presented.

Petri Nets 2021 included a section devoted to Application of Concurrency to System Design, which was in the past a separate event. The final selection of this track was made by Jörg Desel and Alex Yakovlev.

The event was organized by the LoVe (Logics and Verification) team of the computer science laboratory LIPN (Laboratoire d'Informatique de Paris Nord) at the University Sorbonne Paris Nord and CNRS, along with members of the Paris region MeFoSyLoMa (Méthodes Formelles pour les Systèmes Logiciels et Matériels) group. The conference was supposed to take place in the chosen area of Campus Condorcet, the new international research campus for humanities and social sciences in Paris, France.

Unfortunately, because of the coronavirus epidemic, the event was organized as a virtual conference. We would like to express our deepest thanks to the Organizing Committee chaired by Laure Petrucci and Etienne André for the time and effort invested in the organization of this event.

This year, 39 papers were submitted to Petri Nets 2020 by authors from 16 different countries. Each paper was reviewed by at least three reviewers. The discussion phase and final selection process by the Program Committee (PC) were supported by the EasyChair conference system. From regular papers and tool papers, the PC selected 22 papers for presentation: 20 regular papers and 2 tool papers. After the conference, some of these authors were invited to submit an extended version of their contribution for consideration in a special issue of a journal.

We thank the PC members and other reviewers for their careful and timely evaluation of the submissions and the fruitful constructive discussions that resulted in the final selection of papers. The Springer LNCS team (notably Anna Kramer) provided excellent and welcome support in the preparation of this volume.

Due to the virtual format of the event, the keynote presentations of the 2020 edition were postponed until this 2021 edition. They were given by Serge Abiteboul, Luca Bernardinello and Jérôme Leroux. Alongside Petri Nets 2021, the following workshops and events took place: the 11th edition of the Model Checking Contest (MCC 2021), Algorithms and Theories for the Analysis of Event Data (ATAED 2021), and Petri Nets and Software Engineering (PNSE 2021).

We hope you enjoy reading the contributions in this LNCS volume.

June 2021

Didier Buchs
Josep Carmona

Organization

Program Committee

Elvio Gilberto Amparore	University of Turin, Italy
Abel Armas Cervantes	The University of Melbourne, Australia
Paolo Baldan	Dipartimento di Matematica, Università di Padova, Italy
Benoit Barbot	LACL, Université Paris-Est Créteil, France
Beatrice Berard	LIP6, Sorbonne Université and CNRS, France
Didier Buchs	University of Geneva, Switzerland
Josep Carmona	Universitat Politècnica de Catalunya, Spain
Thomas Chatain	LSV, ENS Paris-Saclay, France
Jörg Desel	Fernuniversität in Hagen, Germany
Raymond Devillers	Université de Bruxelles, Belgium
Susanna Donatelli	Dipartimento di Informatica, Università di Torino, Italy
Javier Esparza	Technical University of Munich, Germany
David Frutos Escrig	Universidad Complutense de Madrid, Spain
Stefan Haar	Inria/LSV, ENS Paris-Saclay, France
Xudong He	Florida International University, USA
Loic Helouet	Inria, France
Marieke Huisman	University of Twente, Netherlands
Ryszard Janicki	McMaster University, Canada
Anna Kalenkova	The University of Melbourne, Australia
Jörg Keller	FernUniversität in Hagen, Germany
Ekkart Kindler	Technical University of Denmark, Denmark
Michael Köhler-Bußmeier	University of Applied Science Hamburg, Germany
Irina Lomazova	National Research University Higher School of Economics, Russia
Robert Lorenz	University of Augsburg, Germany
Roland Meyer	TU Braunschweig, Germany
Lukasz Mikulski	Nicolaus Copernicus University, Poland
Andrew Miner	Iowa State University, USA
Andrey Mokhov	Jane Street, UK
Marco Montali	KRDB Research Centre, Free University of Bozen-Bolzano, Italy
Dumitru Potop Butucaru	Inria, France
Pierre-Alain Reynier	Aix-Marseille Université, France
Arnaud Sangnier	IRIF, Université Paris Diderot CNRS, France
Natalia Sidorova	Department of Mathematics and Computer Science, Technische Universiteit Eindhoven, Netherlands

Jaco van de Pol   Aarhus University, Denmark
Boudewijn Van Dongen   Eindhoven University of Technology, Netherlands
Alex Yakovlev   Newcastle University, UK
Wlodek Zuberek   Memorial University, Canada

Additional Reviewers

Badouel, Eric
Balasubramanian, A. R.
Bashkin, Vladimir
Blondin, Michael
Boltenhagen, Mathilde
Busatto-Gaston, Damien
Dal Zilio, Silvano
Desel, Jörg
Furbach, Florian
Gogolinska, Anna
Keskin, Eren
Kurpiewski, Damian

Litzinger, Sebastian
Metzger, Johannes
Oualhadj, Youssouf
Petrak, Lisa
Raskin, Mikhail
Rivkin, Andrey
Rosa-Velardo, Fernando
Schalk, Patrizia
van der Wall, Sören
Welzel, Christoph
Wolff, Sebastian
Zakharov, Vladimir

Contents

Keynotes

Topics in Region Theory and Synthesis Problems 3
   Luca Bernardinello

Flat Petri Nets (Invited Talk) 17
   Jérôme Leroux

Application of Concurrency to System Design

Cost and Quality in Crowdsourcing Workflows 33
   Loïc Hélouët, Zoltan Miklos, and Rituraj Singh

Timed Petri Nets with Reset for Pipelined Synchronous Circuit Design 55
   Rémi Parrot, Mikaël Briday, and Olivier H. Roux

A Turn-Based Approach for Qualitative Time Concurrent Games 76
   Serge Haddad, Didier Lime, and Olivier H. Roux

Games

Canonical Representations for Direct Generation of Strategies
in High-Level Petri Games 95
   Manuel Gieseking and Nick Würdemann

Automatic Synthesis of Transiently Correct Network Updates via
Petri Games 118
   Martin Didriksen, Peter G. Jensen, Jonathan F. Jønler,
   Andrei-Ioan Katona, Sangey D. L. Lama, Frederik B. Lottrup,
   Shahab Shajarat, and Jiří Srba

Verification

Computing Parameterized Invariants of Parameterized Petri Nets 141
   Javier Esparza, Mikhail Raskin, and Christoph Welzel

On the Combination of Polyhedral Abstraction and SMT-Based Model
Checking for Petri Nets 164
   Nicolas Amat, Bernard Berthomieu, and Silvano Dal Zilio

Skeleton Abstraction for Universal Temporal Properties 186
   Sophie Wallner and Karsten Wolf

Reduction Using Induced Subnets to Systematically Prove Properties
for Free-Choice Nets   208
   Wil M. P. van der Aalst

Model Checking of Synchronized Domain-Specific Multi-formalism
Models Using High-Level Petri Nets   230
   Michael Haustermann, David Mosteller, and Daniel Moldt

Synthesis and Mining

Edge, Event and State Removal: The Complexity of Some Basic
Techniques that Make Transition Systems Petri Net Implementable   253
   Ronny Tredup

Synthesis of (Choice-Free) Reset Nets   274
   Raymond Devillers

Synthesis of Petri Nets with Restricted Place-Environments: Classical
and Parameterized   292
   Ronny Tredup

Discovering Stochastic Process Models by Reduction and Abstraction   312
   Adam Burke, Sander J. J. Leemans, and Moe Thandar Wynn

Reachability and Partial Order

Efficient Algorithms for Three Reachability Problems in Safe Petri Nets   339
   Pierre Bouvier and Hubert Garavel

A Lazy Query Scheme for Reachability Analysis in Petri Nets   360
   Loïg Jezequel, Didier Lime, and Bastien Sérée

Abstraction-Based Incremental Inductive Coverability for Petri Nets   379
   Jiawen Kang, Yunjun Bai, and Li Jiao

Firing Partial Orders in a Petri Net   399
   Robin Bergenthum

Semantics

Deterministic Concurrent Systems   423
   Samy Abbes

Deciphering the Co-Car Anomaly of Circular Traffic Queues Using
Petri Nets   443
   Rüdiger Valk

Tools

Cortado—An Interactive Tool for Data-Driven Process Discovery
and Modeling   465
   Daniel Schuster, Sebastiaan J. van Zelst, and Wil M. P. van der Aalst

PROVED: A Tool for Graph Representation and Analysis of Uncertain
Event Data   476
   Marco Pegoraro, Merih Seran Uysal, and Wil M. P. van der Aalst

Author Index   487

Keynotes

Topics in Region Theory and Synthesis Problems

Luca Bernardinello[(✉)]

Dipartimento di Informatica, Sistemistica e Comunicazione, Università Degli Studi di Milano-Bicocca, Viale Sarca 336 U14, Milano, Italy
`luca.bernardinello@unimib.it`

Abstract. Regions, as introduced by Ehrenfeucht and Rozenberg more than thirty years ago, have been used as a fundamental tool in synthesis problems, where a Petri net of a specific type must be built from a specification given in terms of a transition system. Some topics emerged in the research on regions are discussed, and a few open problems are stated. In particular, the paper focuses on three areas: (1) the notion of 'type of nets' as a tool for unifying the theory of regions, and as a notion leading to new variants of Petri nets; (2) the algebraic aspects of region theory; (3) the proposal of a new type of regions, inspired by reaction systems, and the potential for studying problems of synthesis of reaction systems.

1 Introduction

In this paper, I discuss some topics in region theory. The choice of the topics is biased by my personal past and current interests, and aims at pointing out a few open problems, on the theoretical side, which I consider interesting and deserving attention.

Regions were introduced by Ehrenfeucht and Rozenberg in [8] and [9] as a tool within the theory of 2-structures, a general theory of a class of equivalence relations. A 2-structure can be defined as a graph with an equivalence relation on its edges. We can then take the equivalence classes on edges as *labels*, and label each edge accordingly.

In that context, regions were used in order to solve a representation problem: given a labelled 2-structure, find an isomorphic labelled *set* 2-structure, where nodes are subsets of a ground set.

A labelled 2-structure can be seen as a labelled transition system, and from now on, I will directly use labelled transition systems.

In the original definition, regions are subsets of the set of states of a transition system, satisfying a uniformity property with respect to labels which cross the border of that subset. This notion of region corresponds to the conditions (or places) of Elementary Net Systems; a region is in fact the extension of a (potential) place, namely the set of states in which the place is marked, or the corresponding condition is true. Regions of this type will be called *elementary*,

D. Buchs and J. Carmona (Eds.): PETRI NETS 2021, LNCS 12734, pp. 3–16, 2021.
https://doi.org/10.1007/978-3-030-76983-3_1

and are the main ingredient needed to solve the following synthesis problem: given a labelled transition system A, decide whether there exists an Elementary Net System Σ, such that the set of events coincides with the set of labels of A, and the marking graph of Σ is isomorphic to A.

Pushing this idea beyond the original context, new types of regions have been defined, corresponding to places in different kinds of Petri nets, starting from Place/Transition nets, where regions are functions from states to non-negative integers (these will be called *PT regions* in the following).

This led to solve the synthesis problem for several types of Petri nets. In 1996, Badouel and Darondeau gave a systematic presentation of the theory of regions (see [4]), introducing the notion of 'type of nets'. A type of nets is a labelled transition system, modelling the values that a single place can take on, and the possible transitions that it can go through as an effect of transition firings. So, for example, the type of nets corresponding to Elementary Net Systems has two states, corresponding to a place being marked or not (or to the corresponding condition being true or false); the type corresponding to Place/Transition nets is an infinite transition system with states corresponding to natural numbers.

In this general frame, if we define a type of nets τ, a τ-region of a transition system A can be defined as a morphism from A to τ.

This gives an elegant presentation of regions, and of synthesis problems, with a remarkable uniformity in the characterization of transition systems for which the synthesis problem can be solved. Further generalizations have led to define regions on languages and on other formalisms.

The main applications of the synthesis problem so far include process discovery, circuit design, and supervisory control. In all these fields, one starts with a specification relying on an interleaving semantics, with no explicit information on the dependence relations between actions or events; the computation of regions reveals, so to speak, potential concurrency.

Apart from their application in synthesis problems, regions are interesting objects in themselves. Different types of regions bear different algebraic structures. So far, this topic has not raised much interest, but two cases have been studied in some depth: elementary regions and PT regions.

The set of PT regions of a transition system forms a finitely generated group, as was observed in [5]. Elementary regions form instead an orthomodular poset, an algebraic structure, originally conceived to represent the "logical" structure of observables in quantum mechanics.

From these results, one can naturally define a new class of synthesis problems: given an abstract algebraic structure (for instance, a finitely generated group or an orthomodular poset) construct a transition system with an isomorphic regional structure (this would then lead to constructing a corresponding Petri net, of course). This comes with the related problem of characterizing the exact class of those algebraic structures which admit a solution.

The case of elementary regions, giving rise to orthomodular posets, opens the way to speculations towards logic, since those posets are a generalization of Boolean algebras. These ideas are discussed in Sect. 3.

The only original contribution of this paper, in Sect. 4 is the attempt to define a notion of region useful in the study of reaction systems as defined by Ehrenfeucht and Rozenberg in [10]. Reaction systems are intended to model biochemical reactions in living cells. A reaction is defined by three sets, interpreted as the set of reactants, needed for the reaction to take place; the set of inhibitors, which, if present, prevent the reaction; and the set of products of the reaction. As an abstract model of an evolving system, they are based on a few principles, quite different from the conceptual basis of Petri nets. In spite of this difference, I conjecture that a notion of region can be used in order to tackle synthesis problems for reaction systems.

In the following sections, I assume the reader knows the basics of Petri net theory; in particular, I give for granted the definitions of Elementary Net System and of Place/Transition net. Apart from this, I have tried to include all the formal definitions needed to understand my arguments. However, I have not been fully rigorous.

2 Regions and Synthesis of Net Systems

As recalled in the introduction, within region theory a type of nets is a labelled transition system.

Definition 1. *A labelled transition system is a structure $A = (Q, E, T)$, where Q is the set of states, E is the set of labels, and $T \subseteq Q \times E \times Q$ is the set of transitions.*

When (q, e, q') is a transition in T, I will also write $q \xrightarrow{e} q'$. If such a transition exists, e is said to be *enabled* in q.

If the set of states is finite, then A is said to be finite; otherwise, A is infinite.

Given a type of nets $\tau = (Q_\tau, E_\tau, T_\tau)$, and a finite transition system $A = (Q, E, T)$, the set of τ-regions of A is defined as the set of morphisms from A to τ. A morphism is a map on the set of states together with a map on the set of labels which preserve transitions, while respecting labels, in this sense:

Definition 2. *Let $A_i = (Q_1, E_i, T_i)$ be a transition system for $i = 1, 2$. A morphism from A_1 to A_2 is a pair of maps $r_Q : Q_1 \to Q_2$ and $r_E : E_1 \to E_2$ (in the following, the subscripts Q and E will be omitted; the context will always make clear which map is being used), such that, if $q \xrightarrow{e} q'$ in A_1, then $r(q) \xrightarrow{r(e)} r(q')$ in A_2.*

The definition of regions as morphisms into a type of nets allows for a unified treatment of synthesis problems. The key to the solution consists in the appropriate generalization of the so-called state separation and event-state separation problems, which had been first identified for elementary regions.

The state separation problem for a pair of distinct states q and q' of a transition system consists in finding a τ-region r which allows one to distinguish the two states: q and q' are separated by r if $r(q) \neq r(q')$.

The event-state separation problem for an event (a label) e and a state q from which e cannot occur consists in finding a region r which "explains" the fact that e cannot occur in q: a region r separates e from q if $r(e)$ is not enabled in $r(q)$.

For any type of nets τ, the corresponding synthesis problem for a labelled transition system $A = (Q, E, T)$ is solvable if, and only if, for each pair of distinct states (q, q') there is a region r which separates them, and for each state q and each label e not enabled at q, there is a region r separating e from q.

The separation problems are indeed the key to the synthesis problem: a set of regions is sufficient to construct the desired net if it solves all the separation problems.

Given the general solution to the synthesis problems, one can turn to face complexity issues of the procedure to compute a sufficient set of regions, or to decide that there is no solution. A first result on this subject concerned Place/Transition nets. In [1], it was proved that this problem is polynomial in the size of the given transition system. This was proved by exploiting the algebraic structure of PT regions, which allows to devise a procedure which combines algorithms from graph theory and algorithms for computing rational solutions of linear systems.

Lately, in [2], it was proved that the synthesis problem for Elementary Net Systems is instead NP-complete. The reason for the substantial difference in complexity seems to consist in the lack of a rich algebraic structure of elementary regions (but see the next section for a more detailed discussion of the algebraic aspects of elementary regions). As a matter of fact, a simple variant of Elementary Net Systems, in which a third kind of arc is added, so that an event can swap the value of a place (from false to true or viceversa), admits a polynomial algorithm for the corresponding synthesis problem, by exploiting the richest algebraic structure of the corresponding regions. Nets of this type have been called "flip-flop nets". For details, see [3].

More recently, Ronny Tredup has started the ambitious project of systematically determining the complexity of the synthesis problem for all classes of "Boolean nets", namely nets in which places can take on two values, like Elementary Net Systems and flip-flop nets. This undertaking is still ongoing, but a large number of results are already available. In most cases, the problem is NP-complete, with some exceptions. So far, the results are scattered in several papers, but the interested reader can look at [13], and the references therein.

Apart from Boolean nets, there is still room for research in devising more general new types of nets. A convincing argument in this direction can be found in [11], where some types of nets are introduced, and related open problems are stated. Constructing regions with rich algebraic properties is, I believe, another worthwhile direction for research.

3   Regions and Algebra

Besides their use in solving synthesis problems, regions can be interesting objects in themselves, and in relation with the full set of regions of a given transition system.

In general, the set of regions of a given type will bear some algebraic structure, depending on that type. To my knowledge, so far the only cases that have been studied extensively are the case of elementary regions, and the case of PT-regions. In this section, I will briefly recall the main results on elementary regions, and discuss a few directions of research.

3.1   Elementary Regions

Consider a system described by a transition system A. In general, we can define a *property* of A as a subset of its states, with the interpretation that the property holds exactly in those states, and does not hold in any other state. The subset can then be called the *extension* of the property. In principle, any subset of states corresponds to a property. In this view, a state of the system is characterized by the set of all properties that hold in it. We have then a duality between states and properties: (the extension of) a property is a set of states, and a state is a set of properties.

The set of all properties forms a Boolean algebra (here, I will only consider systems with a finite set of states), and the set-theoretical operations of union, intersection, and complement correspond to the logical connectives of disjunction, conjunction, and negation. We can in this way build a logic of properties.

In a distributed system, a global state is not a monolithic object, but rather a combination of local states of its components. Moreover, we should assume that the global state at a given instant is not actually observable (the notion of 'given instant' is actually critical, if we believe in the special theory of relativity).

An Elementary Net System is a model of a distributed system, where places model local states of components, and the global state (or *marking*) is a set of local states. Elementary regions are subsets of states. In an elementary transition system, we can then consider the set of all elementary regions of a given transition system as a partially ordered set. This set has a minimum (the empty set), and a maximum (the set of all states), and is closed by complementation and union of disjoint regions.

In [6], it was proved that the partially ordered set of elementary regions of a transition system is an orthomodular poset (or quantum logic). Orthomodular posets can be characterized as families of partially overlapping Boolean algebras. All the Boolean subalgebras of an orthomodular set share the minimum and the maximum element. Other elements can be shared by all or only some of the Boolean subalgebras.

For convenience, I recall here the definition of elementary region, and of quantum logic.

Definition 3. *A region of a transition system $A = (Q, E, T)$ is a subset r of Q such that $\forall e \in E,\ \forall (q_1, e, q_2), (q_3, e, q_4) \in T$:*

1. $(q_1 \in r$ and $q_2 \notin r) \Rightarrow (q_3 \in r$ and $q_4 \notin r)$; and
2. $(q_1 \notin r$ and $q_2 \in r) \Rightarrow (q_3 \notin r$ and $q_4 \in r)$.

Definition 4. *A* quantum logic *(or* logic*)* $L = (L, \leq, 0, 1, (\,.\,)')$ *is a partially ordered finite set* $(L, \leq)$ *endowed with a least and a greatest element, denoted by* 0 *and* 1*, respectively, and a unary operation* $(\,.\,)'$ *(called* orthocomplement*), such that the following conditions are satisfied:*
$\forall\, x, y \in L$

1. $x \leq y \Rightarrow y' \leq x'$;
2. $(x')' = x$;
3. $x \leq y' \Rightarrow x \vee y \in L$;
4. $x \leq y \Rightarrow y = x \vee (x' \wedge y)$.

This latter condition is sometimes referred to as orthomodular law.

Two elements x and y are said to be *orthogonal* (denoted by $x \perp y$) when $x \leq y'$. A simple quantum logic is shown in Fig. 1. A quantum logic arising from the set of regions of a transition system can always be seen as a family of partially overlapping Boolean algebras. Every Boolean subalgebra corresponds to a partition of the set of states of the transition system formed by regions. In the net system constructed by solving the synthesis problem, Boolean subalgebras correspond to state machine components of the net, or sequential components.

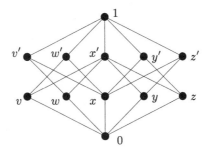

Fig. 1. A simple quantum logic.

The quantum logic shown in Fig. 1 has two maximal Boolean subalgebras; one is formed by the elements in $\{0, v, w, x\}$ and their orthocomplements; the other by the elements $\{x, y, z\}$ and their orthocomplements.

Let us call *regional* a quantum logic if it is isomorphic to the set of regions of an elementary transition system.

Now, given an abstract quantum logic, we can ask whether it is isomorphic to the partially ordered set of regions of some transition system, and if so, how to construct such a transition system. Solving this synthesis problem requires one to determine the set of states, the set of labels, and the set of transitions of the solution.

Quantum logics already provide a notion of state. If we interpret the elements of a quantum logic as propositions, the (partial) meet and join operations as logical connectives, and the orthocomplement as negation, then a state is a maximal set of consistent propositions, a sort of generalization of maximal filter in Boolean algebras.

Definition 5. *A* two-valued state *on a quantum logic L is a mapping $s : L \to \{0,1\}$ such that:*

1. $s(1) = 1$;
2. $\forall x, y \in L \quad x \perp y \Rightarrow s(x \vee y) = s(x) + s(y)$

$\mathcal{S}(L)$ will denote the set of two-valued states of L.

The white circles in Fig. 2 indicate elements of a two-valued state. Then, the ordered symmetric difference of a pair of distinct states is a natural candidate to label a state transition. The idea is that the label specifies which propositions cease to hold, and which start to hold as an effect of this state change. Note the close resemblance with the firing rule of Elementary Net Systems.

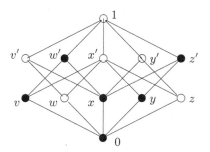

Fig. 2. A two-valued state.

$$E(L) = \{\langle s_1 \setminus s_2, s_2 \setminus s_1 \rangle \mid s_1, s_2 \in \mathcal{S}(L), s_1 \neq s_2\}. \tag{1}$$

The set of transitions is now naturally defined: for each pair of distinct states, s_1 and s_2, there is a transition from s_1 to s_2, labelled by the corresponding ordered symmetric difference.

In the following, $[s_1, s_2]$ will denote $\langle s_1 \setminus s_2, s_2 \setminus s_1 \rangle$.

$$T(L) = \{(s_1, [s_1, s_2], s_2) \mid s_1, s_2 \in \mathcal{S}(L), s_1 \neq s_2\} \tag{2}$$

Of course, the same label can have several occurrences. We can now define the transition system derived from a quantum logic L.

$$\mathcal{A}(\mathcal{L}) = (\mathcal{S}(\mathcal{L}), \mathcal{E}(\mathcal{L}), \mathcal{T}(\mathcal{L})). \tag{3}$$

The transition system $\mathcal{A}(\mathcal{L})$ includes a transition for each ordered pair of states; hence we call it *saturated* of transitions. $\mathcal{A}(\mathcal{L})$ is elementary, as shown in [6], which means that it is isomorphic to the marking graph of an Elementary Net System.

What happens if we now compute the set of regions of $\mathcal{A}(\mathcal{L})$? In general, L embeds into the regional quantum logic of $\mathcal{A}(\mathcal{L})$, as shown in [6].

Several open problems remain in this line of research. Probably the main issue is the characterization of regional quantum logics. We know that there are finite quantum logics that are not regional, and recently we did some progress towards a characterization by giving some necessary conditions for a logic to be regional. A related conjecture states that any regional logic is isomorphic to the logic of its transitions system (regional logics are stable). The interested reader can find details in [7].

On the other hand, the synthesis procedure described above constructs, as noted, a "saturated" transition system. This is not necessary: given a regional quantum logic, there are non saturated solutions. We are working on the problem of characterizing subsets of states and of labels which are sufficient to solve the synthesis problem.

4   Regions for Reaction Systems

Reaction systems were introduced in [10] as a formal tool to model biochemical reactions within living cells. The mathematical basis of reaction systems is simple. A *reaction* is specified by three sets of elements taken from a background set: (1) the set of *reactants*, which intuitively corresponds to the set of substances that must be present for the reaction to take place; (2) the set of *inhibitors*, namely those substances which, if present, prevent the reaction from taking place; and (3) the set of *products*, namely the substances that are produced when the reaction takes place.

A state of a reaction system is fully specified by a subset of the background set, interpreted as the set of substances available at a given instant. A state transition is determined by the occurrence of all the "enabled" reactions: a reaction is enabled at a state if the state contains all its reactants and none of the inhibitors.

In a reaction system, there are no quantities: in a state, a substance is either present or absent. When it is present, all reactions using it as a reactant can take place, without any notion of conflict.

Another fundamental principle underlying this model is the principle of non-persistence of substances. If a substance, present in a state, is not produced by one of the enabled reactions, it disappears in the next state. So the state reached after a state transition comprises the union of the products of all the reactions which were enabled in the previous state.

Usually, a reaction system is considered as living within an environment, which, at each new state, can inject some substances. Here, I will consider isolated reaction systems.

From the previous discussion, it should be clear that reaction systems are quite different from Petri nets, in which tokens are resources which can be used by just one transition, and persist if they are not used. There have been several attempts to define variants of Petri nets with the aim of modelling reaction systems (see, for instance, [12]).

I will take here a different approach, and propose a notion of region corresponding to the notion of substance in a reaction system, and discuss its potential use in the problem of synthesizing a reaction system from a labelled transition system.

Let us first briefly recall the main formal definitions, taken from [10], with minor changes.

Definition 6. *Let S be a finite set (called the* background set*). A reaction on S is a tuple $a = (R, I, P)$, where R, I, P are subsets of S.*

If S' is a subset of S, the result of a *on S', denoted by $res_a(S')$, is defined as $res_a(S') = P$ if $R \subseteq S'$ and $I \cap S' = \emptyset$; $res_a(S') = \emptyset$ otherwise. The result of a subset of reactions Δ on S', denoted by $res_\Delta(S')$, is the union of the results of all the reactions in Δ.*

Definition 7. *A* reaction system *is a pair $\mathcal{A} = (S, Z)$, where S is a finite set (the* background set *of $\mathcal{A}$), and Z is a set of reactions on S.*

Definition 8. *Let $\mathcal{A} = (S, Z)$ be a reaction system. A* transition system on $\mathcal{A}$ *is a labelled transition system $(Q, 2^Z, T)$, where $Q \subseteq 2^S$, with the following properties:*

- *Q is forward closed: if $q \in Q$, and Δ is the set of all reactions enabled at q, then $res_\Delta(q) \in Q$.*
- *$(q, \Delta, q') \in T$ if each a in Δ is enabled at q, and $q' = res_\Delta(q)$.*

We now turn to defining a synthesis problem for reaction systems. The starting point is a labelled transition system TS, where labels are taken from the set of subsets of a given set. In other words, we suppose to know the names of the reactions, and in which global states they can take place. The problem consists in computing a feasible background set S, and the sets of reactants, inhibitors, and products for each reaction. The procedure should specify, for each element in S, in which states it is present. Then, taking the set representations of the states of TS, and the computed reactions, the transition system of the resulting reaction system, with the given set of states, should be isomorphic to TS.

Let Z be a finite set of reaction names (for short, of *reactions*). Define $\Gamma = 2^Z$; Γ is the set of potential labels for a transition system. Let TS $= (Q, \Gamma, T)$ be a labelled transition system, where Q is a finite set of *states*, and $T \subseteq Q \times \Gamma \times Q$ is the set of *transitions*. If $(q, \Delta, q') \in T$, we write $q \xrightarrow{\Delta} q'$.

For each reaction a, define $\bullet a$ as the set of states with at least one outgoing transition whose label includes a:

$$\bullet a = \{q \in Q \mid \exists (q, \Delta, q') \in T, a \in \Delta\}$$

and $a\bullet$, symmetrically:

$$a\bullet = \{q \in Q \mid \exists (q', \Delta, q) \in T, a \in \Delta\}$$

We now try to characterize which subsets of states can be taken as elements of the background set of a reaction system. Such a subset will be interpreted as the set of states in which the corresponding element of the background set is present.

The first step consists in computing the potential relations of a subset of states with a reaction. Let a be a reaction. A subset of states D is

- a *reactant* of a if $\bullet a \subseteq D$
- an *inhibitor* of a if $\bullet a \cap D = \emptyset$
- a *product* of a if $a\bullet \subseteq D$

Then D is a *region* if, for all $q \xrightarrow{\Delta} q'$ such that q' is in D, there is a in Z such that D is a product of a, and a belongs to Δ.

Regions of the given transition system can be interpreted as elements of the background set of a reaction system. For each element z of Z we can determine the corresponding set of reactants, inhibitors, and products: $R(z), I(z), P(z)$, by applying the definitions above.

Example 1. Consider the transition system in Fig. 3. The following table shows the relevant sets of states associated to each reaction.

$\bullet A = \{1,3\}$ $A\bullet = \{2,4\}$	$\bullet B = \{1,2,4,5\}$ $B\bullet = \{2,3,5\}$	
$\bullet C = \{2\}$ $C\bullet = \{3\}$	$\bullet D = \{2,4\}$ $D\bullet = \{3,5\}$	
$\bullet E = \{3,5\}$ $E\bullet = \{3,4\}$	$\bullet F = \{5\}$ $F\bullet = \{3\}$	

Then, for example, $\{3,4,5\}$ is a region, which acts as a reactant of E and F, as an inhibitor of C, and as a product of C, D, E, and F. On the other hand, $\{2,4,5\}$ is not a region, since it is a product of A only, and its presence in state 5 is not justified by any reaction.

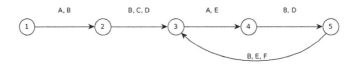

Fig. 3. A transition system.

With this definition of regions, we can tackle the usual problems related to synthesis: characterize the class of transition systems for which the synthesis problem is solvable with this notion of regions; characterize sufficient sets of regions for constructing a solution; studying synthesis problems with stronger constraints, and so on.

The separation problems introduced earlier apply also in this case. For ease of notation, when D is a region, the symbol D will be used also to denote the characteristic function of the corresponding subset of states.

Let $\mathcal{R}(\text{TS})$ be the set of regions of TS. We say that TS has the *state separation property* if, for each pair of distinct states, q, q', there is a region D such that $D(q) \neq D(q')$.

We say that TS has the property of *state-reaction separation* if, whenever z is not enabled in q, there is either a region D in $R(z)$ such that $D(q) = 0$, or there is a region D in $I(z)$ such that $D(q) = 1$.

We say that TS has the *production* property if, for each region D, and for each state q in D, either q is an initial state, or for each transition $q' \xrightarrow{\Delta} q$, there is $z \in \Delta$ such that $D \in P(z)$.

From now on, suppose that $\text{TS} = (Q, 2^Z, T)$ has the three properties above. For each reaction name, z, define

$$R(z) = \{s \in \mathcal{R}(\text{TS}) \mid z \in R(s)\} \tag{4}$$

$$I(z) = \{s \in \mathcal{R}(\text{TS}) \mid z \in I(s)\} \tag{5}$$

$$P(z) = \{s \in \mathcal{R}(\text{TS}) \mid z \in P(s)\} \tag{6}$$

For each $q \in Q$, define $\hat{q} = \{s \in \mathcal{R}(\mathcal{A}) \mid q \in s\}$. This gives a concrete representation of states of $\mathcal{A}$ as sets of regions. Define $\hat{Q} = \{\hat{q} \mid q \in Q\}$.

For each $z \in Z$, define a reaction on the background set $\mathcal{R}(\mathcal{A})$, as follows:

$$\hat{z} = (R(z), I(z), P(z))$$

The reaction system derived from $\mathcal{A}$, denoted $\text{RS}(\mathcal{A})$, is defined by the background set $\mathcal{R}(\mathcal{A})$ and by the set of reactions $\{\hat{z} \mid z \in Z\}$.

Similarly to what happens for usual regions for Petri nets, also in this case the synthesis problem is solved if the given transition system satisfies the separation properties. We can in fact prove that for each $q \in Q$, and for each $z \in Z$, z is enabled in q if, and only if, $\hat{z}$ is enabled in $\hat{q}$.

The set of regions computed in this way is, in a sense, redundant. To solve the synthesis problem, one must find a set of regions which solves all the separation problems.

In the definition of region given above, the set of states determines the realtions with all reactions. However, in reaction systems, this is not necessarily the case. Two elements of the background set can be present in the same states, but bear different relations with reactions. We can then give an alternative definition, which gives a larger set of regions.

Definition 9. *A tuple (r, J, K, L), where r is a subset of states, and J, K, and L are three subsets of Z, is a region if*

- *for all z in J, $\bullet z \subseteq r$*
- *for all z in K, $r \cap \bullet z = \emptyset$*
- *for all z in L, $z\bullet \subseteq r$*
- *for all (q, Δ, q') in T, if $q' \in r$, then there is z in $L \cap \Delta$*

The notions of state separation, and state-reaction separation apply of course, to this new kind of region, and one can use this new kind to solve the synthesis problem. Encoding explicitly in the notion of region the relations with reactions allows us to characterize regions as morphisms of set-labelled transition systems, as shown in the next section.

5   Regions for Reaction Systems and Morphisms

In Sect. 2, we saw that regions can be defined as morphism from a transition system to a 'type of nets', namely a transition system encoding the possible values of a local state, and the possible transitions of state, for a specific kind of Petri nets. By extending the notion of 'type of nets', we now give a definition of region as morphism of set-labelled transition systems.

Definition 10. *A set-labelled transition system $A = (Q, 2^Z, T)$ is given by a set Q of states, a set Z of basic actions, and a set T of transitions, where a transition is a triple (q, Δ, q'), with $q, q' \in Q$, and $\Delta \subseteq Z$.*

A morphism between set-labelled transition systems is given by a map on states, and a map on actions, as follows.

Let $A_i = (Q_i, 2^{Z_i}, T_i)$ be a set-labelled transition system for $i = 1, 2$. Let $f : Q_1 \to Q_2$ and $g : Z_1 \to 2^{Z_2}$ be total maps. For Δ subset of Z_1, define $g(\Delta)$ as the union of $g(z)$ for all z in Δ. Then (f, g) is a morphism from A_1 to A_2 if, whenever (q_1, Δ, q_1') is a transition of A_1, $f(q_1), g(\Delta), f(q_1')$ is a transition of A_2.

5.1   The Type of Reaction Systems

In the general theory of regions, a type of nets is defined as a transition system, modelling the possible states of a single place, and the possible effects on a place of transition firings.

In the same line, we define a set-labelled transition system describing the possible states of a substance in a reaction system. These states correspond to the presence or absence of the substance in a global state. The basic actions correspond to the possible relations holding between a substance and a reaction: reactant (R), inhibitor (I), product (P), or independence (N). In general, more than one relation can hold between a substance and a reaction. This is reflected in the definition of map g of a morphism, which maps a label to a set of labels. Some combinations are meaningless in the context of reaction systems, and could be prohibited.

The labels associated to transitions are the following (note that there are several distinct transitions with the same source and destination states).

$$0 \longrightarrow 0: \quad \{N\}, \{I\}, \{N, I\} \tag{7}$$

$$0 \longrightarrow 1: \quad \{P\}, \{P, I\}, \{P, N\}, \{P, I, N\} \tag{8}$$

$$1 \longrightarrow 0: \quad \{R\}, \{N\}, \{R, N\} \tag{9}$$

$$1 \longrightarrow 1: \quad \{P\}, \{P, R\}, \{P, N\}, \{P, R, N\} \tag{10}$$

With this definition, a morphism from a transition system TS to is a region according to Definition 9, and for each region there is a corresponding morphism.

Some attempts have been made to define variants of Petri nets, with the aim of representing reaction systems. This characterization of regions as morphisms might be a starting point for a new approach.

6   Conclusion

Region theory is now thirty years old and has a fairly large literature. I have tried in these pages to argue that it is still a live field of research, with several interesting open problems.

Acknowledgments. Many colleagues and friends have contributed to shape my views on concurrency theory and on region theory during more than twenty-five years. I gratefully acknowledge the pleasure I took working with them, even if they are too many to mention. I feel I can make an exception for Philippe Darondeau, who will not be able to apply his rigour and scientific perspicacity to comment on these pages.

References

1. Badouel, E., Bernardinello, L., Darondeau, P.: Polynomial algorithms for the synthesis of bounded nets. In: Mosses, P.D., Nielsen, M., Schwartzbach, M.I. (eds.) CAAP 1995. LNCS, vol. 915, pp. 364–378. Springer, Heidelberg (1995). https://doi.org/10.1007/3-540-59293-8_207
2. Badouel, E., Bernardinello, L., Darondeau, P.: The synthesis problem for elementary net systems is np-complete. Theor. Comput. Sci. **186**(1–2), 107–134 (1997)
3. Badouel, Eric., Bernardinello, Luca, Darondeau, Philippe: Petri Net Synthesis. TTCSAES. Springer, Heidelberg (2015). https://doi.org/10.1007/978-3-662-47967-4
4. Badouel, E., Darondeau, P.: Theory of regions. In: Reisig, W., Rozenberg, G. (eds.) ACPN 1996. LNCS, vol. 1491, pp. 529–586. Springer, Heidelberg (1998). https://doi.org/10.1007/3-540-65306-6_22
5. Bernardinello, L., De Michelis, G., Petruni, K., Vigna, S.: On the synchronic structure of transition systems. In: Desel, J., (ed.) Proceedings of the International Workshop on Structures in Concurrency Theory, STRICT 1995, Berlin, Germany, 11–13 May 1995, Workshops in Computing, pp. 69–84. Springer, London (1995) https://doi.org/10.1007/978-1-4471-3078-9_5
6. Bernardinello, L., Ferigato, C., Pomello, L.: An algebraic model of observable properties in distributed systems. Theor. Comput. Sci. **290**(1), 637–668 (2003)

7. Bernardinello, L., Ferigato, C., Pomello, L., Aubel, A.P.: On stability of regional orthomodular posets. Trans. Petri Nets Other Model. Concurr. **13**, 52–72 (2018)
8. Ehrenfeucht, A., Rozenberg, G.: Partial (set) 2-structures. part I: basic notions and the representation problem. Acta Inf. **27**(4), 315–342 (1990)
9. Ehrenfeucht, A., Rozenberg, G.: Partial (set) 2-structures. part II: state spaces of concurrent systems. Acta Inf. **27**(4), 343–368 (1990)
10. Ehrenfeucht, A., Rozenberg, G.: Reaction systems. Fundam. Inf.Inform. **75**(1–4), 263–280 (2007)
11. Kleijn, J., Koutny, M., Pietkiewicz-Koutny, M., Rozenberg, G.: Applying regions. Theor. Comput. Sci. **658**, 205–215 (2017). Formal Languages and Automata: Models, Methods and Application In honour of the 70th birthday of Antonio Restivo
12. Kleijn, J., Koutny, M., Rozenberg, G.: Petri nets for biologically motivated computing. Sci. Ann. Comp. Sci. **21**(2), 199–225 (2011)
13. Rosenke, C., Tredup, R.: The complexity of synthesizing elementary net systems relative to natural parameters. J. Comput. Syst. Sci. **110**, 37–54 (2020)

Flat Petri Nets (Invited Talk)

Jérôme Leroux$^{(\boxtimes)}$ (iD)

University Bordeaux, CNRS, Bordeaux INP, LaBRI, UMR 5800,
Talence 33405, France
jerome.leroux@labri.fr
https://www.labri.fr/~leroux/

Abstract. Vector addition systems with states (VASS for short), or equivalently Petri nets are one of the most popular formal methods for the representation and the analysis of parallel processes. The central algorithmic problem is reachability: whether from a given initial configuration there exists a sequence of valid execution steps that reaches a given final configuration. This paper provides an overview of results about the reachability problem for VASS related to Presburger arithmetic, by presenting 1) a simple algorithm for deciding the reachability problem based on invariants definable in Presburger arithmetic, 2) the class of flat VASS for computing reachability sets in Presburger arithmetic, and 3) complexity results about the reachability problem for flat VASS.

Keywords: Formal methods · Petri nets · Flat systems · Presburger arithmetic

1 Introduction

Vector addition systems with states [30], or equivalently vector addition systems [31], or Petri nets are one of the most popular formal methods for the representation and the analysis of parallel processes [24]. The central algorithmic problem is reachability: whether from a given initial configuration there exists a sequence of valid execution steps that reaches a given final configuration. Many computational problems reduce to this reachability problem in logic, complexity, real-time systems, protocols [28, 49].

A *d-dimensional vector addition system* (*d-VASS*, or just *VASS* when the dimension d is not relevant) is a pair $V = (Q, T)$ where Q is a non empty finite set of elements called *states*, and T is a finite set of triples in $Q \times \mathbb{Z}^d \times Q$ called *transitions*. A *configuration* is a pair $(q, x) \in Q \times \mathbb{N}^d$ also denoted as $q(x)$ in the sequel, and an *action* is a vector in $\mathbb{Z}^d$. The semantics is defined by introducing for each transition t the binary relation $\overset{t}{\to}$ over the configurations defined by $p(x) \overset{t}{\to} q(y)$ if $t = (p, y - x, q)$. We also associate to a word $\sigma = t_1 \ldots t_k$ of

The author is supported by the grant ANR-17-CE40-0028 of the French National Research Agency ANR (project BRAVAS).

D. Buchs and J. Carmona (Eds.): PETRI NETS 2021, LNCS 12734, pp. 17–30, 2021.
https://doi.org/10.1007/978-3-030-76983-3_2

transitions $t_1, \ldots, t_k$ the binary relation $\xrightarrow{\sigma}$ over the configurations defined by $p(x) \xrightarrow{\sigma} q(y)$ if there exists a sequence $c_0, \ldots, c_k$ of configurations such that:

$$p(x) = c_0 \xrightarrow{t_1} c_1 \cdots \xrightarrow{t_k} c_k = q(y)$$

The *reachability set* from a set C_{in} of configurations is the set $\text{Reach}_V(C_{in})$ of configurations c such that $c_{in} \xrightarrow{\sigma} c$ for some configuration $c_{in} \in C_{in}$ and some word σ of transitions.

Example 1. Let us consider the VASS V depicted bellow. This VASS has a loop on state p and another loop on state q. Intuitively, iterating the loop on state p transfers the content of the first counter to the second counter whereas iterating the loop on state q transfers and multiplies by two the content of the second counter to the first counter. Let us denote by t_1, t_2, t_3 and t_4 the transitions $(p, (-1, 1), p)$, $(p, (0, 0), q)$, $(q, (2, -1), q)$ and $(q, 0, 0, p)$. We can prove that the reachability set from $\{p(1, 0)\}$ is equal to $\{p, q\} \times \{(n, m) \mid n + m \geq 1\}$ by observing that if $n, m \in \mathbb{N}$ satisfy $n + m \geq 1$ then:

$$p(1, 0) \xrightarrow{(t_1 t_2 t_3 t_4)^{n+m-1}} p(n+m, 0) \xrightarrow{t_1^m} p(n, m) \xrightarrow{t_2} q(n, m)$$

$$(0, 0)$$

$$(-1, 1) \circlearrowleft \underset{(0,0)}{\boxed{p}} \qquad \boxed{q} \circlearrowright (2, -1)$$

The reachability problem takes as input a VASS V and two configurations c_{in}, c_{out} and it decides if there exists a word σ of transitions such that $c_{in} \xrightarrow{\sigma} c_{out}$. After an incomplete proof by Sacerdote and Tenney [48], decidability of the problem was established by Mayr [44,45], whose proof was then simplified by Kosaraju [32]. Building on the further refinements made by Lambert in the 1990s [34], in 2015, a first complexity upperbound of the reachability problem was provided [39] more than thirty years after the presentation of the algorithm introduced by Mayr [32,34,44,45]. The upperbound given in that paper is cubic Ackermannian. This complexity is obtained by analyzing the computation complexity of the Mayr algorithm. By refining this algorithm and by introducing a new ranking function proving the termination of this refinement, an Ackermannian complexity upperbound was obtained in [40]. This paper also showed that the reachability problem in fixed dimension is primitive recursive by bounding the length of executions thanks to the Grzegorczyk hierarchy. Based on this bound, in [43], the reachability problem for general VASS is shown to be interreducible in log-space to the reachability problem for structurally bounded VASS when numbers are encoded in unary or in binary. Let us recall that a VASS is said to be *structurally bounded* if the reachability set is finite from any initial configuration, and this property is decidable in polynomial time (even when numbers are encoded in binary) thanks to the Kosaraju-Sullivan algorithm [33].

The reachability problem for structurally bounded VASS can be decided by a deterministic brute-force exploration in an obvious way. The computational complexity of such an algorithm is known to be Ackermannian [47]. Moreover, due to the family of VASS introduced in [46] this bound is tight. It follows that the reachability problem for general VASS can be solved with a simple deterministic brute-force algorithm, and last but not least, the reachability problem for structurally bounded VASS is a central problem.

In this paper, we present results about the reachability problem for VASS related to *Presburger arithmetic* fo$(\mathbb{N}, +)$. In this context, a set C of configurations in $Q \times \mathbb{N}^d$ is said to be *Presburger* if there exists a sequence $(\phi_q)_{q \in Q}$ of formulas ϕ_q in Presburger arithmetic denoting sets $X_q \subseteq \mathbb{N}^d$ such that $C = \bigcup_{q \in Q} \{q\} \times X_q$. In Sect. 2 we present a simple algorithm for deciding the reachability problem for VASS based on Presburger inductive invariant that shows that the Presburger sets of configurations are central for deciding the reachability problem for VASS even if, as shown in Example 2, there exists VASS with non Presburger reachability sets. In Sect. 3 we shows that the reachability set of a VASS is Presburger, if and only if, it is flattable, i.e. the VASS can be unfolded into a VASS without nested cycles called *flat VASS*. In Sect. 4 we present complexity results about the reachability problem for flat VASS.

Example 2. In 1979, Hopcroft and Pansiot [30] introduced the VASS depicted bellow. This VASS exhibit a non Presburger reachability set from the initial configuration $p(1, 0, 0)$. Intuitively, on the first and the second counters, the behaviour of that VASS is the same as the one introduced in Example 1. The third counter is incremented each time we come back to state p from q. In [30] the reachability set from the initial configuration $p(1, 0, 0)$ is proved equal to the following set:

$$\{p(x_1, x_2, x_3) \mid x_1 + x_2 \leq 2^{x_3}\} \cup \{q(x_1, x_2, x_3) \mid x_1 + 2x_2 \leq 2^{x_3+1}\}$$

2   Presburger Inductive Invariants

We present in this section a simple algorithm for deciding the reachability problem based on Presburger inductive invariants [35–37] that may have an optimal complexity (this is an open problem). A set C of configurations is called an *inductive invariant* for a VASS V if for every configurations c, c' and every transition t such that $c \xrightarrow{t} c'$, then $c \in C$ implies $c' \in C$.

Theorem 1. ([35]). *For every VASS V, for every Presburger sets of configurations C_{in}, C_{out}, either $c_{in} \xrightarrow{\sigma} c_{out}$ for some configurations $c_{in} \in C_{in}$ and*

$c_{out} \in C_{out}$ and some word σ of transitions, or there exists a Presburger inductive invariant C that contains C_{in} and disjoints from C_{out}.

Since we can decide if a sequence of Presburger formulas denotes an inductive invariant with classical algorithms deciding Presburger arithmetic, the previous theorem shows that a brute-force non-deterministic exploration of the reachability set and sequences of Presburger formulas provides a simple algorithm for deciding the reachability problem. Whereas the proof in [35] was based on a refinement of Lambert's algorithm, in [36] a direct proof based on a well quasi order over the executions is provided. This proof was then simplified a bit more in a paper [37] that received a best paper award at Alan Turing centenary conference in 2012. In those two last papers, Presburger formulas denoting inductive invariants are obtained by proving that reachability sets are "asymptotically" definable in Presburger arithmetic.

Example 3. Let us come back to Example 2. Notice that the reachability set from the initial configuration $p(1, 0, 0)$ is not Presburger. Let us introduce the non-decreasing sequence $(C_n)_{n \in \mathbb{N}}$ of Presburger sets defined as follows:

$$C_n = \{p(x_1, x_2, x_3) \mid \bigvee_{i=0}^{n} (x_1 + x_2 \leq 2^i \wedge x_3 = i) \vee x_3 > n\} \cup$$

$$\{q(x_1, x_2, x_3) \mid \bigvee_{i=0}^{n} (x_1 + 2x_2 \leq 2^{i+1} \wedge x_3 = i) \vee x_3 > n\}$$

Notice that C_n is an inductive invariant that contains the initial configuration $p(1, 0, 0)$ and since $\bigcap_{n \in \mathbb{N}} C_n$ is the reachability set from $p(1, 0, 0)$, it follows that for every configuration c_{out} outside of this reachability set, there exists n such that $c_{out} \notin C_n$.

3   Flat and Flattable VASS

When the reachability set of a VASS is infinite from an initial configuration, a brute-force exploration of the reachability set fails. However, even in that case the computation of the reachability sets may still be possible by using Presburger arithmetic for symbolically representing infinite sets of configurations and by using acceleration techniques to discover infinite sets of reachable configurations. Intuitively, acceleration techniques consist in computing symbolically the effect of iterating cycles of the system. Those techniques were studied for several models: systems with FIFO channels [9–11,15,16], time [2,3,12,13], other data structures [17], and systems manipulating counters including the VASS model [4–6,8,14,18,19,27].

Acceleration techniques for VASS are related to the class of *flat VASS*. Formally, a VASS V is said to be *flat* if for every state q, there exists at most one simple cycle on q (intuitively no nested cycles).

Example 4. The two VASS depicted in Example 1 and Example 2 are not flat. The VASS depicted below is flat.

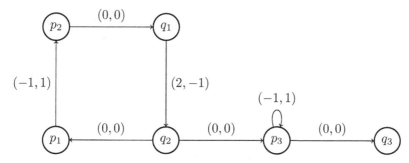

The reachability set of a flat VASS is clearly Presburger by compiling in Presburger arithmetic the effect of iterating simple cycles [27]. The problem of deciding if the reachability set of a general (non flat) VASS from an initial Presburger set of configurations is Presburger was studied thirty years ago independently by Dirk Hauschildt during his PhD [29] and Jean-Luc Lambert. Unfortunately, these two works were never published. Moreover, from these works, it is difficult to derive a simple algorithm for computing Presburger formulas denoting the reachability set. In [38] a simple algorithm for computing such a formula based on flat VASS is given. Intuitively when a VASS V is not flat, one can try to unfold it into a flat VASS V' such that the reachability set of V from a Presburger set C_{in} can be derived from the reachability set of V' from a Presburger set C'_{in} derived from C_{in}.

More formally, an *unfolding* of a VASS $V = (Q, T)$ is a pair (V', f) where $V' = (Q', T')$ is a VASS and $f : Q' \to Q$ is a total mapping such that $(f(p'), a, f(q'))$ is in T for every transition $(p', a, q') \in T'$. We observe that for every set of configurations C_{in} of V, we have where f is extended over the configurations by $f(q', x) = (f(q'), x)$ for every configuration $(q', x) \in Q' \times \mathbb{N}^d$:

$$f(\text{Reach}_{V'}(f^{-1}(C_{\text{in}}))) \subseteq \text{Reach}_V(C_{\text{in}})$$

When the previous inclusion is an equality, the unfolding is said to be *complete* from C_{in}. An unfolding is called a flattening when V' is flat [7]. A VASS V is said to be *flattable* from a set C_{in} of initial configurations if there exists a flattening of V complete from C_{in} (see [42] for various examples of flattable VASS).

Theorem 2. ([38]). *For every VASS V, for every Presburger set C_{in} of configurations, the reachability set $\text{Reach}_V(C_{\text{in}})$ is Presburger if, and only if, V is flattable from C_{in}.*

It follows that if the reachability set from an initial Presburger set of configurations is Presburger, a sequence of Presburger formulas denoting the reachability set can be computed by finding the right flattening. In [26] heuristics and algorithms for finding such a flattening are presented. Those heuristics are implemented in the tool FAST [4–6] for analyzing Minsky machines, a class of systems strictly extending VASS with undecidable reachability problem.

Remark 1. In [38], a stronger version of Theorem 2 is proved. More precisely, it is shown that for every VASS V, for every Presburger set C_{in} of configurations, and for every Presburger set $C \subseteq \text{Reach}_V(C_{in})$, there exists a flattening (V', f) of V such that $C \subseteq f(\text{Reach}_{V'}(f^{-1}(C_{in})))$. This extension is used in [25] in order to provide witnesses of well-specification for population protocols [1].

Example 5. Let $V = (Q, T)$ be the VASS introduced in Example 1, $V' = (Q', T')$ be the flat VASS introduced in Example 4, and $f : Q' \rightarrow Q$ defined by $f(p_i) = p$ and $f(q_i) = q$ for every $i \in \{1, 2, 3\}$. Observe that (V', f) is a flattening of V. Moreover, we derive from Example 1 that this flattening is complete from $\{p(1, 0)\}$. It follows that V is flattable from $\{p(1, 0)\}$.

Flattening are also used for deriving fine complexity results for the reachability problem for 2-VASS. Recall that the reachability sets from an initial configuration have been shown to be effectively Presburger for 2-VASS in [30]. In [41], it was proved that for every 2-VASS V there exists a flattening (V', f) of V effectively computable such that for every configuration $p(x), q(y)$, we have $p(x) \xrightarrow{*}_V q(y)$ if, and only if, there exist two states $p' \in f^{-1}(p)$ and $q' \in f^{-1}(q)$ such that $p'(x) \xrightarrow{*}_{V'} q'(y)$. Based on a similar proof, ten years later, it was proved that for every 2-VASS V, there exists a family $\mathcal{F}$ of flattening (V', f) of V of "small sizes" such that for every configurations $p(x), q(y)$, we have $p(x) \xrightarrow{*}_V q(y)$ if, and only if, there exists a flattening (V', f) in $\mathcal{F}$ and states $p' \in f^{-1}(p)$ and $q' \in f^{-1}(q)$ such that $p'(x) \xrightarrow{*}_{V'} q'(y)$. From this result the reachability problem for 2-VASS encoded in binary was proved to be PSPACE-complete in the same paper. Finally, thanks to the family $\mathcal{F}$, and the fact that the reachability problem for flat 2-VASS encoded in unary is NL-complete [23], the reachability problem for general 2-VASS encoded in unary was proved NL-complete in [23].

4   Reachability Problem for Flat VASS

In this section we present some complexity lowerbounds of the reachability problem for flat VASS. In that context, the subclass of ultraflat VASS will play a central role. Formally, an *ultraflat VASS* is a VASS $V = (Q, T)$ such that $Q = \{q_1, \ldots, q_n\}$ with $n = |Q|$, and $T = \{(q_{j-1}, (0, \ldots, 0), q_j) \mid 2 \leq j \leq n\} \cup \{(q_j, a_j, q_j) \mid 1 \leq j \leq n\}$ for some actions $a_1, \ldots, a_n$. An ultraflat VASS is clearly flat since (q_j, a_j, q_j) is the unique simple cycle on q_j for every j.

The reachability problem for flat 1-VASS with numbers encoded in binary can be easily proved NP-hard by reduction of the subset sum problem. Let us recall that the subset sum problem takes as input a sequence $s, s_1, \ldots, s_k$ of natural numbers encoded in binary and it decides if there exists a finite set $I \subseteq \{1, \ldots, k\}$ such that $s = \sum_{i \in I} s_i$. The following lemma shows that this lowerbound also holds for ultraflat 1-VASS in binary.

Lemma 1. *The reachability problem for ultraflat 1-VASS with numbers encoded in binary is NP-hard.*

Proof. Let us consider an instance $s, s_1, \ldots, s_k$ of the subset sum problem. We can assume that $0 < s < \sum_{j=1}^{k} s_j$ and $k \geq 2$ since other instances are trivially accepting or non accepting. We introduce $x = \sum_{j=1}^{k}(1 + s_j)$. Notice that $x \geq 4$, $s + k < x$, and $1 + s_j < x$ for every $1 \leq j \leq k$.

We introduce the ultraflat 1-VASS $V = (Q, T)$ defined by the set of states $Q = \{p_1, q_1, \ldots, p_k, q_k\}$, and the set of transitions T that contains the transitions labeled by 0 that connect the states of Q to form an ultraflat VASS, and the transitions $\alpha_j = (p_j, u_j, p_j)$ and $\beta_j = (q_j, v_j, q_j)$ with $v_j = 1 - (x-1)x^{2k+1-j}$ and $u_j = s_j + v_j$ for every $1 \leq j \leq k$. Let us prove that $p_1(x^{2k+1}) \xrightarrow{*} q_k(x^{k+1} + s + k)$ if, and only if, there exists $J \subseteq \{1, \ldots, k\}$ such that $s = \sum_{j \in J} s_j$.

Assume first that $p_1(x^{2k+1}) \xrightarrow{*} q_k(x^{k+1} + s + k)$. Since V is an ultraflat VASS, there exist sequences $(n_j)_{1 \leq j \leq k}$, $(m_j)_{1 \leq j \leq k}$, $(a_j)_{1 \leq j \leq k+1}$, and $(b_j)_{1 \leq j \leq k}$ of natural numbers with $a_1 = x^{2k+1}$, $a_{k+1} = x^{k+1} + s + k$, and such that for every $1 \leq j \leq k$, we have:

$$p_j(a_j) \xrightarrow{\alpha_j^{m_j}} p_j(b_j) \quad q_j(b_j) \xrightarrow{\beta_j^{n_j}} q_j(a_{j+1})$$

It follows that $a_{j+1} = a_j + m_j u_j + n_j v_j$ for every $1 \leq j \leq k$.

From $1 + s_j < x$, we derive $u_j < x - (x-1)x^{k+1} \leq (2-x)x^{k+1} \leq -x^{k+1}$ since $x \geq 3$. As $v_j \leq u_j$, we have proved that $u_j, v_j < -x^{k+1}$. Since $a_{k+1} = a_1 + \sum_{j=1}^{k} m_j u_j + n_j v_j \leq x^{2k+1} - (x^{k+1} + 1)\sum_{j=1}^{k}(m_j + n_j)$, and $a_{k+1} \geq 0$, we get $\sum_{j=1}^{k} m_j + n_j < x^k$. Moreover, since $u_j = 1 + s_j \mod x^{k+1}$ and $v_j = 1 \mod x^{k+1}$, we deduce from $a_{j+1} = a_j + m_j u_j + n_j v_j$ that $a_{j+1} = a_j + m_j(1 + s_j) + n_j \mod x^{k+1}$. It follows that $a_{k+1} = a_1 + r \mod x^{k+1}$ where $r = \sum_{j=1}^{k}(m_j(1 + s_j) + n_j)$. As $a_{k+1} = s + k \mod x^{k+1}$ and $a_1 = 0 \mod x^{k+1}$ we get $r = s + k \mod x^{k+1}$. Since $r \leq (x^k - 1)\sum_{j=1}^{k}(1 + s_j) < x^{k+1}$ and $s + k < x \leq x^{k+1}$ we deduce that $r = s + k$. In particular $\sum_{j=1}^{k}(m_j + n_j) \leq s + k$.

Assume by contradiction that there exists $\ell \in \{1, \ldots, k\}$ such that $m_\ell + n_\ell \neq 1$ and let ℓ be the minimal one. By induction we deduce that $a_\ell = x^{2k+2-\ell} + \ell - 1 + \sum_{j \in J} s_j$ where $J = \{j \in \{1, \ldots, \ell-1\} \mid m_j = 1\}$. It follows that $a_\ell < x^{2k+2-\ell} + x$. Notice that if $m_\ell + n_\ell \geq 2$ then:

$$\begin{aligned}
a_{\ell+1} &= a_\ell + m_\ell u_\ell + n_\ell v_\ell \\
&\leq a_\ell + (m_\ell + n_\ell)u_\ell \\
&< x^{2k+2-\ell} + x + 2(x - (x-1)x^{2k+1-\ell}) \\
&\leq x^{2k+1-\ell}(2 - x) + 3x \\
&\leq -x^{k+1} + x^2 \\
&\leq 0
\end{aligned}$$

And we get a contradiction with $a_{\ell+1} \geq 0$. Therefore $m_\ell + n_\ell \leq 1$ and since $m_\ell + n_\ell \neq 1$ we deduce that $m_\ell = n_\ell = 0$. It follows that $a_{\ell+1} = a_\ell$. In particular $a_{\ell+1} \geq x^{2k+2-\ell}$.

Now, observe that $a_{k+1} = a_{\ell+1} + \sum_{j=\ell+1}^{k} n_j u_j + m_j v_j \geq a_{\ell+1} + \sum_{j=\ell+1}^{k}(n_j + m_j) v_{\ell+1} \geq a_{\ell+1} + (s+k) v_{\ell+1}$. It follows that $a_{k+1} \geq x^{2k+2-\ell} + (s+k)(1-(x-1)x^{2k-\ell}) = x^{2k-\ell}(x^2 - (s+k)(x-1)) + s + k$. Since $s+k \leq x-1$, we deduce that $x^2 - (s+k)(x-1) \geq x^2 - (x-1)^2 = 2x - 1 > x$ since $x \geq 2$. In particular $a_{k+1} > x^{k+1} + s + k = a_{k+1}$ and we get a contradiction.

It follows that $m_j + n_j = 1$ for every $1 \leq j \leq k$. Let us introduce $J = \{j \in \{1, \ldots, k\} \mid m_j = 1\}$. An immediate induction shows that $a_{k+1} = x^{k+1} + k + \sum_{j \in J} s_j$. Since $a_{k+1} = x^{k+1} + k + s$, we get $s = \sum_{j \in J} s_j$.

Conversely, observe that if there exists $J \subseteq \{1, \ldots, k\}$ such that $s = \sum_{j \in J} s_j$ then $p_1(x^{2k+1}) \xrightarrow{*} q_k(x^{k+1} + s + k)$ by considering the sequence $(m_j)_{1 \leq j \leq k}$ and $(n_j)_{1 \leq j \leq k}$ satisfying $(m_j, n_j) = (1, 0)$ if $j \in J$ and $(m_j, n_j) = (0, 1)$ otherwise. □

When the dimension is part of the input, the following lemma shows that the reachability problem for ultraflat VASS in unary is also NP-hard.

Lemma 2. *The reachability problem for ultraflat VASS with numbers encoded in unary is NP-hard.*

Proof. Let us consider an instance $s, s_1, \ldots, s_k$ of the subset sum problem. We can assume that $s \leq \sum_{j=1}^{k} s_j$. We consider the minimal $\ell \in \mathbb{N}$ such that $\sum_{j=1}^{k} s_j < 2^\ell$, and we let $d = \ell + k$. We denote by z the zero vector of $\mathbb{N}^d$, and we denote by e_i the ith unit vector of $\mathbb{N}^d$ defined by $e_i(i) = 1$ and $e_i(j) = 0$ if $j \neq i$. We denote for a natural number $n < 2^\ell$ the vector $\text{bin}(n) \in \mathbb{N}^d$ defined as $\sum_{i=1}^{\ell} b_i e_i$ where $b_1, \ldots, b_\ell \in \{0, 1\}$ are such that $n = \sum_{i=1}^{\ell} b_i 2^{i-1}$.

We introduce the ultraflat d-VASS $V = (Q, T)$ where Q is the set of states $q_1, p_1 \ldots, q_k, p_k, q_{k+1}, \ldots, q_{k+\ell}$, and T is the set of transitions that contains the transitions labeled by z that connect the states of Q to form an ultraflat VASS, the transitions $(q_j, -e_{\ell+k}, q_j)$ and $(p_j, -e_{\ell+k} + \text{bin}(s_j), p_j)$ for every $1 \leq j \leq k$, transitions $(q_{k+i}, -2e_i + e_{i+1}, q_{k+i})$ for every $1 \leq i < \ell$.

Just observe that $q_1(\sum_{j=1}^{k} e_{\ell+j}) \xrightarrow{*} q_{k+\ell}(\text{bin}(s))$ if, and only if, there exists $J \subseteq \{1, \ldots, k\}$ such that $s = \sum_{j=1}^{k} s_j$. □

Finally, let us consider the reachability problem for flat d-VASS where d is fixed and numbers are encoded in unary. In this context, the complexity of the problem is difficult to determined since we need to compute with a fix number of counters large numbers with actions that involves only small numbers. This intuition is confirmed up to the dimension 2. In fact, the reachability problem for (not necessarily flat) 1-VASS encoded in unary is NL-complete by using a classical hill-cutting argument. For flat 2-VASS encoded in unary, the reachability problem was also proved to be NL-complete in [23] by observing that if there exists a word σ such that $c_{\text{in}} \xrightarrow{\sigma} c_{\text{out}}$ for a flat 2-VASS, then there exists another one with a length polynomially bounded in the size of the VASS and the configurations $c_{\text{in}}, c_{\text{out}}$ encoded in unary. Such a property is not trivial since an hill-cutting argument can no longer be applied in that context as shown by the following example.

Example 6. Let us introduce the family $(V_n)_{n \in \mathbb{N}}$ of ultraflat 2-VASS parameterized by a natural number n and depicted below (zero vectors are not depicted). Intuitively, iterating the loop on a state p_i transfers and multiplies by two the content of the first counter to the second counter, iterating the loop on a state r_i transfers back the content of the second counter to the first counter, iterating the loop on a state s_i transfers the content of the first counter to the second counter, and iterating the loop on a state q_i transfers back and divides by two the content of the second counter to the first counter. Observe that $p_1(1,0) \xrightarrow{\sigma_n} q_1(1,0)$ where σ_n is a run obtained by executing each loop a maximal number of times. It follows that σ_n is exponentially long in n. Moreover, the set C_n of configurations c such that $p_1(1,0) \xrightarrow{u} c \xrightarrow{v} q_1(1,0)$ where u, v are such that $\sigma_n = uv$ is an exponential set of incomparable configurations for the relation $\sqsubseteq$ defined by $p(x_1, x_2) \sqsubseteq q(y_1, y_2)$ if $p = q$, $x_1 \leq y_1$ and $x_2 \leq y_2$.

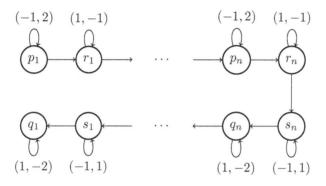

A complexity lowerbound better than NL is open for ultraflat 3-VASS. Starting from that dimension, the minimal length of a word σ such that $c_{in} \xrightarrow{\sigma} c_{out}$ can be exponentially long in the size of the VASS and the configurations c_{in}, c_{out} encoded in unary as shown by the following example.

Example 7. Let us introduce the family $(V_n)_{n \in \mathbb{N}}$ of ultraflat 3-VASS parameterized by a natural number n and depicted below (zero vectors are not depicted). Those VASS are presented as counter programs in [21]. Intuitively, the loop on state p initialized the VASS from $p(1,0,1)$ to a configuration $p_0(x,0,x)$ where x is any positive natural number. Iterating the loop on a state p_i transfers and multiplies by $\frac{n+2-i}{n+1-i}$ the content of the first counter to the second counter, iterating the loop on a state r_i transfers back the content of the second counter to the first counter. By iterating all those loops, starting from $p_0(x,0,x)$, if all the multiplications are performed exactly, we get the configuration $q(x(n+1),0,x)$ since $\prod_{i=1}^{n} \frac{n+2-i}{n+1-i} = n+1$. From such a configuration, by iterating the loop on state q we get the configuration $q(0,0,0)$. In order to obtain such an execution, we prove in [21] that x is necessarily a non zero multiple of $\frac{\text{lcm}(2,\ldots,n+1)}{n+1}$ which is exponential in n. It follows that the minimal word σ such that $p(1,0,1) \xrightarrow{\sigma} q(0,0,0)$ is exponential in n.

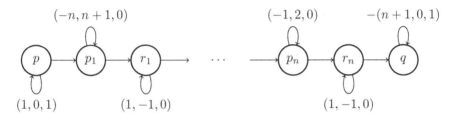

Based on Example 7, the reachability problem for flat 7-VASS in unary is proved to be NP-complete in [21]. In a work in progress, we recently proved that the problem is also NP-complete in dimension 5. It follows that the best complexity lowerbound of the reachability problem for flat d-VASS is NL for $d \in \{3, 4\}$. Proving that the reachability problem is NP-hard in those dimensions requires to find a way to use counters in a non trivial way. Such a trick, if it exists, could pave the way for increasing the best known complexity lower bound [20] of the general (non flat) reachability problem for VASS in order to match the Ackermannian complexity upper bound [40]. On the other side, the best complexity upperbound of the reachability problem for ultraflat d-VASS encoded in unary is NP for $d \geq 3$. Proving that this problem is in P when $d \in \{3, 4\}$ could provide a way to prove that the reachability problem for flat d-VASS in unary is in P.

5   Conclusion

We presented in this paper an overview of results about the reachability problem for VASS related to Presburger arithmetic. Those results came from several peer reviewed publications, except Lemma 1 and Lemma 2 (we are confident that those results are correct, but, in case of, notice that when replacing ultraflat by flat in those lemmas, proofs are almost immediate).

The complexity of the reachability problem for general VASS is still open. The best known complexity lowerbound is a tower function with an height limited by the dimension of the VASS. A possible way to lift up this bound to match the Ackermannian upperbound is to find a trick to reuse some counters in a non trivial way. We think that looking at the complexity of flat 3-VASS encoded in unary may pave the way to get such a trick since if this problem is NP-hard, in order to prove such a bound, we need to encode an NP-hard problem like the subsetsum problem with only three counters and small actions. Intuitively some counters must be reused along the computation. On the other hand, if the problem is in P, we may obtain from the proof of that result a way to prove that the reachability problem for general 3-VASS is elementary (the best known upperbound is tower for structurally bounded 3-VASS).

In this paper, we did not present the vast set of results related to Petri net extensions. Anyway, let us mention that the complexity of the reachability problem for flat 3-VASS encoded in unary seems to be related to the complexity of the coverability problem for pushdown 1-VASS [22].

References

1. Angluin, D., Aspnes, J., Diamadi, Z., Fischer, M.J., Peralta, R.: Computation in networks of passively mobile finite-state sensors. In: Chaudhuri, S., Kutten, S. (eds.) Proceedings of the Twenty-Third Annual ACM Symposium on Principles of Distributed Computing, PODC 2004, St. John's, Newfoundland, Canada, July 25–28, 2004, pp. 290–299. ACM (2004). https://doi.org/10.1145/1011767.1011810

2. Annichini, A., Asarin, E., Bouajjani, A.: Symbolic techniques for parametric reasoning about counter and clock systems. In: Emerson, E.A., Sistla, A.P. (eds.) CAV 2000. LNCS, vol. 1855, pp. 419–434. Springer, Heidelberg (2000). https://doi.org/10.1007/10722167_32

3. Annichini, A., Bouajjani, A., Sighireanu, M.: TReX: a tool for reachability analysis of complex systems. In: Berry, G., Comon, H., Finkel, A. (eds.) CAV 2001. LNCS, vol. 2102, pp. 368–372. Springer, Heidelberg (2001). https://doi.org/10.1007/3-540-44585-4_34

4. Bardin, S., Finkel, A., Leroux, J.: FASTer acceleration of counter automata in practice. In: Jensen, K., Podelski, A. (eds.) TACAS 2004. LNCS, vol. 2988, pp. 576–590. Springer, Heidelberg (2004). https://doi.org/10.1007/978-3-540-24730-2_42

5. Bardin, S., Finkel, A., Leroux, J., Petrucci, L.: FAST: fast acceleration of symbolic transition systems. In: Hunt, W.A., Somenzi, F. (eds.) CAV 2003. LNCS, vol. 2725, pp. 118–121. Springer, Heidelberg (2003). https://doi.org/10.1007/978-3-540-45069-6_12

6. Bardin, S., Leroux, J., Point, G.: FAST extended release. In: Ball, T., Jones, R.B. (eds.) CAV 2006. LNCS, vol. 4144, pp. 63–66. Springer, Heidelberg (2006). https://doi.org/10.1007/11817963_9

7. Bardin, S., Finkel, A., Leroux, J., Schnoebelen, P.: Flat acceleration in symbolic model checking. In: Peled, D.A., Tsay, Y.-K. (eds.) ATVA 2005. LNCS, vol. 3707, pp. 474–488. Springer, Heidelberg (2005). https://doi.org/10.1007/11562948_35

8. Boigelot, B.: On iterating linear transformations over recognizable sets of integers. Theor. Comput. Sci. **309**(1–3), 413–468 (2003). https://doi.org/10.1016/S0304-3975(03)00314-1

9. Boigelot, B.: Domain-specific regular acceleration. Int. J. Softw. Tools Technol. Transf. **14**(2), 193–206 (2012). https://doi.org/10.1007/s10009-011-0206-x

10. Boigelot, B., Godefroid, P.: Symbolic verification of communication protocols with infinite state spaces using qdds. Formal Methods Syst. Des. **14**(3), 237–255 (1999). https://doi.org/10.1023/A:1008719024240

11. Boigelot, B., Godefroid, P., Willems, B., Wolper, P.: The power of QDDs (extended abstract). In: Van Hentenryck, P. (ed.) SAS 1997. LNCS, vol. 1302, pp. 172–186. Springer, Heidelberg (1997). https://doi.org/10.1007/BFb0032741

12. Boigelot, B., Herbreteau, F.: The power of hybrid acceleration. In: Ball, T., Jones, R.B. (eds.) CAV 2006. LNCS, vol. 4144, pp. 438–451. Springer, Heidelberg (2006). https://doi.org/10.1007/11817963_40

13. Boigelot, B., Herbreteau, F., Mainz, I.: Acceleration of affine hybrid transformations. In: Cassez, F., Raskin, J.-F. (eds.) ATVA 2014. LNCS, vol. 8837, pp. 31–46. Springer, Cham (2014). https://doi.org/10.1007/978-3-319-11936-6_4

14. Boigelot, B., Wolper, P.: Symbolic verification with periodic sets. In: Dill, D.L. (ed.) CAV 1994. LNCS, vol. 818, pp. 55–67. Springer, Heidelberg (1994). https://doi.org/10.1007/3-540-58179-0_43

15. Bouajjani, A., Habermehl, P.: Symbolic reachability analysis of FIFO-channel systems with nonregular sets of configurations. In: Degano, P., Gorrieri, R., Marchetti-Spaccamela, A. (eds.) ICALP 1997. LNCS, vol. 1256, pp. 560–570. Springer, Heidelberg (1997). https://doi.org/10.1007/3-540-63165-8_211

16. Bouajjani, A., Habermehl, P.: Symbolic reachability analysis of fifo-channel systems with nonregular sets of configurations. Theor. Comput. Sci. **221**(1–2), 211–250 (1999). https://doi.org/10.1016/S0304-3975(99)00033-X

17. Bozga, M., Iosif, R.: On flat programs with lists. In: Cook, B., Podelski, A. (eds.) VMCAI 2007. LNCS, vol. 4349, pp. 122–136. Springer, Heidelberg (2007). https://doi.org/10.1007/978-3-540-69738-1_9

18. Bozga, M., Iosif, R., Lakhnech, Y.: Flat parametric counter automata. In: Bugliesi, M., Preneel, B., Sassone, V., Wegener, I. (eds.) ICALP 2006. LNCS, vol. 4052, pp. 577–588. Springer, Heidelberg (2006). https://doi.org/10.1007/11787006_49

19. Bozga, M., Iosif, R., Lakhnech, Y.: Flat parametric counter automata. Fundam. Informaticae **91**(2), 275–303 (2009). https://doi.org/10.3233/FI-2009-0044

20. Czerwiński, W., Lasota, S., Lazić, R., Leroux, J., Mazowiecki, F.: The reachability problem for petri nets is not elementary. In: Proceedings of the 51st Annual ACM SIGACT Symposium on Theory of Computing, STOC 2019, Phoenix, AZ, USA, June 23–26, 2019, pp. 24–33. ACM (2019). https://doi.org/10.1145/3313276.3316369

21. Czerwinski, W., Lasota, S., Lazic, R., Leroux, J., Mazowiecki, F.: Reachability in fixed dimension vector addition systems with states. In: Konnov, I., Kovács, L. (eds.) 31st International Conference on Concurrency Theory, CONCUR 2020, September 1–4, 2020, Vienna, Austria (Virtual Conference). LIPIcs, vol. 171, pp. 48:1–48:21. Schloss Dagstuhl - Leibniz-Zentrum für Informatik (2020). https://doi.org/10.4230/LIPIcs.CONCUR.2020.48

22. Englert, M., Hofman, P., Lasota, S., Lazic, R., Leroux, J., Straszynski, J.: A lower bound for the coverability problem in acyclic pushdown VAS. Inf. Process. Lett. **167**, 106079 (2021). https://doi.org/10.1016/j.ipl.2020.106079

23. Englert, M., Lazic, R., Totzke, P.: Reachability in two-dimensional unary vector addition systems with states is nl-complete. In: Grohe, M., Koskinen, E., Shankar, N. (eds.) Proceedings of the 31st Annual ACM/IEEE Symposium on Logic in Computer Science, LICS 2016, New York, NY, USA, July 5–8, 2016, pp. 477–484. ACM (2016). https://doi.org/10.1145/2933575.2933577

24. Esparza, J., Nielsen, M.: Decidability issues for petri nets - a survey. Bull. Eur. Assoc. Theor. Comput. Sci. **52**, 245–262 (1994)

25. Esparza, J., Ganty, P., Leroux, J., Majumdar, R.: Verification of population protocols. In: Aceto, L., de Frutos-Escrig, D. (eds.) 26th International Conference on Concurrency Theory, CONCUR 2015, Madrid, Spain, September 1.4, 2015. LIPIcs, vol. 42, pp. 470–482. Schloss Dagstuhl - Leibniz-Zentrum für Informatik (2015). https://doi.org/10.4230/LIPIcs.CONCUR.2015.470

26. Finkel, A., Leroux, J.: How to compose presburger-accelerations: applications to broadcast protocols. In: Agrawal, M., Seth, A. (eds.) FSTTCS 2002. LNCS, vol. 2556, pp. 145–156. Springer, Heidelberg (2002). https://doi.org/10.1007/3-540-36206-1_14

27. Fribourg, L., Olsén, H.: Proving safety properties of infinite state systems by compilation into Presburger arithmetic. In: Mazurkiewicz, A., Winkowski, J. (eds.) CONCUR 1997. LNCS, vol. 1243, pp. 213–227. Springer, Heidelberg (1997). https://doi.org/10.1007/3-540-63141-0_15

28. Hack, M.H.T.: Decidability questions for Petri nets. Ph.D. thesis, MIT (1975). http://publications.csail.mit.edu/lcs/pubs/pdf/MIT-LCS-TR-161.pdf

29. Hauschildt, D.: Semilinearity of the Reachability Set is Decidable for Petri Nets. Ph.D. thesis, University of Hamburg (1990)
30. Hopcroft, J.E., Pansiot, J.J.: On the reachability problem for 5-dimensional vector addition systems. Theor. Comput. Sci. **8**, 135–159 (1979)
31. Karp, R.M., Miller, R.E.: Parallel program schemata. J. Comput. Syst. Sci. **3**(2), 147–195 (1969). https://doi.org/10.1016/S0022-0000(69)80011-5
32. Kosaraju, S.R.: Decidability of reachability in vector addition systems (preliminary version). In: STOC, pp. 267–281. ACM (1982). https://doi.org/10.1145/800070.802201
33. Kosaraju, S.R., Sullivan, G.F.: Detecting cycles in dynamic graphs in polynomial time (preliminary version). In: Simon, J. (ed.) Proceedings of the 20th Annual ACM Symposium on Theory of Computing, May 2–4, 1988, Chicago, Illinois, USA, pp. 398–406. ACM (1988). https://doi.org/10.1145/62212.62251
34. Lambert, J.: A structure to decide reachability in Petri nets. Theor. Comput. Sci. **99**(1), 79–104 (1992). https://doi.org/10.1016/0304-3975(92)90173-D
35. Leroux, J.: The general vector addition system reachability problem by Presburger inductive invariants. In: Logical Methods in Computer Science, vol. 6, no. 3 (2010). https://doi.org/10.2168/LMCS-6(3:22)2010
36. Leroux, J.: Vector addition system reachability problem: a short self-contained proof. In: POPL, pp. 307–316. ACM (2011). https://doi.org/10.1145/1926385.1926421
37. Leroux, J.: Vector addition systems reachability problem (A simpler solution). In: Turing-100. EPiC Series in Computing, vol. 10, pp. 214–228. EasyChair (2012). http://www.easychair.org/publications/paper/106497
38. Leroux, J.: Presburger vector addition systems. In: 28th Annual ACM/IEEE Symposium on Logic in Computer Science, LICS 2013, New Orleans, LA, USA, June 25–28, 2013, pp. 23–32. IEEE Computer Society (2013). https://doi.org/10.1109/LICS.2013.7
39. Leroux, J., Schmitz, S.: Demystifying reachability in vector addition systems. In: 30th Annual ACM/IEEE Symposium on Logic in Computer Science, LICS 2015, Kyoto, Japan, July 6–10, 2015, pp. 56–67. IEEE Computer Society (2015). https://doi.org/10.1109/LICS.2015.16
40. Leroux, J., Schmitz, S.: Reachability in vector addition systems is primitive-recursive in fixed dimension. In: 34th Annual ACM/IEEE Symposium on Logic in Computer Science, LICS 2019, Vancouver, BC, Canada, June 24–27, 2019, pp. 1–13. IEEE (2019). https://doi.org/10.1109/LICS.2019.8785796
41. Leroux, J., Sutre, G.: On flatness for 2-dimensional vector addition systems with states. In: Gardner, P., Yoshida, N. (eds.) CONCUR 2004. LNCS, vol. 3170, pp. 402–416. Springer, Heidelberg (2004). https://doi.org/10.1007/978-3-540-28644-8_26
42. Leroux, J., Sutre, G.: Flat counter automata almost everywhere! In: Software Verification: Infinite-State Model Checking and Static Program Analysis, 19.02. - 24.02.2006. Dagstuhl Seminar Proceedings, vol. 06081. Internationales Begegnungs- und Forschungszentrum fuer Informatik (IBFI), Schloss Dagstuhl, Germany (2006). http://drops.dagstuhl.de/opus/volltexte/2006/729
43. Leroux, J.: When reachability meets grzegorczyk. In: Hermanns, H., Zhang, L., Kobayashi, N., Miller, D. (eds.) LICS 2020: 35th Annual ACM/IEEE Symposium on Logic in Computer Science, Saarbrücken, Germany, July 8–11, 2020, pp. 1–6. ACM (2020). https://doi.org/10.1145/3373718.3394732

44. Mayr, E.W.: An algorithm for the general petri net reachability problem. In: Proceedings of the 13th Annual ACM Symposium on Theory of Computing, May 11–13, 1981, Milwaukee, Wisconsin, USA, pp. 238–246. ACM (1981). https://doi.org/10.1145/800076.802477
45. Mayr, E.W.: An algorithm for the general Petri net reachability problem. SIAM J. Comput. **13**(3), 441–460 (1984). https://doi.org/10.1137/0213029
46. Mayr, E.W., Meyer, A.R.: The complexity of the finite containment problem for petri nets. J. ACM **28**(3), 561–576 (1981). https://doi.org/10.1145/322261.322271
47. McAloon, K.: Petri nets and large finite sets. Theor. Comput. Sci. **32**, 173–183 (1984). https://doi.org/10.1016/0304-3975(84)90029-X
48. Sacerdote, G.S., Tenney, R.L.: The decidability of the reachability problem for vector addition systems (preliminary version). In: Proceedings of the 9th Annual ACM Symposium on Theory of Computing, May 4–6, 1977, Boulder, Colorado, USA, pp. 61–76. ACM (1977). https://doi.org/10.1145/800105.803396
49. Schmitz, S.: The complexity of reachability in vector addition systems. SIGLOG News **3**(1), 4–21 (2016). https://dl.acm.org/citation.cfm?id=2893585

Application of Concurrency to System Design

Cost and Quality in Crowdsourcing Workflows

Loïc Hélouët[1(✉)], Zoltan Miklos[2], and Rituraj Singh[2]

[1] INRIA Rennes, University of Rennes 1, Rennes, France
`loic.helouet@inria.fr`
[2] University of Rennes 1, Rennes, France
{`zoltan.miklos,rituraj.singh`}`@irisa.fr`

Abstract. Crowdsourcing platforms provide tools to replicate and distribute micro tasks (simple, independent work units) to crowds and assemble results. However, real-life problems are often complex: they require to collect, organize or transform data, with quality and costs constraints. This work considers dynamic realization policies for complex crowdsourcing tasks. Workflows provide ways to organize a complex task in phases and guide its realization. The challenge is then to deploy a workflow on a crowd, i.e., allocate workers to phases so that the overall workflow terminates, with good accuracy of results and at a reasonable cost. Standard "static" allocation of work in crowdsourcing affects a fixed number of workers per micro-task to realize and aggregates the results. We define new *dynamic* worker allocation techniques that consider progress in a workflow, quality of synthesized data, and remaining budget. Evaluation on a benchmark shows that dynamic approaches outperform static ones in terms of cost and accuracy.

Keywords: Crowdsourcing · Data-centric workflows

1 Introduction

Despite recent advances in artificial intelligence and machine learning, many tasks still require human contributions. With the growing availability of Internet, it is now possible to hire workers all around the world on crowdsourcing marketplaces. Many crowdsourcing platforms have emerged in the last decade: Amazon Mechanical Turk[1], Figure Eight[2], Wirk[3], etc. They hire workers from a crowd to solve problems [23]. A platform allows employers to post tasks, that are then realized by workers in exchange for some incentives [3]. Common tasks include image annotation, surveys, classification, recommendation, sentiment analysis, etc. [10].

Work supported by the Headwork ANR.

[1] www.mturk.com.
[2] www.appen.com.
[3] www.wirk.com.

© Springer Nature Switzerland AG 2021
D. Buchs and J. Carmona (Eds.): PETRI NETS 2021, LNCS 12734, pp. 33–54, 2021.
https://doi.org/10.1007/978-3-030-76983-3_3

The existing platforms support simple, repetitive and independent *micro-tasks* which require a few minutes to an hour to complete.

However, many real-world problems are not simple micro-tasks, but rather complex orchestrations of dependent tasks, that process input data and collect workers answers for tasks requiring human expertize. Existing crowdsourcing platforms provide interfaces to execute micro-tasks and access crowd, but lack ways to specify and execute complex tasks. The next stage of crowdsourcing is to design systems to specify more involved tasks over existing crowd platforms. A natural solution is to define complex tasks as workflows, i.e., orchestrations of phases that exchange data to achieve a final objective [27]. The data output by an individual phase is passed to the next one(s) according to the workflow rules.

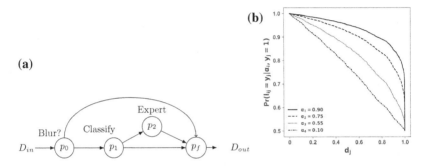

Fig. 1. *a*) A workflow from SPIPOLL, *b*) Generating functions $Pr(l_{ij}=y_j|d_j,\alpha_i,y_j=1)$

We illustrate complex workflows in Fig. 1-*a*). This workflow is an image annotation process on SPIPOLL [5], a platform to survey populations of pollinating insects. Contributors take pictures of insects that are then classified by crowdworkers. Pictures are grouped in a dataset D_{in}, input to node p_0. The process is the following. First, received images are filtered to eliminate bad pictures (fuzzy or blurred ones) in phase p_0. The remaining pictures are sent to workers who try to classify them with the help of the SPIPOLL website. If classification is too difficult, the image is sent to an expert. Initial classification is represented by phase p_1 in the workflow, and expert classification by p_2. Pictures that were discarded, classified easily or studied by experts are then assembled in a result dataset D_{out} in phase p_f, to do statistics on insect populations.

Workflows alone are not sufficient to handle complex tasks with crowdsourcing. Many data-centric applications come with budget and quality constraints: As human workers are prone to errors, one has to hire several workers to aggregate a final answer with sufficient confidence. An unlimited budget allows hiring large pools of workers to assemble reliable answers for each micro-task, but in general, a client for a complex task has a limited budget. This forces to replicate micro-tasks in an optimal way to achieve the best possible quality, but without exhausting the given budget. The objective is hence to obtain a *reliable* result, forged through a *complex orchestration*, at a *reasonable cost*.

This paper proposes a solution for the efficient realization of complex tasks. We define a workflow model, which orchestrates tasks and work distribution according to a dynamic policy that considers confidence in aggregated data and the cost to increase this confidence. A workflow can be seen as an orchestration of phases, where the goal of each phase is to tag records from its input dataset. The output of a phase is used as input for the next ones in the workflow. A complex task terminates when the last of its phases has completed its tagging. For simplicity, we consider simple Boolean tagging tasks that associate a tag in $\{0, 1\}$ to every record in a dataset. Each tagging task on each record is performed by several workers to reduce errors, and the answers are assembled using an aggregation technique. We assume that workers are uniformly paid. For each record, one of the possible answers (called the *ground truth*) is correct, and an aggregated answer is considered as reliable if its probability to be the ground truth (computed by the aggregation technique) is high. Hiring more workers to tag records increases the reliability of the aggregated answer. The overall challenge is hence to realize a workflow within a given budget B_0, while guaranteeing that the final dataset forged during the last phase of the workflow has a high probability to be the ground truth.

Design choices influence realization and quality of workflows realization. First, the chosen aggregation technique influences the quality of the final results. Furthermore, the mechanisms used to hire workers impacts costs and accuracy of answers. The simplest way to replicate micro-tasks is *static execution*, i.e., affect an identical fixed number of workers to each micro-task in the orchestration without exceeding budget B_0. On the other hand, one can allocate workers to tasks *dynamically*. One can wait in each phase to achieve a sufficient reliability of answers for all records of the input before forwarding data. This is called a *synchronous* execution of a workflow. Last, one can eagerly forward records with reliable tags to the next phases without waiting for the total completion of a phase. This is called an *asynchronous* execution.

We then study execution strategies for complex workflows in different contexts. We consider several types of workflows, different aggregation mechanisms (namely Majority Voting (MV) and Expectation Maximization (EM) [12]), several distributions of data, difficulty of tasks and workers expertize. We evaluate the cost and accuracy of workflows execution in these contexts under static, synchronous and asynchronous assignment of workers to tasks. Unsurprisingly, dynamic distribution of work saves costs in all cases. A more surprising result is that synchronous realization of complex tasks is in general more efficient than asynchronous realization.

Related Work: Several works consider data centric models, deployment on crowdsourcing platforms, and aggregation techniques to improve data quality. Due to lack of space, we only mention some of them and refer readers to the long version of this work [14] for a more complete bibliography.

Coordination of tasks has been considered in many languages such as BPMN [22], ORC [16], BPEL [21], or workflow nets [28], a variant of Petri nets dedicated to business processes. They allow parallel or sequential execution of tasks, fork and join operations to create or merge a finite number of parallel threads. Some works propose empirical solutions for complex data acquisition, mainly at the level of micro-tasks [10,19]. Crowdforge uses Map-Reduce techniques to solve complex tasks [17]. Turkit [20] is a crash and rerun programming model. It builds on an imperative language, that allows for repeated calls to services provided by a crowdsourcing platform. Turkomatic [18] is a tool that recruits crowd workers to help clients planning and solving complex jobs. It implements a Price, Divide and Solve (PDS) loop, that asks crowd workers to divide a task into orchestrations of subtasks, and repeats this operation up to the level of micro-tasks. A PDS scheme is also used by [31] in a model based on hierarchical state machines that orchestrates sub-tasks.

In this work, we assemble answers returned by workers using aggregation techniques. Basic aggregation is majority voting (MV), i.e., a mechanism that takes the most returned answer as final result for a tagging task. Several approaches have improved MV by giving more weight to competent workers. Other approaches use aggregation mechanisms based on Expectation Maximization (EM), and consider workers competences, expressed in terms of accuracy (ratio of correct answers) or in terms of recall and specificity (that considers correct classification for each possible type of answer). It is usually admitted [32] that recall and specificity give a finer picture of worker's competence than accuracy. We only highlight works that focus on EM or MV to aggregate data, and refer interested readers to [32] for a more complete survey of the domain. Zencrowd [6] considers workers competences in terms of accuracy and aggregates answers using EM. Workers accuracy and ground truth are hidden variables that must be discovered in order to minimize the deviations between workers answers and aggregated conclusion. D&S [4] uses EM to synthesize answers that minimize error rates from a set of patient records. It considers recall and specificity, but not the difficulty of tasks. [15] proposes an algorithm to assign tasks to workers, synthesize answers, and reduce the cost of crowdsourcing. It assumes that all tasks have the same difficulty, and that workers reliability is a static probability to return a correct value (i.e., the ground truth) that applies to all types of tasks. EM is used by [24] to discover recall and specificity of workers and propose a maximum-likelihood estimator that jointly learns a classifier, discovers the best experts, and estimates the ground truth. Most of the works cited above consider expertise of workers but do not address tasks difficulty. Approaches such as GLAD [30] or [2] also estimate tasks difficulty to improve quality of answers aggregation on a single batch of Boolean tagging tasks.

A few papers on data aggregation focus on costs optimization. CrowdBudget [26] is an approach that divides a budget B among K existing tasks to replicate them and then aggregate answers with MV. Crowdinc [25] is an EMbased aggregation technique that considers task difficulty, recall and specificity of workers to realize a single batch of micro tasks with a good trade-off between costs and data quality. It computes accuracy of an aggregation, and launches new tasks dynamically. The model proposed in this paper is a workflow that orchestrates tasks, replicates them, distributes them and aggregates the returned results before passing the forged dataset to the next tasks. It is a variant of the complex workflow model proposed in [1], and it uses the aggregation technique of Crowdinc [25] to forge reliable answers.

Some works consider deployment of tasks, i.e., synthesis of strategies to hire workers and parallelize realization of batches of tasks. The objective is to improve costs and latency, i.e., the time needed to treat a complete batch with an optimal deployment. CLAMSHELL [13] focuses on latency improvement. It affects workers to batches of tagging tasks and detects staggers. To speed up tasks completion, some batches are replicated. Pools are assembled and maintained by rewarding workers for waiting. This approach improves latency, but increases costs. [9] uses Markov decision processes to dynamically adapt a pricing policy so that batches of tasks are completed with the lowest latency within a fixed budget, or at the lowest price given some time constraint. [11] proposes a solution to compute the best static deployment policies in order to achieve an optimal utility (i.e., a weighted sum of overall cost and accuracy) using sequencing or parallelization of tasks. This approach is an exhaustive search which limits the number of workers and orchestrations that can be considered. [29] is a recommendation technique for deployments, that allows parallelization of tasks, sequential composition, and use of machines to solve open tasks such as translation or text writing. This approach builds on optimization techniques to find deployments that reduces latency and improves quality of data.

2   Complex Workflows with Aggregation

Complex Workflows are inspired by data centric workflows [1], but allow tasks replication, and consider aggregation and budget management. The context of the workflow is the following: A client wants to realize a complex task that needs the knowledge and skills of human workers. Complex tasks are divided into several dependent phases. Each phase processes records from an input dataset or merges different inputs to a single one, and forwards the result to its successor. Datasets are collections of records, i.e., relations of the form $r(a_1, \ldots a_k)$ where each a_i is a value for a field of the record, chosen from a domain Dom_i. One can use First-Order statements with variables denoting fields values to address properties of a record (e.g. write $v_i == true$), or of a set of records in a dataset (e.g. $\exists r(v_1, \ldots v_k) \in D, v_k == true$). We will denote by FO^R the FO formulas for records, and by FO^D the FO formulas for datasets. For simplicity, we assume that processing a record is a micro-task that simply consists in adding a new Boolean field (called a *tag*) to this record. Hence a micro-task can be seen as an operation that transforms a record $r(v_1, \ldots, v_k)$ into a new record $r'(v_1, \ldots, v_k, v_{k+1})$ where v_{k+1} is a Boolean value. This setting can be easily adapted to let v_{k+1} take values from a discrete domain.

As humans are prone to errors, phases are not unique micro tagging tasks, but rather replications of batches of tagging tasks allocated to several workers.

The returned answers are then aggregated before proceeding to the next phase. Hence, an aggregation mechanism is required to combine the answers and forward the results to the next phases. When a phase has several successors, the contents of records is used to decide to which successor(s) it should be forwarded. This allows to split datasets according to the value of a particular field, process differently records depending on their contents, create concurrent threads, etc.

Definition 1 (Complex Workflow). *A complex workflow is a tuple* $W = (\mathcal{P}, \longrightarrow, G, \bigotimes, p_0, p_f)$ *where* $\mathcal{P}$ *is a finite set of phases,* p_0 *is a particular phase without predecessor,* p_f *a phase without successor,* $\longrightarrow \subseteq \mathcal{P} \times FO^R \times \mathcal{P}$ *is a flow relation and* $G : \mathcal{P} \to FO^D$ *associates a guard to every phase, and for every* $p_x \in \mathcal{P}$, $\bigotimes^x$ *is an operator used to merge datasets input to* p_x.

Intuitively, a phase performs a batch of tagging tasks (one for each record in a dataset), but replicates and distributes them to several workers. The answers returned by all hired workers are then aggregated to get a final trusted answer. We assume that workers answers are independent. For a triple $(p_x, g_{x,y}, p_y)$ in $\longrightarrow$, we will say that p_x is a predecessor of p_y. We denote by $Succ(p_x) = \{p_y \mid p_x \longrightarrow^* p_y\}$ the set of phases that must occur after p_x, and by $Pred(p_x) = \{p_y \mid p_y \longrightarrow^* p_x\}$ the set of phases that must occur before p_x. The meaning of guard $g_{x,y}$ is that every record produced by phase p_x that satisfies guard $g_{x,y}$ is forwarded to p_y. We will see in the rest of this section that "producing a record" is not done in a single shot, and requires to duplicate a tagging micro-task, aggregate answers, and decide if the confidence in the aggregated answer is sufficient. When a phase p_x has several successors $p_y^1, \ldots p_y^k$ and the guards $g_{x,y^1}, \ldots g_{x,y^1}$ are exclusive, each record processed by p_x is sent to at most one successor. We will say that p_x is an *exclusive fork* phase. On the contrary, when guards are not exclusive, a copy of each record processed in p_x can be sent to each successor (hence increasing the amount of data processed in the workflow), and p_x is called a *non-exclusive fork* phase. For a phase $p_x \in \mathcal{P}$, we denote by $G_x \in FO^D$ the guard attached to phase p_x. G_x addresses properties of the datasets input to p_x by its predecessors. This allows in particular to require that all records in preceding phases have been processed (we will then say that phase p_x is *synchronous*), that at least one record exists in a dataset produced by a predecessor (the task is then fully *asynchronous*), or more generally satisfaction of any FO expressible property on datasets produced by predecessors of p_x. The operator $\bigotimes^x$ can be either a simple union of datasets, or a more complex join operation. If $\bigotimes^x$ is a join operation, we impose that p_x is synchronous. This is reasonable, as one cannot start processing data produced by a join operation when the final set of records is not known. When $\bigotimes^x$ is a simple union of datasets, as tasks are independent, any record processed on a predecessor of p_x can be forwarded individually without waiting for other results to be available. This allows asynchronous executions in which two phases p_x, p_y can be concurrently active (i.e., have started processing records), even if p_x precedes p_y. On the contrary, if the execution of a phase p_x is synchronous, and p_y is a successor of p_x all records input to p_x must be processed before starting phase p_y.

The semantics of a complex workflow is defined in terms of moves from a configuration to the next one, organized in *rounds*. Configurations memorize the data received by phases, a remaining budget, the answers of workers, aggregated answers quality and workers competences.

Definition 2. *A configuration is a tuple* $C = (\mathcal{D}_{in}, W_{in}, W_{out}, conf, B)$ *where*

- $\mathcal{D}_{in} : \mathcal{P} \to Dsets$ *associates a (possibly empty) dataset to every phase* $p_x \in \mathcal{P}$.
- $W_{in} : \mathcal{P} \times \mathbb{N} \times \to 2^W$ *is a partial map that associates a set of workers to each record in* $\mathcal{D}_{in}(p_x)$.
- $W_{out} : \mathcal{P} \times \mathbb{N} \times W \to \{0,1\} \cup \emptyset$ *is a partial map that associates a tag or the empty set to a worker, a phase and a record.* $W_{out}(p, n, w)$ *is defined only if* $w \in W_{in}(p, n)$. *We denote by* $l_{i,j}^x$ *the answer returned by worker* w_i *when tagging record* r_j *during phase* p_x.
- $conf : \mathcal{P} \times \mathbb{N} \to [0,1]$ *is a map that associates to each record in* $\mathcal{D}_{in}(p_x)$ *a confidence score in* $[0,1]$ *computed from answers in* W_{out}.
- $B \in \mathbb{N}$ *is the remaining budget.*

$W_{out}(p_x, k, w) = \emptyset$ indicates that a worker w in a phase p_x has not yet processed record r_k. We say that phase p_x is *completed* for a record r_k from $\mathcal{D}_{in}(p_x)$ if there is no worker w such that $W_{out}(p_x, w, r_k) = \emptyset$. As soon as phase p_x is completed for r_k, we can derive an *aggregated answer* $r_k'(v_1, \ldots, v_n, y_k^x)$ for each record $r_k(v_1, \ldots v_n)$ from the set of all answers returned by the workers in $W_{in}(p_x, k)$. Similarly, we can compute a confidence score $conf(p_x, k)$ for value y_k^x and the expertise of each worker (we will see how these values are evaluated in Sect. 4). We say that a record r_i in a phase p_x is *inactive* if no more workers are assigned to it. It is *active* otherwise. Given a threshold value Th, we will say that p_x is *finished* for a record r_k from $\mathcal{D}_{in}(p_x)$ if p_x is completed for r_k and $conf(p_x, k) > Th$. Record $r_k'(v_1, \ldots, v_n, y_k^x)$ will then be part of the input of phase p_y if $(p_x, g_{x,y}, p_y) \in \longrightarrow$ and $r_k'(v_1, \ldots, v_n, y_k^x)$ satisfies guard $g_{x,y}$.

We can now detail how rounds change the configuration of a workflow. The key idea is that each round aggregates available answers, and then decides whether the confidence in aggregated results is sufficient. If confidence in a record is high enough, this record is forwarded to the successor phases, if not new workers are hired for the next round, which decreases the remaining budget. The threshold for the confidence decreases accordingly. Then new workers are hired for freshly forwarded data, leaving the system ready for the next round. From a configuration $C = (\mathcal{D}_{in}, W_{in}, W_{out}, conf, exp, B)$, a round produces a new configuration $C' = (\mathcal{D}_{in}, W_{in}, W_{out}, conf, exp, B)$ as follows:

- **Answers:** Workers hired in preceding round produce new data. For every phase p_x, every record $r_n \in \mathcal{D}_{in}(p_x)$, and every worker w_i such that $w_i \in W_{in}(p_x, n)$ and $W_{out}(p_x, w_i, n) = \emptyset$, we produce a new output $l_{i,n}^x \in \{0,1\}$ and set $W_{out}(p_x, w_i, n) = l_{i,n}^x$.
- **Aggregation:** The system aggregates answers in every active phase p_x. For every record r_k in $\mathcal{D}_{in}(p_x)$, we compute an aggregated answer y_k^x from the set of answers $A_k = \{l_{i,k}^x \mid w_i \in W_{in}(p_x, n)\}$. We also compute a new confidence

score $conf'(p_x, n)$ for the aggregated answer (this confidence depends on the aggregation technique), and evaluate workers expertize and the difficulty of tagging each record(with the algorithm shown in Sect. 3).

- **Data forwarding:** We distinguish asynchronous and synchronous phases. Let p_y be an asynchronous phase ($\bigotimes^y$ can only be a union of records). Then p_y accept every new record $r'(v_1, \ldots v_k, y_n^x)$ that was not yet among its inputs from a predecessor p_x provided r' satisfies guard $g_{x,y}$, and the confidence in the aggregated answer y_n^x is high enough. Formally, $\mathcal{D}'_{in}(p_y) = \mathcal{D}_{in}(p_y) \cup \{r'(v_1, \ldots v_k, y_n^x)\}$ if $(p_x, g_{x,y}, p_y) \in \longrightarrow$, $conf'(p_x, n) \geq Th$ and $r'(v_1, \ldots v_k, y_n^x) \models g_{x,y}$. Let p_y be a phase such that $\bigotimes^y$ is synchronous. We will say that a phase is closed if all its predecessors are closed, and for every $n, r_n \in \mathcal{D}_{in}(p_x), conf(p_x, n) \geq Th$. If there exists a predecessor p_x of p_y that is not closed, then $\mathcal{D}'_{in}(p_y) = \emptyset$. Otherwise we can compute an input for phase p_y as a join over datasets computed by all preceding phases. Formally, $\mathcal{D}'_{in}(p_y) = \bigotimes^y \{D_x \mid p_x \longrightarrow p_y\}$, where $D_x = \{r'(v_1, \ldots v_k, y_n^x) \mid r(v_1, \ldots v_k) \in \mathcal{D}_{in}(p_x) \land r'(v_1, \ldots v_k, y_n^x) \models g_{x,y}\}$ i.e., $\mathcal{D}'_{in}(p_y)$ merges data produced by all predecessors of p_y. Hence, for a synchronous phase p_y, the input dataset is obtained by a join operation computed over datasets filtered by guards, and realized only once preceding tasks have produced all their results. In synchronous and asynchronous settings, a phase p_y becomes active if $\mathcal{D}'_{in}(p_y) \models G(p_y)$. We set $conf'(p_y, n) = 0$ for every new record in $\mathcal{D}'_{in}(p_y)$.
- **Worker allocation:** For every p_x that is active and every record $r_n = r(v_1, \ldots v_k) \in \mathcal{D}'_{in}(p_x)$ such that $conf'(p_x, n) < Th$, we allocate k new workers $w_1, \ldots w_k$ to record r_n for phase p_x, i.e., $W_{in}(p_x, n) = W_{in}(p_x, n) \cup \{w_1, \ldots, w_k\}$. This number k of workers depend on the chosen policy (see details in Sect. 4). Accordingly, for every new worker w_i affected to a tagging task for a record r_n in phase p_x, we set $W'_{out}(p_x, n, i) = \emptyset$.
- **Budget update:** We then update the budget. The overall number of workers hired is $nw = \sum_{p_x \in P} \sum_{r_n \in \mathcal{D}'_{in}(p_x)} |W'_{in}(p_x, n) \setminus W_{in}(p_x, n)|$. We consider, for simplicity, that all workers and tasks have identical costs, we hence set $B' = B - nw$.

An execution starts from an initial configuration C_0 in which only p_0 is active, with an input dataset affected to p_0, and begins with workers allocation. Executions end successfully in a configuration C_f where all records in $\mathcal{D}_{in}(p_f)$ are tagged with a sufficient threshold, or fail if they reach a configuration $C \neq C_f$ with a remaining budget $B = 0$. Notice that several factors influence the overall execution of a workflow. First of all, the way workers answers are aggregated influence the number of workers that must be hired to achieve a decent confidence in the synthesized answer. We propose to consider two main aggregation policies. The first one is majority voting (MV), where a fixed static number of workers is hired for each record in each phase. A second policy is the expectation maximization (EM) based technique proposed in [25], in which workers are hired on demand to increase confidence in aggregated answers. With this policy, the confidence in answers is computed taking into account the estimated expertise of workers, and the difficulty of records tagging. The number of workers hired per record in a phase is not fixed, but rather computed considering the difficulty of tagging records, and the remaining budget.

Recall that for a phase p_x, asynchronous execution allows to start processing records as soon as $\mathcal{D}_{in}(p_x) \neq \emptyset$. Conversely, synchronous execution forces p_x to wait for the termination of its predecessors. Choosing a synchronous or asynchronous execution policy may hence influence the time and budget spent to realize a complex task. In Sect. 5, we study the impact of synchronous/asynchronous guards on the overall execution of a workflow.

3 Aggregation Model

As mentioned in previous section, crowdsourcing requires replication of micro-tasks, and aggregation mechanisms for the answers returned by the crowd. For simplicity, we consider Boolean tasks, i.e., with answer 0 or 1. However, the model easily extends to a more general setting with a discrete set of answers.

Consider a phase p_x which input is a set of records $D_x = \{r_1, r_2, \ldots, r_n\}$, and which goal is to associate a Boolean tag to each record of D_x. We assume a set of k independent workers that return Boolean answers, and denote by l_{ij} the answer returned by worker j for a record r_i. $L_i = \bigcup_{j \in 1...k} l_{ij}$ denotes the set of answers returned by k workers for a record r_i and $L = \bigcup_{j \in 1...n} L_j$ denotes the set of all answers. We assume that workers are independent (there is no collaboration and their answers are hence independent), and *faithful* (they do not give wrong answers intentionally). The objective of aggregation is to derive a set of *final answers* $Y = \{y_j, 1 \leq j \leq n\}$ from the set of answers L. Once a final answer y_j is computed, it can be appended as a new field to record r_j. The set of produced results can be forwarded to successor phases of p_x, which may launch new phases.

We consider several parameters to model tasks and workers, namely the difficulty to tag a record, and the expertise of workers. The *difficulty* to tag a record r_j is modeled by a real valued parameter $d_j \in [0, 1]$. Value 0 means that tagging r_j is very easy, and $d_j = 1$ means that it is extremely difficult. Expertise of a worker is often quantified in terms of *accuracy*, i.e., as the ratio of correct answers. However, accuracy can lead to bias in the case of datasets with unbalanced ground truth. Indeed, consider a case where the number of records with ground truth 1 is much higher than the number of records with ground truth 0. If a worker annotates most of records with ground truth 1 as 1 but makes errors when tagging records with ground truth 0, her accuracy will still be very high. We hence prefer a more precise model, where expertise of a worker is given as a pair $\xi_i = \{\alpha_i, \beta_i\}$, where α_i is the **recall** and β_i the **specificity** of worker i. The *recall* α_i is the probability that worker i answers 1 when the ground truth is 1, i.e., $\alpha_i = Pr(l_{ij} = 1 | y_j = 1)$. The *specificity* β_i is the probability that worker i answers 0 when the ground truth is 0, i.e., $\beta_i = Pr(l_{ij} = 0 | y_j = 0)$. We do not have a priori knowledge of the behavior of workers, so we define a generative model to determine the probability of correct answers when α_i, β_i are known.

This probability depends on the difficulty of a task, on recall and specificity of the considered worker, and on the ground truth. We set $Pr(l_{ij} = y_j | d_j, \alpha_i, y_j = 1) = (1 + (1 - d_j)^{(1-\alpha_i)})/2$ and $Pr(l_{ij} = y_j | d_j, \beta_i, y_j = 0) = (1 + (1 - d_j)^{(1-\beta_i)})/2$.

Figure 1-b) shows probability to get $l_{ij} = 1$ when $y_i = 1$. The horizontal axis represents the difficulty of a task, the vertical axis denotes the probability to get answer $l_{ij} = 1$. Each curve represents this probability for a particular value of *recall*. Note that the vertical axis ranges from 0.5 to 1.0 as a random guess by a worker can still provide a correct answer with probability 0.5. As the difficulty of task increases, the probability of giving a correct answer decreases and when the task difficulty is 1 workers only make random guesses. For a fixed difficulty of a task, the higher recall is, the more accurate answers are.

We equip complex workflows with an aggregation technique that uses Expectation Maximization (EM) [12]. EM is an iterative method that alternates between an expectation (E) step and a maximization (M) step. For a pool of k workers processing n records, we estimate jointly latent variables $(\alpha_i)_{i \in 1 \ldots k}$, $(\beta_i)_{i \in 1 \ldots k}, (d_j)_{j \in 1 \ldots n}$ and derive a set of *final answers* $Y = y_1 \ldots y_n$. We denote by θ the values of $(\alpha_i)_{i \in 1 \ldots k}, (\beta_i)_{i \in 1 \ldots k}, (d_j)_{j \in 1 \ldots n}$. In the E-step, we compute for each record r_j the posterior probability of $y_j = 0$ and $y_j = 1$, given the difficulty d_j, workers expertise $(\alpha_i, \beta_i)_{(i \in 1 \ldots k)}$ and the answers $L_j = \{l_{i,j} \mid i \in 1 \ldots k\}$. In the M-Step, we compute the parameters θ that maximize $Q(\theta, \theta^t)$, the expected value of the log likelihood function, with respect to the estimated posterior probabilities of Y computed during the E-step of the algorithm. Let θ^t be the value of parameters computed at step t of the algorithm. We use the observed values of L, and the previous expectation for Y. We find parameters θ that maximize $Q'(\theta, \theta^t) = \mathbb{E}[log Pr(L, Y \mid \theta) \mid L, \theta^t]$ (we refer interested readers to [8]-Chap. 9 and [7] for explanations showing why this is equivalent to maximizing $Q(\theta, \theta^t)$). We take as next value for parameters $\theta^{t+1} = \arg\max_\theta Q'(\theta, \theta^t)$.

This maximization is done with optimization techniques provided by the scipy[4] library. We iterate E and M steps, computing at each iteration t the posterior probability and the parameters θ^{t+1} maximizing $Q'(\theta, \theta^t)$. The algorithm converges, and stops when the difference between two successive joint log-likelihood values is below a threshold (set in our case to $1 \cdot e^{-7}$). It returns values for parameters $(\alpha_i)_{i \in 1 \ldots k}, (\beta_i)_{i \in 1 \ldots k}, (d_j)_{j \in 1 \ldots n}$. The final answers are the most probable y_j's.

4 Cost Model for Workflow

The objective of a complex workflow W over a set of phases $P = \{p_0, \ldots, p_f\}$ is to transform a dataset input to the initial phase p_0 and eventually produce an output dataset. The final answer is the result of the last processed phase p_f. The simplest scenario is a workflow that adds several binary tags to input records. The realization of a micro-task by a worker is paid, and workflows come with a fixed maximal budget B_0 provided by the client. For simplicity, we consider

---

[4] docs.scipy.org/doc/scipy/reference/generated/scipy.optimize.minimize.html.

that each worker receives one unit of credit per realized task. As explained in Sect. 2, each phase receives records, each record is tagged by one or several workers. Answers are then aggregated, and the records produced by a phase p_x are distributed to its successors if they meet some conditions on their data. A consequence of this filtering is that records have different lifetimes and follow different paths in the workflow. Further, one can hire more workers to increase confidence in an aggregated result if needed and if a sufficient budget remains available. Several factors influence the realization of a workflow and its cost: the number of tagging tasks that have to be realized, the available initial budget, the confidence in produced results, workers expertise, the size and nature of input data, the difficulty of tagging, and the policies chosen to realize a workflow and to hire workers. Existing crowdsourcing platforms often use static allocation, i.e., fix a number K_s of workers to hire for each micro-task. An obvious drawback of this approach is that the same effort is spent on easy and difficult tasks.

In Sect. 2, we have defined synchronous and asynchronous schemes to allocate workers on-the-fly to tasks. In this section, we define the cost model associated with these schemes, and in particular, the threshold measure used to decide whether more workers should be hired. We show in Sect. 5 that the algorithm achieves a good trade-off between cost and accuracy. Recall that at each round, we allocate new micro-tagging tasks to workers, to obtain answers for records that are still open. EM aggregation is used to compute a plausible aggregated tag y_j^x for each record r_j from a set of answers L_j^x obtained in each active phase p_x. The algorithm also gives an estimation of difficulty d_j^x (the difficult of the micro-task that consists in tagging record $r_j \in p_x$), and evaluates the expertise level of every worker w_i, i.e., its recall α_i and its specificity β_i. We also obtain a *confidence score* $\hat{c}_j^x$ for the aggregated answer y_j^x. This score is used to decide whether one needs more answers or conversely has to consider y_j^x as a definitive result. Let $k_j^x = |L_j^x|$ denote the number of answers for record $r_j \in p_x$ at a given instant. The *confidence* $\hat{c}_j^x$ in final label y_j^x is defined as:

$$\hat{c}_j^x = \begin{cases} \frac{1}{k_j^x} \cdot \sum_{i=1}^{k_j^x} \{l_{ij}^x \times (\frac{1+(1-d_j^x)^{(1-\alpha_i)}}{2}) + (1 - l_{ij}^x) \times (1 - \frac{1+(1-d_j^x)^{(1-\alpha_i)}}{2})\} & \text{if } y_j^x = 1 \\ \frac{1}{k_j^x} \cdot \sum_{i=1}^{k_j^x} \{(1 - l_{ij}^x) \times (\frac{1+(1-d_j^x)^{(1-\beta_i)}}{2}) + (l_{ij}^x) \times (1 - \frac{1+(1-d_j^x)^{(1-\beta_i)}}{2})\} & \text{if } y_j^x = 0 \end{cases}$$

Confidence $\hat{c}_j^x$ is a weighted sum of individual confidence of workers in the aggregated result. Each worker adds its probability of answering correctly (i.e., choose $l_{ij}^x = y_j^x$) when aggregating the final answer. This probability depends on y_j^x, but also on worker's competences. If confidence $\hat{c}_j^x$ is greater than a current threshold Th, then answer y_j^x is considered as definitive and the record r_j is closed. Otherwise, the record remains active. We fix a maximal number $\tau \geq 1$ of workers that can be hired during a round for a particular record. Let T_{ar} denote the set of active records after aggregation and D_{max}^x the maximal difficulty for an active record in T_{ar} tagged by phase p_x. For every record $r_j^x \in T_{ar}$ with difficulty d_j^x, we allocate $\mathbf{a}_j^x = \lceil (d_j^x / D_{max}^x) \times \tau \rceil$ new workers for the next round. Intuitively, we allocate more workers to difficult tasks. Now, T_{ar} and hence $\mathbf{a}_j^x$ depend on the threshold computed at each round. An appropriate threshold must consider the remaining budget, the remaining work to do, that depends on the number of records to be processed, on the structure of the workflow, and on the chosen policy. The first parameter to fix for the realization of a workflow is the initial budget B_0. The height and width of a workflow can be used to find a coarse overapproximation of the budget allowing to complete an execution of a workflow. To obtain sharper bound, we first bound the number of remaining phases that a record have to go through to the final phase p_f when it is processed in a phase p_x. We call this number the *foreseeable workload* at phase p_x and denote it by $fw(p_x)$.

Definition 3 (Foreseeable workload). *The foreseeable workload $fw(p_x)$ at phase p_x is the maximal number of phases visited by a record processed in p_x.*

We give an algorithm to compute the foreseeable workload in [14]. Intuitively, it considers the structure of the workflow to compute the number of phases visited between a fork node n_1 and the corresponding merge node n_2: it is the longest path in case of an exclusive fork node, and the total number of nodes between n_1 and n_2 otherwise.

Definition 4 (Foreseeable task number). *Let C be a configuration, and n_x denote the total number of active records at a phase p_x in C. The foreseeable task number from p_x in C is denoted $ft_C(p_x)$ and defined as $ft_C(p_x) = n_x \times fw(p_x)$. The foreseeable task number in C is the sum $FTN(C) = \sum_{p_x \in \mathcal{P}} ft_C(p_x)$.*

Let us now define a threshold function based on the current configuration of a workflow. This function must consider all records that still need processing, the remaining budget, and an upper bound on the number of tagging tasks that will have to be realized to complete the workflow. Further, the execution policy will influence the way workers are hired, and hence the budget spent. In a synchronous execution, records in a phase p_x can be processed only when *all* records in preceding phases have been processed. On the contrary, in asynchronous execution mode, processing of records input to a phase p_x can start without waiting for the closure of all records input to preceding phases. A consequence is that in synchronous modes, the decision to hire workers to improve accuracy of answers for a task can be taken *locally* to each phase, while in an asynchronous mode, this decision depends on a *global* view of the remaining work in the workflow. Hence, for a synchronous execution policy, we will define a local threshold computed for each phase, and for an asynchronous execution policy, we will consider a global threshold, computed for the whole workflow.

Asynchronous Execution: The execution of a workflow starts from a configuration C_0 with an expected workload $FTN(C_0)$. It is an upper bound, as all records do not necessarily visit this maximal number of phases. We define a global ratio $\Gamma_C \in [0,1]$ of already executed or avoided work in configuration C as $\Gamma_C = \frac{(FTN(C_0) - FTN(C))}{FTN(C_0)}$. Note that at the beginning of an execution, $\Gamma_{C_0} = 0$ as no record is processed yet. When records are processed and moved to successor phases, Γ increases, and we necessarily have $\Gamma_{C_f} = 1$ when no record remains to process in a final configuration C_f. Now, the threshold value has to account for the remaining budget to force the progress of records processing. Let B_0 denote the initial budget at the beginning of execution, and B_C be the budget consumed in configuration C. We denote by β_C the fraction of B_0 consumed in configuration C, i.e., $\beta_C = \frac{B_C}{B_0}$. In the initial configuration, $\beta_{C_0} = 0$. The value of β increases at every round of the execution, and takes value $\beta = 1$ when the whole budget is spent. However, our objective is to end executions with $\beta < 1$. We now define a global threshold value $Th_C \in [0.5, 1.0]$ that accounts for the remaining work and budget.

$$Th_C = \frac{1 + (1 - \beta_C)^{\Gamma_C}}{2} \tag{1}$$

We remind that in a phase p_x, a record r_j with confidence level $\hat{c}_j^x > Th_C$ is considered as processed for phase p_x. In an asynchronous execution policy, the threshold is a global value and applies to all records in the workflow at a given instant. The intuition for Th_C is simple: when only a few records remain to be processed, and the remaining budget is sufficiently high, then one can hire more workers. With more contributions, the confidence in aggregated final answers is expected to increase for several records. Conversely, if the number of records to be processed is high and the remaining budget is low, then the threshold decreases, and even records which current answer have a low confidence level are considered as processed and moved to the next phase(s).

Synchronous Execution: In asynchronous execution, a phase does not wait for the completion of its predecessors to start. As a consequence, records can be processed in all phases, and we consider a global threshold Th_C, and hence a global policy to hire workers. However, in synchronous execution, records are processed phase by phase, i.e., a phase does not start processing its input dataset until all records in the preceding phases have been processed. Using our global threshold Th_C may produce data with poor quality: as a phase is not launched as long as a preceding phase has an unprocessed record, one can easily meet situations where the larger part of the budget B_{in} is spent to hire workers in the first phases of the workflow, forcing to accept final answers with low confidence in the next phases. To avoid this problem, we propose to allocate the budget phase by phase. The idea is to divide the budget among phases based on the number of records processed.

We will say that a task becomes *active* when it starts processing records, i.e., once preceding phases have tagged *all* their records with a sufficient confidence on aggregated answer and the obtained datasets meet guard G_x. We denote by $init(p_x)$ the number of records input to p_x when the phase becomes active. As for asynchronous execution, synchronous execution of a workflow starts from an initial configuration C_0 with an initial budget B_0, and in each configuration C, the remaining budget is denoted by $B_r(C)$. The key idea in synchronous execution is to compute resources needed for each active phase, and to maintain after each round a ratio of input records that still need additional answers to forge a trusted answer, and a local threshold per phase. Let p_x be a phase that becomes active when the execution reaches configuration C. The initial budget allocated to p_x with $init(p_x)$ records in a configuration C is:

$$B_{in}^x = \frac{B_r(C)}{\sum_{p_i \text{ active phase}} FTN(p_i)} \times init(p_x) \qquad (2)$$

Intuitively, the remaining budget is shared among active phases to allow termination of the workflow from each phase. Then for each active phase p_x, we maintain the consumed budget B_c^x, and the ratio $\beta^x = \frac{B_c^x}{B_{in}^x}$ of consumed budget. At the end of each round, for each active phase p_x, we compute the ratio of processed tasks

$$\Gamma_C^x = \frac{|\{r_i \mid \hat{c}_i \le Th^x\}|}{Init(p_x)} \qquad (3)$$

where Th^x is the threshold computed at previous round. Obviously, if $\Gamma_C^x = 1$, phase p_x becomes inactive. Otherwise, a local threshold Th'^x for p_x to be used in the next round is computed, using the formula:

$$Th'^x = \frac{1 + (1 - \beta^x)^{\Gamma_C^x}}{2} \qquad (4)$$

With the convention that the initial threshold Th_x for a starting active phase, as no record is processed yet is $Th_x = \frac{1+(1-\beta^x)}{2}$.

Realization of Workflows: Regardless of the chosen policy, the execution of a workflow always follows the same principles. The structure of workflow W is static and does not change with time. It describes a set of phases $P = \{p_0, \ldots, p_f\}$, their dependencies, and guarded data flows from one phase to the next one. A set of n records $R = \{r_1, \ldots, r_n\}$ is used as input to W, i.e., is passed to initial phase p_0, and must be processed with a budget smaller than a given initial budget B_0. As no information about the difficulty of a task d_j^x is available at the beginning of phase p_0, τ workers are allocated to each record for an initial estimation round. The same principle is followed for each record when it enters a new phase $p_x \in W$. After collection of τ answers, at each round we first apply EM aggregation to estimate the difficulty d_j^x of active records $r_j \in p_x$, $\hat{c}_j^x$ the confidence in the final aggregated answer y_j^x and the recall α_i and specificity β_i of each worker w_i. Then we use a stopping threshold to decide whether we need more answers for each record. In asynchronous execution, the threshold Th is a global threshold, and in synchronous mode, the confidence of each record r_j in p_x is compared to the local threshold Th_x. Records with sufficient confidence are passed to the next phase(s). We hire new workers to obtain more answers for other records. This can increase the confidence level, but also decrease the threshold, as a part of the remaining budget is consumed. Executions stop when the whole budget B_{in} is exhausted or when there is no additional record left to process. Last, the final phase p_f returns the aggregated answer for each record.

Termination: Obviously, when the remaining budget decreases, the threshold(s) decrease too. However, there are situations where the confidence in some answers remains low, and the remaining budget reaches 0 before the threshold attains the lower bound 0.5 (that forces moving any record to the next phase(s)). Similarly, when records do not progress in the workflow, the ratio of realized work Γ_C remains unchanged for many rounds. As a consequence, synchronous and asynchronous realization of a workflow may fail. We will see in the experimental results section that even with poor accuracy of workers, this situation was never met. Failure corresponds to situations where the weighted answers of workers remain balanced for a long time. The threshold decreases slowly, and the confidence in aggregated answers remains lower. In that case, when threshold and confidence values coincide (in the worst case at value 0.5), the remaining budget is too low to realize the remaining work. Solutions to solve this issue and guarantee termination is to bound the sojourn time of a record in a phase, or to keep a sufficient budget to terminate the workflow with a static worker allocation policy hiring only a small number of workers per record. Another solution is to limit allocation of workers to tasks with the highest remaining workload. Yet, realization can still fail if records remain stuck in the last but one phase.

5   Experiments and Results

In this section, we evaluate execution policies on typical workflows. We consider a standard situation, where a client wants to realize a complex task defined by a workflow on a crowdsourcing platform. The client provides input data, and has a budget B_0. Crowd workers do not collaborate and hence realize their micro-tasks independently. As there exists no platform to realize complex tasks, there is no available data to compare the realization of a workflow with our approach to existing complex task executions. To address this issue, we design several typical workflows, synthetic data, and consider realizations of these workflows for various execution policies, characteristics of data, and accuracy of workers.

We consider 5 different workflows, represented in Fig. 2. Workflow W_1 is a sequence of tasks, W_2 is a standard fork-join pattern i.e., parallel processing of data followed by a merge of branches results, W_3 and W_4 are fork-join patterns with equal and different lengths on branches, and W_5 is a more complex workflow with two consecutive forks followed by merges on each branch. We consider micro-tasks that simply add Boolean tags to records. Guards from one phase p_x to the next phase p_y are simple exclusive guards sending each record to one successor, depending on the tag obtained at phase p_x. Formally, guards are FO formulas of the form $f == l_0$ or $f == l_1$, where f is the new field produced by the phase. To avoid unnecessary blocking of workflow progressions, we set $G_x == true$ for every phase $p_x \in \mathcal{P}$. In Fig. 2 we depict these choices by pairs of letters (l_0, l_1) representing the binary decision taken on each phase, For example, in workflow W_1, phase p_0 considers two possible tags denoted A and B. Phases p_7, p_8 in workflow W_5 are simple aggregations, and hence are not labeled by choices. After realization of the tasks, if the records are tagged as A by the workers then records are moved to the phase p_f and if tagged with B the records are assigned to phase p_1 for further processing. Each phase of workflows implements similar tagging and decision.

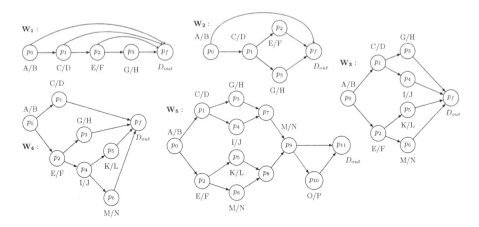

Fig. 2. Five different workflows. W_1: sequence of phases, W_2: parallel data transformations followed by an aggregation of results, W_3: fork-join patterns with uniform lengths of branches, W_4, W_5: fork-join patterns with nonuniform lengths of branches.

We evaluate average costs and accuracies achieved by workflows realizations with the following parameters. First, the input of each complex task is a dataset of 80 records. Notice that despite this fixed size, the number of micro-tasks realized during executions depend on workers competences, on the execution policy, on the value of data fields produced by workers, but also on the initial dataset, on the initial budget, etc. Each record in the original dataset has initially known data fields, and new fields are added by aggregation of workers answers during the execution of the workflow. For these fields, we assume a prior ground truth, which influences the probability that a worker answers 0 or 1 when filling this field. We generate balanced (equal numbers of 0 and 1 in fields) and unbalanced datasets (unbalanced numbers of 0 and 1).

We run the experiment with 4 randomly generated pools of 50 crowd workers, making their accuracy range from low to high expertise. For each pool, we sampled accuracies of workers according to normal distributions ranging respectively in intervals $[0.2, 0.7]$ ($Expertise_0$, low expertise of workers), $[0.4, 0.9]$ ($Expertise_1$, low to average expertise), $[0.6, 0.99]$ ($Expertise_3$, average expertise) and $[0.8, 0.99]$ ($Expertise_4$, high expertise).

The last parameter to set is the initial budget B_0. We first evaluated the cost for the realization of workflows with a static allocation policy that associates a fixed number of k_{smv} workers to each record in each phase, and aggregates their answers with Majority Voting. We call this policy *Static Majority Voting (SMV)*. A priori, running SMV with a chosen value for k_{smv} should consume a budget lower than $k_{smv}.ft_{C_0}(p_0)$, i.e., $80.k_{smv}.fw(p_0)$. This is however a coarse upper bound for the total budget B_{mv} consumed during the realization of a workflow with an SMV policy, as B_{mv} depends on the execution path followed by records during execution, and hence on random answers of workers. Yet, SMV was shown to be a naive and costly approach in most benchmarks (see for instance [25]), so starting with a budget $B_0 = B_{mv}$ for realization techniques tailored to save costs when accuracy is sufficient is a sensible approach. For each workflow, and for three different values $k_{smv} = 10/20/30$, we performed random runs of SMV to evaluate the maximal budget B_{mv} needed.

In a second step, we used the total budget B_{mv} spent by the SMV approach as initial budget for realization of the same workflow with synchronous and asynchronous policies. The objective was to achieve at least the same accuracy as SMV with synchronous and asynchronous execution policies with the same initial budget $B_0 = B_{mv}$, while spending a smaller fraction of this budget. Overall, our experiments cover realization of 5 different workflows with different values for initial budget, workers accuracy, characteristics of data, and realization policy. This means 72 different contexts, represented in Table 1 (one type of experiment represents a selection of one entry in each row). We ran each experiment 15 times to get rid of bias. This represents a sample of 1080 workflow realizations.

Table 1. Evaluation parameters

Workflow	W_1	W_2	W_3	W_4	W_5
Parameter	Value				
Worker accuracy	Low	Mid	Average	High	
Value of k_{smv}	10	20	30		
Data type	Balanced	Unbalanced			
Mechanisms	Static MV	Synchronous	Asynchronous		

We can now analyze the outcomes of our experiments. A first interesting result is that all workflow executions terminated without exhausting their given initial budget, even with low competences of workers. A second interesting (but rather expected) result is that for all realization policies, and for all workflows, executions end with poor accuracy when expertize is low. Consider for instance the results of Fig. 3. This Figure gives the consumed budget and achieved accuracy for a given workflow and a given initial budget when workers have a low expertize. The first series of results concern Workflow 1 with a parameter k_{smv} set to 10, 20, 30 workers per record in each phase. The overall expended budget with an SMV approach is around 1200, 2800, 4000, respectively. Regardless of the initial budget, synchronous and asynchronous approaches spend only a fraction of the budget allowed by SMV. Accuracy is not conclusive, as the best realization policy varies with each experiment: for instance, for W_1 with 10 workers per record to tag in each phase, SMV seems to be the best approach, while with a budget of 20, the synchronous approach is the best. However, most of the experiments achieve accuracies below 0.2, which is quite low. An explanation is that, as shown in Fig. 1−b), with low expertise, workers answers are almost random choices. Hence when all workers have a low expertise, individual errors are not corrected by other answers, and the ground truth does not influence the results. At each phase, the algorithms take their decisions mostly based on wrong answers provided by the workers and as a consequence errors accumulate. The system's behavior is then completely random, which results in poor performance. This tendency shown for all workflows and initial budgets with balanced data is confirmed on unbalanced data (the results of the experiments with unbalanced data are available in [14]).

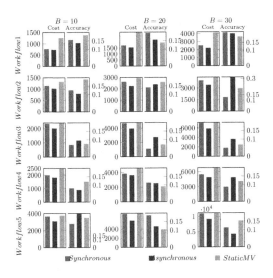

Fig. 3. Budgets and accuracies with low expertize

Next experiments consider mid-level to high expertise, which is a common setting in crowdsourcing. The experiments with competent workers and *synchronous* and *asynchronous* execution policies clearly show that dynamic allocation schemes outperform the SMV approach both in terms of cost and accuracy. One can easily see these results Figs. 4 and 5, that represent executions of workflows W_1 with three levels of expertise, 3 values of k_{smv} (and hence 3 different initial budgets), and all execution policies, respectively for balanced and unbalanced data. We show similar results in [14] for workflows W_2, W_3, W_4, W_5.

In the worst cases, *synchronous* and *asynchronous* executions achieve accuracies that are almost identical to that of SMV, but often give answers with better accuracy. With a sufficient initial budget, dynamic approaches achieve an accuracy greater than 0.9. An explanation for this improvement of synchronous and asynchronous executions w.r.t. SMV is that in SMV, one does not consider the expertise of the worker, whereas the *synchronous* and *asynchronous*

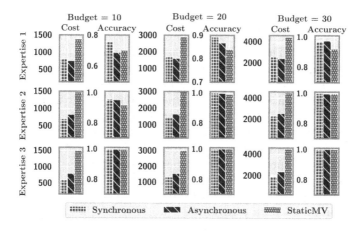

Fig. 4. Workflow 1 on balanced data

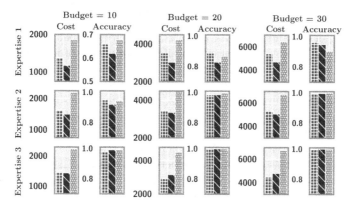

Fig. 5. Workflow 1 on unbalanced data

executions are EM based algorithms that compute the final answers by weighting individual answers according to worker's expertise. This makes EM-based evaluation of final answers more accurate than SMV. This improvement already occurs at the level of a single phase execution (this was also the conclusion of [25]). The reasons for cost improvement with respect to SMV are also easy to figure. SMV allocates a fixed number of workers to every record in every phase of a workflow, whereas synchronous and asynchronous execution schemes allocate workers on-the-fly based on a confidence level which depends on the difficulty of tasks, workers expertise, and returned answers. By comparing confidence levels with a dynamic threshold, workers allocation considers the remaining budget and workload as well. This clever allocation of workers saves costs, as easy tasks call for the help of fewer workers than the fixed number imposed by SMV. The resources that are not used on easy tasks can be reused later for difficult tasks, hence improving accuracy.

These results were expected. A more surprising outcome of the experiment is that in most cases synchronous execution outperforms asynchronous execution in terms of accuracy. The intuitive reason behind this result is that the way records are spread in the workflow execution affects the evaluation of expertize and difficulty. The synchronous execution realizes tasks in phases, while asynchronous execution starts tasks independently in the whole workflow. A consequence is that evaluation of hidden variables such as the difficulty of tasks and workers expertise in the EM aggregation improves with a larger number of records per phase in synchronous execution, while it might remain imprecise when the records are spread in different phases during an asynchronous execution. This precise estimation helps synchronous execution to allocate workers as well as to derive the final answers more efficiently and hence outperform asynchronous execution. A third general observation is that both synchronous and asynchronous executions need a greater budget to complete a workflow when data is unbalanced. Observe the results in Fig. 4 and Fig. 5: the budgets spent are always greater with unbalanced data. A possible explanation is that with balanced data, records are sent uniformly to all phases, which helps evaluation of workers expertize and difficulty of tasks, while with unbalanced data, some phases receive only a few records, which affects evaluation of hidden variables.

Unsurprisingly (see for instance Fig. 4), for a fixed budget, when worker expertise increases, accuracy increases too, and consumed budget decreases. Competent workers return correct answers, reach a consensus earlier, and hence achieve better accuracy faster. Similarly for a fixed expertise level, increasing the initial budget increases the overall accuracy of the workflow. Again, the explanation is straightforward: a higher budget increases the threshold used to consider an aggregated answer as correct, giving better accuracies. To summarize, for a fixed initial budget and high enough expertize, synchronous and asynchronous policies usually improve both cost and accuracy.

6   Conclusion

This work has proposed a model to realize complex tasks with the help of a crowd of workers. It fosters on the advantages of crowdsourcing systems and workflow. A particular attention is paid to quality of the data produced, and to the overall cost of complex tasks realization. We have compared several task distribution strategies through experiments and showed that dynamic distribution of work outperforms static allocation in terms of cost and accuracy.

A short-term extension is to consider termination of complex tasks realization with dynamic policies. Indeed, workflows realized with dynamic policies may not terminate: this happens when the guard associated with a phase is never satisfied, or when for some record, all workers agree to return the answers that do not increase the confidence. However, this latter situation was never met during our experiments, even with low expertize of workers. The probability of non-terminating executions with synchronous/asynchronous policies seems negligible. In our future work, we plan to demonstrate formally that $\mathbb{P}(B_r = 0 \wedge FTN(C) > 0)$, the probability of reaching a configuration with exhausted budget and remaining work to do is very low.

This work opens the way to new challenges. The next step is to test our approach with existing crowdsourcing platforms on a real case study. We are targeting citizen science initiatives, that typically require orchestration of various competence to reach a final objective. Now that our model is settled, another objective is to consider various strategies to hire workers in the most efficient way. A possibility to address this challenge is to see complex workflows as stochastic games, in which one player tries to maximize accuracy and reduce costs, while its opponent tries to achieve the opposite objectives.

References

1. Bourhis, P., Hélouët, L., Miklos, Z., Singh, R.: Data centric workflows for crowd-sourcing. Proc. Petri Nets **2020**, 46–61 (2020)
2. Dai, P., Lin, C.H., Weld, D.S.: Pomdp-based control of workflows for crowdsourcing. Artif. Intell. **202**, 52–85 (2013)
3. Daniel, F., Kucherbaev, P., Cappiello, C., Benatallah, B., Allahbakhsh, M.: Quality control in crowdsourcing: a survey of quality attributes, assessment techniques, and assurance actions. ACM Comput. Surv. **51**(1), 7 (2018)
4. Dawid, A., Skene, A.: Maximum likelihood estimation of observer error-rates using the EM algorithm. J. Roy. Stat. Soc. Ser. C (Appl. Stat.) **28**(1), 20–28 (1979)
5. Deguines, N., Julliard, R., De Flores, M., Fontaine, C.: The whereabouts offlower visitors: contrasting land-use preferences revealed by a country-widesurvey based on citizen science. PLOS ONE **7**(9), e45822 (2012)
6. Demartini, G., Difallah, D., Cudré-Mauroux, P.: ZenCrowd: leveraging probabilistic reasoning and crowdsourcing techniques for large-scale entity linking. In: Proceedings of the WWW 2012, pp. 469–478. ACM (2012)
7. Dempster, A., Laird, N.M., Rubin, D.B.: Maximum likelihood from incomplete data via the EM algorithm. J. Roy. Stat. Soc. Ser. B (Methodol.) **39**(1), 1–22 (1977)
8. Flach, P.: Machine Learning - The Art and Science of Algorithms that Make Senseof Data. Cambridge University Press (2012)
9. Gao, Y., Parameswaran, A.G.: Finish them!: pricing algorithms for human computation. Proc. VLDB Endow. **7**(14), 1965–1976 (2014)
10. Garcia-Molina, H., Joglekar, M., Marcus, A., Parameswaran, A., Verroios, V.: Challenges in data crowdsourcing. Trans. Knowl. Data Eng. **28**(4), 901–911 (2016)
11. Goto, S., Ishida, T., Lin, D.: Understanding crowdsourcing workflow: modeling and optimizing iterative and parallel processes. In: Proceedings of the HCOMP 2016, pp. 52–58. AAAI Press (2016)
12. Gupta, M., Chen, Y.: Theory and use of the EM algorithm. Found. Trends Sig. Process. **4**(3), 223–296 (2011)
13. Haas, D., Wang, J., Wu, E., Franklin, M.J.: CLAMShell: speeding up crowds for low-latency data labeling. Proc. VLDB Endow. **9**(4), 372–383 (2015)
14. Hélouët, L., Miklos, Z., Singh, R.: Cost and Quality Assurance in Crowdsourcing Workflows (October 2020). Extended version. https://hal.inria.fr/hal-02964736
15. Karger, D., Oh, S., Shah, D.: Iterative learning for reliable crowdsourcing systems. In: Proceedings of the NIPS 2011, pp. 1953–1961 (2011)

16. Kitchin, D., Cook, W., Misra, J.: A language for task orchestration and its semantic properties. In: Proceedings of the CONCUR 2006, pp. 477–491 (2006)

17. Kittur, A., Smus, B., Khamkar, S., Kraut, R.: CrowdForge: crowdsourcing complex work. In: Proceedings of the UIST 2011, pp. 43–52. ACM (2011)

18. Kulkarni, A., Can, M., Hartmann, B.: Collaboratively crowdsourcing workflows with Turkomatic. In: Proceedings of the CSCW 2012, pp. 1003–1012. ACM (2012)

19. Li, G., Wang, J., Zheng, Y., Franklin, M.: Crowdsourced data management: a survey. Trans. Knowl. Data Eng. **28**(9), 2296–2319 (2016)

20. Little, G., Chilton, L., Goldman, M., Miller, R.: TurKit: tools for iterative tasks on Mechanical Turk. In: Proceedings of the HCOMP 2009, pp. 29–30. ACM (2009)

21. OASIS: Web Services Business Process Execution Language. Technical report, OASIS (2007)

22. OMG: Business Process Model and Notation (BPMN). OMG (2011)

23. Quinn, A., Bederson, B.: Human computation: a survey and taxonomy of a growing field. In: Proceedings of the SIGCHI Conference on Human Factors in Computing Systems, pp. 1403–1412 (2011)

24. Raykar, V.C., et al.: Learning from crowds. J. Mach. Learn. Res. **11**, 1297–1322 (2010)

25. Singh, R., Hélouët, L., Miklós, Z.: Reducing the cost of aggregation in crowdsourcing. In: Proceedings of the ICWS 2020 (2020)

26. Tran-Thanh, L., Venanzi, M., Rogers, A., Jennings, N.: Efficient budget allocation with accuracy guarantees for crowdsourcing classification tasks. In: Proceedings of the AAMAS 2013, pp. 901–908 (2013)

27. Tsai, C.H., Luo, H.J., Wang, F.J.: Constructing a BPM environment with BPMN. In: 11th IEEE International Workshop on Future Trends of Distributed Computing Systems, FTDCS 2007, pp. 164–172. IEEE (2007)

28. Van Der Aalst, W., et al.: Soundness of workflow nets: classification, decidability, and analysis. Formal Aspects Comput. **23**(3), 333–363 (2011)

29. Wei, D., Roy, S., Amer-Yahia, S.: Recommending deployment strategies for collaborative tasks. In: Proceedings of the 2020 International Conference on Management of Data, SIGMOD Conference 2020, pp. 3–17. ACM (2020)

30. Whitehill, J., Wu, T., Bergsma, J., Movellan, J., Ruvolo, P.: Whose vote should count more: optimal integration of labels from labelers of unknown expertise. In: Proceedings of the NIPS 2009, pp. 2035–2043 (2009)

31. Zheng, Q., Wang, W., Yu, Y., Pan, M., Shi, X.: Crowdsourcing complex task automatically by workflow technology. In: MiPAC 2016 Workshop, pp. 17–30 (2016)

32. Zheng, Y., Li, G., Li, Y., Shan, C., Cheng, R.: Truth inference in crowdsourcing: is the problem solved? Proc. VLDB Endow. **10**(5), 541–552 (2017)

Timed Petri Nets with Reset
for Pipelined Synchronous Circuit Design

Rémi Parrot[(✉)], Mikaël Briday, and Olivier H. Roux

École Centrale de Nantes, LS2N UMR CNRS, 6004 Nantes, France
`remi.parrot@ec-nantes.fr`

Abstract. This paper introduces an extension of Timed Petri Nets for the modeling of synchronous electronic circuits, addressing pipeline design problems. Petri Nets have been widely used for the modeling of electronic circuits. In particular, Timed Petri Nets which capture timing properties are perfectly suited for scheduling problems. Our extension, through *reset* that model the pipeline stages, and through *delayable* transitions that relax timing constraints, allows to widen the conception space of pipelined systems.

After discussing about maximal-step firing rule and the semantics of Timed Petri Nets "à la Ramchandani", we define our Timed Petri Nets with reset and delayable (non-asap) transitions.

We then study the decidability and the complexity of the main problems of interest. We propose an abstraction of the state space. We then establish a translation of this model into a single-clock timed automata, which preserves the language. This translation settles the decidability on language inclusion and universality problems.

Finally, an algorithm for the exploration of the state space is provided, and can be driven by the optimisation of various properties of the pipeline.

1 Introduction

The field of hardware verification seems to have been started in a little-known 1957 paper by Alonzo Church, 1903–1995, in which he described the use of logic to specify sequential circuits [11]. Today's semiconductor designs are still dominated by synchronous circuits. In these circuits, clock signals synchronize the logic, providing the designer with a simple operational model.

A major step in the design of synchronous circuits concerns the automatic generation of the pipeline. The pipeline does not functionally modify the circuit, but allows to split a process into several steps in order to increase the operating frequency (throughput). Its implementation can be seen as an optimisation problem whose aim is to reach a target operating frequency while minimizing the hardware cost of the pipeline stages (registers).

This work is supported by the Renault-Centrale Nantes chair dedicated to the propulsion performance of electric vehicles.

© Springer Nature Switzerland AG 2021
D. Buchs and J. Carmona (Eds.): PETRI NETS 2021, LNCS 12734, pp. 55–75, 2021.
https://doi.org/10.1007/978-3-030-76983-3_4

As introduced in [15], a circuit can be abstracted by a weighted directed graph, where the vertices are the *operators* of the circuits and the edges are the *connections* in between. Weights are added to edges representing the number of registers, and to vertices representing the *propagation delays* of operators. The authors then proposed an operation called *retiming*, which consists in moving registers from one place to another (an operation on the edge-weights) without altering the circuit's behaviour, in order to explore various pipeline solutions.

A more suitable formalism of this approach is actually the (Timed) Marked Graph (or Event Graph), which is a subclass of Petri Net where each place has one incoming arc, and one outgoing arc.

Petri Nets to Model Circuits: Due to their concurrency nature, Petri Net have been extensively used to analyse and optimise timing properties of both synchronous and asynchronous circuits [6,7,17,21].

For example, it has shown particularly effective for building resource-optimal pipeline on Latency-insensitive systems [8], in [6], and with *control-flow* structures in [14], which is of particular interest in the *High-Level Synthesis* (HLS) approach. Furthermore, Marked Graph have also been used to pipeline asynchronous systems, through *slack matching* in [21]. More recently, in [17] the authors manage to pipeline mode-based asynchronous circuits, where there are given probabilities to switch between modes, using a combination of Markov chains and Marked Graph.

All those works share the same method of resolution: deduce the timing constraints from the Petri Net structure, and get back to an Integer Linear Problem. In contrast, we propose to encapsulate the time in our model, and to explore the states of the circuit using directly the semantics of our model. In other words, we suggest a novel modeling of the classical timing closure problem, on synchronous dataflow circuits without loops.

Petri Nets with Time: The two main time extensions of Petri Nets are Time Petri Nets [16] and Timed Petri Nets [20]. While a transition can be fired within a given interval for Time Petri Nets, deterministic (or constant) "firing duration" are assigned to transitions of Timed Petri Nets.

For Timed Petri Nets [20], each transition takes a positive time (duration) to fire according to a three-phases firing. The tokens are consumed when the transitions are enabled, then as soon as the delays have elapsed, the tokens corresponding to the firing of the transitions are produced. The model is restricted to decision-free nets (as Timed marked graph) and zero time delay is prohibited.

Enhanced timed nets are proposed in [22] that combine "immediate" nets which are in fact ordinary (i.e., timeless) free-choice acyclic Petri nets, and free-choice bounded timed Petri nets.

In [19], Popova generalised Timed Petri Nets and proposed a semantic based on the same three-phases firing, allowing null duration. Transitions are fired as soon as possible (asap) according to the maximal-step rule, *i.e.* in each marking, a maximal set of firable transitions fires at once.

Contributions and Outline of the Paper

We first rewrite in Sect. 2 the semantics of Timed Petri Nets with an atomic maximal step firing rule i.e. without three-phases firing. We then propose in Sect. 3, a Timed Petri Net extension closer to real synchronous circuits, which embeds the effect of registers on the circuit's timing with a particular *reset* action, and which permits to relax some timing constraints (allow lag on operations of the circuit) with *delayable* transitions. This new model is proved in Sect. 4 to have a PSPACE-Complete complexity for the reachability (and TCTL model checking) problem. Moreover, we provide in Sect. 5 a symbolic states space exploration algorithm, using simplified zones. This latter allows us in Sect. 6 to build a translation into a single-clock timed automata, preserving timed behaviour. Then it gives the decidability on timed language inclusion and universality problems. Finally, we present in Sect. 7 a use case of this model: to build a pipeline of a circuit, minimising the total number of flip-flops (registers of one bit), while ensuring the operating frequency to be into an interval. To do so, we provide a heuristic for the state space exploration, by adding *costs* which measure the total number of flip-flops of a state.

2   Maximal Step Firing Rule and Timed Petri Net

The introduction of deterministic time into Petri nets was first attempted by Ramchandani [20]. The time labels (duration) were assigned to each transition, denoting the fact that actions take time to complete.

$\mathbb{N}$ and $\mathbb{R}_{\geq 0}$ are respectively the sets of integer and non-negative real numbers. For vectors of size n, the usual operators $+, -, \times, <, \leq, >, \geq$ and $=$ are used on vectors of $\mathbb{N}^n$ and $\mathbb{R}_{\geq 0}^n$ and are the point-wise extensions of their counterparts in $\mathbb{N}$ and $\mathbb{R}_{\geq 0}$. Let $\bar{0}$ be the null vector of size n.

2.1   Three-Phases Firing

Ramchandani proposed a three-phases firing semantics: delete the input tokens of the transition (consumption), wait until the firing time is reached (delay) and create the output tokens of the transition (production). This firing process when initiated cannot be interrupted or stopped, therefore the consumption phase can be seen as a *reservation* (in particular in case of conflict). Moreover the transitions in the process of firing are synchronized to a global clock, through a *token balance equation* linking the tokens added and removed to the tokens present in a place between two instants. Zero time firing is prohibited, preventing the same transition from being fired twice when other transitions are in conflict.

More recently, Popova proposed a semantics based on the same three-phases firing, allowing null duration but selecting beforehand a maximal-step (a set) of transitions to be fired in the same action [19]. In other words, instead of being reserved one after the other, the transitions are selected and then reserved all at the same time (consumption phase).

2.2 Maximal-Step Firing

The classical semantics of timeless Petri Nets is the interleaving semantics. From a practical point of view, the maximal-step semantics avoids interleaving and is very interesting for synchronous system modelling.

Given a Petri Nets, the maximal-step firing compared to interleaving semantics puts more strain on the firing, increases expressiveness and removes reachable markings.

Popova shows how a counter machine can be simulated by Timed Petri nets [19]. But she also shows that this reduction can be done by timeless Petri nets with maximal-step firing. In particular the so-called zero-test, can be simulated by a timeless net thanks to the maximal-step firing rule. It means that Timeless as Timed Petri nets firing in maximal-step are Turing equivalent.

2.3 Timed Petri Net

We propose to rewrite Timed Petri Nets semantics without any reservation: waiting is done while keeping the tokens in their place, then when at least one transition is fireable we select the maximal step and fire (consumption and production) all the transitions in one atomic action. The maximal step contains, in our case, enabled transitions which have been enabled for a period of time equal to their delays.

Informally, with each transition of the Net is associated a clock and a delay. The clock measures the time since the transition has been enabled and the delay is interpreted as a firing condition: the transition may and must fire if the value of its clock is equal to the delay.

Formally:

Definition 1 (TPN). *A Timed Petri Net is a tuple* $(P, T, {}^\bullet(.), (.)^\bullet, \delta, M_0)$ *defined by:*

- $P = \{p_1, p_2, \ldots, p_m\}$ *is a non-empty set of places,*
- $T = \{t_1, t_2, \ldots, t_n\}$ *is a non-empty set of transitions,*
- ${}^\bullet(.) : T \to \mathbb{N}^P$ *is the backward incidence function,*
- $(.)^\bullet : T \to \mathbb{N}^P$ *is the forward incidence function,*
- $M_0 \in \mathbb{N}^P$ *is the initial marking of the Petri Net,*
- $\delta : T \to \mathbb{N}$ *is the function giving the firing times (delays) of transitions.*

A marking M is an element of $\mathbb{N}^P$ such that $\forall p \in P$, $M(p)$ is the number of tokens in place p.

A marking M enables a transition $t \in T$ if: $M \geq^\bullet t$. The set of transitions enabled by a marking M is $enab(M) = \{t \in T \mid M \geq^\bullet t\}$.

Firable transitions are fired according to the maximal-step firing rule and thus must fire simultaneously. For marked graph where every place has one incoming arc, and one outgoing arc, there can not be any conflict and the firing of a transition cannot disable another transition. In the general case, there can be conflict and, from a given state, there can be several maximal steps τ.

From a marking M, the simultaneous firing of a set τ of transitions leads to a marking $M' = M + \Sigma_{t \in \tau}(t^\bullet - {}^\bullet t)$.

A transition t' is said to be *newly* enabled by the firing of a set of transitions τ if $M + \Sigma_{t \in \tau}(t^\bullet - {}^\bullet t)$ enables t' and $(M - \Sigma_{t \in \tau} {}^\bullet t)$ did not enable t'. If t remains enabled after its firing then t is newly enabled. The set of transitions newly enabled by a set of transitions τ for a marking M is noted $\uparrow enab\,(M, \tau)$.

A state is a pair (M, v) where M is a marking and $v \in \mathbb{R}^T_{\geq 0}$ is a time valuation of the system (*i.e.* the value of the clocks). $v(t)$ is the time elapsed since the transition $t \in T$ has been newly enabled. $\bar{0}$ is the valuation assigning 0 to every transitions.

Definition 2 (Maximal Step). *Let $q = (M, v)$ be a state of the Timed Petri Net $(P, T, {}^\bullet(.), (.)^\bullet, \delta, M_0)$, $\tau \subseteq T$ is a maximal step from q iff:*

1. $\forall t \in \tau,\ v(t) = \delta(t)$
2. $\Sigma_{t \in \tau} {}^\bullet t \leq M$
3. $\forall t' \in T,\ (v(t') = \delta(t')\ and\ {}^\bullet t' \leq M\ and\ t' \notin \tau) \Rightarrow \Sigma_{t \in \tau} {}^\bullet t + {}^\bullet t' \not\leq M$

The set of maximal steps from q is noted $maxStep(q)$

The first condition ensures that the transitions are ready to fire, *i.e.* the clocks are equal to the delays. The second condition ensures that the transition are firable, *i.e.* enabled and not in conflict with another transition of τ. The third condition disallows the existence of a proper superset of τ which fulfils the previous two conditions.

The semantics of TPN is defined as a Timed Transition System (TTS). Waiting in a marking is a delay transition of the TTS and firing a transition of the TPN is a discrete transition of the TTS.

Definition 3 (Semantics of a TPN). *The semantics of a TPN is defined by the Timed Transition System $\mathcal{S} = (Q, q_0, \rightarrow)$:*

- $Q = \mathbb{N}^P \times \mathbb{R}^T_{\geq 0}$ *is the set of states,*
- $q_0 = (M_0, \bar{0})$ *is the initial state,*
- $\rightarrow \in Q \times (\mathbb{R}_{\geq 0} \cup 2^T) \times Q$ *is the transition relation including a discrete transition and a delay transition.*
  - *The delay transition is defined $\forall d \in \mathbb{R}_{\geq 0}$ by:*

$$(M, v) \xrightarrow{d} (M, v')\ iff\ \forall t \in enab\,(M),\ v'(t) = v(t) + d\ and\ v'(t) \leq \delta(t)$$

  - *The discrete transition is defined $\forall \tau \in maxStep((M, v))$ by:*

$$(M, v) \xrightarrow{\tau} (M', v')\ iff\ \begin{cases} M' = M + \Sigma_{t \in \tau}(t^\bullet - {}^\bullet t) \\ v'(t) = \begin{cases} 0 & if\ t \in \uparrow enab\,(M, \tau)\ or\ t \notin enab\,(M') \\ v(t) & otherwise \end{cases} \end{cases}$$

A run in a Timed Petri Net is a sequence $q_0 \xrightarrow{\alpha_1} q_1 \xrightarrow{\alpha_2} \ldots$, such that for all i, $q_i \xrightarrow{\alpha_{i+1}} q_{i+1}$ is a transition in the semantics.

2.4   Comparison with Ramchandani's Semantics

In the absence of conflict, the atomic semantics of Definition 3 is equivalent to the three-phases one of Ramchandani (extended with zero firing delay [19]): it exists only one execution, no indeterminism.

In case of conflict, it is possible to construct the three-phases firing in our semantics: just add a zero time transition before each transition, in order to simulate the reservation action as illustrated in Figs. 1 and 2. Notice that in Fig. 1b, the maximal step $\{t_0, t_1, t_2\}$ is only the consumption phase, while the production is done implicitly after the delay. Our semantics is then at least as expressive as the Ramchandani's one.

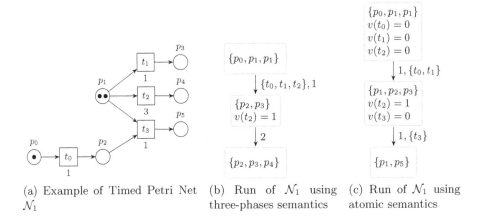

(a) Example of Timed Petri Net $\mathcal{N}_1$

(b) Run of $\mathcal{N}_1$ using three-phases semantics

(c) Run of $\mathcal{N}_1$ using atomic semantics

Fig. 1. Comparison with three-phases firing semantics[1].

3   TPN with Reset and Delayable Transitions

We now extend TPN. A transitions can be of two types: either it is fired as soon as possible, as in Definition 1, or it is delayable (non-asap) i.e. may fire either if the value of its clock is equal to the delay or if the value of its clock is greater than the delay and if it is associated with another transition whose clock is equal to its delay. Moreover, the clocks can be reset (let *reset* be the corresponding action) and the delay between two successive resets is given by an interval I_{reset}.

Formally:

Definition 4 (RTPN). *A Timed Petri Net with reset and delayable transitions (RTPN) $\mathcal{N}$ is a tuple $(P, T, T_D, {}^\bullet(.), (.)^\bullet, \delta, I_{reset}, M_0)$ defined by:*

---

[1] For the sake of brevity, in all the following figures, we note a marking M as a set of marked places instead of a vector and we give the valuation v only for the enabled transitions.

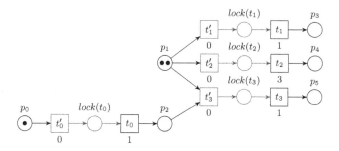

Fig. 2. Timed Petri Net $\mathcal{N}_2$ that simulates $\mathcal{N}_1$ with the three-phases firing semantics.

- $(P, T, ^\bullet(.), (.)^\bullet, \delta, M_0)$ is a Timed Petri Net,
- $T_D \subseteq T$ is the set of delayable transitions,
- I_{reset} is the reset time interval with lower ($\underline{I_{reset}}$) and upper ($\overline{I_{reset}}$) bounds in $\mathbb{N}$.

From a state (M, v), a transition is firable if it is enabled and its clock is greater or equal to its delay. As for Timed Petri Net, the clock of asap transition $t \notin T_D$ cannot exceed $\delta(t)$. Hence $v(t) \leq \delta(t)$ and t must fire when its clock is equal to its delay.

A delayable transition $t \in T_D$, can fire either when $v(t) = \delta(t)$ (not delayed in this case) or when $v(t) > \delta(t)$, but in this second case, t must be associated with at least one (or more if any) other firable transition t' such that $v(t') = \delta(t')$.

The maximal step is now maximal only from the asap transitions point of view as follows:

Definition 5 (Maximal Step w.r.t. T_D). *Let $q = (M, v)$ be a state of $\mathcal{N}$. $\tau \subseteq T$ is a maximal step w.r.t. T_D from q iff:*

1. $\forall t \in \tau,\ v(t) \geq \delta(t)$
2. $\exists t \in \tau\ s.t.\ v(t) = \delta(t)$
3. $\sum_{t \in \tau} {}^\bullet t \leq M$
4. $\forall t' \in T \setminus T_D,\ (v(t') = \delta(t')\ and\ {}^\bullet t' \leq M\ and\ t' \notin \tau) \Rightarrow \sum_{t \in \tau} {}^\bullet t + {}^\bullet t' \nleq M$

The set of maximal steps w.r.t. T_D from q is noted $maxStep_{\setminus T_D}(q)$.

A state is now a pair (M, v) where $v \in \mathbb{R}_{\geq 0}^{T \cup \{reset\}}$ is extended with a value for the *reset, i.e.* the time elapsed since the last action *reset*. The *reset* action resets all the clocks of the model. It is possible when the clock of the reset is in the reset time interval $v(reset) \in I_{reset}$.

The semantics of RTPN is defined as a Timed Transition System (TTS). Waiting in a marking is a delay transition of the TTS and firing a set of transitions of the RTPN or resetting the clocks is a discrete transition of the TTS.

Definition 6 (Semantics of a RTPN). *The semantics of a RTPN $\mathcal{N}$ is defined by the Timed Transition System $\mathcal{S}_\mathcal{N} = (Q, q_0, \rightarrow)$:*

- $Q = \mathbb{N}^P \times \mathbb{R}_{\geq 0}^{T \cup \{reset\}}$ *is the set of states,*

- $q_0 = (M_0, \bar{0})$ is the initial state,
- $\rightarrow \in Q \times (\mathbb{R}_{\geq 0} \cup 2^T \cup \{reset\}) \times Q$ is the transition relation including a discrete transition and a delay transition.
  - The delay transition is defined $\forall d \in \mathbb{R}_{\geq 0}$ by:

$$(M, v) \xrightarrow{d} (M, v') \text{ iff } \begin{cases} \forall t \in enab\,(M) \cup \{reset\}, \; v'(t) = v(t) + d \\ v'(reset) \leq \overline{I_{reset}} \\ \forall t \in enab\,(M) \setminus T_D, \; v'(t) \leq \delta(t) \end{cases}$$

  - The discrete transition is defined by:
    * $\forall \tau \in maxStep_{\setminus T_D}\big((M, v)\big)$,

$$(M, v) \xrightarrow{\tau} (M', v') \text{ iff } \begin{cases} M' = M + \Sigma_{t \in \tau}\left(t^\bullet - {}^\bullet t\right) \\ v'(t) = \begin{cases} 0 & \text{if } t \in \uparrow enab\,(M, \tau) \;\; or \;\; t \notin enab\,(M') \\ v(t) & otherwise \end{cases} \end{cases}$$

    * $(M, v) \xrightarrow{\{reset\}} (M, v') \text{ iff } \begin{cases} v(reset) \in I_{reset} \\ v' = \bar{0} \end{cases}$

Definition 7 (Runs). *Let $\mathcal{N}$ be a RTPN and $\mathcal{S}_{\mathcal{N}}$ its semantics. A run of $\mathcal{N}$ from q_1 is a finite or infinite sequence $\rho = q_r \xrightarrow{d_1} q_{d_1} \xrightarrow{\tau_1} q_{\tau_1} \ldots \xrightarrow{d_n} q_{d_n} \xrightarrow{\tau_n} q_{\tau_n}$ of alternating d_i delay (possibly null) and τ_i discrete transition where either $\tau_i \subseteq T$ or $\tau_i = \{reset\}$.*

4   Complexity of Reachability Problem

First we have the following theorem:

Theorem 1. *Reachability problem for RTPN is undecidable.*

Proof. The behaviour of a timeless Petri Net with maximal step firing rule is simulated by a *RTPN* with the same structure and initial marking, and such that $T_D = \emptyset$, $\forall t \in T$, $\delta(t) = 0$ and $I_{reset} > 0$. Moreover, timeless Petri Nets with maximal step firing rule are Turing powerful [19].   □

In the sequel we then consider bounded Nets.

Lemma 1. *Reachability for safe timeless Petri Nets with maximal-step firing rule, for safe TPN and for safe RTPN is PSPACE-hard.*

Proof. We first consider 1-safe timeless Petri Net. We reduce the reachability problem for a 1-safe Petri Net with interleaving semantics to reachability for a 1-safe Petri Net with maximal-step firing rule. Let $\mathcal{N} = (P, T, {}^\bullet(.), (.)^\bullet, m_0)$ a 1-safe Petri Net with interleaving semantics. We translate $\mathcal{N}$ into $\mathcal{N}' = (P, T', pre, post, m_0)$ with maximal-step firing rule such that $T \subseteq T'$, $\forall t \in T$, $pre(t) = {}^\bullet t$ and $post(t) = t^\bullet$. Moreover $T' = T \cup T_p$ where T_p is a set of transitions defined by $T_p \cap T = \emptyset$ and $\forall p \in P$ there exists a transition $t_p \in T_p$ such that $pre(t_p) = post(t_p) = p$. Informally speaking, we add a self loop from all places of P.

Hence we create a conflict between all transitions of $\mathcal{N}'$ with a transition of T_p. Since the firing of a transition t_p of T_p preserves the marking of p, this translation allows to simulate the interleaving semantics from the maximal-step firing rule.

Since reachability in timeless 1-safe Petri net with interleaving semantics is a PSPACE-complete problem [10], it follows that reachability for 1-safe Petri net with Maximal-step firing rules is PSPACE-hard. Moreover, as for the proof of Theorem 1, we can now consider that $\mathcal{N}'$ is a TPN with $\forall t \in T'$, $\delta(t) = 0$ or a $RTPN$ with $T_D = \emptyset$, $\forall t \in T'$, $\delta(t) = 0$ and $\underline{I_{reset}} > 0$ proving the lemma. $\square$

TCTL, introduced in [2], is a real-time extension of the branching-time temporal logic CTL. It has been trivially adapted in [9] for Time Petri Nets where atomic propositions are linear constraints over markings such as Generalized Mutual Exclusion Constraints [12].

To construct a finite structure in order to employ usual discrete model checking techniques, we can use the region equivalence relation $\simeq$ over clock interpretations defined for timed automata [2,3]. This region equivalence can be easily adapted for Timed Petri Nets with or without reset as in [5]. To compute the region graph, we now just change the computation of the firing step (*i.e.* the discrete step) by applying the maximal firing rule. This region graph is exponential in the size of the input $T + P$. However, we can proceed like in [2,5] to check TCTL formulas by a recursive procedure $label(vertex, \varphi)$ called for each sub-formula leading to a polynomial space algorithm. Finally thanks to the PSPACE-hardness (Lemma 1), we obtain the following theorem and corollaries.

Theorem 2. *Reachability and TCTL model checking for bounded Timed Petri Nets with or without reset is PSPACE-complete.*

Corollary 1. *The result holds for Timed Petri Nets "à la Ramchandani".*

Corollary 2. *Reachability and CTL model checking for bounded timeless Petri Nets with maximal-step firing rule is PSPACE-complete.*

Note that this PSPACE complexity is theoretical and, for Timed Automata and Time Petri Nets, no effective PSPACE algorithm has been proposed and all real implementations are with exponential algorithms.

5   State Space Computation

The semantics of RTPN is a transition system in which each state is a pair of a marking and a clock valuation. Observe that there are only finitely many markings, but there are uncountably many values for clocks due to the denseness of time (in particular for the states from which a reset can be done). Hence, the semantics of RTPN has an uncountably infinite state space.

The region graph partitions the space of valuations into a finite number of regions. However, the region graph approach turns out to be impractical. A more efficient solution is to work with convex sets of valuations called zones described by constraints between clocks.

Using zones, a symbolic semantics graph of RTPN, can be defined. A symbolic state of a RTPN is a pair (M, Z) representing a set of states of the RTPN, where M is a marking and Z is a zone. A symbolic transition describes all the possible concrete transitions from the set of states.

Definition 8. *A symbolic state is a pair (M, Z) where M is a marking and the zone Z is a set of valuations v on $T \cup \{reset\}$ represented by a conjunction of:*

- *rectangular constraints over valuations: $(v(x) \sim c)$ where $x \in T \cup \{reset\}$ and $\sim \in \{\leq, =, \geq\}$ and $c \in \mathbb{N}$, with*
- *$\forall t \in enab(M)$ diagonal constraints on pairs: $(v(reset) - v(t) = c)$ where $c \in \mathbb{N}$.*

We said that a valuation v_i is in a zone: $v_i \in Z$, if it verifies all its constraints. We note $\bar{0}$ the zone containing only the valuation $\bar{0}$.

5.1   Operations over Symbolic States

Since diagonal constraints are equalities, by setting the value of a single variable $v(x)$ we obtain a point in the zone. To ensure this, we set to zero the valuations of non enabled transitions.

Let M be a marking and Z a zone. The computation of the reachable markings from M according to the zone Z is done by using the following operations:

1. Compute the possible evolution of time (future): $\overrightarrow{Z} = \{v' \mid v \in Z \text{ and } v'(x) = v(x) + d \text{ with } d \geq 0, x \in enab(M) \cup \{reset\}\}$. This is obtained by setting all upper bounds of $v(x)$ to infinity for $x \in enab(M) \cup \{reset\}$.
2. Select only the possible valuations for which M could exist, *i.e.* valuations of enabled transitions $t \notin T_D$ must not be greater than $\delta(t)$ and valuation of the *reset* must not be greater than $\overline{I_{reset}}$:

$$Z' = \overrightarrow{Z} \wedge \left(v(reset) \leq \overline{I_{reset}}\right) \bigwedge_{t \in enab(M) \backslash T_D} (v(t) \leq \delta(t))$$

So, Z' is the maximal zone starting from Z for which the marking M is legal according to the semantics.

3. Determine the set of firable transitions sets $fireable_z(M, Z') = \{(\tau, z) \mid z \subseteq Z', \tau \in 2^T \cup \{reset\}$ is firable from $(M, z)\}$ defined by:
   - $\tau \subseteq T$ is firable from the firing point $(M, \{v_p\})$ if $\tau \in maxStep_{\backslash T_D}((M, v_p))$ and $\exists t \in \tau$ such that $Z' \wedge (v(t) = \delta(t)) = \{v_p\}$,
   - *reset* is firable from (M, z_{reset}) with $z_{reset} = Z' \wedge \left(v(reset) \geq I_{reset}\right)$ if z_{reset} is a non empty zone.
4. Fire transitions

- Firing a firable set of transitions $\tau_i \subseteq T$ from the firing point $v_p \in Z'$ leads to the new marking $M' = M + \Sigma_{t \in \tau}(t^\bullet - {}^\bullet t)$ and the point zone v_i such that:

$$\forall t \in T, v_i(t) = \begin{cases} 0 & \text{if } t \in\uparrow enab\,(M, \tau) \text{ or } t \notin enab\,(M') \\ v_p(t) & \text{otherwise} \end{cases}$$

  and $v_i(reset) = v_p(reset)$.
- Firing a *reset* leads to the point zone $\bar{\mathbf{0}}$ and then to $(M, \bar{\mathbf{0}})$

A set of transitions $\tau \subseteq T$ is always fired from a point of a zone Z and the zone obtained after the firing of τ from Z is also a point. A reset is fired from a part of a zone but since it resets all the clocks, the zone obtained by the firing of *reset* from a given zone Z is also a point.

An integer point v_i of a zone Z is a valuation such that for all $x \in T \cup \{reset\}$, $v(x) \in \mathbb{N}$.

Lemma 2. $\forall \tau \subseteq T$, $(\tau, z_\tau) \in fireable_z(M, Z) \Rightarrow z_\tau$ is an integer point.

Proof. Each reset leads to an integer point zone $\bar{\mathbf{0}}$. Between two reset, only firings of transitions $\tau \subseteq T$ can occur. By definition of the semantics (and of operation 3), the firing of a set of transitions $\tau \subseteq T$ can occur only if at least one transition $t \in \tau$ is such that $v(t) = \delta(t)$ in $\mathbb{N}$. Hence a set of transitions $\tau \subseteq T$ is always fired from an integer point of a zone Z and the zone obtained after the firing of τ from Z is also an integer point. $\qquad\square$

Hence, the set of zones is closed under these 4 operations, in the sense that the result of the operations is also a zone as defined in Definition 8.

The successor operator $succ\,((M, Z), \tau)$ gives the symbolic state obtained from (M, Z) by applying successively operations 4 with $\tau \in 2^T \cup \{reset\}$, 1 and then 2.

5.2   State Graph

For a RTPN $\mathcal{N}$, the initial symbolic state (M_0, Z_0) is obtained from $(M_0, \bar{\mathbf{0}})$ by applying operations 1 and 2.

We compute forward the reachable symbolic states from (M_0, Z_0) by iteratively applying the successor operator for all the firable transitions. The set of reachable symbolic states from the initial symbolic state is $Reach(\mathcal{N})$.

Lemma 3. $Reach(\mathcal{N})$ is finite.

Proof. First, we consider bounded nets therefore the number of marking is bounded. Lemma 2 gives that applying operation 4 leads to integer points. The coordinates of those points are bounded by $\overline{I_{reset}}$, then there is a finite number of reachable integer points, and then a finite number of zones after applying operations 1 and 2. $\qquad\square$

Definition 9. *The state graph of $\mathcal{N}$ is the graph $SG(\mathcal{N}) = (Reach(\mathcal{N}), (M_0, Z_0), \hookrightarrow, \Sigma)$ such that $\Sigma = 2^T \cup \{reset\}$ and $\forall s \in Reach(\mathcal{N}),\ \tau \in \Sigma$, $s \overset{\tau}{\hookrightarrow} s'$ if $s' = succ(s, \tau)$.*

This symbolic semantics corresponds closely to the operational semantics in the sense that $(M, Z) \overset{\tau}{\hookrightarrow} (M', Z')$ implies for all $v' \in Z'$, $(M, v) \overset{\tau}{\rightarrow} (M', v')$ for some $v \in Z$. The symbolic semantics is a correct and full characterisation of the operational semantics given in Definition 6.

Example 1. An example of RTPN is presented in Fig. 3a, in its initial state, where the delays are in red, and the delayable transitions in gray (only t_0 here). The corresponding part of state graph $SG(\mathcal{N})$ obtained in only one step is given in Fig. 3b.

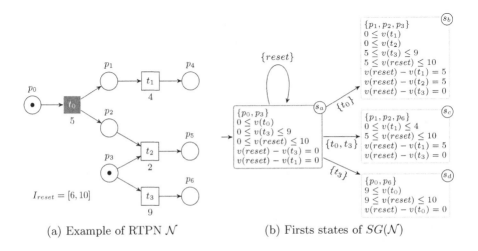

(a) Example of RTPN $\mathcal{N}$ (b) Firsts states of $SG(\mathcal{N})$

Fig. 3. Example of RTPN and part of its state graph

6   Decidability of Some Timed Language Problems

A timed word of $\mathcal{N}$ is a finite or infinite sequence $w = (d_1, \tau_1)(d_1 + d_2, \tau_2) \ldots$ $(\Sigma_{i=1\ldots n} d_n, \tau_n)$ such that $\rho = q_0 \overset{d_1}{\rightarrow} q_{d_1} \overset{\tau_1}{\rightarrow} q_{\tau_1} \ldots \overset{d_n}{\rightarrow} q_{d_n} \overset{\tau_n}{\rightarrow} q_{\tau_n}$ is a run of $\mathcal{N}$ from q_0. The timed language $L(\mathcal{N})$ recognized by $\mathcal{N}$ is the set of words w of $\mathcal{N}$.

The language of a Petri Net is generally prefix-closed but it is easy to extend Petri Nets with final or repeated markings as in [4] in order to have non-prefix-closed languages over finite or infinite words.

Language inclusion and universality problems are known to be undecidable for Timed Automata and Time Petri Nets. However these problems are decidable on finite words for one clock Timed Automata. We then propose, from any bounded RTPN $\mathcal{N}$, to build a single-clock timed automaton which recognizes the same timed language as $\mathcal{N}$.

6.1   From Bounded RTPN to Single-Clock Timed Automata

Timed Automata. Timed automata were first introduced by Alur and Dill in [2,3] and extend finite automata with a finite number of clocks.

An *atomic constraint* is a formula of the form $x \bowtie c$ for $x \in X$, $c \in \mathbb{N}$ and $\bowtie \in \{<, \leq, \geq, >, =\}$. The set of *constraints* over a set X of variables is denoted by $\xi(X)$ and consists of conjunctions of atomic constraints.

Definition 10 (Timed Automaton). *A timed automaton $\mathcal{A}$ is a tuple $(L, l_0, X, \Sigma, E, Inv)$ where L is a finite set of locations, $l_0 \in L$ is the initial location, X is a finite set of clocks, Σ is a finite set of actions, $E \subseteq L \times \xi(X) \times \Sigma \times 2^X \times L$ is a finite set of edges where $e = (l, \gamma, a, R, l') \in E$ represents an edge from the location l to the location l' with the guard $\gamma \in \xi(X)$, the label $a \in \Sigma$ and the reset set $R \subseteq X$, $Inv \in \xi(X)^L$ assigns an invariant to any location; we restrict the invariants to conjunctions of terms of the form $x \leq k$ for $x \in X$ and $k \in \mathbb{N}$.*

A clock valuation is a function $\nu : X \to \mathbb{R}_{\geq 0}$. If $R \subseteq X$ then $v[R \mapsto 0]$ denotes the valuation such that $\forall x \in X \setminus R$, $\nu[R \mapsto 0](x) = \nu(x)$ and $\forall x \in R$, $\nu[R \mapsto 0](x) = 0$. The satisfaction relation $\nu \models c$ for $c \in \xi(X)$ is defined in the natural way.

Definition 11 (Semantics of a Timed Automaton). *The semantics of the timed automaton $\mathcal{A} = (L, l_0, X, \Sigma, E, Inv)$ is the timed transition system $\mathcal{S}_\mathcal{A} = (Q, q_0, \to)$ with $Q = \{(l, \nu) \in L \times (\mathbb{R}_{\geq 0})^X \mid \nu \models Inv(l)\}$, $q_0 = (l_0, \bar{0})$ is the initial state and $\to$ is defined by:*

- *the discrete transitions relation $(l, \nu) \xrightarrow{a} (l', \nu')$ iff $\exists (l, \gamma, a, R, l') \in E$ s.t. $\nu \models \gamma$, $\nu' = \nu[R \mapsto 0]$ and $\nu' \models Inv(l')$;*
- *the continuous transition relation $(l, \nu) \xrightarrow{d} (l', \nu')$ iff $l = l'$, $\nu' = \nu + d$ and $\nu' \models Inv(l)$.*

Translation from RTPN to Single-Clock Timed Automaton. By definition of zone, all enabled transition t verify the diagonal contraint $(v(reset) - v(t) = c)$ with $c \in \mathbb{N}$ and all transition t' not enabled verify $v(t') = 0$. Hence a point of a zone is fully characterised by $v(reset)$. Note that $v(reset) - v(t) = c$ means that the transition t has been enabled c time units after the last reset.

From any RTPN $\mathcal{N}$, we wish to build a single-clock timed automaton $\mathcal{A}_\mathcal{N}$ in which the single clock x has the value of $v(reset)$.

We construct the single-clock timed automaton $\mathcal{A}_\mathcal{N} = (L, l_0, X, \Sigma, E, Inv)$ from the state graph $SG(\mathcal{N}) = (Reach(\mathcal{N}), (M_0, Z_0), \hookrightarrow, \Sigma)$, as follows:

- $\phi : Reach(\mathcal{N}) \mapsto L$ is a bijection
- $L = \{\phi(s) \mid s \in Reach(\mathcal{N})\}$
- $X = \{x\}$
- The initial location is $l_0 = \phi((M_0, Z_0))$
- For each $l \in L$, set the invariant $(x \leq \overline{I_{reset}})$

- For each $s \in Reach(\mathcal{N})$,
  add $\Big(\phi(s), x \geq \underline{I_{reset}}, \{reset\}, \{x\}, \phi\big(succ(s, \{reset\})\big)\Big)$ to E
- For all $(\tau, v_\tau) = fireable_z(M, Z)$ such that $(M, Z) \xrightarrow{\tau} (M', Z')$ do both statements:
  - add $\Big(\phi((M, Z)), x = v_\tau(reset), \tau, \emptyset, \phi((M', Z'))\Big)$ to E
  - if $\exists t \in \tau$ such that $t \notin T_D$ then add the constraint (by conjunction) $\big(x \leq v_\tau(reset)\big)$ to $Inv\big(\phi((M, Z))\big)$.

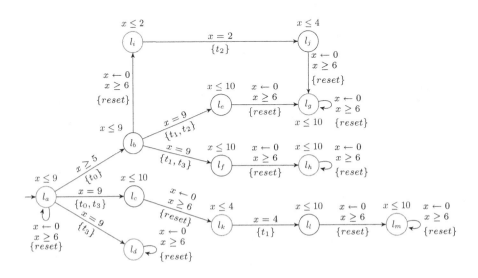

Fig. 4. Translation from the RTPN $\mathcal{N}$ into a single-clock TA $\mathcal{A}_\mathcal{N}$

Theorem 3. *The single-clock timed automaton $\mathcal{A}_\mathcal{N}$ and the RTPN $\mathcal{N}$ recognize the same timed language: $L(\mathcal{N}) = L(\mathcal{A}_\mathcal{N})$.*

Proof. Let $(M_\mathcal{N}, v)$ a state of $\mathcal{N}$ and $\big(\phi((M_\mathcal{A}, Z)), \nu\big)$ a state of $\mathcal{A}_\mathcal{N}$. The relation $\simeq$, defined by $(M_\mathcal{N}, v) \simeq \big(\phi((M_\mathcal{A}, Z)), \nu\big)$ iff $M_\mathcal{N} = M_\mathcal{A}$, $v \in Z$ and $v(reset) = \nu(x)$ is a timed bisimulation between $\mathcal{N}$ and $\mathcal{A}_\mathcal{N}$. From this timed bisimulation we can state that $L(\mathcal{A}_\mathcal{N}) = L(\mathcal{N})$. $\qquad\square$

Example 2. Figure 4 presents the timed automata constructed from the RTPN $\mathcal{N}$ of Fig. 3a, where the guards and clock resets (denoted $x \leftarrow 0$) are in pink, the invariants in orange, and the locations reachable by a *reset* are in cyan. In the TA, we omit *reset* when it is not possible *i.e.* from a location with an invariant $x \leq c$ with $c < \underline{I_{reset}}$.

6.2   Corollaries

Thanks to the translation from RTPN to one-clock Timed Automata, we inherit these decidability results established on one-clock Timed Automata.

Given two timed model A and B, asking if all the timed words recognised by B also recognised by A (language inclusion problem) is known to be undecidable for Timed Automata. However it becomes decidable on finite words if A is restricted to having at most one clock [18].

Corollary 3. *Language inclusion problem is decidable for finite words for RTPN.*

The universality problem for timed model is: given a timed model A, does A accept all timed words? Alur and Dill have shown that the universality problem is undecidable for timed automata with two clocks. However, for one-clock timed automata over finite words, the one-clock universality problem is decidable [1].

Corollary 4. *Universality problem is decidable for finite words for RTPN.*

7   Application to the Pipeline Problem

Now that we have a model that closely represents pipelined synchronous circuits, we are able to perform model checking of TCTL properties, for example to ensure the sequentiality of some operations. An interesting use case would be the sharing of resources (sections of the circuits), indeed thanks to some simple TCTL properties, one can build a pipeline that prevents conflicts with shared resources. But as it is an high stakes issue for the designing of synchronous systems, we wish to focus on the building of an optimised pipeline w.r.t. the number of registers.

The problem of building a pipeline that minimises the number of registers, while ensuring a minimal throughput have already been solved in [15]. However, the minimisation of the number of registers does not imply the minimisation of the number of flip-flops: it depends on the size of signals. Even though the signal's size can be taken into account in the model of [15] by adding edges in parallel, there is still a problem that cannot be solved with this approach: the placement of registers when there is one signal used by multiple operators, what we call *branch points*. We claim to solve this particular issue by using RTPN.

For a circuit representation, where the transitions illustrate the operators, and the places illustrate the connections, the Petri Net is actually a Marked Graph, thus there is no conflict. However the state space still has an exponentiel size w.r.t. the size of the RTPN, then we will add features that allows us to cut branches in the exploration according to an optimisation goal. In this particular case, we aim at the minimisation of the total number of flip-flops (1-bit register), thus we extend our model with costs which represent the number of flip-flop of a given pipeline. Remind that the considered circuits are finite with unfolded loops, so we only focus on finite runs of the RTPN, then adding an only increasing cost won't affect the termination.

7.1 RTPN with Cost

We extend RTPN with a cost associated with each place and a marking cost function.

Definition 12 (RTPN with Cost). *A RTPN extended with cost (RTPN with Cost) is a tuple* $(\mathcal{N}, \mathcal{C}, \omega)$ *where* $\mathcal{N} = (P, T, T_D, {}^\bullet(.), (.)^\bullet, \delta, I_{reset}, M_0)$ *is a RTPN and*

- $\mathcal{C} : P \to \mathbb{N}$ *is the place cost function.*
- $\omega : \mathbb{N}^P \to \mathbb{N}$ *is the marking cost function (recall that a marking* $M \in \mathbb{N}^P$*).*

In Marked Graphs, the marking of a place $M(p)$ *can only take its value in* $\{0, 1\}$*, which can be interpreted both as a boolean and as an integer value. Therefore, we allow to use both arithmetical operators (in* $\{+, *\}$*) and the logical* or *operator* $\vee$ *in the definition of the marking cost function* ω*.*

Example for $\omega(M) = (M(p_1) \vee M(p_2)) * 4 + M(p_2) * 10$. Assume $M_1(p_1) = M_1(p_2) = 1$ then $\omega(M_1) = (1 \vee 1) * 4 + 1 * 10 = 14$.

A classical marking cost function is $\omega(M) = \sum_{p \in P} M(p) * \mathcal{C}(p)$ which is the sum of marked places weighted by their cost.

Definition 13 (Cost of a run). *The cost* $\Omega(\rho)$ *of a run* ρ *is the cumulated marking cost of the states after each reset transition over the run, starting with the cost of the initial marking. It is inductively defined on a run* $\rho_n = \rho_{n-1} \xrightarrow{\alpha_n} q_n$*, with* $\alpha_n \in \mathbb{R}_{\geq 0} \cup 2^T \cup \{reset\}$ *and* $q_n = (M_n, v_n)$ *by:*

- $\Omega(q_0) = \omega(M_0)$
- $\Omega(\rho_n) = \begin{cases} \Omega(\rho_{n-1}) + \omega(M_n) & \text{if } \alpha_n = \{reset\} \\ \Omega(\rho_{n-1}) & \text{otherwise} \end{cases}$

7.2 From a Pipelining Problem to a RTPN with Cost

As stated before, Marked Graphs have been extensively used to model circuits, where transitions stand for the atomic operators, places for the connections in between, and where tokens represent the registers on each connection. This can be improved by considering branch points (points where a signal is used by multiple operators) as operators with a null propagation delay, and then integrating them into the model with more transitions.

We propose to use our model of Timed Petri Net with reset and delayable transitions, in order to build a pipeline of a synchronous circuit, which minimises the number of registers, while ensuring that the throughput is in a target interval $[f_{min}, f_{max}]$.

We build the RTPN with Cost $((P, T, T_D, {}^\bullet(.), (.)^\bullet, \delta, I_{reset}, M_0), \mathcal{C}, \omega)$ from the circuit by creating a transition $t \in T$ for each operator and branch point, with its delay equal to the propagation delay, a place $p \in P$ for each connection of the circuit, and with ${}^\bullet(.)$ and $(.)^\bullet$ preserving the network structure of the circuit. The initial marking M_0 sets a token in all the places corresponding to input connections of the circuit. The placement of tokens models the placement of registers, so our model won't hold the fully pipelined circuit in its state, but only one stage at a time, a complete pipeline is built from a run. The *reset* action settles the registers placement of each pipeline stage, then $I_{reset} = [\frac{1}{f_{max}}, \frac{1}{f_{min}}]$ guarantees the throughput to be in $[f_{min}, f_{max}]$. The cost of each place $\mathcal{C}(p)$ will be the size of the signal (in bits) held by the corresponding connection in the circuit. The marking cost function is such that it gives the number of flip-flops corresponding to a marking M: $\omega(M) = \sum_{p_k \in P_{Op}} \mathcal{C}(p_k) \cdot (M(p_k) \vee \bigvee_{p_{kl} \in P_B(p_k)} M(p_{kl}))$, where P_{Op} is the set of places respectives to connections outgoing from operators, and $P_B(p_k)$ is the set of places respectives to connections outgoing from the branch point after the connection corresponding to p_k. In this manner, the cost of a run will be equal to the number of flip-flops in the pipeline so far. Finally, all the transitions t corresponding to operators with a larger bus width at the output than at the input, are set to be delayable $t \in T_D$. Thus, we relax the constraints on those transitions, and allow to explore states where the register is *before* the operator, and so with less flip-flops. It is actually possible to make all transitions delayable, but this will obviously lead to an explosion of the state space of the model.

An example of circuit is presented on Fig. 5a, involving some operators op_i with propagation delays in red and some signals s_j transmitted by connections with sizes in green.

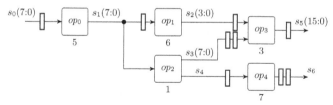

(a) Pipelined circuit (with frequency $\frac{1}{8} \leq f \leq \frac{1}{4}$)

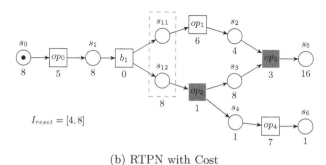

(b) RTPN with Cost

Fig. 5. A synchronous circuit example (Color figure online)

The RTPN with Cost produced from this circuit is represented on Fig. 5b, with the delays of transitions in red, the costs of places in green, and the delayable transitions in gray. The two places s_{11} and s_{12} are represented inside a dotted green box, because they "share" their cost, as it models signals outgoing from the same branch point. The cost function is $\omega(M) = 8 \cdot M(s_0) + 8 \cdot (M(s_1) \vee (M(s_{11}) \vee M(s_{12}))) + 4 \cdot M(s_2) + 8 \cdot M(s_3) + M(s_4) + 16 \cdot M(s_5) + M(s_6)$.

Firstly the benefit of this approach is that it builds the pipeline from a non-pipelined circuit. Secondly, the stage produced can be compared on-the-fly, as they are added to the pipeline. Therefore the exploration can be lead by some heuristics.

Finally, the reset interval offers flexibility over a fixed value and shorter pipeline stages can be defined to allow exploration of other configurations. However if the stages are too short, this increases the number of stages (and thus the cost in registers). A good trade-off is to restrict I_{reset} to $\left[\frac{1}{2f}, \frac{1}{f}\right]$ with some target frequency f.

7.3 Pipeline Exploration

Each reachable state of the model represents a possible pipeline stage of the real circuit. A *reset* operation defines a transition from one pipeline stage to the next

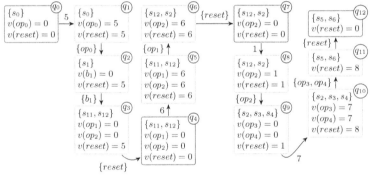

(a) One run of the RTPN with Cost of Fig. 5b. States after a *reset* are framed in cyan (q_0, q_4, q_7 and q_{12})

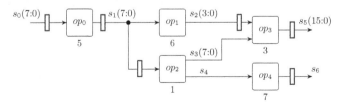

(b) One possible pipeline of the circuit of Fig. 5a

Fig. 6. Example of the extraction of a pipeline from a run

one. The full pipeline is retrieved by a walk along a branch of the state graph, collecting *reset* operations.

One run ρ of the RTPN with Cost of Fig. 5b, is represented on Fig. 6a. It is the best run achievable by our model, *i.e.* the one that minimises the cost. The corresponding pipeline on the circuit is presented on Fig. 6b.

The marking of every state after a *reset* (framed in cyan in Fig. 6a) gives the position of the registers in the pipelined circuit. Although, if all the signals outgoing from a branch point are marked, then only one register is needed for the unique signal that they represent. For example the marking $M_4 = \{s_{11}, s_{12}\}$, leads to only one register on s_1.

Let $q_i = (M_i, v_i)$ $(0 \leq i \leq 12)$ be the states of this run ρ. The run cost is $\Omega(\rho) = \omega(M_0) + \omega(M_4) + \omega(M_7) + \omega(M_{12}) = \mathcal{C}(s_0) + \mathcal{C}(s_1) + \mathcal{C}(s_1) + \mathcal{C}(s_2) + \mathcal{C}(s_5) + \mathcal{C}(s_6) = 45$. This cost matches with the number of flip-flops in the pipeline of Fig. 6a. Note that on this example, a classical greedy algorithm as implemented in FloPoCo [13] (a well-known generator of arithmetic operators with pipeline for FPGAs), produces the result in Fig. 5a, with a total of 55 flip-flops.

8   Conclusion

We have proposed an extension of Timed Petri Nets for the modeling of synchronous electronic circuits, addressing pipelined design problems.

Through a translation from RTPN into a single-clock timed automata, we have proved the decidability of language inclusion and universality problems for bounded RTPN. We have proved that the complexity of the reachability problem for bounded RTPN is PSPACE-Complete. This induces the same complexity for Timed Petri Nets "à la Ramchandani" and for timeless Petri Nets with maximal-step firing rule.

We have given a symbolic abstraction of the state space for RTPN. Thanks to two degrees of freedom through delayable transitions and reset interval of RTPN, the state space computation allows to generate multiple pipeline configurations. This makes it possible to address a wide range of interesting problems such as checking the absence of conflict between sharing resources (sections of the circuits).

We then have shown that we can also deal with the problem of the construction of a pipeline optimised w.r.t. the number of registers. We have proposed a cost extension leading to a state space exploration algorithm guided by cost, allowing to choose among all combinations those that minimizes the resources allocated to the pipeline, while ensuring a frequency objective. While this use case is interesting on its own, we believe that RTPN can handle the design of pipelined circuits in many ways: for instance to address timed division multiplexing problem, or to manage behavioural registers by adding explicit reset transitions in the model.

References

1. Abdulla, P.A., Deneux, J., Ouaknine, J., Quaas, K., Worrell, J.: Universality analysis for one-clock timed automata. Fundam. Informaticae **89**(4), 419–450 (2008)
2. Alur, R., Courcoubetis, C., Dill, D.: Model-checking in dense real-time. Inf. Comput. **104**(1), 2–34 (1993)
3. Alur, R., Dill, D.L.: A theory of timed automata. Theoret. Comput. Sci. **126**(2), 183–235 (1994)
4. Bérard, B., Cassez, F., Haddad, S., Lime, D., Roux, O.H.: The expressive power of time Petri nets. Theoretical Comput. Sci. (TCS) **474**, 1–20 (2013)
5. Boucheneb, H., Gardey, G., Roux, O.H.: TCTL model checking of time Petri nets. J. Log. Comput. **19**(6), 1509–1540 (2009)
6. Bufistov, D., Cortadella, J., Kishinevsky, M., Sapatnekar, S.: A general model for performance optimization of sequential systems. In: 2007 IEEE/ACM International Conference on Computer-Aided Design, pp. 362–369 (2007)
7. Campos, J., Chiola, G., Colom, J.M., Silva, M.: Properties and performance bounds for timed marked graphs. IEEE Trans. Circuits Syst. I: Fund. Theory Appl. **39**(5), 386–401 (1992)
8. Carloni, L.P., McMillan, K.L., Saldanha, A., Sangiovanni-Vincentelli, A.L.: A methodology for correct-by-construction latency insensitive design. In: 1999 IEEE/ACM International Conference on Computer-Aided Design. Digest of Technical Papers (Cat. No.99CH37051), pp. 309–315 (1999)
9. Cassez, F., Roux, O.H.: Structural translation from Time Petri Nets to Timed Automata - Model-Checking Time Petri Nets via Timed Automata. J. Syst. Softw. **79**(10), 1456–1468 (2006)
10. Cheng, A., Esparza, J., Palsberg, J.: Complexity results for 1-safe nets. Theoret. Comput. Sci. **147**, 117–136 (1995)
11. Church, A.: Application of recursive arithmetic to the problem of circuit synthesis, pp. 3–50 (1957)
12. Giua, A., DiCesare, F., Silva, M.: Generalized mutual exclusion constraints on nets with uncontrollable transitions. In: IEEE International Conference on SMC (1992)
13. Istoan, M., de Dinechin, F.: Automating the pipeline of arithmetic datapaths. In: Design, Automation & Test in Europe Conference & Exhibition (DATE 2017), pp. 704–709, Lausanne, Switzerland (2017)
14. Josipović, L., Sheikhha, S., Guerrieri, A., Ienne, P., Cortadella, J.: Buffer placement and sizing for high-performance dataflow circuits. In: Proceedings of the 2020 ACM/SIGDA Int. Symposium on Field-Programmable Gate Arrays, FPGA 2020, pp. 186–196. Association for Computing Machinery, New York (2020)
15. Leiserson, C.E., Saxe, J.B.: Retiming synchronous circuitry. Algorithmica **6**(1–6), 5–35 (1991)
16. Merlin, P.M.: A study of the recoverability of computing systems. Ph.D. thesis, Dep. of Information and Computer Science, University of California, Irvine, CA (1974)
17. Najibi, M., Beerel, P.A.: Slack matching mode-based asynchronous circuits for average-case performance. In: Proceedings of the International Conference on Computer-Aided Design, ICCAD 2013, pp. 219–225. IEEE Press (2013)
18. Ouaknine, J., Worrell, J.: On the language inclusion problem for timed automata: closing a decidability gap. In: Proceedings of the 19th Annual IEEE Symposium on Logic in Computer Science, 2004, pp. 54–63 (2004)
19. Popova-Zeugmann, L.: Time and Petri Nets. Springer (2013)

20. Ramchandani, C.: Analysis of asynchronous concurrent systems by timed Petri nets. Ph.D. thesis, Massachusetts Institute of Technology, Cambridge, MA (1974)
21. Kim, S., Beerel, P.A.: Pipeline optimization for asynchronous circuits: complexity analysis and an efficient optimal algorithm. IEEE Trans. Comput.-Aided Des. Integrated Circuits Syst. **25**(3), 389–4022 (2006)
22. Zuberek, W.: D-timed petri nets and modeling of timeouts and protocols. Trans. Soc. Comput. Simul. **4**(4), 331–357 (1987)

A Turn-Based Approach for Qualitative Time Concurrent Games

Serge Haddad[1], Didier Lime[2(✉)] (iD), and Olivier H. Roux[2]

[1] ENS Paris-Saclay, LMF, CNRS, INRIA, Université Paris-Saclay,
Gif-sur-Yvette, France
[2] École Centrale de Nantes, LS2N, CNRS, UMR, 6004 Nantes, France
`Didier.Lime@ec-nantes.fr`

Abstract. We address concurrent games with a qualitative notion of time with parity objectives. This setting allows to express how potential controllers interact with their environment and more specifically includes relevant features: transient states where the environment will eventually act, controller avoiding of an environment action either by an immediate controller action or by masking it, etc. In order to solve the controller synthesis in this framework, we design a linear-time building of a timeless turn-based game and show a close connection between strategies of the controller in the two games. Thus we reduce the synthesis problem to a standard problem of turn-based game with parity objectives establishing as a side effect that pure memoryless strategies are enough for winning. Moreover we introduce permissiveness for safety and reachability games as a criterion to choose between winning strategies and prove that one can compute a most permissive strategy (when it exists) in linear time.

1 Introduction

Games and Controller Synthesis. Finite games on graphs [13] are widely recognized as an adequate formalism to address problems such as controller synthesis on discrete event systems, originally expressed and studied within the theory of supervision [9,12,14]. The control problem can indeed be expressed as a game between two players representing respectively the controller and the environment. A controller for the system can be synthesized as a winning strategy for the controller player, when it exists.

Real-Time Controller Synthesis. For real-time systems, a strict turned-based game is an unnatural model, since the controller and the environment may play concurrently leading to concurrent games [5,7,8]. Adding quantitative delays before playing actions is a way to select which action should be played. This results in formalisms called (concurrent) timed games [6,11], for which tools like UPPAAL-Tiga are available [4]. Nevertheless, the algorithmics of timed games is costly, and for instance, the mere existence of a controller is an EXPTIME-complete problem [10] and the resulting strategies can be very

D. Buchs and J. Carmona (Eds.): PETRI NETS 2021, LNCS 12734, pp. 76–92, 2021.
https://doi.org/10.1007/978-3-030-76983-3_5

large [1]. Furthermore from a modelling point of view, often the exact timing constraints are unknown and indeed not needed to ensure the existence of a winning strategy.

Qualitative Time Concurrent Games. To overcome these issues the authors of [2,3] have introduced a model of qualitative time concurrent game with the following features: actions of the environment may occur immediately or require some non null unknown delay and transient states where when the controller chooses not to (or cannot) play, an action of the environment is guaranteed to eventually occur. Then they have designed polynomial time (ad-hoc) algorithms synthesising (when it exists) a controller for reachability and safety goals.

Our Contribution. Our contribution is threefold. First we extend the model of [2] by allowing some actions of the environment to be blocked by the controller and considering parity objectives. Then we design a linear-time building of a timeless turn-based game and show a close (but not one-to-one) connection between strategies of the controller in the two games. Thus we reduce the synthesis problem to a standard problem of turn-based game with parity objectives establishing as a side effect that pure memoryless strategies are enough for winning. Finally we introduce permissiveness for safety and reachability games as a criterion to choose between winning strategies and prove that one can compute a most permissive strategy (when it exists) in linear time.

Organisation. We illustrate by a relevant example the interest of our framework in Sect. 2. Section 3 gives the basic definitions and terminology used in this paper. Section 4 provides the translation to turn-based games and establishes the connections between strategies. Section 5 introduces the notion of permissiveness and shows how to compute most permissive strategies. Finally, we conclude in Sect. 6.

2   A Motivating Example

A device driver is the interface between the hardware device and the application or the operating system. Being executed with supervisor privileges, any error in a driver may have a serious impact on the integrity of the entire system. A specific driver (as opposed to a generic driver) is a driver dedicated to an application, i.e., with a smaller memory footprint.

In the context of driver synthesis, the environment is both the hardware device and the application using the driver. Then uncontrollable actions are interrupts that are triggered by the hardware and the requests made by the application.

Let us consider an analog-to-digital converter (ADC) inspired by the one of the MPC5xx microcontrollers family. An ADC cell has multiplexed acquisition channels (only one channel at a time). In order to allow the conversion of several channels, the conversions are combined into a conversion chain.

There are two types of conversion: normal or injected. In a normal conversion, it is possible to make a chain of conversion uniquely (*oneShot*) or continuously (*scan*). In *oneShot* mode, the cell stops acquisition at the end of the conversion, while in *scan* mode it repeats the chain ad infinitum (until a stop action is performed).

If one wants to make a *oneShot* acquisition in the middle of a conversion in *scan* mode, it is possible to use the injected conversions. An injected conversion is analogous to a software interrupt (`inject`) that can be maskable, i.e., it can be disabled. When an injected conversion is started, any conversion in progress is interrupted. The injected conversion is then carried out. At the end of the injected conversion, the chain which was interrupted resumes where it had been stopped (see Fig. 1 for channel CH5). If a conversion is interrupted twice by two injected conversions, then it is lost and the *scan* goes to the next one (see Fig. 1 for channel CH6).

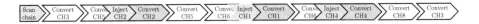

Fig. 1. Conversion chain Scan for channels 3, 5, 6, 8 with some injected conversions.

A conversion takes a non-null time that is not known precisely. At the end of a conversion the hardware generates an interrupt EOC (end of conversion). This interruption is ineluctable, i.e., it is guaranteed to happen eventually.

Finally, the converter can sleep or be awake but if a conversion request occurs while sleeping, the driver must return an error to the application. Let us take stock of what we need to model:

– controllable actions of the driver
– uncontrollable actions of the environment where:
  - some uncontrollable actions take a non-null time and cannot happen immediately (*oneShot, scan, eoc*);
  - some uncontrollable actions are guaranteed to happen eventually. The input state of such a transition is then a transient state (*oneShot, scan, eoc*);
  - some uncontrollable actions are *maskable* (`inject`): they can be disabled by the controller.

3   Definitions

The following definitions introduce a kind of concurrent game between the controller (denoted by C) and the environment (denoted by U). In all states $q \in Q$, C (resp. U) selects an action in $Avail_C(q) \subseteq A_C$ (resp. $Avail_U(q) \subseteq A_U$). As seen in Definition 2, it selects a qualitative delay for performing its action and it can block a subset of the maskable actions of the environment (A_U^m). C may also be inactive while the environment has to act in *transient states* (QT). Thus in all $q \in QT$, an unmaskable action is available, i.e. $Avail_U(q) \setminus A_U^m \neq \emptyset$.

Definition 1 (Game structure). *A game structure is a tuple*
$\mathcal{G} = (Q, A_C, Avail_C, A_U, Avail_U, \delta)$ *where:*

- $Q = QT \uplus QI$ *is a set of states partitioned in transient states* QT *and idle states* QI *with* $q_0 \in Q$, *the initial state;*
- A_C *is the set of actions of the controller and* $Avail_C : Q \to 2^{A_C}$ *defines its available actions depending on states.*
- A_U *is the set of actions of the environment with* $A_C \cap A_U = \emptyset$ *and* $Avail_U : Q \to 2^{A_U}$ *defines its available actions depending on states.* A_U *includes the set of avoidable actions* A_U^a *and the set of maskable actions* A_U^m.
- $\delta : Q \times A_C \cup A_U \to Q$ *is the transition function such that* $\delta(q, a)$ *is defined if and only if* $a \in Avail_C(q) \cup Avail_U(q)$. *For all* $q \in QT$, $Avail_U(q) \setminus A_U^m \neq \emptyset$.

Example 1. The game structure of the case study presented in Sect. 2 is depicted in Fig. 2. We use the following graphical notations: Idle (resp. transient) states states are represented by (resp. double) circles, controller (resp. environment) transitions are represented by solid (resp. dashed) arrows, avoidable transitions start with a small circle and maskable actions are written in TrueType font.

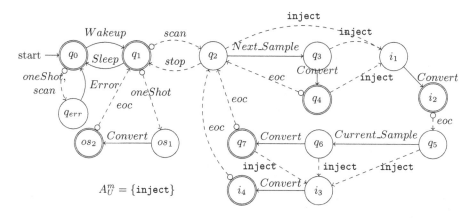

Fig. 2. Game structure of the case study

Given a current state q, the controller decides whether it intends to act (**0** or $\overline{\mathbf{0}}$) or not (ε), which action it intends to perform, when it will act (immediately: **0**; or later: $\overline{\mathbf{0}}$) and which actions it will block.

Definition 2 (Decision). *Let* $\mathcal{G}$ *be a game structure and* $q \in Q$. *Then the set of decisions of the controller* $Dec(q)$ *is defined as follows.* $(\gamma, \tau, B) \in Dec(q)$ *if:*

- *the action* $\gamma \in Avail_C(q) \cup \{\varepsilon\}$ *where* ε *denotes* inaction;
- *the delay* $\tau \in \{\mathbf{0}, \overline{\mathbf{0}}, \varepsilon\}$ *fulfills* $\tau = \varepsilon$ *iff* $\gamma = \varepsilon$;
- *the actions to be masked* $B \subseteq A_U^m \cap Avail_U(q)$.

Given a state q and a decision d of the controller, the next state to be reached may be either (1) q itself if it is idle and the controller is inactive, either (2) the state reached by the action selected by the controller (if any), or (3) a state reached by an environment action that has not be preempted by the action of the controller played without delay or masked by the controller.

Definition 3 (Play transitions). *Let $\mathcal{G}$ be a game structure, $q \in Q$ and $d = (\gamma, \tau, B) \in Dec(q)$, the set of play transitions $Next(q, d)$ is defined by:*

- *If $\gamma = \varepsilon$ and $q \in QI$ then $q \overset{d,\varepsilon}{\Longrightarrow} q \in Next(q, d)$;*
- *If $\gamma \neq \varepsilon$ then $q \overset{d,\gamma}{\Longrightarrow} \delta(q, \gamma) \in Next(q, d)$;*
- *For all $a \in Avail_U(q) \setminus (A_U^a \cup B)$, $q \overset{d,a}{\Longrightarrow} \delta(q, a) \in Next(q, d)$;*
- *If $\tau \neq \mathbf{0}$ then for all $a \in Avail_U(q) \cap A_U^a \setminus B$, $q \overset{d,a}{\Longrightarrow} \delta(q, a) \in Next(q, d)$.*

By construction, $Next(q, d)$ is never empty: if $\gamma \neq \varepsilon$ then $q \overset{d,\gamma}{\Longrightarrow} \delta(q, \gamma) \in Next(q, d)$ else if $q \in QI$ then $q \overset{d,\varepsilon}{\Longrightarrow} q \in Next(q, d)$ else there is some $a \in Avail_U(q) \setminus A_U^m$ such that $q \overset{d,a}{\Longrightarrow} \delta(q, a) \in Next(q, d)$.

A play is a finite or infinite sequence of play transitions such that the source of a non initial transition is the destination of the transition that precedes it.

Definition 4 (Play). *Let $\mathcal{G}$ be a game structure. Then $r = (q_n \overset{d_n,a_n}{\Longrightarrow} q_{n+1})_{n \in \mathbb{N}}$ where for all $n \in \mathbb{N}$, $d_n \in Dec(q_n)$ and $q_n \overset{d_n,a_n}{\Longrightarrow} q_{n+1} \in Next(q_n, d_n)$ is an infinite play. A finite play r is a finite prefix of an infinite play, ending in a state denoted $\mathrm{Last}(r)$. $\overline{\mathcal{R}}$ (resp. $\mathcal{R}$) denotes the set of infinite (resp. finite) plays. The empty play is denoted by λ with $\mathrm{Last}(\lambda) = q_0$.*

A strategy of the controller restricts the underlying transition system of the concurrent game by selecting a decision for all finite plays allowed by the strategy. Thus this mapping is inductively defined, simultaneously with its domain.

Definition 5 (Strategy). *A controller strategy s_C is a partial mapping from $\mathcal{R}$ to $\bigcup_{q \in Q} Dec(q)$ with its domain denoted $Dom(s_C)$ inductively defined by:*

- *$\lambda \in Dom(s_C)$;*
- *for all $r \in Dom(s_C)$, $s_C(r) \in Dec(\mathrm{Last}(r))$ and for all $\mathrm{Last}(r) \overset{s_C(r),a}{\Longrightarrow} q \in Next(\mathrm{Last}(r), s_C(r))$, $r(\mathrm{Last}(r) \overset{s_C(r),a}{\Longrightarrow} q) \in Dom(s_C)$.*

A play $r = (q_n \overset{d_n,a_n}{\Longrightarrow} q_{n+1})_{n \in \mathbb{N}}$ complies with s_C if for all n, $d_n = s_C((q_m \overset{d_m,a_m}{\Longrightarrow} q_{m+1})_{m < n})$. The outcome of s_C, denoted $\mathrm{Outcome}(s_C)$, is the set of infinite plays complying with it.

Any decision of a *positional* (also called *memoryless*) strategy s_C only depends on the last state of the play. For such a strategy, given some $q \in Q$, $s_C(q)$ denotes the decision of s_C for any finite play with last state q.

A *goal* W for the controller is a subset of Q^ω. W.r.t. W, a strategy s_C is *winning* for q_0 if for all plays $(q_n \xrightarrow{d_n, a_n} q_{n+1})_{n \in \mathbb{N}}$ complying with s_C, $(q_n)_{n \in \mathbb{N}} \in W$. A *parity goal* W_μ is defined by a mapping μ from Q to $\mathbb{N}$. Let $s = (q_n)_{n \in \mathbb{N}}$ define $m_\mu(s) = \max(i \mid \forall n\ \exists n' \geq n\ i = \mu(q_{n'}))$. Then $s \in W_\mu$ iff $m_\mu(s)$ is even. Parity goals include several kinds of goals like *safety* and *reachability* goals. A *game* is a pair $(\mathcal{G}, W)$.

4 From Concurrent Games to Turn-Based Games

A Turn-Based Game Interpretation. In order to apply the theory and algorithms of parity turn-based games, we propose below a linear-time translation of a concurrent game structure $\mathcal{G}$ into a turn-based one $\widehat{\mathcal{G}}$ such that given some parity goal W: (1) the controller has a winning strategy for $(\mathcal{G}, W)$ iff it has a winning strategy $(\widehat{\mathcal{G}}, W)$ and (2) from a positional winning strategy in $(\widehat{\mathcal{G}}, W)$, one can build in linear time a positional winning strategy in $(\mathcal{G}, W)$.

Definition 6. $\widehat{\mathcal{G}} = (\widehat{Q}, \rightarrow)$, *a turn-based game structure is defined by:*

- $\widehat{Q} = \widehat{Q}_C \uplus \widehat{Q}_U$, *the set of states with $q_0 \in \widehat{Q}_C$, the initial state;*
- $\rightarrow \subset \widehat{Q} \times \widehat{Q}$ *the transition relation fulfilling:* $\forall q \in \widehat{Q}\ \exists q' \in \widehat{Q}\ q \rightarrow q'$.

We denote Own the mapping from $\widehat{Q}$ to $\{C, U\}$ defined for all $q \in \widehat{Q}$ by $Own(q) = C$ if and only if $q \in \widehat{Q}_C$. $(q_n)_{n \in \mathbb{N}} \in \widehat{Q}^\omega$ is an infinite *play* of $\widehat{\mathcal{G}}$ if for all $n \in \mathbb{N}$, $q_n \rightarrow q_{n+1}$. A sequence $(q_m)_{m \leq n} \in \widehat{Q}^* \widehat{Q}_C$ is a finite play if for all $m < n$, $q_m \rightarrow q_{m+1}$. Note that we only define finite plays ending in states owned by C, as only those will be useful to define strategies for C.

Definition 7 (Strategy). *A strategy s_C of $\widehat{\mathcal{G}}$ is a partial mapping from the set of of finite plays to $\widehat{Q}$ with its domain denoted $Dom(s_C)$ inductively defined by:*

- $q_0 \in Dom(s_C)$;
- *for all $\rho = (q_m)_{m \leq n} \in Dom(s_C)$, $s_C(\rho) \in \widehat{Q}$ with $q_n \rightarrow s_C(\rho)$ and for all $\rho'' = \rho s_C(\rho)\rho'$, such that $s_C(\rho)\rho' \in \widehat{Q}_U^* \widehat{Q}_C\ \rho'' \in Dom(s_C)$.*

A play $\rho = (q_n)_{n \in \mathbb{N}}$ complies with s_C if for all n, such that $q_n \in \widehat{Q}_C$, $q_{n+1} = s_C((q_m)_{m \leq n})$. The outcome of s_C, denoted $\mathrm{Outcome}(s_C)$, is the set of infinite plays complying with it.

Let $\widehat{W} \subseteq \widehat{Q}^\omega$ be a goal, s_C is winning in $(\widehat{\mathcal{G}}, \widehat{W})$, if $\mathrm{Outcome}(s_C) \subseteq \widehat{W}$.

In order to obtain a canonical translation we assume an enumeration order of A_C and A_U. It should be clear that the results hold whatever the chosen order. According to this order, define for all $q \in Q$:

- maskable uncontrollable actions:

$$Avail_U(q) \cap A_U^m = \{\alpha_1^q, \ldots, \alpha_{\ell_q}^q\};$$

- unavoidable maskable uncontrollable actions:

$$(Avail_U(q) \cap A_U^m) \setminus A_U^a = \{\alpha_1^q, \ldots, \alpha_{k_q}^q\} \text{ with } k_q \leq \ell_q;$$

- unmaskable uncontrollable actions:

$$Avail_U(q) \setminus A_U^m = \{\beta_1^q, \ldots, \beta_{n_q}^q\};$$

- unavoidable unmaskable uncontrollable actions:

$$Avail_U(q) \setminus (A_U^a \cup A_U^m) = \{\beta_1^q, \ldots, \beta_{m_q}^q\} \text{ with } m_q \leq n_q;$$

- controllable actions:

$$Avail_C(q) = \{\gamma_1^q, \ldots, \gamma_{p_q}^q\}.$$

In order to avoid handling particular cases, Definition 8 assumes that for all q, $k_q \geq 1$, $n_q \geq 1$, and $p_q \geq 1$. Afterwards, we explain how to adapt the translation when this is not the case. Let us first informally describe how the turn-based version $\widehat{\mathcal{G}}$ of game $\mathcal{G}$ is specified:

- The states of $\mathcal{G}$ are also states of $\widehat{\mathcal{G}}$ and belong to the controller. In such a state q, the controller has three choices (see Fig. 3):
  - either it decides to (try to) play immediately going to state (q_0^C, α_1^q);
  - either it decides to (try to) play not immediately going to state $(q_{\overline{0}}^C, \alpha_1^q)$;
  - or it decides to be inactive, going to state $(q_\varepsilon^C, \alpha_1^q)$.

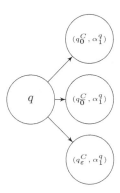

Fig. 3. Choices of the controller

- From state (q_0^C, α_1^q), the controller successively either lets the environment the availability of action α_1^q by going to $(q_\varepsilon^U, \alpha_1^q)$ or masks this action by going to (q_0^C, α_2^q). After all maskable unavoidable actions have been enumerated (and masked or not played), in q_0^U the environment can play any unavoidable and unmaskable action or, by going in q_0^C, let the controller play an action (see Fig. 4).

– The situation from state $(q_{\overline{0}}^C, \alpha_1^q)$ is similar to the previous one except that the controller first enumerates *all* the maskable actions and in $q_{\overline{0}}^U$ the environment can play *any* unmaskable action (see Fig. 5).
– The situation from state $(q_{\varepsilon}^C, \alpha_1^q)$ is similar to the previous one except that in q_{ε}^U, if q is transient the environment must play *some* unmaskable action while if q is idle it can decide to be inactive going back to q (see Fig. 6).

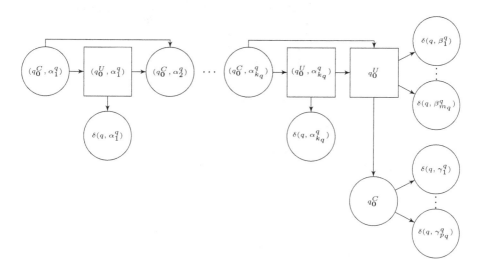

Fig. 4. The controller tries to play immediately

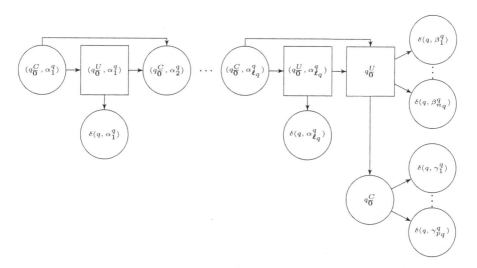

Fig. 5. The controller tries to play but not immediately

Definition 8. *Let $\mathcal{G}$ be a game structure. Then $\widehat{\mathcal{G}} = (\widehat{Q}, Own, \rightarrow)$, a turn-based game structure, is defined as follows:*

- $\widehat{Q} = Q \cup \{q_x^y \mid q \in Q \wedge ((x \in \{\mathbf{0}, \overline{\mathbf{0}}\} \wedge y \in \{C, U\}) \vee (x = \varepsilon \wedge y = U))\}$
  $\cup \{(q_x^y, \alpha_i^q)) \mid q \in Q \wedge y \in \{C, U\} \wedge (x = \mathbf{0} \wedge (i \leq k_q)) \vee (x \in \{\overline{\mathbf{0}}, \varepsilon\} \wedge (i \leq \ell_q))\};$
- $\widehat{Q}_C = Q \cup \{q_x^C\}_{q,x} \cup \{(q_x^C, z)\}_{q,x,z}, \ \widehat{Q}_U = Q \cup \{q_x^U\}_{q,x} \cup \{(q_x^U, z)\}_{q,x,z};$
- $\rightarrow$ *is defined by for all $q \in Q$ and all $x \in \{\mathbf{0}, \overline{\mathbf{0}}, \varepsilon\}$, $q \rightarrow (q_x^C, \alpha_1^q)$ and:*

Case of immediate play controller choice.
- *For all q and $i \leq k_q$, $(q_{\mathbf{0}}^C, \alpha_i^q) \rightarrow (q_{\mathbf{0}}^U, \alpha_i^q)$ and $(q_{\mathbf{0}}^U, \alpha_i^q) \rightarrow \delta(q, \alpha_i^q);$*
- *For all q and $i < k_q$, $(q_{\mathbf{0}}^C, \alpha_i^q) \rightarrow (q_{\mathbf{0}}^C, \alpha_{i+1}^q)$ and $(q_{\mathbf{0}}^U, \alpha_i^q) \rightarrow (q_{\mathbf{0}}^C, \alpha_{i+1}^q);$*
- *For all q, $(q_{\mathbf{0}}^C, \alpha_{k_q}^q) \rightarrow q_{\mathbf{0}}^U$ and $(q_{\mathbf{0}}^U, \alpha_{k_q}^q) \rightarrow q_{\mathbf{0}}^C;$*
- *For all q and $i \leq m_q$, $q_{\mathbf{0}}^U \rightarrow \delta(q, \beta_i^q)$ and $q_{\mathbf{0}}^U \rightarrow q_{\mathbf{0}}^C;$*
- *For all q and $i \leq p_q$, $q_{\mathbf{0}}^U \rightarrow \delta(q, \gamma_i^q).$*

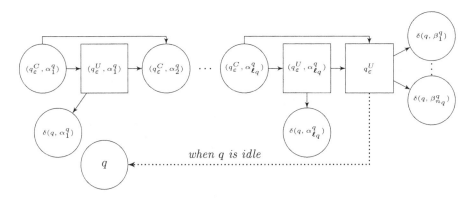

Fig. 6. The controller decides not to play

Other Cases $(x \in \{\overline{\mathbf{0}}, \varepsilon\})$.
- *For all q and $i \leq \ell_q$, $(q_x^C, \alpha_i^q) \rightarrow (q_x^U, \alpha_i^q)$ and $(q_x^U, \alpha_i^q) \rightarrow \delta(q, \alpha_i^q);$*
- *For all q and $i < \ell_q$, $(q_x^C, \alpha_i^q) \rightarrow (q_x^C, \alpha_{i+1}^q)$ and $(q_x^U, \alpha_i^q) \rightarrow (q_x^C, \alpha_{i+1}^q);$*
- *For all q, $(q_x^C, \alpha_{\ell_q}^q) \rightarrow q_x^U$ and $(q_x^U, \alpha_{\ell_q}^q) \rightarrow q_x^U;$*
- *For all q and $i \leq n_q$, $q_x^U \rightarrow \delta(q, \beta_i^q);$*
- *$q_{\overline{\mathbf{0}}}^U \rightarrow q_{\overline{\mathbf{0}}}^C$ and for all $q \in Q$ and $i \leq p_q$, $q_{\overline{\mathbf{0}}}^C \rightarrow \delta(q, \gamma_i^q);$*
- *When $q \in QI$, $q_\varepsilon^U \rightarrow q.$*

Let us explain how to address the particular cases. For instance when $\ell_q=0$, the three transitions outgoing from q target the states $q_{\mathbf{0}}^U$, $q_{\overline{\mathbf{0}}}^U$ and q_ε^U. The other particular cases are similarly handled.

Example 2. Figure 7 partly illustrates this translation for the game structure of Fig. 2, starting from transient state q_0. In all figures illustrating $\widehat{\mathcal{G}}$, states owned by the controller are represented by circles while states owned by the environment are represented by rectangles.

$Avail_U(q_0) \setminus A_U^m = \{oneShot, scan\}$

$Avail_C(q_0) = \{Wakeup\}$

$k_{q_0} = 0, \ell_{q_0} = 0, m_{q_0} = 0, n_{q_0} = 2$

$\beta_1^{q_0} = \beta_{m+1}^{q_0} = oneShot$

$\beta_2^{q_0} = \beta_n^{q_0} = scan$

$\gamma_1^{q_0} = Wakeup$

$\delta(q_0, \beta_1^{q_0}) = \delta(q_0, \beta_2^{q_0}) = q_{err}, \delta(q_0, \gamma_1^{q_0}) = q_1$

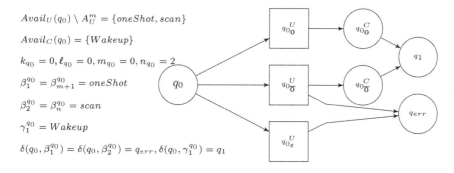

Fig. 7. The controller choices from q_0

Correspondence Between Plays. We want to establish a correspondence between plays of $\mathcal{G}$ and plays of $\widehat{\mathcal{G}}$. With this aim, we introduce *connecting paths* of $\widehat{\mathcal{G}}$.

Definition 9. *A* connecting path *of* $\widehat{\mathcal{G}}$ *is a path from some* $q \in Q$ *to some* $q' \in Q$ *that does not visit in between other states of* Q.

Let ρ be a connecting path. We denote $src(\rho)$ the source state of ρ and $dst(\rho)$ its destination state. Given two connecting paths ρ, ρ' such that q, the target state of ρ, is the source state of ρ', $\rho\rho'$ denotes the concatenation of ρ and ρ' without the repetition of q. As usual, this operation is extended to finite or countable paths. Observe that any infinite play of $\widehat{\mathcal{G}}$ is a countable concatenation of connecting paths.

The following lemma lists all possible connecting paths. We omit its proof as it straightforwardly comes from examination of the local structure of $\widehat{\mathcal{G}}$. Observe that we associate an action with a connecting path.

Lemma 1. *Let* $q \in Q$. *Then all connecting paths starting from* q *can be concisely specified (with an associated action* a_ρ) *as follows:*

- $\rho(q, y, B, \alpha_i^q)$, *the path that (1) starts from* q *to* (q_y^C, α_1^q), *(2) goes to* (q_y^U, α_i^q) *avoiding the set* $\{(q_y^U, \alpha_j^q) \mid \alpha_j^q \in B\}$, *and (3) reaches* $\delta(q, \alpha_i^q)$. $a_\rho = \alpha_i^q$;
- $\rho(q, y, B, \beta_i^q)$, *the path that (1) starts from* q *to* (q_y^C, α_1^q), *(2) goes to* q_y^U *avoiding the set* $\{(q_y^U, \alpha_j^q) \mid \alpha_j^q \in B\}$, *and (3) reaches* $\delta(q, \beta_i^q)$. $a_\rho = \beta_i^q$;
- $\rho(q, y, B, \gamma_i^q)$, *the path that (1) starts from* q *to* (q_y^C, α_1^q), *(2) goes to* q_y^C *avoiding the set* $\{(q_y^U, \alpha_j^q) \mid \alpha_j^q \in B\}$, *and (3) reaches* $\delta(q, \gamma_i^q)$. $a_\rho = \gamma_i^q$;
- *when* q *is idle,* $\rho(q, \varepsilon, B, \varepsilon)$, *the path that (1) starts from* q *to* $(q_\varepsilon^C, \alpha_1^q)$, *(2) goes to* q_ε^U *avoiding the set* $\{(q_y^U, \alpha_j^q) \mid \alpha_j^q \in B\}$, *and (3) returns to* q. $a_\rho = \varepsilon$.

We now relate play transitions of $\mathcal{G}$ to connecting paths of $\widehat{\mathcal{G}}$.

Definition 10. *Let* $e = q \overset{d,a}{\Longrightarrow} q'$ *with* $d = (\gamma, \tau, B)$ *be a play transition of* $\mathcal{G}$. *Then the connecting path* $\hat{e}$ *is defined as follows:*

- *If* $a = \alpha_i^q$ *for some* α_i^q *then* $\hat{e} = \rho(q, \tau, B', \alpha_i^q)$ *with* $B' = \{\alpha_j^q \mid j < i\} \cap B$;

- If $a = \beta_i^q$ for some β_i^q then $\hat{e} = \rho(q, \tau, B, \beta_i^q)$
- If $a = \gamma_i^q$ for some γ_i^q then $\hat{e} = \rho(q, \tau, B, \gamma_i^q)$;
- If $a = \varepsilon$ then $\hat{e} = \rho(q, \tau, B, \varepsilon)$;

We now extend in a natural way the correspondence between transitions to plays.

Definition 11 (Relation between plays). Let $r = (e_n)_{n \leq N}$ (resp. $r = (e_n)_{n \in \mathbb{N}}$) be a finite (resp. infinite) play of $\mathcal{G}$. Then $\hat{r}$ a finite (resp. infinite) play of $\widehat{\mathcal{G}}$, is defined by $\hat{r} = (\hat{e}_n)_{n \leq N}$ (resp. $\hat{r} = (\hat{e}_n)_{n \in \mathbb{N}}$).

Correspondence Between Strategies. Let $\mathcal{G}$ be a concurrent game structure and $\rho = (q_n)_{n \in \mathbb{N}} \in \widehat{Q}^\omega$. Define $\pi(\rho)$ as $(q_{\alpha(n)})_{n \in \mathbb{N}}$ where α is a strictly increasing mapping from $\mathbb{N}$ to $\mathbb{N}$ with the range of α being $\{n \mid q_n \in Q\}$. This means, π extracts the subsequence of states from the original game from any play in the turn-based game. Let W be a goal of $\mathcal{G}$. Then $\widehat{W} \subseteq \widehat{Q}^\omega$, a goal of $\widehat{\mathcal{G}}$, is defined by $\widehat{W} = \{\rho \mid \pi(\rho) \in W\}$.

In order to obtain a correspondence between strategies, we focus on translating a decision $d \in Dec(q)$ in $\mathcal{G}$ into a *local positional strategy* $\hat{d}$ of $\widehat{\mathcal{G}}$. Given a state q and a decision $d = (\gamma, \tau, B)$ induced by a strategy in $\mathcal{G}$, the 'local' corresponding strategy in $\widehat{\mathcal{G}}$ consists in allowing exactly the connecting paths $\hat{e}$ such that $e \in Next(q, d)$. Thus, (1) in q the strategy selects (q_τ^C, α_1^q), (2) in all states (q_τ^C, α_i^q), it selects (q_τ^U, α_i^q) if $\alpha_i^q \notin B$ and avoids it otherwise, and (3) when $\gamma \neq \varepsilon$, it selects in q_τ^C the state $\delta(q, \gamma)$.

Definition 12 (From decisions to local strategies). Let $q \in Q$ and $d = (\gamma, \tau, B) \in Dec(q)$. The partial mapping $\hat{d} : \widehat{Q} \to \widehat{Q}$ is defined as follows:

- $\hat{d}(q) = (q_\tau^C, \alpha_1^q)$;
- for all (defined) (q_τ^C, α_i^q) if $\alpha_i^q \notin B$ then $\hat{d}(q_\tau^C, \alpha_i^q) = (q_\tau^U, \alpha_i^q)$;
- for all (defined) (q_τ^C, α_i^q) if $\alpha_i^q \in B$ and $(i, \tau) \notin \{(k_q, \mathbf{0}), (\ell_q, \overline{\mathbf{0}}), (\ell_q, \varepsilon)\}$
  then $\hat{d}(q_\tau^C, \alpha_i^q) = (q_\tau^C, \alpha_{i+1}^q)$;
- for all (defined) (q_τ^C, α_i^q) if $\alpha_i^q \in B$ and $(i, \tau) \in \{(k_q, \mathbf{0}), (\ell_q, \overline{\mathbf{0}}), (\ell_q, \varepsilon)\}$
  then $\hat{d}(q_\tau^C, \alpha_i^q) = q_\tau^U$;
- If $\gamma \neq \varepsilon$ then $\hat{d}(q_\tau^C)) = \delta(q, \gamma)$.

Definition 13. Let $d \in Dec(q)$. A connecting path ρ starting from q complies with d if for all transitions $q_1 \to q_2$ of ρ with $q_1 \in \widehat{Q}_C$, $q_2 = \hat{d}(q_1)$.

By examining all possible cases, one gets the following lemma.

Lemma 2. Let $q \in Q$ and $d = (\gamma, \tau, B) \in Dec(q)$. Then:

$$Next(q, d) = \{q \overset{d, a_\rho}{\Longrightarrow} dst(\rho) \mid \rho \text{ is a connecting path with } src(\rho) = q \text{ complying with } d\}$$

Example 3. Let us consider the example of Fig. 2. Figure 8 shows the result of the decision $d = (\gamma_1^{q_2}, \mathbf{0}, \{\alpha_1^{q_2}\}) = (Next\_Sample, \mathbf{0}, \{\texttt{inject}\})$ from q_2.

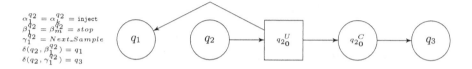

$$\begin{aligned}
\alpha_1^{q_2} &= \alpha_k^{q_2} = inject\\
\beta_1^{q_2} &= \beta_m^{q_2} = stop\\
\gamma_1^{q_2} &= Next\_Sample\\
\delta(q_2, \beta_1^{q_2}) &= q_1\\
\delta(q_2, \gamma_1^{q_2}) &= q_3
\end{aligned}$$

Fig. 8. The connecting paths from q_2 complying with $d = (Next\_Sample, \mathbf{0}, \{\texttt{inject}\})$

Given the decisions selected by a strategy s_C of $\mathcal{G}$ and applying the corresponding local strategies, one gets a strategy $\hat{s}_C$ of $\widehat{\mathcal{G}}$. The next definition is sound since, by induction and using Lemma 2, for every finite play ρ that complies with $\hat{s}_C$ and ends in Q, there is some r_ρ that complies with s_C such that $\rho = \hat{r}$.

Let $q \in Q$, $\tau \in \{\mathbf{0}, \overline{\mathbf{0}}, \varepsilon\}$, $\alpha \in A_U^m$ such that $(q_\tau^U, \alpha) \in \widehat{Q}$ and $B \subseteq A_U^m \cap Avail_U(q)$. Then $(q_\tau^U, \alpha)_{\downarrow B}$ denotes (q_τ^U, α) if $\alpha \notin B$ and the empty word of $\widehat{Q}^*$ otherwise.

Definition 14. *Let s_C be a controller strategy of $\mathcal{G}$. Then $\hat{s}_C$, a controller strategy of $\widehat{\mathcal{G}}$ is defined by induction on the length of $\rho = \hat{r}_\rho$ where r_ρ is a finite play of $\mathcal{G}$ complying with s_C and ending in some $q \in Q$ as follows when denoting $s_C(r)$ by $d = (\gamma, \tau, B)$.*

- $\hat{s}_C(\rho) = \hat{d}(q)$;
- *for all (defined) (q_τ^C, α_i^q), $\hat{s}_C(\rho(q_\tau^C, \alpha_1^q)(q_\tau^U, \alpha_1^q)_{\downarrow B} \ldots (q_\tau^C, \alpha_i^q)) = \hat{d}(q_\tau^C, \alpha_i^q)$*
- *If $\gamma \neq \varepsilon$ then $\hat{s}_C(\rho(q_\tau^C, \alpha_1^q)(q_\tau^U, \alpha_1^q)_{\downarrow B} \ldots (q_\tau^C, \alpha_{k_q}^q)(q_\tau^U, \alpha_{k_q}^q)_{\downarrow B}q_\tau^U q_\tau^C)) = \hat{d}(q_\tau^C))$.*

Let $\rho^* = \rho\rho'$ where ρ' is a a connecting path complying with d and associated action $a_{\rho'}$. Then $r_{\rho^*} = r_\rho(Last(r_\rho) \xRightarrow{d, a_{\rho'}} dst(\rho'))$.

Applying inductively Lemma 2, one immediately gets the next proposition and corollary:

Proposition 1. *Let s_C be a controller strategy of $\mathcal{G}$. Then:*

$$\text{Outcome}(\hat{s}_C) = \{\hat{r} \mid r \in \text{Outcome}(s_C)\}$$

Corollary 1. *Let $Goal \subseteq Q^\omega$ be a goal and s_C be a winning strategy for Goal in $\mathcal{G}$. Then $\hat{s}_C$ is a winning strategy for Goal in $\widehat{\mathcal{G}}$.*

In order to get the converse result we exploit the fact that positional strategies are sufficient for turn-based parity games.

Definition 15. *Let s_C be a positional controller strategy of $\widehat{\mathcal{G}}$. Then $\overline{s}_C$ is a positional controller strategy of $\mathcal{G}$ defined as follows. Let $q \in Q$ and denote $\overline{s}_C(q) = (\gamma, \tau, B)$.*

- If $s_C(q) = (q_{\overline{0}}^C, \alpha_1^q)$ and $s_C(q_{\overline{0}}^C) = \delta(a, \gamma_i^q)$ then
  $\gamma = \gamma_i^q$, $\tau = \mathbf{0}$ and $B = \{i \leq k_q \mid s_C((q_{\overline{0}}^C, \alpha_i^q)) \neq (q_{\overline{0}}^U, \alpha_i^q)\}$;
- If $s_C(q) = (q_{\overline{0}}^C, \alpha_1^q)$ and $s_C(q_{\overline{0}}^C) = \delta(a, \gamma_i^q)$ then
  $\gamma = \gamma_i^q$, $\tau = \overline{\mathbf{0}}$ and $B = \{i \leq \ell_q \mid s_C((q_{\overline{0}}^C, \alpha_i^q)) \neq (q_{\overline{0}}^U, \alpha_i^q)\}$;
- If $s_C(q) = (q_{\varepsilon}^C, \alpha_1^q)$ then
  $\gamma = \varepsilon$, $\tau = \varepsilon$ and $B = \{i \leq \ell_q \mid s_C((q_{\overline{0}}^C, \alpha_i^q)) \neq (q_{\overline{0}}^U, \alpha_i^q)\}$;

Observe that $\overline{s}_C(q)$ is an item of $Dec(q)$, say d, and that s_C restricted to the subset of states of $\widehat{Q}$ related to q can be viewed as a local strategy, say d'. By construction, $d' = \hat{d}$. Thus applying inductively Lemma 2, one immediately gets the next proposition and corollary:

Proposition 2. *Let s_C be a positional controller strategy of $\widehat{\mathcal{G}}$. Then:*

$$\text{Outcome}(s_C) = \{\hat{r} \mid r \in \text{Outcome}(\overline{s}_C)\}.$$

Corollary 2. *Let $Goal \subset Q^\omega$ be a parity goal and s_C be a winning positional strategy for $Goal$ in $\widehat{\mathcal{G}}$. Then $\overline{s}_C$ is a winning positional strategy for $Goal$ in $\mathcal{G}$.*

Combining Corollaries 1 and 2, one gets:

Theorem 1. *Let $\mathcal{G}$ be a concurrent game, $Goal \subseteq Q^\omega$ be a parity goal.*

- *Let s_C be a controller strategy. Then s_C is a winning strategy for $Goal$ if and only if $\hat{s}_C$ is a winning strategy for $Goal$;*
- *If there is a winning strategy for $\mathcal{G}$, there is a positional one, i.e. $\overline{s}'_C$ where s'_C is any positional winning strategy of $\widehat{\mathcal{G}}$.*

5   Permissivity of the Controller

Permissivity is a criterion that has to be taken into account for choosing between possible controllers. In our context and due to the results of the previous section, we limit ourselves to positional controllers. First we introduce an order between controller decisions. In words, a decision d' is more permissive than d if (1) either d' is inactive or intends to play the same action as d, (2) the delay before acting of d' is greater or equal than the delay before acting of d, and (3) the set of actions masked by d' is a subset of actions masked by d. Here permissivity should be interpreted as the controller avoiding to restrict the behaviour of the environment.

Definition 16. *Let q be a state and $d = (\gamma, \tau, B)$ and $d' = (\gamma', \tau, B)$ be two decisions, one says that d' is more (or equally) permissive than d denoted by $d \preceq d'$ if: $\gamma' = \gamma$ or $\gamma' = \varepsilon$, $\tau \leq \tau'$ and $B' \subseteq B$.*

Assuming that there exists a winning strategy for q_0, our goal is to define and synthesise a maximally permissive winning strategy for q_0 w.r.t. safety and reachability goals. As explained later we introduce slightly different notions of maximally permissive winning strategy depending on the kind of goal. Observe that in the next definition, we do not care about losing states of the game since winning strategies do not enter losing states.

Definition 17. *Let s_C and s'_C be positional winning controller strategies of $(\mathcal{G}, Goal)$ where Goal is a safety goal. Then s'_C is more permissive than s_C w.r.t. Goal, denoted $s_C \preceq s'_C$, if for all q winning state of $(\mathcal{G}, Goal)$, $s_C(q) \preceq s'_C(q)$. Additionally, $s_C \prec s'_C$ if $s_C \preceq s'_C$ and $s'_C \not\preceq s_C$.*
s_C is a maximally permissive winning strategy w.r.t. Goal if there is no s'_C such that $s_C \prec s'_C$.

The synthesis of a maximally permissive winning strategy w.r.t. a safety goal is easy. Since it is enough to stay in the winning states, maximally permissive decisions can be combined without any restriction to get a maximally permissive winning strategy.

Theorem 2. *Let $\mathcal{G}$ be a concurrent game and Goal be a safety goal. Once the winning states of $(\widehat{\mathcal{G}}, Goal)$ have been computed, a maximally permissive winning strategy of $(\mathcal{G}, Goal)$ for q_0 can be computed in linear time.*

Proof. We compute for all q, winning state of $(\mathcal{G}, Goal)$ (and thus of $(\widehat{\mathcal{G}}, Goal)$), a most permissive decision. Since we deal with a safety goal, any combination of decisions works. A maximally permissive decision for q is computed as follows.

- If $(q_\varepsilon^C, \alpha_1^q)$ is winning then define $B = \{\alpha_i^q \mid i \leq \ell_q \wedge \delta(q, \alpha_i^q) \text{ is not winning}\}$. By construction, $(\varepsilon, \varepsilon, B)$ is a maximally permissive decision;
- If $(q_\varepsilon^C, \alpha_1^q)$ is not winning and $(q_{\overline{0}}^C, \alpha_1^q)$ is winning then define $B = \{\alpha_i^q \mid i \leq \ell_q \wedge \delta(q, \alpha_i^q) \text{ is not winning}\}$. The set $\{\gamma_i^q \mid i \leq p_q \wedge \delta(q, \gamma_i^q) \text{ is winning}\}$ is not empty since $(q_{\overline{0}}^C, \alpha_1^q)$ is winning. Pick some γ_i^q inside. By construction, $(\gamma_i^q, \overline{0}, B)$ is a maximally permissive decision;
- If $(q_\varepsilon^C, \alpha_1^q)$ and $(q_{\overline{0}}^C, \alpha_1^q)$ are not winning then (q_0^C, α_1^q) is winning. Define $B = \{\alpha_i^q \mid i \leq k_q \wedge \delta(q, \alpha_i^q) \text{ is not winning}\}$. The set $\{\gamma_i^q \mid i \leq p_q \wedge \delta(q, \gamma_i^q) \text{ is winning}\}$ is not empty since (q_0^C, α_1^q) is winning. Pick some γ_i^q inside. By construction, $(\gamma_i^q, 0, B)$ is a maximally permissive decision.

$\square$

The definition of a maximally permissive winning strategy w.r.t. a reachability goal is more involved since we want to take into account the *delay* before reaching the target state q_f assumed w.l.o.g. to be absorbing (i.e., its only outgoing transition is a self-loop).

Definition 18. *Let $\mathcal{G}$ be a concurrent game structure and Goal be a reachability goal defined by an absorbing target state q_f. Let s_C be a positional winning strategy and $q \in Q$ be a winning state. Then $delay_{s_C}(q)$, the delay of $q \neq q_f$ w.r.t. s_C is defined by:*

$$delay_{s_C}(q) = \max(n \mid (q_m \xrightarrow{d_m, a_m} q_{m+1})_{1 \leq m < n} \text{ complying to } s_C, q = q_1, \forall i \; q_i \neq q_f)$$

By convention, $delay_{s_C}(q_f) = 0$.
The minimal delay of a q, denoted $delay^*(q)$ is defined by:

$$delay^*(q) = \min(delay_{s_C}(q) \mid s_C \text{ is a winning strategy})$$

We also restrict the constraint for maximality for decisions of a strategy s_C to states *visited* by s_C: a state q is visited by s_C if there is a finite play complying to s_C starting in q_0 and ending in q.

Definition 19. *Let s_C and s'_C be positional winning controller strategies of $(\mathcal{G}, Goal)$ where Goal is a reachability goal defined by an absorbing target state q_f. Then s'_C is more permissive than s_C w.r.t. Goal, denoted $s_C \preceq s'_C$, if for all q state visited by s_C, $s_C(q) \preceq s'_C(q)$ and $delay_{s'_C}(q) \leq delay_{s_C}(q)$.*
$s_C \prec s'_C$ if $s_C \preceq s'_C$ and $s'_C \npreceq s_C$.
s_C is a maximally permissive *winning strategy w.r.t. Goal if there is no s'_C such that $s_C \prec s'_C$.*

Let us informally explain how the proof of the next theorem proceeds to synthesise a maximally permissive winning strategy. If $q_0 = q_f$ we are done. Otherwise define $Q_0 = \{q_f\}$ and Q_1 to be the set of states $q \neq q_f$ for which there is a decision d such that all the transitions of $Next(q, d)$ lead to Q_0. Observe that this set is non empty. Otherwise q_0 could not be a winning state. For such q select d maximally permissive among these kinds of decisions. If $q_0 \in Q_1$ we are done. We iterate this process until q_0 is found. Assume $Q_0, \ldots, Q_i$ have been built with $q_0 \notin \biguplus_{j \leq i} Q_j$. Define Q_{i+1} to be the set of states $q \notin \biguplus_{j \leq i} Q_j$ for which there is decision d such that all the transitions of $Next(q, d)$ lead to $\biguplus_{j \leq i} Q_j$. Observe that this set is non empty. Otherwise q_0 could not be a winning state. For such q select d maximally permissive among these kinds of decisions. Since Q is finite this process must stop with q_0 belonging to some Q_{i^*}.

Theorem 3. *Let $\mathcal{G}$ be a concurrent game structure and Goal be a reachability goal defined by an absorbing target state q_f. Then a maximally permissive winning strategy of $(\mathcal{G}, Goal)$ for q_0 can be computed in linear time.*

Proof. As explained we proceed by iteratively building disjoint sets of states $Q_0, Q_1, \ldots$ as long as q_0 does not belong to these states. Furthermore we associate decisions with every state belonging to these sets, that define the strategy s_C. We set $Q_0 = \{q_f\}$ and since q_f is absorbing, the associated decision is $(\varepsilon, \varepsilon, \emptyset)$. Assume that $Q_0, Q_1, \ldots, Q_i$ have been built with $q_0 \notin \biguplus_{j < i} Q_j$. For sake of readability, we write $Q' = \biguplus_{j \leq i} Q_j$. Let $q \notin Q'$.

- If q is transient and for all $k \leq n_q, \delta(q, \beta_k^q) \in Q'$ then $q \in Q_{i+1}$ and defining $B = \{\alpha_k^q \mid k \leq \ell_q \wedge \delta(q, \alpha_k^q) \notin Q'\}$, $(\varepsilon, \varepsilon, B)$ is the decision associated with q;
- Else if for all $k \leq n_q, \delta(q, \beta_k^q) \in Q'$ and there exists γ_p^q such that $\delta(q, \gamma_p^q) \in Q'$ then $q \in Q_{i+1}$ and defining $B = \{\alpha_k^q \mid k \leq \ell_q \wedge \delta(q, \alpha_k^q) \notin Q'\}$, $(\gamma_p^q, \overline{\mathbf{0}}, B)$ is the decision associated with q;
- Else if for all $k \leq m_q, \delta(q, \beta_k^q) \in Q'$ and there exists γ_p^q such that $\delta(q, \gamma_p^q) \in Q'$ then $q \in Q_{i+1}$ and defining $B = \{\alpha_k^q \mid k \leq k_q \wedge \delta(q, \alpha_k^q) \notin Q'\}$, $(\gamma_p^q, \mathbf{0}, B)$ is the decision associated with q.

By construction, for all $q \in Q$, one has $q \in Q_i$ if and only if $delay^*(q) = i$. Furthermore $delay_{s_C}(q) = delay^*(q)$. Assume there exists a winning strategy s'_C

such that $s_C \prec s'_C$. Since s_C achieves the minimality of delay for states visited by s_C, there must exist a state q visited by s_C that belongs to some Q_i such that $s'_C(q)$ is strictly more permissive than $s_C(q)$. By definition of $s_C(q)$, this implies that there exists a state $q'' \notin \biguplus_{j<i} Q_j$ which is the destination of a transition of $Next(q, d)$. Thus $delay_{s'_C}(q) > delay_{s_C}(q)$, establishing a contradiction.

In order to obtain a linear time algorithm, one proceeds as follows. In the sequel i denotes the current iteration when the sets $Q_0, Q_1, \ldots, Q_{i-1}$ have been built. Here Q' denotes $Q' = \biguplus_{j \leq i} Q_j$.

- Initially one builds a "reverse graph" whose vertices are the states and there is an edge $q' \xrightarrow{\beta^q_k} q$ (resp. $q' \xrightarrow{\gamma^q_p} q$) with $k \leq m_q$ (resp. $p \leq p_q$) if $\delta(q, \beta^q_k) = q'$ (resp. $\delta(q, \gamma^q_p) = q'$).
- At beginning of the ith iteration, the set Q_i has already been computed using the variables described below.
- One associates a counter c_q and a boolean b_q with every state q and a boolean b_q with every idle state q. Boolean b_q is true if there exists some action γ^q_p such that $\delta(q, \gamma^q_p) \in Q'$ and c_q is the size of the set $\{k \leq m_q, \delta(q, \beta^q_k) \notin Q'\}$. A state q belongs to Q_i if (1) it does not belong to Q', (2) its counter c_q is null, and (3) either it is transient or its boolean b_q is true.
- The ith iteration consists in two stages for all $q \in Q_i$. First one determines $s_C(q)$ using the rules described above. Then one updates the counters and the booleans using the edges of the reverse graph: for all $q \xrightarrow{\beta^{q'}_k} q'$, $c_{q'}$ is decremented and if there exists some $q \xrightarrow{\gamma^{q'}_p} q'$ $b_{q'}$ is set to true. If q' satisfies the three conditions w.r.t. iteration $i + 1$, it enters the set Q_{i+1}.

It is routine to check that this procedures operates in linear time.   □

6   Conclusion

We have introduced a model of qualitative time concurrent game between a controller and an environment. We have designed a linear-time translation from such a game to an untimed turn-based game and shown that given any parity goal, the concurrent game is winning for the controller if and only if the turn-based game is winning for it. This allows us, taking as input a positional winning strategy of the turn-based game, to build in linear time a positional winning strategy in the original game. Furthermore we have introduced a notion of permissivity for strategies and we have established that one can compute in linear time a maximally permissive winning strategy for safety and reachability goals. For future work, we want to design an algorithm for computing in polynomial time a maximally permissive winning strategy for repeated reachability goals. We also plan to modify the notion of delay for reachability goals by integrating the delay of actions when computing a maximally permissive winning strategy for reachability. Finally our notion of qualitative delay for actions could be refined in order to take into account more precise (but still qualitative) information.

References

1. Ashok, P., Křetínský, J., Larsen, K.G., Le Coënt, A., Taankvist, J.H., Weininger, M.: SOS: safe, optimal and small strategies for hybrid Markov decision processes. In: Parker, D., Wolf, V. (eds.) QEST 2019. LNCS, vol. 11785, pp. 147–164. Springer, Cham (2019). https://doi.org/10.1007/978-3-030-30281-8_9
2. Béchennec, J.-L., Lime, D., Roux, O.H.: Control of DES with urgency, avoidability and ineluctability. In: Keller, J., Penczek, W. (eds.) ACSD 2019. pp, pp. 92–101. IEEE Computer Society, Aachen, Germany (2019)
3. Béchennec, J.-L., Lime, D., Roux, O.H.: Logical time control of concurrent DES. Discrete Event Dynamic Systems 1–33 (2021). https://doi.org/10.1007/s10626-020-00333-x
4. Behrmann, G., Cougnard, A., David, A., Fleury, E., Larsen, K.G., Lime, D.: UPPAAL-tiga: time for playing games! In: Damm, W., Hermanns, H. (eds.) CAV 2007. LNCS, vol. 4590, pp. 121–125. Springer, Heidelberg (2007). https://doi.org/10.1007/978-3-540-73368-3_14
5. Chatterjee, K., de Alfaro, L., Henzinger., T.A. Strategy improvement for concurrent reachability and safety games. CoRR http://arxiv.org/abs/1201.2834 (2012)
6. de Alfaro, L., Faella, M., Henzinger, T.A., Majumdar, R., Stoelinga, M.: The element of surprise in timed games. In: Amadio, R., Lugiez, D. (eds.) CONCUR 2003. LNCS, vol. 2761, pp. 144–158. Springer, Heidelberg (2003). https://doi.org/10.1007/978-3-540-45187-7_9
7. de Alfaro, L., Henzinger, T.A., Kupferman, O.: Concurrent reachability games. Theoret. Comput. Sci. **386**(3), 188–217 (2007)
8. de Alfaro, L., Henzinger, T.A., Majumdar, R.: Symbolic algorithms for infinite-state games. In: Larsen, K.G., Nielsen, M. (eds.) CONCUR 2001. LNCS, vol. 2154, pp. 536–550. Springer, Heidelberg (2001). https://doi.org/10.1007/3-540-44685-0_36
9. Golaszewski, C., Ramadge, P.: Control of discrete event processes with forced events. In: Proceedings of the 26th Conference on Decision and Control, pp. 247–251 (1987)
10. Jurdziński, M., Trivedi, A.: Reachability-time games on timed automata. In: Arge, L., Cachin, C., Jurdziński, T., Tarlecki, A. (eds.) ICALP 2007. LNCS, vol. 4596, pp. 838–849. Springer, Heidelberg (2007). https://doi.org/10.1007/978-3-540-73420-8_72
11. Maler, O., Pnueli, A., Sifakis, J.: On the synthesis of discrete controllers for timed systems. In: Mayr, E.W., Puech, C. (eds.) STACS 1995. LNCS, vol. 900, pp. 229–242. Springer, Heidelberg (1995). https://doi.org/10.1007/3-540-59042-0_76
12. Ramadge, P.J., Wonham, W.M.: Supervisory control of a class of discrete event processes. SIAM J. Control. Optim. **25**(1), 206–230 (1987)
13. Thomas, W.: On the synthesis of strategies in infinite games. In: Mayr, E.W., Puech, C. (eds.) STACS 1995. LNCS, vol. 900, pp. 1–13. Springer, Heidelberg (1995). https://doi.org/10.1007/3-540-59042-0_57
14. Wonham, W.M., Ramadge, P.J.: On the supremal controllable sublanguage of a given language. In: The 23rd IEEE Conference on Decision and Control, pp. 1073–1080 (1984)

Games

Canonical Representations for Direct Generation of Strategies in High-Level Petri Games

Manuel Gieseking⬤ and Nick Würdemann(✉)⬤

Department of Computing Science, University of Oldenburg, Oldenburg, Germany
{gieseking,wuerdemann}@informatik.uni-oldenburg.de

Abstract. Petri games are a multi-player game model for the synthesis problem in distributed systems, i.e., the automatic generation of local controllers. The model represents causal memory of the players, which are tokens on a Petri net and divided into two teams: the controllable system and the uncontrollable environment. For one environment player and a bounded number of system players, the problem of solving Petri games can be reduced to that of solving Büchi games.

High-level Petri games are a concise representation of ordinary Petri games. Symmetries, derived from a high-level representation, can be exploited to significantly reduce the state space in the corresponding Büchi game. We present a new construction for solving high-level Petri games. It involves the definition of a unique, canonical representation of the reduced Büchi game. This allows us to translate a strategy in the Büchi game directly into a strategy in the Petri game. An implementation applied on six structurally different benchmark families shows in most cases a performance increase for larger state spaces.

1 Introduction

Whether telecommunication networks, electronic banking, or the world wide web, *distributed systems* are all around us and are becoming increasingly more widespread. Though an entire system may appear as one unit, the local controllers in a network often act autonomously on only incomplete information to avoid constant communication. These independent agents must behave correctly under all possible uncontrollable behavior of the environment. *Synthesis* [7] avoids the error-prone task of manually implementing such local controllers by automatically generating correct ones from a given specification (or stating the nonexistence of such controllers). In case of a single process in the underlying model, synthesis approaches have been successfully applied in nontrivial applications (e.g., [3,25]). Due to the incomplete information in systems with multiple

This work was supported by the German Research Foundation (DFG) through the Research Training Group (DFG GRK 1765) SCARE and through Grant Petri Games (No. 392735815).

D. Buchs and J. Carmona (Eds.): PETRI NETS 2021, LNCS 12734, pp. 95–117, 2021.
https://doi.org/10.1007/978-3-030-76983-3_6

processes progressing on their individual rate, modeling *asynchronous distributed systems* is even more cumbersome and particularly benefits from a synthesis approach.

Petri games [12] (based on an underlying Petri net [31] where the tokens are the players in the game) are a well-suited multi-player game model for the *synthesis of asynchronous distributed systems* because of its subclasses with comparably low complexity results. For Petri games with a single *environment* (*uncontrollable*) player, a bounded number of *system* (*controllable*) players, and a *safety objective*, i.e., all players have to avoid designated *bad places*, deciding the existence of a winning strategy for the system players is EXPTIME-complete [13]. This problem is called the *realizability problem*. The result is obtained via a reduction to a two-player Büchi game with enriched markings, so called *decision sets*, as states.

High-level Petri nets [24] can concisely model large distributed systems. Correspondingly, *high-level Petri games* [17] are a concise high-level representation of ordinary Petri games. For solving high-level Petri games, the *symmetries* [33] of the system can be exploited to build a *symbolic Büchi game* with a significantly smaller number of states [18]. The states are *equivalence classes* of decision sets and called *symbolic decision sets*. For generating a Petri game strategy for a high-level Petri game the approach proposed in [18] resorts to the original strategy construction in [13], i.e., the equivalence classes of a symbolic two-player strategy are dissolved and a strategy for the standard two-player game is generated. Figure 1 shows the relation of the elements just described.

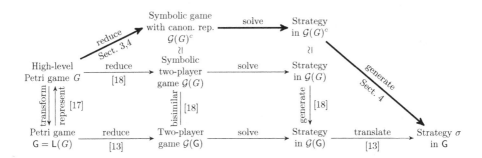

Fig. 1. An overview of the scope of this paper. The connections between the different elements describe their interplay and where these methods are introduced. The connections labeled with "solve" mean that a two-player (Büchi) game can be solved by standard algorithms in game theory (e.g., [21]). The bottom level corresponds to the original reduction in [13], the level above corresponds to the high-level counterparts described in [17,18], and the top level contains the elements introduced in this paper. The new reduction is marked by thick edges.

In this paper, we propose a new construction for solving high-level Petri games to avoid this detour while generating the strategy. In [18] the symbolic Büchi game is generated by comparing each newly added state with all already added ones for equivalence, i.e., the *orbit problem* must be answered. The new approach calculates a *canonical representation* for each newly added state (the *constructive orbit problem* [8]), and only stores these representations. This generation of a symbolic Büchi game with canonical representations is based on the corresponding ideas for reachability graphs from [5]. As in [18], we consider safe Petri games with a high-level representation, and exclude Petri games where the system can proceed infinitely without the environment. For the decidability result we consider, as in [13], Petri games with only one environment player, i.e., in every reachable marking there is at most one token on an environment place.

One of the main advantages of the new approach is that the canonical representations allow to directly generate a Petri game strategy from a symbolic Büchi game strategy without explicitly resolving all symmetries (cp. thick edges in Fig. 1). Another advantage is the complexity for constructing the symbolic Büchi game. Even though, the calculation of the canonical representation comes with a fixed cost, less comparisons can be necessary, depending on the input system. We implemented the new algorithm and applied our tool on the benchmark families used in [18] and Example 1. The results show in general a performance increase with an increasing number of states for most of the benchmark families.

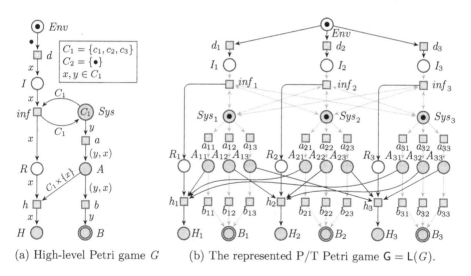

(a) High-level Petri game G (b) The represented P/T Petri game $\mathsf{G} = \mathsf{L}(G)$.

Fig. 2. *Client/Server*: The environment decides on one out of three computers to host a server. The system players (computers) can win the game by getting informed on the decision of the environment and connecting correctly.

We now introduce the example on which we demonstrate the successive development stages of the presented techniques throughout the paper.

Example 1. The high-level Petri game G depicted in Fig. 2a models a simplified scenario where one out of three computers must host a server for the others to connect to. The environment nondeterministically decides which computer must be the host. The places in the net are partitioned into *system places* (gray) and *environment places* (white). An object on a place is a player in the corresponding team. *Bad places* are double-bordered. The variables x, y on arcs are bound only locally to the transitions, and an assignment of objects to these variables is called a *mode* of the transition.

The environment player $\bullet$, initially residing on place Env, decides via transition d in mode $x = \tilde{c}$ on a computer $\tilde{c}$ that should host the server. The system players (computers $c_1, c_2, c_3 \in C_1$), initially residing on place Sys, can either directly, individually connect themselves to another computer via transition a, or wait for transition inf to be enabled. When they choose to connect themselves directly, after firing transition a in different modes, the corresponding pairs of computers reside on place A. Since the players always have to give the possibility to proceed in the game, and transition h cannot get enabled any more, they must take transition b to the bad place B. So instead, all players should initially only allow transition inf (in every possible mode x). After the decision of the environment, transition inf can be fired in mode $x = \tilde{c}$, placing $\tilde{c}$ on R. In this firing, the system players get informed on the environment's decision. Back on place Sys they can, equipped with this knowledge, each connect to the computer $\tilde{c}$ via transition a, putting the three objects $(c_1, \tilde{c}), (c_2, \tilde{c}), (c_3, \tilde{c})$ on place A. Thus, transition h can be fired in mode $\tilde{c}$, and the game terminates with $\tilde{c}$ in H. Since the system players avoided reaching the bad place B, they win the play. This scenario is highly symmetric, since it does not matter which computer is chosen to be the host, as long as the others connect themselves correctly.

The remainder of this paper is structured as follows: In Sect. 2 we recall the definitions of (high-level) Petri nets and (high-level) Petri games. In Sect. 3 we present the idea, formalization, and construction of canonical representations. In Sect. 4 we show the application of these canonical representations in the symbolic two-player Büchi game, and how to directly generate a Petri game strategy. In Sect. 5, experimental results of the presented techniques are shown. Section 6 presents the related work and Sect. 7 concludes the paper.

Further details can be found in the full version of the paper [19].

2   Petri Nets and Petri Games

This section recalls (high-level) Petri nets and -games, and the associated concept of strategies established in [13,17,18]. Figure 2 serves as an illustration.

2.1   P/T Petri Nets

A (marked P/T) *Petri net* is a tuple $N = (P, T, F, M_0)$, with the disjoint sets of *places* P and *transitions* T, a *flow function* $F : (P \times T) \cup (T \times P) \to \mathbb{N}$, and an *initial marking* M_0, where a *marking* is a multi-set $M : P \to \mathbb{N}$ that indicates the number of tokens on each place. $F(x, y) = n > 0$ means there is an *arc* of *weight* n from node x to y describing the flow of tokens in the net.

A transition $t \in T$ is *enabled* in a marking M if $\forall p \in P : F(p, t) \leq M(p)$. If t is enabled then t can *fire* in M, leading to a new marking M' calculated by $\forall p \in P : M'(p) = M(p) - F(p, t) + F(t, p)$. This is denoted by $M[t\rangle M'$. N is called *safe* if for all markings M that can be reached from M_0 by firing a sequence of transitions we have $\forall p \in P : M(p) \leq 1$. For each transition $t \in T$ we define the *pre-* and *postset* of t as the multi-sets $pre(t) = F(\cdot, t)$ and $post(t) = F(t, \cdot)$ over P.

An example for a Petri net can be seen in Fig. 2b. Ignoring the different shades and potential double borders for now, the net's places are depicted as circles, transitions as squares. Dots represent the number of tokens on each place in the initial marking of the net. The flow is depicted as weighted arcs between places and transitions. Missing weights are interpreted as arcs of weight 1. In the initial marking, all transitions a_{ij} and d_i are enabled. Firing, e.g., d_1 results in the marking with one token on I_1, Sys_1, Sys_2, and Sys_3, each.

2.2 P/T Petri Games

Petri games are an extension of Petri nets to incomplete information games between two teams of players: the controllable system vs. the uncontrollable environment. The tokens on places in a Petri net represent the individual players. The place a player resides on determines their team membership. Particularly, a player can switch teams. For that, the places are divided into system places and environment places. A play of the game is a concurrent execution of transitions in the net. During a play, the *knowledge* of each player is represented by their *causal history*, i.e., all visited places and used transitions to reach to current place. Players enrich this local knowledge when synchronizing in a joint transition. Then the complete knowledge of all participating players are exchanged. Based on this, players allow or forbid transitions in their postset. A transition can only fire if every player in its preset allows the execution. The system players in a Petri game win a play if they satisfy a safety-condition, given by a designated set of bad places they must not reach.

Formally, a (P/T) *Petri game* is a tuple $G = (P_S, P_E, T, F, M_0, P_B)$, with a set of *system places* P_S, a set of *environment places* P_E, and a set of *bad places* $P_B \subseteq P_S$. The set of all places is denoted by $P = P_S \dot\cup P_E$, and T, F, M_0 are the remaining components of a Petri net $N = (P, T, F, M_0)$, called the *underlying net* of G. We consider only Petri games with finitely many places and transitions.

In Fig. 2b, a Petri game is depicted. We just introduced the underlying net of the game. The system places are shaded gray, the environment places are white. Bad places are marked by a double border. This Petri game is the P/T-version of the high-level Petri game described in the introduction. The three tokens/system players residing on Sys_i represent the computers. The environment player residing on *Env* makes their decision which computer should host a server by taking a transition d_i. The system players can then get informed of the decision and react accordingly as described above.

A *strategy* for the system players in a Petri game G can be formally expressed as a *sub-process* of the *unfolding* [10]: in the unfolding of a Petri net, every loop is unrolled and every backward branching place is expanded by duplicating the place, so that every transition represents the unique occurrence of a transition during an execution of the net. The causal dependencies in G (and thus, the knowledge of the players) are naturally represented in its unfolding, which is the unfolding of the underlying net with system-, environment-, and bad places marked correspondingly.

A strategy is obtained from the unfolding by deleting some of the branches that are under control of the system players. This sub-process has to meet three conditions: (i) The strategy must be *deadlock-free*, to avoid trivial solutions; it must allow the possibility to continue, whenever the system can proceed. Otherwise the system players could win with the respect to the safety objective (bad places) if they decide to do nothing. (ii) The system players must act in a *deterministic* way, i.e., in no reachable marking of the strategy two transitions involving the same system player are enabled. (iii) *Justified refusal:* if a transition is not in the strategy, then the reason is that a system player in its preset forbids all occurrences of this transition in the strategy. Thus, no pure environment decisions are restricted, and system players can only allow or forbid a transition of the original net, based on only their knowledge. In a *winning* strategy, the system players cannot reach bad places.

In Fig. 3, we see the already informally described winning strategy for the system players in the Petri game G. For clarity, we only show the case in which the environment chose the first computer to be the host completely. All computers, *after* getting informed of the environment's decision, act correspondingly and connect to the first computer. The remaining branches in the unfolding are cut off in the strategy. The other two cases (after firing inf_2 or inf_3) are analogous. We include the formal definitions of unfoldings and strategies in the full version [19].

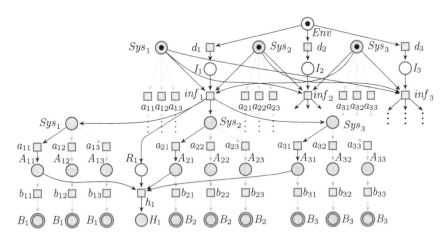

Fig. 3. Part of a winning strategy for the system players in G (solid), obtained by deleting some of the branches of the unfolding (solid and greyed out).

2.3   High-Level Petri Nets

While in P/T Petri nets only tokens can reside on places, in high-level Petri nets each place is equipped with a *type* that describes the form of data (also called *colors*) the place can hold. Instead of weights, each arc between a place p and a transition t is equipped with an *expression*, indicating which of these colors are taken from or laid on p when firing t. Additionally, each transition t is equipped with a *guard* that restricts when t can fire.

Formally, a *high-level* Petri net is a tuple $N = (P, T, F, ty, e, g, M_0)$, with a set of *places* P, a set of *transitions* T satisfying $P \cap T = \emptyset$, a *flow relation* $F \subseteq (P \times T) \cup (T \times P)$, a *type function* ty from P such that for each place p, $ty(p)$ is the set of colors that can lie on p, a mapping e that, for every transition t, assigns to each arc (p, t) (or (t, p)) in F an expression $e(p, t)$ (or $e(t, p)$) indicating which colors are withdrawn from p (or laid on p) when t is fired, a *guard function* g that equips each transition t with a Boolean expression $g(t)$, an *initial marking* M_0, where a *marking* in N is a function M with domain P indicating what colors reside on each place, i.e., $\forall p \in P : M(p) \in [ty(p) \to \mathbb{N}]$.

Figure 2a, a high-level Petri net is depicted. As in the P/T case, we ignore the different shadings and borders of places for now. The types of the places can be deducted from the surrounding arcs. For example, the place E has the type $ty(E) = C_2 = \{\bullet\}$, and the place A has the type $ty(A) = C_1 \times C_1$. Each arc is equipped with an expression, e.g., $e(Sys, a) = y$, and $e(a, A) = (y, x)$. In the given net, all guards of transitions are *true* and therefore not depicted.

Typically, expressions and guards will contain variables. A *mode* (or valuation) v of a transition $t \in T$ assigns to each variable x occurring in $g(t)$, or any expression $e(p, t)$ or $e(t, p)$, a value $v(x)$. The set $Val(t)$ contains all modes of t. Each $v \in Val(t)$ assigns a Boolean value, denoted by $v(t)$, to $g(t)$, and to each arc expression $e(p, t)$ or $e(t, p)$ a multi-set over $ty(p)$, denoted by $v(p, t)$ or $v(t, p)$. A transition t is *enabled in a mode* $v \in Val(t)$ in a marking M if $v(t) = true$ and for each arc $(p, t) \in F$ and every $c \in ty(p)$ we have $v(p, t)(c) \leq M(p)(c)$. The marking M' reached by firing t in mode v from M (denoted by $M[t.v\rangle M'$) is calculated by $\forall p \in P \; \forall c \in ty(p) : M'(p)(c) = M(p)(c) - v(p, t)(c) + v(t, p)(c)$.

A high-level Petri net N can be *transformed* into a P/T Petri net $\mathsf{L}(N)$ with $\mathsf{P} = \{p.c \mid p \in P, c \in ty(p)\}$, $\mathsf{T} = \{t.v \mid t \in T, v \in Val(t), v(t) = true\}$, the flow F defined by $\forall p.c \in \mathsf{P} \; \forall t.v \in \mathsf{T} : \mathsf{F}(p.c, t.v) = v(p, t)(c) \wedge \mathsf{F}(t.v, p.c) = v(t, p)(c)$, and initial marking $\mathsf{M_0}$ defined by $\forall p.c \in \mathsf{P} : \mathsf{M_0}(p.c) = M_0(p)(c)$. The two nets then have the same semantics: the number of tokens on a place $p.c$ in a marking in $\mathsf{L}(N)$ indicates the number of colors c on place p in the corresponding marking in N. Firing a transition $t.v$ in $\mathsf{L}(N)$ corresponds to firing transition t in mode v in N. We say a high-level Petri net N *represents* the P/T Petri net $\mathsf{L}(N)$.

2.4   High-Level Petri Games

Just as P/T Petri games are structurally based on P/T Petri nets, a *high-level Petri game* $G = (P_S, P_E, T, F, ty, e, g, M_0, P_B)$ with *underlying high-level net*

$N = (P, T, F, ty, e, g, M_0)$ divides the places P into *system places* P_S and *environment places* P_E. The set $P_B \subseteq P_S$ indicates the bad places. High-level Petri games represent P/T Petri games: a high-level Petri game G (with underlying high-level net N) *represents* a P/T Petri game $\mathsf{L}(G)$ with underlying P/T Petri net $\mathsf{L}(N)$. The classification of places $p.c$ in $\mathsf{L}(G)$ into system-, environment-, and bad places corresponds to the places p in the high-level game.

In Fig. 2a, a high-level Petri game G and its represented Petri game $\mathsf{G} = \mathsf{L}(G)$ are depicted. For the sake of clarity, we abbreviated the nodes in $\mathsf{L}(G)$. Thus, e.g., the transition $a.[x = c_1, y = c_2]$ is renamed to a_{12}. We often use notation from the represented P/T Petri game to express situations in a high-level game.

3   Canonical Representations of Symbolic Decision Sets

In this paper, we investigate for a given high-level Petri game G with one environment player whether the system players in $\mathsf{L}(G)$ have a winning strategy (and possibly generate one). This problem is solved via a reduction to a symbolic two-player Büchi game $\mathcal{G}(G)^c$. The general idea of this reduction is, as in [13], to equip the markings of the Petri game with a set of transitions for each system player (called *commitment sets*) which allows the players to fix their next move. In the generated Büchi game, only a subset of all interleavings is taken into account, in the way that the moves of the environment player are delayed until no system player can progress without interacting with the environment. By that, each system player gets informed about the environment's last position during their next move. This means that in every state, every system player knows the current position of the environment or learns it in the next step, before determining their next move. Thus, the system players can be considered to be completely informed about the game. This is only possible due to the existence of only one environment player. For more environment players such interleavings would not ensure that each system player is informed (or gets informed in their next move) about all environment positions. The nodes of the game are called *decision sets*. In [18], symmetries in the Petri net are exploited to define equivalence classes of decision sets, called *symbolic decision sets*. These are used to create an equivalent, but significantly smaller, Büchi game.

In this section we introduce the new canonical representations of symbolic decision sets which serve as nodes for the new Büchi game. We transfer relations between and properties of (symbolic) decision sets to the established canonical representations. We start by recalling the definitions of symmetries in Petri nets [33] and of (symbolic) decision sets [18].

From now on we consider high-level Petri games G representing a safe P/T Petri game $\mathsf{L}(G)$ that has one environment player, a bounded number of system players with a safety objective, and where the system cannot proceed infinitely without the environment.

3.1 Symmetric Nets

High-level representations are often created using *symmetries* [33] in a Petri net. Conversely, in some high-level nets, symmetries can be read directly from the given specification. A class of nets which allow this are the so called *symmetric nets* (SN) [4].[1] In symmetric nets, the types of places are selected from given (finite) *basic color classes* $C_1, \ldots, C_n$. For every place $p \in P$, we have $ty(p) = C_1^{p_1} \times \cdots \times C_n^{p_n}$ for natural numbers $p_1, \ldots, p_n \in \mathbb{N}$, where $C_i^{p_i}$ denotes the p_i-fold Cartesian product of C_i.[2] The possible values of variables contained in guards and arc expressions are also basic classes. Thus, the modes of each transition $t \in T$ are also given by a Cartesian product $Val(t) = C_1^{t_1} \times \cdots \times C_n^{t_n}$. Guards and arc expressions treat all elements in a color class equally.

Example 2. The underlying high-level net N in Fig. 2a is a symmetric net with basic color classes $C_1 = \{c_1, c_2, c_3\}$ and $C_2 = \{\bullet\}$. We have, e.g., $Val(a) = C_1 \times C_1$ (the two coordinates representing y and x), and therefore, $a_1 = 2, a_2 = 0$.

Remark 1. In general, each basic color class C_i is possibly partitioned into *subclasses* $C_i = \bigcup_{q=1}^{n_i} C_{i,q}$. In this paper, we omit this partition. The detailed proofs in the full version [19] take the general case into account.

Proposition 1. *Any high-level Petri net can be transformed into a SN with the same basic structure, same place types, and equivalent arc labeling (cf. [4]).*

The *symmetries* ξ_N in a symmetric net N are all tuples $s = (s_1, \ldots, s_n)$ such that each s_i is a permutation on C_i. A symmetry s can be applied to a single color $c \in C_i$ by $s(c) = s_i(c)$. The application to tuples, e.g., colors on places or transition modes, is defined by the application in each entry. The set ξ_N, together with the function composition $\circ$, forms a group with identity $(id_{C_i})_{i=1}^n$. In the represented P/T Petri net $L(N)$, symmetries can be applied to places $\mathsf{p} = p.c \in \mathsf{P}$ and transitions $\mathsf{t} = t.v \in \mathsf{T}$ by defining $s(p.c) = p.s(c)$ and $s(t.v) = t.s(v)$. The structure of symmetric nets ensures $\forall s \in \xi_N \ \forall \mathsf{t} \in \mathsf{T} : pre(s(\mathsf{t})) = s(pre(\mathsf{t}))$ and $post(s(\mathsf{t})) = s(post(\mathsf{t}))$. Thus, symmetries are compatible with the firing relation; $\forall s \in \xi_N : \mathsf{M}[\mathsf{t}\rangle\mathsf{M}' \Leftrightarrow s(\mathsf{M})[s(\mathsf{t})\rangle s(\mathsf{M}')$. In a symmetric net, we can w.l.o.g. assume the initial marking M_0 to be *symmetric*, i.e., $\forall s \in \xi_N : s(M_0) = M_0$.

3.2 Symbolic Decision Sets

A *decision set* is a set $D \subseteq \mathsf{P} \times (\mathbb{P}(\mathsf{T}) \cup \mathsf{T})$. An element $(\mathsf{p}, \mathsf{C}) \in D$ with $\mathsf{C} \subseteq post(\mathsf{p})$ indicates there is a player on place p who allows all transitions in C to fire. C is then called a *commitment set*. An element $(\mathsf{p}, \mathsf{T}) \in D$ indicates the player on place p has to choose a commitment set in the next step. The step of this decision is called T-*resolution*.

---

[1] Symmetric Nets were formerly known as Well-Formed Nets (WNs). The renaming was part of the ISO standardization [23].

[2] In the Cartesian products $ty(p)$ and $Val(t)$, we omit all C_i^x with $x = 0$ (empty sets).

In a $\top$-resolution, each $\top$-symbol in a decision set D is replaced with a suitable commitment set. This relation is denoted by $D[\top\rangle D'$. If there are no $\top$-symbols in D, a transition t is *enabled*, if $\forall p \in pre(t)\ \exists(p, C) \in D : t \in C$, i.e., there is a token on every place in $pre(t)$ (as for markings) and additionally, t is in every commitment set of such a token. In the process of firing an enabled transition, the tokens are moved accordingly to the flow F. The moved or generated tokens on system places are then equipped with a $\top$-symbol, while the tokens on environment places allow *all* transitions that they are involved in. This relation is denoted by $D[t\rangle D'$. The *initial decision set* is given by $D_0 = \{(p, \{t \in T \mid p \in pre(t)\}) \mid p \in P_E \cap M_0\} \cup \{(p, \top) \mid p \in P_S \cap M_0\}$, i.e., the environment in the initial marking allows all possible transitions, the system players still have to choose a commitment set.

Example 3. Assume in the Petri game in Fig. 2a that the computers initially allow transition inf in every mode. The environment player on Env fires transition d in mode c_1. After that, the system gets informed of the environment's decision via transition inf in mode c_1. The system players, now back on Sys, decide via $\top$-resolution they all want to assign themselves to c_1. This corresponds to the following sequence of decision sets, where we abbreviate Sys by S.

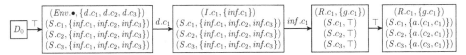

A high-level Petri game G has the same symmetries ξ_N as its underlying symmetric net N. They can be applied to decision sets by applying them to every occurring color c or mode v. For a decision set D, an equivalence class $\{s(D) \mid s \in \xi_N\}$ is called the *symbolic decision set* of D, and contains symmetric situations in the Petri game. In [18], these equivalence classes replace individual decision sets in the two-player Büchi game to achieve a substantial state space reduction.

Example 4. Consider the second to last decision set in the sequence above. This situation is symmetric to the cases where the environment chose computer c_2 or c_3 as the host. In the example G, we have the two color classes C_1 and C_2. Since $|C_2| = 1$, the only permutation on C_2 is id_{C_2}. Thus, the symmetries in G are the permutations on C_1. Symmetries transform the elements in the symbolic decision set into each other. The corresponding symbolic decision set contains the following three elements D, D', and D'':

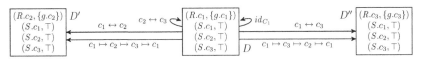

Each edge between two decision sets corresponds to the application of a symmetry. The abbreviated notation $c \mapsto c' \mapsto c'' \mapsto c$ means that each element is mapped to the next in line. Analogously, $c \leftrightarrow c'$ means that c and c' are switched.

3.3   Canonical Representations

In order to exploit symmetries to reduce the size of the state space, one aims to consider only one representative of each of the equivalence classes induced by the symmetries. This can be done either by checking whether a newly generated state is equivalent to any already generated one, or by transforming each newly generated state into an equivalent, canonical representative. In [18] we consider the former approach. The nodes of the symbolic Büchi game are symbolic decision sets. In the construction, an arbitrary representative $\overline{D}$ is chosen for each of these equivalence classes. This means, when reaching a new node D', we must apply every symmetry s to test whether there already is a representative $\overline{D'} = s(D')$, or whether D' is in a new symbolic decision set.

In this section, we now aim at the second approach and define the new *canonical representations* of symbolic decision sets. For that, we first define *dynamic representations*, and then show how to construct a canonical one. We use these instead of (arbitrary representatives of) symbolic decision sets in the construction of the symbolic Büchi game in Sect. 4.

Dynamic Representations. A dynamic representation is an abstract description of a symbolic decision set. It consists of dynamic subclasses of variables, and a dynamic decision set where these dynamic subclasses replace explicit colors. Any (valid) assignment of values to the variables in the dynamic subclasses results in a decision set in the equivalence class.

Formally, a *dynamic representation* is a tuple $\mathcal{R} = (\mathcal{C}, \mathcal{D})$, with the set of *dynamic subclasses* $\mathcal{C} = \{Z_i^j \mid 1 \leq i \leq n, 1 \leq j \leq m_i\}$ for natural numbers m_i, and a *dynamic decision set* $\mathcal{D}$. A dynamic subclass Z_i^j contains a finite number of variables with values in C_i. Each Z_i^j has a cardinality $|Z_i^j|$ that indicates the number of variables. In total, there are as many variables with values in C_i as there are colors, i.e., $\sum_{j=1}^{m_i} |Z_i^j| = |C_i|$. An *assignment* $va : \bigcup_{i=1}^{n} C_i \rightarrow \mathcal{C}$ is *valid* if it respects the cardinality of dynamic subclasses, i.e., $|\{c \in C_i \mid va(c) = Z_i^j\}| = |Z_i^j|$. Every valid assignment of colors $c \in C_i$ to the $Z_i^j, 1 \leq j \leq m_i$, gives a partition of C_i. A dynamic decision set is the same as a decision set, with dynamic subclasses replacing explicit colors. For every decision set D in a symbolic decision set with dynamic representation $\mathcal{R} = (\mathcal{C}, \mathcal{D})$, there is a valid assignment va_D such that $D = va_D^{-1}(\mathcal{D})$. In general, there are several dynamic representations of a symbolic decision set.

Example 5. Consider the symbolic decision set from the last example. We can naively build a dynamic representation by taking one of the decision sets, and replacing each color by a dynamic subclass of cardinality 1. This results in as many dynamic subclasses as there are colors, i.e., $\mathcal{C} = \{Z_1^1, Z_1^2, Z_1^3, Z_2^1\}$ with $|Z_i^j| = 1$ for all i, j. Below, the resulting dynamic decision set $\mathcal{D}$ is depicted, with valid assignments that lead to elements D, D', and D''.

$(R.Z_1^1, \{g.Z_1^1\})$		$(R.c_1, \{g.c_1\})$	$Z_1^2 \hookleftarrow c_1, Z_1^1 \hookleftarrow c_2,$	$(R.c_2, \{g.c_2\})$	D'	D''	$(R.c_3, \{g.c_3\})$
$(S.Z_1^1, \top)$	$Z_1^j \hookleftarrow c_j$	$(S.c_1, \top)$	$Z_1^3 \hookleftarrow c_3$	$(S.c_1, \top)$			$(S.c_1, \top)$
$(S.Z_1^2, \top)$		$(S.c_2, \top)$		$(S.c_2, \top)$	$Z_1^3 \hookleftarrow c_1, Z_1^2 \hookleftarrow c_2,$		$(S.c_2, \top)$
$(S.Z_1^3, \top)$	$\mathcal{D}$	$(S.c_3, \top)$	D	$(S.c_3, \top)$	$Z_1^1 \hookleftarrow c_3$		$(S.c_3, \top)$

The element $(R.Z_1^1, \{g.Z_1^1\})$, e.g., represents one arbitrary color c (since $|Z_1^1| = 1$) on place R with $g.c$ in its commitment set. The same color is on place S, equipped with a $\top$-symbol.

Minimality. We notice in the example above that Z_1^2 and Z_1^3 appear in the same contexts in $\mathcal{D}$. The *context* $\mathrm{con}(Z_i^j)$ of a dynamic subclass Z_i^j is defined as the set of tuples in $\mathcal{D}$ where exactly one appearance of Z_i^j is replaced by a symbol $\triangledown$. So in our example $\mathrm{con}(Z_1^2) = \{(S.\triangledown, \top)\} = \mathrm{con}(Z_1^3)$, and $\mathrm{con}(Z_1^1) = \{(S.\triangledown, \top), (R.\triangledown, g.c_1), (R.c_1, g.\triangledown)\}$. This means Z_1^2 and Z_1^3 can be *merged* into a new dynamic subclass of cardinality 2. The resulting new dynamic representation is given by $\mathcal{C}_{min} = \{Z_1^1, Z_1^2, Z_2^1\}$ with $|Z_1^1| = |Z_2^1| = 1$ and $|Z_1^2| = 2$, and $\mathcal{D}_{min} = \{(R.Z_1^1, \{g.Z_1^1\}), (S.Z_1^1, \top), (S.Z_1^2, \top)\}$.

$(R.Z_1^1, \{g.Z_1^1\})$		$(R.c_1, \{g.c_1\})$	$Z_1^1 \hookleftarrow c_2$	$(R.c_2, \{g.c_2\})$	D'	D''	$(R.c_3, \{g.c_3\})$
$(S.Z_1^1, \top)$	$Z_1^j \hookleftarrow c_1$	$(S.c_1, \top)$	$Z_1^2 \hookleftarrow c_1, c_3,$	$(S.c_1, \top)$			$(S.c_1, \top)$
$(S.Z_1^2, \top)$	$Z_1^2 \hookleftarrow c_2, c_3$	$(S.c_2, \top)$		$(S.c_2, \top)$	$Z_1^1 \hookleftarrow c_3$		$(S.c_2, \top)$
	$\mathcal{D}_{min}$	$(S.c_3, \top)$	D	$(S.c_3, \top)$	$Z_1^2 \hookleftarrow c_1, c_2,$		$(S.c_3, \top)$

Minimal representations do not contain any two dynamic subclasses Z_i^j, Z_i^k with the same context. Given a dynamic representation, it is algorithmically simple to construct a minimal representation of the same symbolic decision set by merging all dynamic subclasses with the respectively same context. The dynamic representation above that resulted from merging the subclasses is therefore minimal. Still, minimality is not enough to obtain a unique canonical representation, since we can permute the indices j of the dynamic subclasses Z_i^j.

Lemma 1. *The minimal representations of a symbolic decision set are unique up to a permutation of the dynamic subclasses.*

This lemma can be proved using the observation that every minimal representation can be reached from a dynamic representation that contains only dynamic subclasses of cardinality 1 (as the one in Example 5) by merging.

Ordering. We can choose one of the minimal representations by *ordering* the dynamic subclasses. In the following, we give a possible way to do that. We display the dynamic decision set $\mathcal{D}$ as a matrix, with rows and columns indicating (tuples of) dynamic subclasses Z. An element of the matrix at entry (Z, Z') returns all tuples (p, t) satisfying $(p.Z, \mathsf{C}) \in \mathcal{D}$ and $t.Z' \in \mathsf{C}$ for a commitment set C. Also, all tuples $(p, \top)$ satisfying $(p.Z, \top) \in \mathcal{D}$ and all tuples $(p, \emptyset)$ satisfying $(p.Z, \emptyset) \in \mathcal{D}$ are returned (in these cases, Z' is neglected).

The elements of the matrix are in $\mathbb{P}(\mathsf{P} \times (\mathsf{T} \cup \{\top, \emptyset\}))$. Since this set is finite, we can give an arbitrary, but fixed, total order on it. This order can be extended to the matrices over the set (the lexicographic order by row-wise traversion through a matrix). Then we can determine a permutation such that, when applied to the dynamic subclasses, the matrix is *minimal with respect to the lexicographic order*. The corresponding dynamic representation is called *ordered*.

Example 6. On the left we see the matrix of the dynamic decision set $\mathcal{D}$ in the minimal representation given above. The first entry, at (Z_1^1, Z_1^1), e.g., is $\{(S, \top), (R, g)\}$ since $(S.Z_1^1, \top)$ and $(R.Z_1^1, g.Z_1^1)$ are in $\mathcal{D}$. Assume $\{(S, \top)\} < \{(S, \top), (R, g)\}$. When the permutation switching Z_1^1 and Z_1^2 is applied, we get the right matrix, which is lexicographically smaller (the first entry is smaller).

$$
\begin{array}{c|cc}
 & Z_1^1 & Z_1^2 \\
\hline
Z_1^1 & \{(S, \top), (R, g)\} & \{(S, \top)\} \\
Z_1^2 & \{(S, \top)\} & \{(S, \top)\}
\end{array}
\qquad
\xleftrightarrow{\; Z_1^1 \leftrightarrow Z_1^2 \;}
\qquad
\begin{array}{c|cc}
 & Z_1^1 & Z_1^2 \\
\hline
Z_1^1 & \{(S, \top)\} & \{(S, \top)\} \\
Z_1^2 & \{(S, \top)\} & \{(S, \top), (R, g)\}
\end{array}
$$

Thus, the minimal representation from above is transformed into a *minimal and ordered* representation $(\mathcal{C}_{ord}, \mathcal{D}_{ord})$ by the permutation $Z_1^1 \leftrightarrow Z_1^2$.

Theorem 1. *For every symbolic decision set there is exactly one minimal and ordered dynamic representation. We call this the* canonical (dynamic) *representation.*

The proof follows from Lemma 1 and the observation that if two ordered dynamic representations can be transformed into each other by applying a permutation of the dynamic subclasses, they must have the same dynamic decision set.

We can algorithmically order a minimal representation by calculating all symmetric representations and finding the one with the lexicographically smallest matrix. These are maximally $|\xi_N|$ comparisons of dynamic representations. So by first making a dynamic representation minimal, and then ordering it, we get the respective canonical representation.

Corollary 1. *We can construct the canonical representation of a given symbolic decision set in $O(|\xi_N|)$.*

3.4   Relations Between Canonical Representations

Between decision sets, we have the two relations $D[\top\rangle D'$ and $D[t.v\rangle D'$ with $v \in Val(t) = C_1^{t_1} \times \cdots \times C_n^{t_n}$. In canonical representations, we abstract from specific colors $c \in C_i$ and replace them by dynamic subclasses Z_i^j of variables. However, in the process of $\top$-resolution or transition firing, two objects represented by the same dynamic subclass can act differently. This means we *instantiate* special objects in the classes that are relevant in the $\top$-resolution or transition firing.

For this, each dynamic subclass Z_i^j in a canonical representation $\mathcal{R}$ is *split* into finitely many $Z_i^{j,k}$ of cardinality $|Z_i^{j,k}| = 1$ with $k > 0$, and a subclass $Z_i^{j,0}$, containing the possibly remaining, non-instantiated, variables. Then, a $\top$ is resolved, or a transition is fired, with the dynamic subclasses $Z_i^{j,k}$ replacing explicit data entries $c \in C_i$. Finally, the canonical representation $\mathcal{R}'$ of the reached dynamic representation is found. These relations are denoted by $\mathcal{R}[\top\rangle\mathcal{R}'$ and $\mathcal{R}[t.Z\rangle\mathcal{R}'$, where Z is a tuple of instantiated $Z_i^{j,k}$.

Below we see an example that corresponds to the last two steps in Example 3. We calculated the second canonical representation in the last section. It is reached from the first canonical representation by firing $inf.Z_1^{2,1}$. In this process one (the only) element in Z_1^2 is instantiated by a dynamic subclass $Z_1^{2,1}$ of cardinality 1. After the actual firing, the reached representation is made canonical. Then, a $\top$ is resolved. Here, Z_1^1 is split into $Z_1^{1,1}$ and $Z_1^{1,2}$ with $|Z_1^{1,1}| = |Z_1^{1,2}| = 1$. In the reached dynamic representation, no two subclasses have the same context, so it is already minimal. After ordering we get the canonical representation.

$$
\begin{array}{c}
|Z_1^1| = 2 \\
|Z_1^2| = 1
\end{array}
\begin{array}{|l|}
\hline
(I.Z_1^2, \{inf.Z_1^2\}) \\
(S.Z_1^1, \{inf.Z_1^1, inf.Z_1^2\}) \\
(S.Z_1^2, \{inf.Z_1^1, inf.Z_1^2\}) \\
\hline
\end{array}
\xrightarrow{inf.(Z_1^{2,1})}
\begin{array}{|l|}
\hline
(R.Z_1^1, \{g.Z_1^2\}) \\
(S.Z_1^1, \top) \\
(S.Z_1^2, \top) \\
\hline
\end{array}
\begin{array}{c}
|Z_1^1| = 2 \\
|Z_1^2| = 1
\end{array}
\xrightarrow[\forall j \in \{1,2,3\}:]{\top}
\begin{array}{c}
\\
|Z_1^j| = 2
\end{array}
\begin{array}{|l|}
\hline
(R.Z_1^3, \{g.Z_1^3\}) \\
(S.Z_1^1, \{a.(Z_1^2, Z_1^3)\}) \\
(S.Z_1^2, \{a.(Z_1^2, Z_1^3)\}) \\
(S.Z_1^3, \{a.(Z_1^2, Z_1^3)\}) \\
\hline
\end{array}
$$

Theorem 2. *Every relation $D[t.v\rangle D'$ or $D[\top\rangle D'$ between two decision sets D and D' is represented by exactly one symbolic relation $\mathcal{R}[t.Z\rangle\mathcal{R}'$ or $\mathcal{R}[\top\rangle\mathcal{R}'$ between the respective canonical representations $\mathcal{R}$ and $\mathcal{R}'$.*

The proof for the case $D[t.v\rangle D'$ follows by applying a valid assignment va_D to v and splitting $\mathcal{R}$ correspondingly. The case $D[\top\rangle D'$ works analogously.

3.5   Properties of Canonical Representations

The goal is to use canonical representations instead of individual decision sets or (arbitrary representatives of) symbolic decision sets as nodes in a two player game. The edges $(\mathcal{R}, \mathcal{R}')$ in this game are built from relations $\mathcal{R}[t.Z\rangle\mathcal{R}'$ and $\mathcal{R}[\top\rangle\mathcal{R}'$, depending on the properties of $\mathcal{R}$. For example, if $\mathcal{R}$ describes nondeterministic situations in the Petri game, then the edges from $\mathcal{R}$ are built in such a way that player 0 (representing the system) cannot win the game from there. In this section, we define the relevant properties of canonical representations.

In [13], the following properties of a decision set D are defined. Let $\mathsf{M}(D) = \{p \in P \mid (p, \top) \in D \vee \exists C \subseteq T : (p, C) \in D\}$ be the *underlying marking* of D. D is *environment-dependent* iff $\neg D[\top\rangle$, i.e., there is no $\top$ symbol in any tuple in D, and $\forall t \in T : D[t\rangle \Rightarrow pre(t) \cap P_E \neq \emptyset$, i.e., all enabled transitions have an environment place in their preset. D *contains a bad place* iff $P_B \cap \mathsf{M}(D) \neq \emptyset$. D is a *deadlock* iff $\neg D[\top\rangle$, and $\exists t' \in T : \mathsf{M}(D)[t'\rangle \wedge \forall t \in T : \neg D[t\rangle$, i.e., there is a transition that is enabled in the underlying marking, but the system forbids all enabled transitions. D is *terminating* iff $\forall t \in T : \neg\mathsf{M}(D)[t\rangle$. D is *nondeterministic* iff $\exists t_1, t_2 : t_1 \neq t_2 \wedge P_S \cap pre(t_1) \cap pre(t_2) \neq \emptyset \wedge D[t_1\rangle \wedge D[t_2\rangle$, i.e., two separate transitions sharing a system place in their presets both are enabled.

In [18], we showed that all decision sets in one equivalence class share the same of the properties defined above. Thus, we say a symbolic decision set has one of the above properties iff its individual members have the respective property.

We now define these properties for canonical representations. Since we do not want to consider individual decision sets, we do that on the level of dynamic representations. Let $\mathcal{R} = (\mathcal{C}, \mathcal{D})$ be a canonical representation. $\mathcal{R}$ is *environment-dependent* iff $\neg\mathcal{R}[\top\rangle$, i.e., there is no $\top$ symbol in any tuple in $\mathcal{D}$, and

$\forall t.Z \; : \; \mathcal{R}[t.Z\rangle \; \Rightarrow \; \exists p \in P_E \; : \; (p,t) \in F$, and $\mathcal{R}$ *contains a bad place* iff $\exists p \in P_B \exists X : (p.X, \top) \in \mathcal{R} \vee \exists C : (p.X, C) \in \mathcal{D}$. Both these properties are rather analogous to the respective property of decision sets. For termination and deadlocks, we introduce for the given $\mathcal{R}$ the representation $\mathcal{R}_{all}$ with the same dynamic subclasses, and the dynamic decision set where every player has all possible transitions $t.Z$ in their commitment set. Since then all transitions that could fire in the underlying marking are enabled, this substitutes for $M(D)$. We say $\mathcal{R}$ is a *deadlock* iff $\neg\mathcal{R}[\top\rangle$, and $\exists t'.Z' : \mathcal{R}_{all}[t'.Z'\rangle \wedge \forall t.Z : \mathcal{R}[t.Z\rangle$. Analogously, $\mathcal{R}$ is *terminating* iff $\forall t.Z : \neg\mathcal{R}_{all}[t.Z\rangle$. For nondeterminism we have to consider two cases. The first one is analogous to the property for individual decision sets. $ndet_1(\mathcal{R}) = \exists t.Z, t'.Z' : t.Z \neq t'.Z' \wedge \exists p \in P_S \exists X, C : (p.X, C) \in \mathcal{D} \wedge t.Z, t'.Z' \in C \wedge \mathcal{R}[t, Z\rangle \wedge \mathcal{R}[t'.Z'\rangle$. The second case considers the situation that *two instances* of one $t.Z$ can both fire with a shared system place in their preset. $ndet_2(\mathcal{R}) = \exists t.Z \exists p \in P_S \exists X, C : (p.X, C) \in \mathcal{D} \wedge t.Z \in C \wedge \exists Z_i^{j,k} \in Z : |Z_i^j| > 1 \wedge \mathcal{R}[t, Z\rangle$. Finally, $\mathcal{R}$ is *nondeterministic* iff $ndet_1(\mathcal{R}) \vee ndet_2(\mathcal{R})$.

Corollary 2. *The properties of a symbolic decision set and its canonical representation coincide.*

For the proof, Theorem 2 is applied to the properties of individual decision sets.

4   Applying Canonical Representations

In this section, we define for a high-level Petri game G the two-player Büchi game $\mathcal{G}(G)^c$ with canonical representations $\mathcal{R}$ of symbolic decision sets, rather than arbitrary representative $\overline{D}$ as in [18]. The edges between nodes are directly implied by the relations $\mathcal{R}[t.Z\rangle\mathcal{R}'$ and $\mathcal{R}[\top\rangle\mathcal{R}'$ between canonical representations. This allows to *directly* generate a winning strategy for the system players in G from a winning strategy for player 0 in $\mathcal{G}(G)^c$ (cf. Fig. 1).

4.1   The Symbolic Two-Player Game

We reduce a Petri game G with high-level representation G to a two-player Büchi game $\mathcal{G}(G)^c$. The goal is to directly create a strategy σ for the system players in G from a strategy f for player 0 in $\mathcal{G}(G)^c$. Recall that a Petri game strategy must be deadlock-free, deterministic, and satisfy the justified refusal condition. Additionally, to be winning, it must not contain bad places.

The nodes in $\mathcal{G}(G)^c$ are canonical representations of symbolic decision sets, equivalence classes of situations in the Petri game. The properties of canonical representations defined in Sect. 3.5 characterize these situations. These properties are used in the construction of the game. As in [13,18], the environment in the game $\mathcal{G}(G)^c$ only moves when the system players cannot continue alone. Thus, they get informed of the environment's decisions in their next steps and the system can therefore be considered as completely informed. Bad situations (nondeterminism, deadlocks, tokens on bad places) result in player 0 not winning.

If player 0 can avoid these situations and always win the game, this strategy can be translated into a winning strategy for the system players in the Petri game.

The *symbolic two-player Büchi game with canonical representations* $\mathcal{G}(G)^c = (V_0, V_1, E, V_F, \mathcal{R}_0)$ for a high-level Petri game G has the following components. The *nodes* $V = V_0 \dot{\cup} V_1$ are all possible canonical representations $\mathcal{R}$ of symbolic decision sets in G. The partition into *player 0's nodes* V_0 and *player 1's nodes* V_1 is given by $V_1 = \{\mathcal{R} \mid \mathcal{R}$ is environment dependent$\}$ and $V_0 = V \setminus V_1$. The *edges* E are constructed as follows. If $\mathcal{R} \in V$ contains a bad place, is a deadlock, is terminating, or is nondeterministic, there is only a self-loop originating from $\mathcal{R}$. If $\mathcal{R} \in V_0$ then $(\mathcal{R}, \mathcal{R}') \in E$ if either $\mathcal{R}[\top\rangle\mathcal{R}'$, or, if no $\top$ can be resolved, $\mathcal{R}[t.Z\rangle\mathcal{R}'$ with only system players participating in $t.Z$. If $\mathcal{R} \in V_1$, then $(\mathcal{R}, \mathcal{R}') \in E$ for every $\mathcal{R}'$ such that $\mathcal{R}[t.Z\rangle\mathcal{R}'$, i.e., transitions involving environment players can only fire if nothing else is possible. The set V_F of *accepting nodes* contains all representations $\mathcal{R}$ that are terminating or environment-dependent, but are not a deadlock, nondeterministic, or contain a bad place. The *initial state* $\mathcal{R}_0$ is the canonical representation of the symbolic decision set containing D_0.

A function $f : V^* V_0 \to V$ s.t. $\forall \mathcal{R}_0' \cdots \mathcal{R}_k' \in V^* V_0 : (\mathcal{R}_k', f(\mathcal{R}_0' \cdots \mathcal{R}_k')) \in E$ is called a *strategy* for player 0. A strategy f is called *winning* iff every *run* $\rho = \mathcal{R}_0 \mathcal{R}_1 \mathcal{R}_2 \cdots$ from $\mathcal{R}_0$ in $\mathcal{G}(G)^c$ (i.e., $\forall k : (\mathcal{R}_k, \mathcal{R}_{k+1}) \in E$) that is *consistent* with f (i.e., $\mathcal{R}_k \in V_0 \Rightarrow \mathcal{R}_{k+1} = f(\mathcal{R}_0 \cdots \mathcal{R}_k)$) satisfies the *Büchi condition* w.r.t. V_F (i.e., $\forall k \; \exists k' \geq k : \mathcal{R}_{k'} \in V_F$).

In the game $\mathcal{G}(G)$ in [18], player 0 has a winning strategy if and only if the system players in $L(G)$ have a winning strategy. As described above, it is built from the relations $\overline{D}[t.c\rangle D'$ and $\overline{D}[\top\rangle D'$ from representatives $\overline{D}$ of symbolic decision sets. The introduced game $\mathcal{G}(G)^c$ is built analogously, with the difference that the nodes are now canonical representations instead of arbitrary representatives of symbolic decision sets, and the edges are built from the relations $\mathcal{R}[t.Z\rangle\mathcal{R}'$ and $\mathcal{R}[\top\rangle\mathcal{R}'$ (cf. Theorem 2 and Corollary 2). The two games are isomorphic, as depicted in Fig. 1. Thus, we get the following result.

Theorem 3. *Given a Petri game G with one environment player, a bounded number of system players with a safety objective, and a high-level representation G, there is a winning strategy for the system players in G if and only if there is a winning strategy for player 0 in $\mathcal{G}(G)^c$.*

The size of $\mathcal{G}(G)^c$ is the same as of $\mathcal{G}(G)$ (exponential in the size of G). This means, using $\mathcal{G}(G)^c$, the question whether a winning strategy in G exists can still be answered in single exponential time [13]. In $\mathcal{G}(G)$ we must, for a newly reached node D', test if it is equivalent to another, already existing, representative. This means we check for all symmetries $s \in \xi$ if $s(D')$ is already a node in the game. In the best case, if we directly find the node, this is 1 comparison. In the worst case, at step i with currently $|V^i|$ nodes, we must make $|\xi_N||V^i|$ comparisons (no symmetric node is in the game so far). To get the canonical representation of a reached node in $\mathcal{G}(G)^c$, we must make less than $|\xi_N|$ comparisons to order the dynamic representation (cf. Corollary 1), and then compare it to all existing nodes. Thus, $|\xi_N| + 1$ comparisons in the best case vs. $|\xi_N| + |V^i|$ in the i-th step in the worst case. We further investigate experimentally on this in Sect. 5.

4.2 Direct Strategy Generation

The solving algorithm in [18] builds a strategy in the Petri game $\mathsf{G} = \mathsf{L}(G)$ from a strategy in $\mathcal{G}(G)$ by first generating a strategy in the low-level equivalent $\mathcal{G}(\mathsf{G})$. Constituting the canonical representations as nodes allows us to directly generate a winning strategy σ for the system players in G from a winning strategy f for player 0 in $\mathcal{G}(G)^c$ (cf. Fig. 1).

The key idea is the same as in [13]. The strategy f is interpreted as a tree T_f with labels in V, and root r_0 labeled with $\mathcal{R}_0$. The tree is traversed in breadth-first order, while the strategy σ is extended with every reached node. To show that this procedure is correct, we must show that the generated strategy σ is satisfying the conditions justified refusal, determinism, and deadlock freedom. Justified refusal is satisfied because of the delay of environment transitions. Assuming nondeterminism or deadlocks in the generated strategy σ leads to the contradiction that there are respective decision sets in T_f. Finally, σ is *winning*, since f also does not reach representations that contain a bad place. For the detailed proof, cf. the full version [19].

Initially, the strategy σ contains places corresponding to the initial marking M_0 in the Petri game, i.e., places labeled with $p.c$ for every $p.c \in \mathsf{M}_0$, each with a token on them. They constitute the initial marking M_0^σ of σ. Every node r in T_f, labeled with $\mathcal{R}$, is now associated with a set K_r of *cuts* – reachable markings in the strategy/unfolding. The set K_{r_0}, associated to the root r_0, contains only the cut $\kappa_0 = \mathsf{M}_0^\sigma$, the initial marking described above.

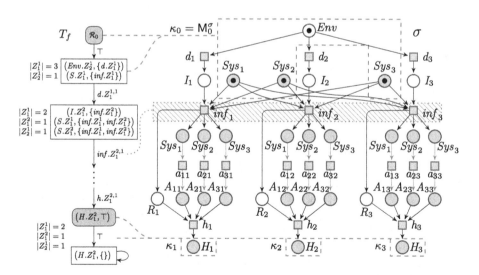

Fig. 4. Parts of a winning strategy for player 0 in $\mathcal{G}(G)^c$ (a tree with gray nodes for player 0), and the generated strategy for the system players in $\mathsf{L}(G)$.

Every edge (r, r') in T_f corresponds to either a relation $\mathcal{R}[t.Z\rangle\mathcal{R}'$ or $\mathcal{R}[\top\rangle\mathcal{R}'$. Suppose now in the breadth-first traversal of T_f we reach a node r with label $\mathcal{R}$

and associated cuts K_r. Further suppose there is an edge in T_f from r to a node r' labeled with $\mathcal{R}'$. If the edge (r, r') in T_f corresponds to a relation $\mathcal{R}[\top\rangle\mathcal{R}'$ in G, then the node r' is associated to the same set of cuts $K_{r'} = K_r$ and nothing is added to the strategy. If the edge (r, r') in T_f corresponds to a relation $\mathcal{R}[t.Z\rangle\mathcal{R}'$ in G and $\mathcal{R} \in V_0$, then there is, for every cut $\kappa \in K_r$, a transition $t.v$ corresponding to $t.Z$ that can be fired from κ (cf. Theorem 2). We add a transition labeled with $t.v$ to the strategy, with its preset in κ. Furthermore, we add places corresponding to its postset to the strategy. The cut κ' that results from firing the new transition from κ is added to $K_{r'}$. If the edge (r, r') in T_f corresponds to a relation $\mathcal{R}[t.Z\rangle\mathcal{R}'$ in G and $\mathcal{R} \in V_1$, then we proceed exactly as in the last case, but with the crucial difference that instead of *one* transition $t.v$ fireable from a cut $\kappa \in K_r$, we consider *all* such transition instances and add them to the strategy. In this step, the number of associated cuts can increase.

In Fig. 4, the strategy tree T_f (consisting of only one branch) in $\mathcal{G}(G)^c$ and the generated strategy σ in (the unfolding of) $\mathsf{L}(G)$ for the running example G are depicted. The strategy σ was already informally described in Sect. 1 and partly shown in Fig. 3. The initial canonical representation $\mathcal{R}_0$ is associated to the cut representing the initial marking in the Petri game. The $\top$-resolution does not change the associated cuts. Firing $d.Z_1^{2,1}$ corresponds to the three firings of d_1, d_2, and d_3 in the strategy. Thus, the third canonical representation is associated to the three cuts $\{Sys_1, Sys_2, Sys_3, I_j\}$, $j = 1, 2, 3$. The strategy T_f terminates in the canonical representation at the bottom, which corresponds to the three situations where all computers connected to the correct host.

5 Experimental Results

In this section we investigate the impact of using *canonical representations* for solving the realizability problem of distributed systems modeled with high-level Petri games with one environment player, an arbitrary number of system players, a safety objective, and an underlying symmetric net.

A prototype [18] for generating the reduced state space of $\mathcal{G}(G)$ for such a high-level Petri game G shows a state space reduction by up to three orders of magnitude compared to $\mathcal{G}(\mathsf{L}(G))$ (cf. Fig. 1) for the considered benchmark families [18]. For this paper we extended this prototype and implemented algorithms to obtain the same state space reduction by using canonical representations in $/\mathcal{G}(G)^c$. Furthermore, we implemented a solving algorithm to exploit the reduced state space for the realizability problem of high-level Petri games. As a reference, we implemented an explicit approach which does not exploit any symmetries of the system. We applied our algorithms on the benchmark families presented in [18] and added a benchmark family for the running example introduced in this paper. An extract of the results for three of these benchmark families are given in Table 1. The complete results are in the full version [19].

The benchmark family *Client/Server (CS)* corresponds to the running example of the paper. With *Document Workflow (DW)* a cyclic document workflow between clerks is modeled. In this benchmark family the symmetries of the systems are only one rotation per clerk. In *Concurrent Machines (CM)* a hostile

Table 1. Comparison of the run times of the *canonical (Canon.)* and *membership (Memb.)* approach solving the realizability problem (✓/✗) for the 3 benchmark families *CS*, *DW*, *CM* with the number of states $|V|$ and number of symmetries $|\xi|$. A gray number of states $|V|$ for the *explicit reference approach* indicates a timeout. Results are obtained on an AMD Ryzen^{TM} 7 3700X CPU, 4.4 GHz, 64 GB RAM and a timeout (TO) of 30 min. The run times are in seconds.

CS	$|V|$	$\models$	$|V|$	$|\xi|$	Memb.	Canon.
1	21	✓	21	1	.38	.36
2	639	✓	326	2	.63	.64
3	45042	✓	7738	6	**5.20**	6.05
4	*7.225e6*	✓	3.100e5	24	151.62	**148.08**
5	*3.154e9*	–	–	120	TO	TO

DW	$|V|$	$\models$	$|V|$	$|\xi|$	Memb.	Canon.
1	57	✓	57	1	.40	**.39**
2	457	✓	241	2	.67	**.62**
...	...	...	...	...	...	...
7	*4.055e6*	✓	5.793e5	7	100.67	**75.24**
8	*2.097e7*	✓	2.621e6	8	986.77	**671.04**
9	*1.053e8*	–	–	9	TO	TO

CM	$|V|$	$\models$	$|V|$	$|\xi|$	Memb.	Canon.
2/1	155	✓	79	2	**.49**	.52
2/2	2883	✗	760	4	**1.07**	1.08
2/3	58501	✗	5548	12	4.38	5.94
2/4	*1.437e6*	✗	33250	48	15.12	**14.40**
2/5	*3.419e7*	✗	1.701e5	240	296.05	**185.81**
2/6	*8.376e8*	–	–	1440	TO	TO
3/1	702	✓	147	6	.71	**.58**
3/2	45071	✓	4048	12	4.46	4.99
3/3	*1.431e6*	✗	91817	36	89.35	**49.90**
3/4	*2.622e8*	–	–	144	TO	TO
4/1	2917	✓	239	24	1.24	1.42
4/2	*6.587e5*	✓	16012	48	25.42	**14.09**
4/3	*1.546e8*	–	–	144	TO	TO

environment can destroy one of the machines processing the orders. Since each machine can only process one order, a positive realizability result is only obtained when the number of orders is smaller than the number of machines. In Table 1 we can see that for those benchmark families the extra effort of computing the canonical representations (Canon.) is worthwhile for most instances compared to the cost of checking the membership of a decision set in an equivalence class (Memb.). This is not the case for all benchmark families.

In Fig. 5 we have plotted the instances of all benchmark families according to their number of symmetries and states. The color of the marker shows the percentaged in- or decrease in performance when using canonical representations while solving high-level Petri games. Blue (unhatched) indicates a performance gain when using the canonical approach. This shows that the benchmarks in general benefit from the canonical approach for an increasing number of states (the right blue (unhatched) area). However, the *DWs* benchmark (a simplified version of *DW*) exhibits the opposite behavior. This is most likely explained by the very simple structure, which favors a quick member check.

The algorithms are integrated in ADAMSYNT[3] [11], open source, and available online[4]. Additionally, we created an artifact with the current version running in a virtual machine for reproducing and checking all experimental data with provided scripts[5].

---

[3] https://github.com/adamtool/adamsynt.
[4] https://github.com/adamtool/high-level.
[5] https://doi.org/10.6084/m9.figshare.13697845.

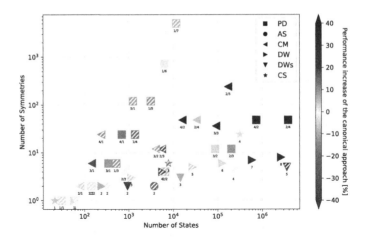

Fig. 5. Comparing the percentage performance gain of the canonical and the membership approach with respect to the number of states and symmetries of the input problem for the benchmark families *Package Delivery (PD)*, *Alarm System (AS)*, *CM*, *DW*, *DWs*, *CS*. Labels are the parameters of the benchmark. A blue (unhatched) marker indicates a performance increase when using canonical representations. (Color figure online)

6   Related Work

For the synthesis of distributed systems other approaches are most prominently the Pnueli/Rosner model [30] and Zielonka's asynchronous automata [34]. The synchronous setting of Pnueli/Rosner is in general undecidable [30], but some interesting architectures exist that have a decision procedure with nonelementary complexity [15,26,32]. For asynchronous automata, the decidability of the control problem is open in general, but again there are several interesting cases which have a decision procedure with nonelementary complexity [16,28,29].

Petri games based on P/T Petri nets are introduced in [12,13]. Solving unbounded Petri games is in general undecidable. However, for Petri games with one environment player, a bounded number of system players, and a safety objective the problem is EXPTIME-complete. The same complexity result holds for interchanged players [14]. High-level Petri games have been introduced in [17]. In [18], such Petri games are solved while exploiting symmetries.

The symbolic Büchi game is inspired by the symbolic reachability graph for high-level nets from [5], and the calculation of canonical representatives [4] from [6]. There are several works on how to obtain symmetries of different subclasses of high-level Petri nets efficiently [4,6,9,27] and for efficiency improvements for systems with different degrees of symmetrical behavior [1,2,22].

7   Conclusions and Outlook

We presented a new construction for the synthesis of distributed systems modeled by high-level Petri games with one environment player, an arbitrary number of system players, and a safety objective. The main idea is the reduction to a symbolic two-player Büchi game, in which the nodes are equivalence classes of symmetric situations in the Petri game. This leads to a significant reduction of the state space. The novelty of this construction is to obtain the reduction by introducing canonical representations. To this end, a theoretically cheaper construction of the Büchi game can be obtained depending on the input system. Additionally, the representations now allow to skip the inflated generation of an explicit Büchi game strategy and to directly generate a Petri game strategy from the symbolic Büchi game strategy. Our implementation, applied on six structurally different benchmark families, shows in general a performance gain in favor of the canonical representatives for larger state spaces.

In future work, we plan to integrate the algorithms in AdamWEB [20], a web interface[6] for the synthesis of distributed systems, to allow for an easy insight in the symbolic games and strategies. Furthermore, we want to continue our investigation on the benefits of canonical representations, e.g., to directly generate high-level representations of Petri game strategies that match the given high-level Petri game.

References

1. Baarir, S., Haddad, S., Ilié, J.: Exploiting partial symmetries in well-formed nets for the reachability and the linear time model checking problems. In: Proceedings of WODES 2004, pp. 219–224 (2004). https://doi.org/10.1016/S1474-6670(17)30749-8
2. Bellettini, C., Capra, L.: A quotient graph for asymmetric distributed systems. In: Proceedings of MASCOTS 2004, pp. 560–568 (2004). https://doi.org/10.1109/MASCOT.2004.1348313
3. Bloem, R., Galler, S.J., Jobstmann, B., Piterman, N., Pnueli, A., Weiglhofer, M.: Interactive presentation: Automatic hardware synthesis from specifications: a case study. In: Proceedings of DATE 2007, pp. 1188–1193 (2007). https://dl.acm.org/citation.cfm?id=1266622
4. Chiola, G., Dutheillet, C., Franceschinis, G., Haddad, S.: On well-formed coloured nets and their symbolic reachability graph. In: Jensen, K., Rozenberg, G. (eds.) High-level Petri Nets: Theory and Application, pp. 373–396. Springer (1991). https://doi.org/10.1007/978-3-642-84524-6_13
5. Chiola, G., Dutheillet, C., Franceschinis, G., Haddad, S.: A symbolic reachability graph for coloured Petri nets. Theor. Comput. Sci. 176(1–2), 39–65 (1997). https://doi.org/10.1016/S0304-3975(96)00010--2
6. Chiola, G., Dutheillet, C., Franceschinis, G., Haddad, S.: Stochastic well-formed colored nets and symmetric modeling applications. IEEE Trans. Comput. 42(11), 1343–1360 (1993)

---

[6] http://adam.informatik.uni-oldenburg.de/.

7. Church, A.: Applications of recursive arithmetic to the problem of circuit synthesis. In: Summaries of the Summer Institute of Symbolic Logic. vol. 1, pp. 3–50. Cornell Univ., Ithaca, NY (1957)

8. Clarke, E.M., Emerson, E.A., Jha, S., Sistla, A.P.: Symmetry reductions in model checking. In: Hu, A.J., Vardi, M.Y. (eds.) CAV 1998. LNCS, vol. 1427, pp. 147–158. Springer, Heidelberg (1998). https://doi.org/10.1007/BFb0028741

9. Dutheillet, C., Haddad, S.: Regular stochastic petri nets. In: Rozenberg, G. (ed.) ICATPN 1989. LNCS, vol. 483, pp. 186–209. Springer, Heidelberg (1991). https://doi.org/10.1007/3-540-53863-1_26

10. Esparza, J., Heljanko, K.: Unfoldings - A Partial-Order Approach to Model Checking. EATCS Monographs in Theoretical Computer Science, Springer (2008). https://doi.org/10.1007/978-3-540-77426-6

11. Finkbeiner, B., Gieseking, M., Olderog, E.-R.: ADAM: causality-based synthesis of distributed systems. In: Kroening, D., Păsăreanu, C.S. (eds.) CAV 2015. LNCS, vol. 9206, pp. 433–439. Springer, Cham (2015). https://doi.org/10.1007/978-3-319-21690-4_25

12. Finkbeiner, B., Olderog, E.: Petri games: Synthesis of distributed systems with causal memory. In: Proceedings of GandALF 2014, pp. 217–230. EPTCS161 (2014). https://doi.org/10.4204/EPTCS.161.19

13. Finkbeiner, B., Olderog, E.: Petri games: synthesis of distributed systems with causal memory. Inf. Comput. **253**, 181–203 (2017). https://doi.org/10.1016/j.ic.2016.07.006

14. Finkbeiner, B., Gölz, P.: Synthesis in distributed environments. In: Proceedings of FSTTCS 2017, pp. 28:1–28:14 (2017). https://doi.org/10.4230/LIPIcs.FSTTCS.2017.28

15. Finkbeiner, B., Schewe, S.: Uniform distributed synthesis. In: Proceedings of LICS 2005, pp. 321–330 (2005). https://doi.org/10.1109/LICS.2005.53

16. Genest, B., Gimbert, H., Muscholl, A., Walukiewicz, I.: Asynchronous games over tree architectures. In: Fomin, F.V., Freivalds, R., Kwiatkowska, M., Peleg, D. (eds.) ICALP 2013. LNCS, vol. 7966, pp. 275–286. Springer, Heidelberg (2013). https://doi.org/10.1007/978-3-642-39212-2_26

17. Gieseking, M., Olderog, E.: High-level representation of benchmark families for Petri games. CoRR abs/1904.05621 (2019). http://arxiv.org/abs/1904.05621

18. Gieseking, M., Olderog, E.-R., Würdemann, N.: Solving high-level Petri games. Acta Inform. **57**(3), 591–626 (2020). https://doi.org/10.1007/s00236-020-00368-5

19. Gieseking, M., Würdemann, N.: Canonical representations for direct generation of strategies in high-level Petri games (full version). CoRR abs/2103.10207 (2021). http://arxiv.org/abs/2103.10207

20. Gieseking, M., Hecking-Harbusch, J., Yanich, A.: A web interface for petri nets with transits and petri games. TACAS 2021. LNCS, vol. 12652, pp. 381–388. Springer, Cham (2021). https://doi.org/10.1007/978-3-030-72013-1_22

21. Grädel, E., Thomas, W., Wilke, T. (eds.): Automata Logics, and Infinite Games. LNCS, vol. 2500. Springer, Heidelberg (2002). https://doi.org/10.1007/3-540-36387-4

22. Haddad, S., Ilié, J.M., Taghelit, M., Zouari, B.: Symbolic reachability graph and partial symmetries. In: De Michelis, G., Diaz, M. (eds.) ICATPN 1995. LNCS, vol. 935, pp. 238–257. Springer, Heidelberg (1995). https://doi.org/10.1007/3-540-60029-9_43

23. Hillah, L., Kordon, F., Petrucci, L., Trèves, N.: PN standardisation: a survey. In: Najm, E., Pradat-Peyre, J.-F., Donzeau-Gouge, V.V. (eds.) FORTE 2006. LNCS, vol. 4229, pp. 307–322. Springer, Heidelberg (2006). https://doi.org/10.1007/11888116_23

24. Jensen, K.: Coloured Petri Nets - Basic Concepts, Analysis Methods and Practical Use - Volume 1. EATCS Monographs on Theoretical Computer Science, Springer (1992). https://doi.org/10.1007/978-3-662-06289-0

25. Kress-Gazit, H., Fainekos, G.E., Pappas, G.J.: Temporal-logic-based reactive mission and motion planning. IEEE Trans. Robot. **25**(6), 1370–1381 (2009). https://doi.org/10.1109/TRO.2009.2030225

26. Kupferman, O., Vardi, M.Y.: Synthesizing distributed systems. In: Proceedings of LICS 2001, pp. 389–398 (2001). https://doi.org/10.1109/LICS.2001.932514

27. Lindqvist, M.: Parameterized reachability trees for predicate/transition nets. In: Rozenberg, G. (ed.) ICATPN 1991. LNCS, vol. 674, pp. 301–324. Springer, Heidelberg (1993). https://doi.org/10.1007/3-540-56689-9_49

28. Madhusudan, P., Thiagarajan, P.S., Yang, S.: The MSO theory of connectedly communicating processes. In: Sarukkai, S., Sen, S. (eds.) FSTTCS 2005. LNCS, vol. 3821, pp. 201–212. Springer, Heidelberg (2005). https://doi.org/10.1007/11590156_16

29. Muscholl, A., Walukiewicz, I.: Distributed synthesis for acyclic architectures. In: Raman, V., Suresh, S.P. (eds.) Proceedings of FSTTCS 2014, pp. 639–651. LIPIcs29 (2014). https://doi.org/10.4230/LIPIcs.FSTTCS.2014.639

30. Pnueli, A., Rosner, R.: Distributed reactive systems are hard to synthesize. In: Proceedings of FOCS, pp. 746–757. IEEE Computer Society Press (1990). https://doi.org/10.1109/FSCS.1990.89597

31. Reisig, W.: Petri Nets: An Introduction, EATCS Monographs on Theoretical Computer Science, vol. 4. Springer (1985). https://doi.org/10.1007/978-3-642-69968-9

32. Rosner, R.: Modular Synthesis of Reactive Systems. Ph.D. thesis, Weizmann Institute of Science, Rehovot, Israel (1992)

33. Starke, P.H.: Reachability analysis of Petri nets using symmetries. Syst. Anal. Model. Simul. **8**(4–5), 293–303 (1991)

34. Zielonka, W.: Asynchronous automata. In: Diekert, V., Rozenberg, G. (eds.) The Book of Traces, pp. 205–247. World Scientific (1995). https://doi.org/10.1142/9789814261456_0007

Automatic Synthesis of Transiently Correct Network Updates via Petri Games

Martin Didriksen, Peter G. Jensen, Jonathan F. Jønler, Andrei-Ioan Katona, Sangey D.L. Lama, Frederik B. Lottrup, Shahab Shajarat, and Jiří Srba[✉]

Department of Computer Science, Aalborg University, Aalborg, Denmark
srba@cs.aau.dk

Abstract. As software-defined networking (SDN) is growing increasingly common within the networking industry, the lack of accessible and reliable automated methods for updating network configurations becomes more apparent. Any computer network is a complex distributed system and changes to its configuration may result in policy violations during the transient phase when the individual routers update their forwarding tables. We present an approach for automatic synthesis of update sequences that ensures correct network functionality throughout the entire update phase. Our approach is based on a novel translation of the update synthesis problem into a Petri game and it is implemented on top of the open-source model checker TAPAAL. On a large benchmark of synthetic and real-world network topologies, we document the efficiency of our approach and compare its performance with state-of-the-art tool NetSynth. Our experiments show that for several networks with up to thousands of nodes, we are able to outperform NetSynth's update schedule generation.

1 Introduction

Modern computer networks are met with increasing demands on scalability, security, reliability and performance. This stipulates the need of frequently updating the network configurations in order to adapt to changes in flow demands, link failures and other disturbances. The complexity of current networks shows the limitation of the traditional manual network maintenance as the risks of introducing faulty behaviour and security leaks become too high. The software defined networking (SDN) paradigm [2] is a recent methodology that aims to combat the increased complexity of network operation by centralizing the network control and hence allowing for fully automatic updates of network configurations. This enables the option of dynamically updating networks with increased frequency in order to optimize their performance, but it also requires a reliable way to govern and schedule updates, so that disruption of service due to forwarding loops or blackholes and security leaks when e.g. a critical firewall is bypassed, can be avoided.

Even though the initial and final configurations are correct and satisfy a number of desirable properties like reachability and waypointing (before a packet

© Springer Nature Switzerland AG 2021
D. Buchs and J. Carmona (Eds.): PETRI NETS 2021, LNCS 12734, pp. 118–137, 2021.
https://doi.org/10.1007/978-3-030-76983-3_7

leaves the network a certain router, e.g. a firewall, must be visited), there is no guarantee that any transient configuration, where individual routers are updated one by one, preserves the required policies. The update synthesis problem [10] asks, in which order to update the routers in the network so that at any moment there is never any policy violation.

As our first contribution, we propose to translate the update synthesis problem into a two-player Petri game between the controller and the environment. The objective of the controller is to reach the updated network routing from the initial one by sequentially scheduling the updates of the individual switches. The environment can at any time interrupt the construction of the update sequence and initialize a check on whether the current partially constructed update sequence satisfies the given security policies. If the policies are satisfied, the controller is the winner, otherwise the environment wins. A winning strategy for the controller then defines a transiently correct update sequence.

As a second contribution, we implement the Petri game translation on top of the open-source model checker TAPAAL [6,12] and update its game engine with efficient state-space exploration strategies for solving the game synthesis problem. Our fully automated tool chain accepts the descriptions of network topologies and the initial and final routings as a JSON file, together with policy properties that include loop-freedom, reachability and waypointing. The tool then outputs that either there does not exist any transiently correct update sequence or it synthesizes such a sequence.

Finally, we conduct a number of experiments on both synthetic and real-world benchmarks and compare the performance of our approach with state-of-the-art tool NetSynth [21] that relies on counterexample-guided search and incremental model checking techniques, and allows for the use of different model checkers including NuSMV [5] as its backend engine. The results confirm that on two, out of the three scalable synthetic networks, we obtain several orders of magnitude speed up. In one case where the synthetic network shares as many possible links in both the initial and final routing, our method is performing slower. Experiments on the real-world Internet topologies, where the routings are constructed using the common method based on shortest paths (see e.g. the Equal-Cost-MultiPath (ECMP) [11] or the Open Shortest Path First (OSPF) [22] routing protocols), demonstrate that both tools are able to solve smaller instances of the update synthesis problem below once second, however, for the larger instances our method wins by a clear margin. As an additional contribution to the research community, we also provide a publicly available reproducibility package [7] including both the code as well as all experimental data.

Related Work. The work on updates in SDN is heavily influenced by the work by Reitblatt et al. [24] that defines the per-packet and per-flow consistency. In per-packet consistency, each packet traverses the network within at most one stable configuration, whereas per-flow guarantees that all packets in a flow traverse the network in the same configuration. The per-packet consistency, which is also the main focus of our work, inspired further research in this area [3,17,20]. In particular, Mahajan and Wattenhofer [20] suggest an approach that eliminates the use of packet header rewriting and the expensive two-phase update. They devise a solution that preserves loop-freedom with weak consistency by examining the dependencies of switches in a network and conclude that half of the updates with around 100 switches only depended on zero or a single critical switch update and in 90% of the cases, updates are only dependent on at most three switches, in contrast to Reitblatt et al. [24] who rely on updating all switches. Their work has since then been refined and extended to support more properties [1,18], including waypointing [19]. However, it is known that the update synthesis problem with waypointing and loop-freedom becomes NP-complete [18] (for a detailed complexity overview see e.g. [10]). More recently, Nate Foster et al. [21] introduce a specialized incremental model checker NetSynth that automatically synthesises correct update sequences from LTL specifications. In [21] the authors argue that their tool is outperforming other existing approaches on a variety of network topologies for ensuring reachability and waypointing policies. NetSynth essentially performs a (heuristic) search through all possible update sequences and relies on the assumption that the routings are loop-free. Our approach can verify also the presence of loops and it uses the general concept of two-player games instead of the explicit enumeration of all possible update sequences. Another approach that allows to present updates in concurrent steps is given in [25].

A recent work by Christensen et al. [4] introduces the tool Latte that models the problem as a timed-arc colored Petri net and its main focus is on reducing the delays between the updates of the individual routers. While it extends the analysis with timing aspects, their work relies on obtaining a correct update sequence from third-party tools (NetSynth in their case) and as such it does not solve the synthesis problem and focuses purely on the timing optimization aspects and the discovery of possible concurrent updates. Another line of work by Finkbeiner et al. [8,9] focuses on verifying concurrent network updates against Flow-LTL specifications, using Petri nets extended with transits as the underlying modelling formalism and circuit model checking as the backend engine. The experiments show that their tool can verify in minutes networks up to a hundred of routers, however, similarly as Latte [4], they can only verify but not synthesise update sequences. To the best of our knowledge, our approach is the first one that employs the game semantics of Petri nets and allows hence for fully automate update synthesis as well as the reuse of generic game model checkers like TAPAAL. As such, the generated update sequences can be then further optimized, e.g. for the timing and concurrent aspects, by the above mentioned approaches.

2   Update Synthesis

We shall now formalize the notion of a network and routing in a network, define some essential routing properties and formulate the update synthesis problem.

Definition 1 (Network). *A network is a directed graph $G = (V, E)$ where V is a finite set of nodes (switches), and $E \subseteq V \times V$ is a set of edges (links) such that $(v, v) \notin E$ for all $v \in V$.*

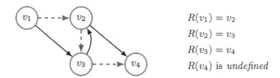

$R(v_1) = v_2$

$R(v_2) = v_3$

$R(v_3) = v_4$

$R(v_4)$ is *undefined*

Fig. 1. A network with routing R depicted by the red dashed arrows (the black arrows represent existing links that are not used in R) (Color figure online)

A network defines the set of links that connect the switches. For a given packet type (given by its header and usually determined by its destination), each switch contains a forwarding table defining the next hop. A switch contains this information for all different packet types and we project on a certain type in order to define its routing in a network. For the rest of this section, let $G = (V, E)$ be a fixed network.

Definition 2 (Routing). *A* routing *in G is a partial function $R : V \hookrightarrow V$ such that $(u, R(u)) \in E$ for all $u \in V$ where $R(u)$ is defined.*

Consider the network in Fig. 1 with a routing R, indicated by the dashed edges. The routing naturally defines a path, which is a (unique) sequence of next hops. A path can be either infinite or finite. Any infinite path, or finite path that ends in a node with undefined next hop, is called a maximal path.

Definition 3 (Path). *A* path *π under a routing R is a sequence of nodes $v_1 v_2 ... v_n ... \in V^* \cup V^\omega$ such that $R(v_i) = v_{i+1}$ for all i. A* maximal path *is either an infinite path or a finite path that ends in a node v where $R(v)$ is undefined.*

Our example in Fig. 1 contains the maximal path $v_1 v_2 v_3 v_4$. This path demonstrates reachability between v_1 and v_4 and the routing does not contain any infinite path. Moreover, the path starting in v_1 contains the switch v_2 before v_4 is reached. A switch with this property is called a waypoint. We shall now formally define these basic properties of a routing function.

Definition 4 (Routing Properties). *Let u, v and w be three different nodes in a network. A routing R satisfies*

– *the* reachability *property $reach(u, v)$ if there is a path $\pi = v_1 v_2 ... v_n$ under R such that $v_1 = u$ and $v_n = v$,*
– *the* waypointing *property $wp(u, v, w)$ if for every path $\pi = v_1 v_2 ... v_n$ under R where $v_1 = u$ and $v_n = v$ there exists an i, $1 < i < n$, such that $v_i = w$, and*
– *the* loop-freedom *property $loopfree(u)$ if the maximal path under R starting in u is finite.*

We shall note that (i) the waypointing property $wp(u, v, w)$ is trivially satisfied whenever there is no path under R from u to v, e.g. in our running example the property $wp(v_3, v_1, v_4)$ holds and (ii) any infinite path under a given routing must form a loop after some finite initial prefix (as the number of nodes is finite).

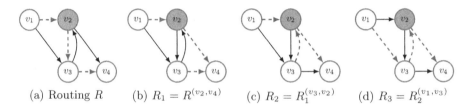

(a) Routing R (b) $R_1 = R^{(v_2, v_4)}$ (c) $R_2 = R_1^{(v_3, v_2)}$ (d) $R_3 = R_2^{(v_1, v_3)}$

Fig. 2. A correct update sequence for $reach(v_1, v_4)$ and $wp(v_1, v_4, v_2)$

2.1   Network Updates

In order to be able to change the given routing (and therefore to influence the routing path) we introduce the notion of an update. An update can either change the forwarding function of a given node to an alternative next hop, or it can undefine the routing function (remove the entry from the forwarding table).

Definition 5 (Update).   *Given a routing R, an* update *is an element $e \in E \cup (V \times \{undefined\})$. For a given update $e = (u, u')$, the updated routing R^e is given by*

$$R^{(u,u')}(v) = \begin{cases} R(v) & \text{if } v \neq u \quad (1a) \\ u' & \text{if } v = u. \quad (1b) \end{cases}$$

In order to update one routing into another, a number of updates must be performed in a sequence (a so-called update sequence), executing the updates from left to right.

Definition 6 (Update Sequence).   *Given a routing R and an update sequence $\omega = e_1 e_2 ... e_n \in (E \cup (V \times \{undefined\}))^*$, we inductively define the final routing R^ω by (i) $R^\epsilon = R$ and (ii) $R^{e\omega} = (R^e)^\omega$ for any $e \in E$ and any $\omega \in (E \cup V \times \{undefined\})^*$.*

Finally, we must guarantee that an update sequence transiently satisfies a given set of properties $\mathcal{P}$, containing e.g. $reach(u, v)$, $wp(u, v, w)$ and $loopfree(u)$.

Definition 7 (Correct Update Sequence).   *We say that ω is a* correct update sequence *for R with respect to a set of properties $\mathcal{P}$ if for every prefix ω' of ω the routing $R^{\omega'}$ satisfies every property from $\mathcal{P}$.*

Figure 2 shows the steps of a correct update sequence $\omega = (v_2, v_4) \circ (v_3, v_2) \circ (v_1, v_3)$ on our running example, for $\mathcal{P} = \{reach(v_1, v_4), wp(v_1, v_4, v_2)\}$. At any moment during the update sequence, the node v_4 is reachable from v_1 and at the same time the node v_2 (in grey) is always present on the path from v_1 to v_4.

The update synthesis problem is, for a given initial routing R and a final routing R', to find an update sequence that transforms R into R' while preserving a given set of path properties $\mathcal{P}$. Moreover, we allow at most one update of every node in the network. As we are modelling one fixed flow in the network, we assume that the path properties always start with the same fixed node $u \in V$.

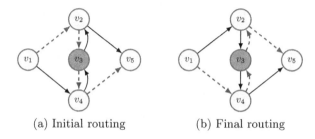

(a) Initial routing   (b) Final routing

Fig. 3. A network with no solution preserving $reach(v_1, v_5)$ and $wp(v_1, v_5, v_3)$

Definition 8 (Update Synthesis Problem). *The* update synthesis problem *is a tuple* $\mathcal{U} = (G, R, R', \mathcal{P}^u)$ *where G is a network, R is an initial routing, R' is a final routing, and $\mathcal{P}^u \subseteq \{wp(u, v, w), reach(u, v), loopfree(u) \mid v, w \in V\}$ for some fixed $u \in V$ is the set of path properties. The question is whether there exists a correct update sequence $\omega \in (E \cup (V \times \{undefined\}))^*$ with respect to $\mathcal{P}^u$ such that $R^\omega = R'$ and where every $v \in V$ appears in ω at most once as a source node. We then say that ω is a* solution *to the update synthesis problem.*

The update synthesis problem in NP-complete [18]. Figure 2 shows a solution to the update synthesis problem transforming the routing R into R_3 while preserving $reach(v_1, v_4)$ and $wp(v_1, v_4, v_2)$. We notice that this is in fact the only solution. If we in the routing R first update the router v_1, we break the way-pointing property and if we instead decide to update first v_3, we create a loop and invalidate the reachability property. Similarly, in the routing R_1 the only choice is to first update v_3, as updating v_1 in R_1 will avoid the waypoint. Lastly, in Fig. 3 we give an example of an update synthesis problem that does not have a solution for preserving the properties $reach(v_1, v_5)$ and $wp(v_1, v_5, v_3)$. If v_3 is updated before v_2, a loop is created and $reach(v_1, v_5)$ is violated and if v_2 is updated before v_3, the property $wp(v_1, v_5, v_3)$ is violated.

3   Petri Games

A Petri net (see e.g. [23]) is a mathematical model of distributed systems that allows us to model concurrent and nondeterministic behaviour. A Petri game extends P/T nets by introducing the controllable and environmental transitions. This kind of games were studied in [13] also including the timing aspects.

Let $\mathbb{N}^0$ be the set of natural numbers including 0 and let $\mathbb{N}^\infty$ be the set of natural numbers extended with the symbol ∞ that is larger than any other natural number.

Definition 9 (Petri Game). *A Petri game is a 4-tuple $N = (P, T, W, I)$ where P is a finite set of* places, *T is a finite set of* transitions *partitioned into the controllable T_{ctrl} and environmental T_{env} ones s.t. $T = T_{ctrl} \uplus T_{env}$ and $P \cap T = \emptyset$, the function $W : (P \times T) \cup (T \times P) \to \mathbb{N}^0$ assigns weights to normal arcs, and the function $I : P \times T \to \mathbb{N}^\infty$ assigns weights to inhibitor arcs.*

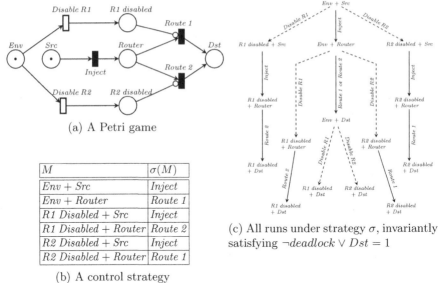

(a) A Petri game

M	$\sigma(M)$
$Env + Src$	$Inject$
$Env + Router$	$Route\ 1$
$R1\ Disabled + Src$	$Inject$
$R1\ Disabled + Router$	$Route\ 2$
$R2\ Disabled + Src$	$Inject$
$R2\ Disabled + Router$	$Route\ 1$

(b) A control strategy

(c) All runs under strategy σ, invariantly
satisfying $\neg deadlock \vee Dst = 1$

Fig. 4. A Petri game example

Let $N = (P, T, W, I)$ be a fixed Petri game. Places in a Petri game are
depicted as circles, controllable transitions by filled rectangles and environmental
by empty rectangles and whenever $W(p, t) > 0$ or $W(t, p) > 0$ we draw an arc
between p and t, resp. t and p, labeled by the corresponding weight (no label
stands for the weight 1, and if $I(p, t) < \infty$ we draw an inhibitor arc depicted
circle-head between p and t and annotated by the weight. An example of Petri
game is given in Fig. 4a and it models a scenario where a packet must travel
through a network from Src to Dst through one of two possible routes $Route\ 1$
or $Route\ 2$. However, the environment can disable one of the two routes, and
depending on which it disables we must choose the remaining route.

A *marking* M on N is a total function $M : P \to \mathbb{N}^0$ that marks each place
with 0 or more tokens. Let $\mathcal{M}(N)$ bet the set of all markings on N. We can
represent a marking as a formal sum $k_0 p_0 + k_1 p_1 + \cdots + k_n p_n$ where k_i denotes
the number of tokens present in p_i. A marking is graphically denoted by dots
in places. A transition $t \in T$ is *enabled* in a marking M if $W(p, t) \le M(p)$
and $I(p, t) > M(p)$ for all $p \in P$. If a transition $t \in T$ is enabled in a marking
M, it can *fire* resulting in the marking M', written $M \xrightarrow{t} M'$, where $M'(p) =
M(p) + W(p, t) - W(t, p)$ for all $p \in P$.

The controller's choice of which controllable transition to select in each marking, is given by the concept of a (memoryless) strategy.

Definition 10 (Strategy). *Let $N = (P, T, W, I)$ be a Petri game. A strategy
$\sigma : \mathcal{M}(N) \hookrightarrow T_{ctrl}$ is a partial function such that if $\sigma(M) = t$ then $M \xrightarrow{t} M'$ for
some M' and $\sigma(M)$ is undefined iff there is no $t \in T_{ctrl}$ enabled in M.*

An example of a strategy for the game net from Fig. 4a is given in Fig. 4b. A strategy determines the set of runs such that from every marking either any environmental transition or the transition proposed by the strategy can be executed. A set of all runs in our running example, organized into a tree where run prefixes are shared, is given in Fig. 4c. Controllable transitions are depicted with solid lines and environmental with dashed lines.

Definition 11 (Run). *Let* $N = (P, T, W, I)$ *be a Petri game with an initial marking* M_0 *and a strategy* σ. *The set of finite and infinite runs under* σ *is given by* $runs^{\sigma}(M_0) = \{M_0 M_1 M_2 \ldots \mid$ *for all i holds* $M_i \xrightarrow{t} M_{i+1}$ *s.t.* $t \in T_{env}$ *or* $M_i \xrightarrow{\sigma(M_i)} M_{i+1}$ *if* $\sigma(M_i)$ *is defined*$\}$.

The goal of the controller in the game is to invariantly preserve a given safety objective, expressed as a *marking predicate* φ given by a Boolean combination of expressions of the form $e \bowtie e$ or *deadlock* where

$$e ::= p \mid n \mid e + e \mid e - e \mid e * e$$

such that $p \in P$, $\bowtie \in \{\leq, <, =, \neq, \geq, >\}$ and $n \in \mathbb{N}_0$. We write $M \models deadlock$ if there are no enabled transitions in M and the semantics of the Boolean connectives as well as of the expressions $e \bowtie e$ is given in a natural way, assuming that p stands for $|M(p)|$. If a marking M satisfies a predicate φ, we write $M \models \varphi$.

Definition 12 (Winning Control Strategy). *Let* $N = (P, T, W, I)$ *be a Petri game with the initial marking* M_0 *and let* φ *be a marking predicate. We write* $M_0 \models control : AG\,\varphi$ *if there exists a strategy* σ *such that for every run* $M_0 M_1 M_2 \ldots \in runs^{\sigma}(M_0)$ *we have* $M_i \models \varphi$ *for all relevant i. If such a strategy exists, we say that* σ *is a* winning strategy.

In Fig. 4a we have $Env + Src \models control : AG\,(\neg deadlock \vee Dst = 1)$ as the strategy defined in Fig. 4b is a winning strategy for the safety objective $\neg deadlock \vee Dst = 1$. This can be verified by exploring the possible runs under the strategy given in Fig. 4c and noticing that all deadlocked markings have a token in the place Dst. In general, for any bounded Petri game the existence of a winning control strategy is decidable (see e.g. [15]).

4   From Update Synthesis Problem to Petri Games

We shall now present a reduction from the update synthesis problem into a Petri game. The reduction idea is that the controller is allowed to step-wise generate any possible sequence of updates and a given control strategy fixes a concrete update sequence. The environment can at any moment decide to stop the generation process and inject a packet into the network in order to verify whether the current prefix of the update sequence satisfies the given properties. If this is not the case, the environment wins, otherwise the controller has a winning strategy that corresponds to a correct update sequence.

The reduction, for a given update synthesis problem $(G, R, R', \mathcal{P}^u)$, is split into the creation of the following components:

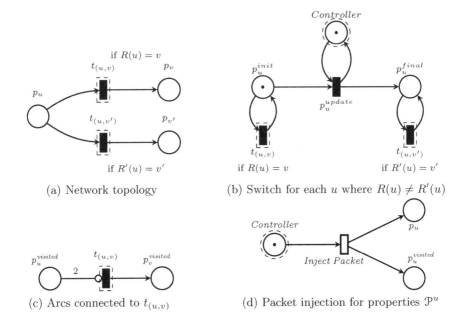

(a) Network topology

(b) Switch for each u where $R(u) \neq R'(u)$

(c) Arcs connected to $t_{(u,v)}$

(d) Packet injection for properties $\mathcal{P}^u$

Fig. 5. Translation components

1. *Network topology component* based on G.
2. *Switch components* based on R and R'.
3. *Visited places* to track how many times a node is visited.
4. *Packet injection component* to start the verification phase.
5. *Satefy objective* based on $\mathcal{P}^u$.

For clarity reasons, we decompose our Petri net construction into separate components. If the same place or transition name appears in multiple components, we refer to them as shared and indicate this by surrounding them with a dashed line. We assume that the shared places and transitions are merged.

4.1   Translation to Petri Games

Let $\mathcal{U} = (G, R, R', \mathcal{P}^u)$ be an update synthesis problem such that $G = (V, E)$ is the underlying network. We create a Petri game $N(\mathcal{U}) = (P, T, I, W)$ where $T = T_{ctrl} \uplus T_{env}$ by defining the different components mentioned above.

1. *Network topology components.* For all $u \in V$ where $R(u)$ or $R'(u)$ is defined create the place p_u and
   – If $R(u) = v$ we create a shared transition $t_{(u,v)} \in T_{ctrl}$ and a place p_v.
   – If $R'(u) = v'$ we create a shared transition $t_{(u,v')} \in T_{ctrl}$ and a place $p_{v'}$.
   The created places and transitions are then connected by normal arcs as illustrated in Fig. 5a.

2. *Switch components.* For every $u \in V$ where $R(u) \neq R'(u)$, we create a switch component with a controllable transition $p_u^{update} \in T_{ctrl}$ connected to a globally shared place *Controller* and two places p_u^{init} (initially marked) and p_u^{final} as in Fig. 5b. Moreover,
   – if $R(u) = v$, we add the arcs p_u^{init} to $t_{(u,v)}$ and $t_{(u,v)}$ to p_u^{init}, and
   – if $R'(u) = v'$, we add the arcs p_u^{final} to $t_{(u,v')}$ and $t_{(u,v')}$ to p_u^{final}.
3. *Visited places.* For each place p_u that was already created, we create a dual place $p_u^{visited}$ as illustrated in Fig. 5c where
   – for all $v \in V$ such that there exists a $u \in V$ where $R(u) = v$ or $R'(u) = v$ we add an arc from $t_{(u,v)}$ to $p_v^{visited}$, and
   – for all $u \in V$ such that $R(u) = v$ or $R'(u) = v$ we add an inhibitor arc with weight of 2 from $p_u^{visited}$ to $t_{u,v}$.
4. *Packet injection component.* Here we add (the only) environmental transition *Inject Packet* $\in T_{env}$ that connects, as depicted in Fig. 5d, the place *Controller* (initially marked with a token) to the places p_u and $p_u^{visited}$ where u is the initial node fixed in the set of properties $\mathcal{P}^u$.
5. *Verification queries.* For the given set of properties $\mathcal{P}^u = \{P_1, P_2, \ldots, P_k\}$ we construct the safety objective *control* : $AG\ \varphi_{P_1} \wedge \varphi_{P_2} \ldots \wedge \varphi_{P_k}$ where:
   – $\varphi_{loopfree(u)} \equiv \bigwedge_{u \in V} p_u^{visited} < 2$
   – $\varphi_{reach(u,v)} \equiv \neg deadlock \vee p_v^{visited} \geq 1$
   – $\varphi_{wp(u,v,w)} \equiv p_w^{visited} \geq 1 \vee p_v^{visited} = 0$

The translation is illustrated by a small example in Fig. 6. For the network in Fig. 6a, we want to update the initial routing via v_1, v_2 and v_4 to the final routing v_1, v_3 and v_4, while preserving the reachability between v_1 and v_4. The translation of the switch components is given in Fig. 6b and the translation of the renaming components in Fig. 6c, where the place *Controller* as well as the transitions $t_{(1,2)}$, $t_{(1,3)}$, $t_{(2,4)}$ and $t_{(3,4)}$ are shared between the two figures. We note that since $R(v_4) = R'(v_4) = undefined$, we do not create a switch component for v_4.

Finally, we notice the once the environmental *Inject Packet* transition fires, a token is removed from the place *Controller* and it is no longer possible to change the current transient routing. The construction guarantees that token injected to p_1 now follows exactly the path corresponding to the current routing and every time a place receives a token, the corresponding visited place also gets a token. Should there be a loop, the first marking where one of the visited places obtains a second token deadlocks due to the introduction of the inhibitor arcs. The constructed Petri game is so guaranteed to be bounded. The property we wish to preserve is $reach(v_1, v_4)$ and it translates to the safety objective $\neg deadlock \vee p_4^{visited} \geq 1$. This guarantees that during the execution of the routing path we do not deadlock (which can be caused either by a blackhole or a loop) before the target place p_4 is reached.

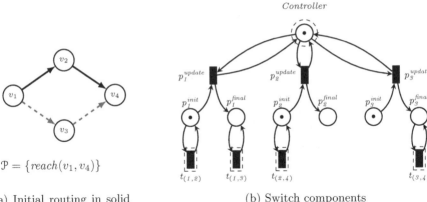

(a) Initial routing in solid   (b) Switch components
lines, final in dashed lines

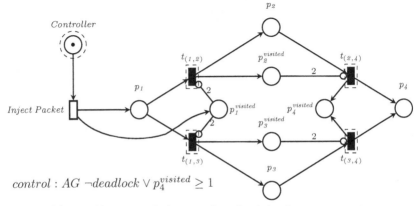

(c) Topology, visited places and packet injection components

Fig. 6. Translation example

4.2 Translation Correctness

Let $\mathcal{U} = (G, R, R', \mathcal{P}^u)$ be an update synthesis problem and let $N(\mathcal{U})$ be the constructed Petri game with the initial marking M_0 and the safety objective φ. The main correctness theorem is stated as follows.

Theorem 1. *The problem $\mathcal{U}$ has a solution iff $M_0 \models control : AG\varphi$ in $N(\mathcal{U})$.*

We shall first observe that all runs in the constructed Petri game are finite.

Lemma 1. *The net $N(\mathcal{U})$ contains no infinite run from the initial marking M_0.*

Proof. Assume by contradiction that there is an infinite run from M_0. Clearly, the firing of each p_u^{update} can happen at most once for each u, so necessarily the transition *Inject Packet* must fire during such a sequence, initializing the execution in the topology component. There must be now a transition $t_{(u,v)}$ that fires at least twice in the infinite run. Each time $t_{(u,v)}$ fires, a new token is deposited into $p_v^{visited}$ and hence the place contains two tokens after the second time $t_{(u,v)}$ is fired. However, all outgoing transitions from the place p_v (containing a token after $t_{(u,v)}$ is fired) are disabled due to the inhibitor arcs of weight 2 from $p_v^{visited}$. Hence the net deadlocks and this contradicts the existence of the infinite run.   □

Now we prove Theorem 1 by establishing each direction of the claim.

Lemma 2. *If ω is a correct update sequence for $\mathcal{U}$ then in $N(\mathcal{U})$ there is a winning control strategy σ for the formula control* : $AG\ \varphi$.

Proof. Let $\omega = e_1e_2...e_n$ be a correct update sequence such that $e_i = (u_i, u_i')$ where $u_i \in V$ and $u_i' \in V \cup \{undefined\}$. We shall now define a winning strategy σ for the controller, given the initial marking M_0, as follows: $\sigma(M_0) = p_{u_1}^{update}$ and we let $M_0 \xrightarrow{p_{u_1}^{update}} M_1$; we continue to define $\sigma(M_{i-1}) = p_{u_i}^{update}$ for all i, $1 < i \leq n$, where we let $M_{i-1} \xrightarrow{p_{u_i}^{update}} M_i$. For all other markings M, we let $\sigma(M) = M'$ for an arbitrary M' such that $M \xrightarrow{M'}_{t}$ and $t \in T_{ctrl}$, if such marking M' exists; otherwise is $\sigma(M)$ undefined. In other words, the controller fires the controllable $p_{u_i}^{update}$ transitions in the order specified in the update sequence ω. We also notice that once the *Inject Packet* transition is fired by the environment, there is in any reachable marking at most one enabled controllable transition, exactly simulating the path under the currently generated sequence of updates. As stated in Lemma 1, this path is finite as the net deadlocks as soon as the same node is visited the second time.

Now we argue that the strategy σ is a winning control strategy by showing that every marking M reachable under the strategy σ satisfies φ. As φ is a conjunction of marking predicates of three different types (depending on the properties in $\mathcal{P}^u$), we discuss the three cases.

- $\varphi_{loopfree(u)} \equiv \bigwedge_{u \in V} p_u^{visited} < 2$. Before the environment fires *Inject Packet*, no token can be placed in any $p_u^{visited}$ and the invariant is satisfied. If *Inject Packet* is fired, because ω is a correct update sequence it follows by Definition 7 that the currently generated prefix of ω yields a loop-free routing and therefore any reachable marking satisfies $p_u^{visited} < 2$.
- $\varphi_{reach(u,v)} \equiv \neg deadlock \vee p_v^{visited} \geq 1$. Before firing *Inject Packet*, the net cannot deadlock as *Inject Packet* is still enabled and hence the property holds. Once *Inject Packet* is fired, due to our assumption that the currently generated prefix of ω is correct and the corresponding routing hence eventually reaches the node v, and because the Petri net faithfully mimics the routing path, we know that the Petri net execution eventually marks the place $p_v^{visited}$ and cannot deadlock before this: there cannot be any blackhole as this contradicts the reachability of v and there cannot be any node visited twice either before reaching v, as the path is deterministic and it will imply the existence

of a loop and contradict the reachability of v. Hence $\neg deadlock \vee p_v^{visited} \geq 1$ invariantly holds.

- $\varphi_{wp(u,v,w)} \equiv p_w^{visited} \geq 1 \vee p_v^{visited} = 0$. Before firing *Inject Packet* the invariant clearly holds as none of the places $p_u^{visited}$ can be marked for all nodes u, including v. After *Inject Packet* is fired, as we assume that any prefix of ω corresponds to a correct routing, meaning that the node v cannot be marked before the node w. Hence the invariant holds also in this case. $\qquad\square$

Lemma 3. *If σ is a winning control strategy in $N(\mathcal{U})$ for the formula control :* $AG\ \varphi$ *then there exists a correct update sequence ω solving $\mathcal{U}$.*

Proof. Assume that σ is a winning strategy for the formula *control* : $AG\ \varphi$. Given the initial marking M_0, the strategy σ generates the sequence of markings $M_0, M_1, \ldots, M_n$ such that $\sigma(M_{i-1}) = p_{u_i}^{update}$ and $M_{i-1} \xrightarrow{p_{u_i}^{update}} M_i$ for all i, $1 \leq i \leq n$. This sequence naturally defines the update sequence $\omega = e_1 e_2 ... e_n$ where $e_i = (u_i, R'(u_i))$ for all i, $1 \leq i \leq n$.

We know that once *Inject Packet* fires at a marking M_i, there exists a unique path in the Petri net, following exactly the routing defined by update sequence $e_1 e_2 \ldots e_i$. As σ is a winning control strategy, we know that φ holds in every marking on such a path. We shall argue by case analysis that the routing after applying the update sequence $e_1 e_2 \ldots e_i$ satisfies every path property $P \in \mathcal{P}^u$.

- $P \equiv loopfree(u)$. Then the path in the Petri net satisfies the marking property $\varphi_{loopfree(u)} \equiv \bigwedge_{u \in V} p_u^{visited} < 2$ which by the net construction implies that the path must be finite.
- $P \equiv reach(u,v)$. Then the path in the Petri net satisfies $\varphi_{reach(u,v)} \equiv \neg deadlock \vee p_v^{visited} \geq 1$. This implies that the net cannot deadlock before marking the place v, which gives that v must be necessarily reached as there is no infinite run due to Lemma 1. The property P hence holds.
- $P \equiv wp(u,v,w)$. Then every marking in the path satisfies $\varphi_{wp(u,v,w)} \equiv p_w^{visited} \geq 1 \vee p_v^{visited} = 0$, exactly formulating the requirement that p_v cannot be marked before p_w gets marked, again implying the property P. $\qquad\square$

4.3 Optimization for Reachability and Waypointing

It is a natural requirement that every transient routing in an update sequence should preserve at least the reachability between the source and the target node. We may also assume that once the target node is reached, any further routing from the target becomes undefined as the packet is considered delivered. In this case, the preservation of reachability also implies loop freedom. Finally, we may also allow to use multiple waypointing properties for different waypoints, as long as they are between the source and the destination that are connected in every transient routing. Formally, we define a set of *reachability and waypointing properties* for a given source $u \in V$ and target $v \in V$ as

$$\mathcal{P}^{(u,v)} = \{reach(u,v)\} \cup \{wp(u,v,w) \mid w \in W\}$$

where $W \subseteq V$ is a given set of waypoints. For this restricted set of path properties, we can notice that the construction of the Petri game solving the update synthesis problem can be optimized as follows.

Let $\mathcal{U} = (G, R, R', \mathcal{P}^{(u,v)})$ be an update synthesis problem with the set of reachability and waypointing properties $\mathcal{P}^{(u,v)}$ such that $G = (V, E)$ is the underlying network. We partition all relevant switches $u \in V$ where $R(u) \neq R'(u)$ into three categories:

- $V^{init} = \{u \in V \mid R(u) \neq R'(u), R(u) \text{ is undefined } \}$,
- $V^{final} = \{u \in V \mid R(u) \neq R'(u), R'(u) \text{ is undefined } \}$, and
- $V^{both} = \{u \in V \mid R(u) \neq R'(u), \text{ both } R(u) \text{ and } R'(u) \text{ are defined } \}$.

We shall now observe that if there is a correct update sequence transforming the routing R into R' while preserving $\mathcal{P}^{(u,v)}$ then there is also one where all switches from V^{init} are placed (in arbitrary order) at the beginning of the update sequence and all switches from V^{final} can be (again in arbitrary order) updated at the very end of the update sequence.

Lemma 4. *Let ω be a correct update sequence for the update synthesis problem $\mathcal{U} = (G, R, R', \mathcal{P}^{(u,v)})$ with reachability and waypointing property set $\mathcal{P}^{(u,v)}$. Let ω' be a subsequence of ω containing only updates of the form (u, u') where $u \in V^{both}$. Let $\omega^{init} = (v_1, R'(v_1)) \circ \ldots \circ (v_k, R'(v_k))$ be a sequence of updates of all nodes from the set $V^{init} = \{v_1, \ldots, v_k\}$. Let $\omega^{final} = (u_1, undefined) \circ \ldots \circ (u_\ell, undefined)$ be a sequence of updates of all nodes from the set $V^{final} = \{u_1, \ldots, u_\ell\}$. Then $\omega^{init} \circ \omega' \circ \omega^{final}$ is also a correct update sequence.*

Proof. Let ω be a correct update sequence such that $\omega = \omega_1 \circ (x, x') \circ \omega_2$ where either (i) $x \in V^{init}$ meaning that $R(x)$ is undefined, or (ii) $x \in V^{final}$ meaning that $R'(x)$ is undefined. We want to show that in case (i) $(x, x') \circ \omega_1 \circ \omega_2$ and in case (ii) $\omega_1 \circ \omega_2 \circ (x, x')$ is also a correct update sequence. These two facts imply the statement in the lemma.

In case (i), we know that for any prefix of ω_1, the corresponding routing path always connects u and v a hence it cannot pass thought the node x for which the next hop is undefined. Hence moving (x, x') to the beginning of the update sequence does not change the routing path and after the sequence of updates $(x, x') \circ \omega_1$ and $\omega_1 \circ (x, x')$ we arrive to the identical switch configuration.

In case (ii), we know that the reachability between u and v is preserved all the time during the update sequence ω. As the update $(x, undefined)$ creates a blackhole, after the sequence ω_1 is applied, the switch x cannot be part of the prefix of the routing path from u until v is reached. This implies that moving the update to the end does not influence the given set of path properties.   □

We can apply this lemma to improve the efficiency of our translation by creating a reduced Petri game $N'(\mathcal{U})$ where from the original net $N(\mathcal{U})$ we remove the whole switch component for every u where $u \in V^{init} \cup V^{final}$. This means that for the switches in V^{init}, the final routing will be enabled already from the start of the net execution and for the switches from V^{final} we never undefine the routing function. We can so conclude with the following theorem.

Theorem 2. *Let* $\mathcal{U} = (G, R, R', \mathcal{P}^{(u,v)})$ *be an update synthesis problem with reachability and waypointing properties* $\mathcal{P}^{(u,v)}$. *Let* $N'(\mathcal{U})$ *be the reduced Petri game with the initial marking* M_0 *defined above and* φ *the safety objective constructed from* $\mathcal{P}^{(u,v)}$. *Then* $\mathcal{U}$ *has a solution iff* $M_0 \models control : AG \; \varphi$ *in* $N'(\mathcal{U})$.

In Fig. 7 we show that Theorem 2 does not hold e.g. for the property *loopfree*(v_1) alone. As usual, the initial routing is in black solid lines and the final routing in dashed red lines. Clearly, $(v_1, v_3) \circ (v_3, v_4) \circ (v_4, v_2) \circ (v_2, undefined)$ is a correct update sequence transforming the initial routing to the final one. However, even though the final routing for the node v_4 is undefined, it is not possible to move this update to the beginning of the update sequence as updating first the node v_4 creates a forwarding loop and hence breaks the *loopfree*(v_1) property.

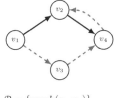

$$\mathcal{P} = \{reach(v_1, v_4)\}$$

Fig. 7. Counter example

5   Implementation and Experiments

We implemented a prototype tool in Python that translates the update synthesis problem into Petri game. Our tool accepts a JSON file with the description of the network topology, initial and final routing as well as the list of required security policies: reachability, loop-freedom and (multiple) waypointing. The tool provides three types of output: (i) update synthesis problem definition in Net-Synth [21] input format, (ii) XML file with Petri game and a safety query to be opened in the GUI of TAPAAL model checker [6], and (iii) XML Petri game model file and query file to be used with the command-line engine `verifypn` (part of the TAPAAL framework).

The game engine `verifypn` is based on the algorithms presented in [13,15], utilizing PTries for efficient state-storage [14], and we designed a new state-space exploration strategy in order to speedup the game synthesis algorithm. The engine `verifypn` is further extended to output the synthetized game strategy from which the update sequence can be derived. In order to experiment with network topologies in the standard `gml` format as used e.g. in the Topology Zoo dataset [16], our tool also facilitates the generation of update synthesis problems (in the TAPAAL and NetSynths formats) directly from the `gml` input files. We evaluate the performance of our tool (using the engine `verifypn` from the TAPAAL model checker) on both synthetic and real-world ISP topologies.

5.1   Synthetic Network Topologies

The synthetic topologies presented in Fig. 8 define scalable network update problems that model two extreme situations where the initial and final routing paths are either *disjoint* or fully *dependent*, as well as a third, more realistic, scenario where the two routing paths *share* a number of waypoints, while still using independent routers in between.

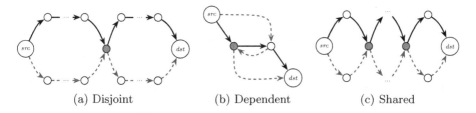

(a) Disjoint   (b) Dependent   (c) Shared

Fig. 8. Synthetic network topologies; initial routing in black solid lines and final routing in red dashed lines (Color figure online)

The disjoint network template from Fig. 8a follows the structure used in NetSynth benchmarks from [21]. The size of the update problem is scaled by increasing the lengths of the disjoint paths. The dependent network type in Fig. 8b aims to minimize the number of correct update sequences as many of the possible update sequences either create a loop or avoid the waypoint drawn in gray. The problem is scaled by sequentially concatenating (repeating) the same structure number of times. Finally, the shared topology from Fig. 8c combines a number of shared nodes that are connected by short disjoint paths of length two. The scaling is achieved by repeating the depicted pattern several times. The verified properties in all synthetic networks are $reach(src, dst)$ in conjunction with $wp(src, dst, w)$ where w is a single selected waypoint that is shared on the initial and final routing path (for disjoint topology there is exactly one such node drawn in gray). For the case of dependent topologies, we also study the variant with multiple waypoint properties.

5.2   Topology Zoo Benchmark

The topology Zoo database [16] contains 261 network topologies (with up to 700 nodes) from real internet service providers. In order to achieve additional scaling and larger instances, we combine existing topologies by further nesting/concatenating them. To create realistic initial and final routing paths, we emulate the standard protocols like OSPF [22] and ECMP [11] that are based on routing along the shortest paths in the weighted network (the weights are typically manually assigned by network operators). For each topology, we compute the diameter of the graph in order to identify the source src and destination dst nodes that maximize the smallest number of hops between them (for large network configurations, only a random subset of nodes are chosen in order to limit the computational effort for generating the test-cases). We randomly assign the weights up to 5 to every edge and set the initial routing path as one of the shortest paths between src and dst. We then increase the weights on the initial path that enforces a change in the set of shortest paths between src and dst in order to determine an alternative final routing path. The verified properties are $reach(src, dst)$ in conjunction with $wp(src, dst, w)$ where w is a randomly chosen waypoint that is shared by both the initial and final path (routing paths with no common waypoint are discarded).

5.3   Results

We compare the performance of our tool translating the update synthesis problem into a Petri game and using the TAPAAL engine as the backend (this tool chain is referred to as TAPAAL in the plots) against the state-of-the-art update synthesis tool NetSynth [21]. The results are presented in the form of *cactus plots* where the synthesis times used by each tool are (independently of each other) sorted in nondecreasing sequence and plotted on the x-axis, while the y-axis shows the concrete runtime for each instance. These graphs do not provide instance-to-instance runtime comparison but instead give an overall picture of the tools performances. All of our experiments are conducted using the Linux 5.8.0 kernel running on AMD EPYC 7551 processors with hyperthreading disabled and limited to 7 GB of memory (the memory limit was though never exceeded). The tool source code as well as the experimental setup that allows us to rerun all experiments is available in [7].

Figure 9a shows the cactus plot for the disjoint network experiment. While NetSynth uses 57 s to solve the largest instance with 1000 nodes, our tool with TAPAAL backend uses only 0.15 s (contributed mainly to the frequent applicability of Lemma 4). For the case of dependent networks in Fig. 9b with a single waypoint, the NetSynth incremental algorithm with build-in loop detection check outperforms TAPAAL, as many update sequence candidates create forwarding loops that TAPAAL is less efficient to detect. As a result, we need almost 386 s to solve the largest instance while NetSynth can synthetise the update sequence in about 16 s. However, once we require that every 10th node must be a waypoint, the relative performance of TAPAAL improves as seen in Fig. 9c and once every 5th node is set as a waypoint, we already outperform NetSynth as shown in Fig. 9d. This demonstrates that our approach scales better with increasing complexity of the required path properties. Moreover, our performance on the more realistic shared network in Fig. 9e is several orders of magnitude faster than NetSynth and NetSynth only solves 21 instances within the 1000 s timeout, while we need less than 37 s to solve even the largest instance.

Finally, Fig. 9f shows the performance on the dataset of existing networks emulating a realistic network operators behaviour. Here the routing paths are computed via a shortest path algorithm. We observe that our approach is in the middle case 8.9 times faster than NetSynth. We manage to solve the majority (933 instances) of problems in less than 1 s, while NetSynth solved only 689 instances within 1 s. We also remark that NetSynth is unable to solve 42 problems within the time limit of 100 s while TAPAAL manages to solve all but the 12 hardest instances (not surprising as already deciding the existence of a correct update sequence is an NP-complete problem [18]). In particular, we notice that NetSynth is noticeably slower at providing answers for problems where no update sequence exists.

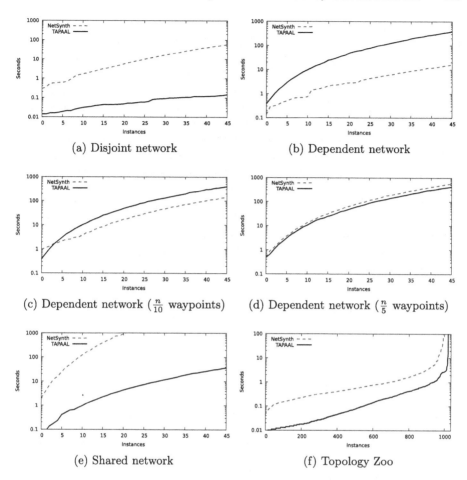

Fig. 9. Cactus plots where x-axis shows the increasing problem instances and y-axis depicts (on logarithmic scale) the synthesis time; n is the number of nodes

6   Conclusion

We presented a fully automatic approach for synthetising correct-by-construction update sequences in programmable networks. The obtained update sequences are guaranteed to satisfy a number of network policies like preservation of reachability, loop-freedom and additional waypointing requirements. Our approach is based on reducing the problem to Petri games and employing the generic model checker TAPAAL for solving the game synthesis problem. The experiments demonstrate that our method is significantly outperforming the state-of-the-art tool NetSynth, except for the case of fully dependent networks where both the initial and final routings share every single node in the network. However, this scenario is unlikely to appear during the operation of real networks as the packets are commonly routed along the shortest paths in the network.

Indeed, the experiments on over one thousand of real network topologies, using the shortest paths routing algorithms like OSPF and ECMP, document that we are consistently (in median 8.9 times) faster in synthetising the correct update sequences compared to NetSynth. In particular, we are able to solve majority of the realistic update synthesis problems in less than 1 s, which is important in nowadays dependable networks that must be promptly updated with increasing frequencies in order to react e.g. to sudden changes in the traffic.

The Petri net model and the generated update sequence is well suited for a further integration with the tool Latte [4], a plug-in to TAPAAL that extends the model with timing features and enables further reduction of the duration of network updates. In the future work, we plan to directly integrate the two tools into a single tool chain in order to reduce the overhead from parsing/exchanging of different file formats. We will also consider adding more network update policies like service chaining (where a given sequence of switches must be visited in a prescribed order); we expect that this will require only smaller modifications to our reduction.

Acknowledgements. We thank to Anders Mariegaard for his help with setting up NetSynth. This work received a support from the DFF project QASNET.

References

1. Amiri S.A., Dudycz, S., Schmid, S., Wiederrecht, S.: Congestion-free rerouting of flows on DAGs. In: ICALP 2018), volume 107 of Leibniz International Proceedings in Informatics (LIPIcs), pp. 143:1–143:13. Dagstuhl (2018)
2. Benzekki, K., El Fergougui, A., Elbelrhiti Elalaoui, A.: Software-defined networking (SDN): a survey. Secur. Comm. Netw. **9**(18), 5803–5833 (2016)
3. Brandt, S., Förster, K., Wattenhofer, R.: On consistent migration of flows in SDNs. In: INFOCOM 2016, pp. 1–9. IEEE (2016)
4. Christesen, N., Glavind, M., Schmid, S., Srba, J.: Latte: improving the latency of transiently consistent network update schedules. In: IFIP PERFORMANCE 2020, vol. 48, no. 3 of Performance Evaluation Review, pp. 14–26. ACM (2020)
5. Cimatti, A., et al.: NuSMV 2: an opensource tool for symbolic model checking. In: Brinksma, E., Larsen, K.G. (eds.) CAV 2002. LNCS, vol. 2404, pp. 359–364. Springer, Heidelberg (2002). https://doi.org/10.1007/3-540-45657-0_29
6. David, A., Jacobsen, L., Jacobsen, M., Jørgensen, K.Y., Møller, M.H., Srba, J.: TAPAAL 2.0: integrated development environment for timed-arc petri nets. In: Flanagan, C., König, B. (eds.) TACAS 2012. LNCS, vol. 7214, pp. 492–497. Springer, Heidelberg (2012). https://doi.org/10.1007/978-3-642-28756-5_36
7. Didriksen, M., et al.: Artefact for: Automatic Synthesis of Transiently Correct Network Updates via Petri Games (2021). https://doi.org/10.5281/zenodo.4497000
8. Finkbeiner, B., Gieseking, M., Hecking-Harbusch, J., Olderog, E.-R.: Model checking data flows in concurrent network updates. In: Chen, Y.-F., Cheng, C.-H., Esparza, J. (eds.) ATVA 2019. LNCS, vol. 11781, pp. 515–533. Springer, Cham (2019). https://doi.org/10.1007/978-3-030-31784-3_30
9. Finkbeiner, B., Gieseking, M., Hecking-Harbusch, J., Olderog, E.-R.: ADAMMC: a model checker for petri nets with transits against flow-LTL. In: Lahiri, S.K., Wang, C. (eds.) CAV 2020. LNCS, vol. 12225, pp. 64–76. Springer, Cham (2020). https://doi.org/10.1007/978-3-030-53291-8_5

10. Foerster, K., Schmid, S., Vissicchio, S.: Survey of consistent software-defined network updates. IEEE Commun. Surv. Tutorials **21**(2), 1435–1461 (2019)

11. Hopps, C., et al.: Analysis of an equal-cost multi-path algorithm. Technical report, RFC 2992, November 2000

12. Jensen, J.F., Nielsen, T., Oestergaard, L.K., Srba, J.: TAPAAL and reachability analysis of P/T nets. Trans. Petri Nets Other Mod. Concurrency (ToPNoC) **9930**, 307–318 (2016)

13. Jensen, P.G., Larsen, K.G., Srba, J.: Real-time strategy synthesis for timed-arc petri net games via discretization. In: Bošnački, D., Wijs, A. (eds.) SPIN 2016. LNCS, vol. 9641, pp. 129–146. Springer, Cham (2016). https://doi.org/10.1007/978-3-319-32582-8_9

14. Jensen, P.G., Larsen, K.G., Srba, J.: Ptrie: data structure for compressing and storing sets via prefix sharing. In: ICTAC 2017, vol. 10580 of LNCS, pp. 248–265. Springer (2017)

15. Jensen, P.G., Larsen, K.G., Srba, J.: Discrete and continuous strategies for timed-arc Petri net games. Int. J. Softw. Tools Technol. Transf. **20**(5), 529–546 (2017). https://doi.org/10.1007/s10009-017-0473-2

16. Knight, S., Nguyen, H.X., Falkner, N., Bowden, R., Roughan, M.: The internet topology Zoo. IEEE J. Select. Areas Comm. **29**(9), 1765–1775 (2011)

17. Liu, H.H., Wu, X., Zhang, M., Yuan, L., Wattenhofer, R., Maltz, D.: Zupdate: updating data center networks with zero loss. SIGCOMM Comput. Commun. Rev. **43**(4), 411–422 (2013)

18. Ludwig, A., Dudycz, S., Rost, M., Schmid, S.: Transiently secure network updates. In: ACM SIGMETRICS, pp. 273–284. ACM (2016)

19. Ludwig, A., Marcinkowski, J., Schmid, S.: Scheduling loop-free network updates: it's good to relax! In: PODC 2015, pp. 13–22. ACM (2015)

20. Mahajan, R., Wattenhofer, R.: On consistent updates in software defined networks. HotNets-XII, New York, NY, USA. ACM (2013)

21. McClurg, J., Hojjat, H., Černy, P., Foster, N.: Efficient synthesis of network updates. ACM Sigplan Not. **50**(6), 196–207 (2015)

22. Moy, J.: RFC2328: OSPF version 2 (1998). https://tools.ietf.org/html/rfc2328

23. Murata, T.: Petri nets: properties, analysis and applications. Proc. IEEE **77**(4), 541–580 (1989)

24. Reitblatt, M., Foster, N., Rexford, J., Schlesinger, C., Walker, D.: Abstractions for network update. In: ACM SIGCOMM 2012, pp. 323–334. ACM (2012)

25. Vissicchio, S., Cittadini, L.: FLIP the (flow) table: fast lightweight policy-preserving SDN updates. In: INFOCOM 2016, pp. 1–9. IEEE (2016)

Verification

Computing Parameterized Invariants
of Parameterized Petri Nets

Javier Esparza, Mikhail Raskin, and Christoph Welzel[(✉)]

Technical University of Munich, Munich, Germany
{esparza,raskin,welzel}@in.tum.de

Abstract. A fundamental advantage of Petri net models is the possibility to automatically compute useful system invariants from the syntax of the net. Classical techniques used for this are place invariants, P-components, siphons or traps. Recently, Bozga et al. have presented a novel technique for the *parameterized* verification of safety properties of systems with a ring or array architecture. They show that the statement "for every instance of the parameterized Petri net, all markings satisfying the linear invariants associated to all the P-components, siphons and traps of the instance are safe" can be encoded in WS1S and checked using tools like MONA. However, while the technique certifies that this infinite set of linear invariants extracted from P-components, siphons or traps are strong enough to prove safety, it does not return an explanation of this fact understandable by humans. We present a CEGAR loop that constructs a *finite* set of *parameterized* P-components, siphons or traps, whose infinitely many instances are strong enough to prove safety. For this we design parameterization procedures for different architectures.

1 Introduction

A fundamental advantage of Petri net system models is the possibility to automatically extract useful system invariants from the syntax of the net at low computational cost. Classical techniques used for this purpose are place invariants, P-components, siphons or traps [19,40,41]. All of them are syntactic objects that can be computed using linear algebra or boolean logic, and from which semantic linear invariants can be extracted. For example, from the fact that a set of places Q is an initially marked trap of the net one extracts the linear invariant $\sum_{p \in Q} M(Q) \geq 1$, which is satisfied for every reachable marking M. This information can be used to prove safety properties: Given a set S of safe markings, if every marking satisfying the invariants extracted from a set of objects is safe, then all reachable markings are safe.

Classical net invariants have been very successfully used in the verification of single systems [10,12,26], or as complement to state-space exploration [45]. Recently, an extension of this idea to the *parameterized* verification of safety properties of systems with a ring or array architecture has been presented in [14,15]. The parameterized verification problem asks whether a system composed of n processes is safe for every $n \geq 2$ [4,11,24]. Bozga *et al.* show in [14,15] that the statement

© Springer Nature Switzerland AG 2021
D. Buchs and J. Carmona (Eds.): PETRI NETS 2021, LNCS 12734, pp. 141–163, 2021.
https://doi.org/10.1007/978-3-030-76983-3_8

"For every instance of the parameterized system, all markings satisfying
the linear invariants associated to all the P-components, siphons and traps
of the corresponding Petri net is safe"

can be encoded in Weak Second-order Logic With One Successor (WS1S), or its
analogous WS2S for two successors. This means that the statement holds iff its
formula encoding is valid. This problem is decidable, and highly optimized tools
exist for it, like MONA [36,43]. The method of [14] is not complete (i.e., there
are safe systems for which the invariants derived from P-components, siphons
and traps are not strong enough to prove safety), but it succeeds for a remarkable set of examples. Further, incompleteness is inherent to every algorithmic
method, since safety of parameterized nets is undecidable even if processes only
manipulate data from a bounded domain [5,11].

While the technique of [14,15] is able to prove interesting properties of numerous systems, it does not yet provide an explanation of why the property holds.
Indeed, when the technique succeeds for a given parameterized Petri net, the
user only knows that the set of all invariants deduced from siphons, traps, and
P-components together are strong enough to prove safety. However, the technique
does not return a minimal set of these invariants. Moreover, since the parameterized Petri net has infinitely many instances, such a set contains infinitely many
invariants. In this paper we show how to overcome this obstacle. We present a
technique that automatically computes a *finite* set of *parameterized invariants*,
readable by humans. This is achieved by lifting a CEGAR (counterexample-
guided abstraction refinement) loop, introduced in [27] and further developed
in [12,26,28], to the parameterized case. Each iteration of the loop of [26,27]
first computes a counterexample, i.e., a marking that violates the desired safety
property but satisfies all invariants computed so far, and then computes a P-
component, siphon, or trap showing that the marking is not reachable. If no
counterexample exists the property is established, and if no P-component, siphon
or trap can be found the method fails. The technique is implemented on top of
an SMT-solver, which receives as input a linear constraint describing the set of
safe markings, and iteratively computes the set of linear invariants derived from
P-components, siphons, and traps.

If we naively lift the CEGAR loop to the parameterized case, the loop never
terminates. Indeed, since the loop computes one new invariant per iteration,
and infinitely many invariants are needed to prove correctness of all instances,
termination is not possible. So we need a procedure to extract from one single invariant for one instance a *parameterized* invariant, i.e., an infinite set of
invariants for all instances, finitely represented as a WS1S-formula. We present
a *semi-automatic* and an *automatic approach*. In the semi-automatic approach
the user guesses the parameterized invariant, and automatically checks it, using
the WS1S-checker. The automatic approach does not need user interaction, but
only works for systems with symmetric structure. We provide automatic procedures for rings and barrier crowds, a class of systems closely related to broadcast

protocols. We present experimental results on a number of systems. The semi-automatic approach requires more human interaction, but produces smaller sets of invariants.

Related Work. The parameterized verification problem has been extensively studied for systems whose associated transition systems are well-structured [1,32,35] (see e.g. [4] for a survey). In this case the verification problem reduces to a coverability problem, for which different algorithms exist [13,31,34,42]; the marking equation (which is roughly equivalent to place invariants) have also been applied [6]. However, the transition systems of parametric rings and arrays are typically not well-structured.

Parameterized verification of ring and array systems has also been studied in a number of papers. Three popular techniques are regular model checking (see e.g. [2,3,39]), abstraction [8,9], and automata learning [17]. All of them apply symbolic state-space exploration to try to compute a finite automaton recognizing the set of reachable markings of all instances, or an abstraction thereof. Our technique avoids any state-space exploration. Also, symbolic state-space exploration techniques are not geared towards providing explanations. Indeed, while the set of reachable markings of all instances is the strongest invariant of the system, it is also one single monolithic invariant, typically difficult to interpret by human users. Our CEGAR loop aims at finding a *collection* of invariants, each of them simple and interpretable.

Many works in parameterized follow the cut-off approach, where one manually proves a *cut-off* bound $c \geq 2$ such that correctness for at most c processes implies correctness for any number of processes (see e.g. [7,16,21,22,37], and [11] for a survey). It then suffices to prove the property for systems of up to c processes, which can be done using finite-state model checking techniques. Compared to this technique, ours is fully automatic.

Full Version. Due to space restrictions we refer to the full version [29] for proofs and extended explanations.

2 Preliminaries

WS1S. Formulas of WS1S over first-order variables $x, y, \ldots$ and second-order variables $X, Y, \ldots$ have the following syntax:

$$t := x \mid 0 \mid succ(t) \qquad \text{(terms)}$$
$$\phi := t_1 \leq t_2 \mid x \in X \mid \phi_1 \wedge \phi_2 \mid \neg \phi_1 \mid \exists x \colon \phi \mid \exists X \colon \phi \quad \text{(formulas)}$$

An interpretation assigns elements of $\mathbb{N}_0 = \{0, 1, 2, 3, \ldots\}$ to first order variables and *finite* subsets of $\mathbb{N}_0$ to second-order variables. Given an interpretation, the semantics that assigns numbers to terms and truth values to formulas is defined in the usual way.

We extend the syntax with constants $0, 1, 2, 3, \ldots$, and terms of the form $\boldsymbol{x}+c$ with $c \in \mathbb{N}_0$. Further, a term $\boldsymbol{x} \oplus_n 1$ in a formula φ stands for

$$(\boldsymbol{x} + 1 < \boldsymbol{n} \wedge \varphi[\boldsymbol{x} \oplus_n 1 \leftarrow \boldsymbol{x} + 1]) \vee (\boldsymbol{n} = \boldsymbol{x} + 1 \wedge \varphi[\boldsymbol{x} \oplus_n 1 \leftarrow 0])$$

where $\varphi[t \leftarrow t']$ denotes the result of substituting t' for t in φ. The terms $\boldsymbol{x} \oplus_n c$ for every $1 \leq c$ are defined similarly. We let $\varphi(\boldsymbol{x}_1, \ldots, \boldsymbol{x}_\ell, \boldsymbol{X}_1, \ldots, \boldsymbol{X}_k)$ denote that φ uses at most $\boldsymbol{x}_1, \ldots, \boldsymbol{x}_\ell$ and $\boldsymbol{X}_1, \ldots, \boldsymbol{X}_k$ as free first-order resp. second-order variables. Finally, we also make liberal use of the following macros:

$\boldsymbol{X} = \emptyset$		$\forall \boldsymbol{x}: \neg(\boldsymbol{x} \in \boldsymbol{X})$
$\boldsymbol{X} = \{\boldsymbol{x}\}$		$\boldsymbol{x} \in \boldsymbol{X} \wedge \forall \boldsymbol{y}: \boldsymbol{y} \in \boldsymbol{X} \rightarrow \boldsymbol{y} = \boldsymbol{x}$
$\boldsymbol{X} = [\boldsymbol{n}]$		$\forall \boldsymbol{x}: \boldsymbol{x} \in \boldsymbol{X} \leftrightarrow \boldsymbol{x} < \boldsymbol{n}$
$\boldsymbol{X} \cap \boldsymbol{Y} = \emptyset$	stands for	$\forall \boldsymbol{x}: \neg(\boldsymbol{x} \in \boldsymbol{X} \vee \boldsymbol{x} \in \boldsymbol{Y})$
$|\boldsymbol{X}| = 1$		$\exists \boldsymbol{x}: \boldsymbol{X} = \{\boldsymbol{x}\}$
$|\boldsymbol{X}| \leq 1$		$\boldsymbol{X} = \emptyset \vee |\boldsymbol{X}| = 1$
$\boldsymbol{X} = \overline{\boldsymbol{Y}}$		$\forall \boldsymbol{x}: \boldsymbol{x} \in \boldsymbol{X} \leftrightarrow \neg(\boldsymbol{x} \in \boldsymbol{Y})$
$\boldsymbol{Y} = \boldsymbol{X} \oplus_n 1$		$\forall \boldsymbol{x}: \boldsymbol{x} \oplus_n 1 \in \boldsymbol{Y} \leftrightarrow \boldsymbol{x} \in \boldsymbol{X}$

Petri Nets. We use a presentation of Petri nets equivalent to but slightly different from the standard one. A *net* is a pair $\langle P, T \rangle$ where P is a nonempty, finite set of *places* and $T \subseteq 2^P \times 2^P$ is a set of *transitions*. Given a transition $t = \langle P_1, P_2 \rangle$, we call P_1 the *preset* and *postset* of t, respectively. We also denote P_1 by ${}^\bullet t$ and P_2 by $t^\bullet$. Given a place p, we denote by ${}^\bullet p$ and $p^\bullet$ the sets of transitions $\langle P_1, P_2 \rangle$ such that $p \in P_2$ and $p \in P_1$, respectively. Given a set X of places or transitions, we let ${}^\bullet X := \bigcup_{x \in X} {}^\bullet x$ and $X^\bullet := \bigcup_{x \in X} x^\bullet$.

A *marking* of $N = \langle P, T \rangle$ is a function $M \colon P \to \mathbb{N}$. A Petri net is a pair $\langle N, M \rangle$, where N is a net and M is the *initial marking* of N. A transition $t = \langle P_1, P_2 \rangle$ is enabled at a marking M if $M(p) \geq 1$ for every $p \in P_1$. If t is enabled at M then it can *fire*, leading to the marking M' given by $M'(p) = M(p) + 1$ for every $p \in P_2 \setminus P_1$, $M'(p) = M(p) - 1$ for every $p \in P_1 \setminus P_2$, and $M'(p) = M(p)$ otherwise. We write $M \xrightarrow{t} M'$, and $M \xrightarrow{\sigma} M'$ for a finite sequence $\sigma = t_1 t_2 \ldots t_n$ if there are markings $M_1, \ldots, M_n$ such that $M \xrightarrow{t_1} M_1 \xrightarrow{t_2} \cdots M_{n-1} \xrightarrow{t_n} M'$. M' is reachable from M if $M \xrightarrow{\sigma} M'$ for some sequence σ.

A marking M is *1-bounded* if $M(p) \leq 1$ for every place p. A Petri net is *1-bounded* if every marking reachable from the initial marking is 1-bounded. A 1-bounded marking M of a Petri net is also defined by the set of marked places; i.e., $\langle M \rangle = \{p \in P \colon M(p) = 1\}$.

3   Parameterized Petri Nets

Definition 1 (Parameterized Nets). *A parameterized net is a pair $\mathcal{N} = \langle \mathcal{P}, Tr \rangle$, where $\mathcal{P}$ is a finite set of place names and $Tr(\boldsymbol{n}, \mathcal{X}, \mathcal{Y})$ is a WS1S-formula over one first-order variable $\boldsymbol{n}$ which represents the considered size of* *the instance and two tuples* $\mathcal{X}$ *and* $\mathcal{Y}$ *of second-order variables for each place name of* $\mathcal{P}$; *i.e., for a fixed enumeration* $p_1, \ldots, p_k$ *of the elements of* $\mathcal{P}$ *we get* $\mathcal{X} = \langle \mathcal{X}_{p_i} \rangle_{i=1}^k$ *and* $\mathcal{Y} = \langle \mathcal{Y}_{p_i} \rangle_{i=1}^k$. *We call such tuples of variables* placeset *variables.*

Let $[n] = \{0, \ldots, n-1\}$. A parameterized net $\mathcal{N}$ induces a net $\mathcal{N}(n) = \langle P_n, T_n \rangle$ for every $n \geq 1$, where $P_n = \mathcal{P} \times [n]$ (i.e., P_n consists of n copies of $\mathcal{P}$), and T_n contains a transition $\langle P_1, P_2 \rangle$ for every pair $P_1, P_2 \subseteq P_n$ of sets of places such that "$Tr(n, P_1, P_2)$" holds. More formally, this means that $\mu \models Tr$ for an interpretation μ s.t. $\mu(n) = n$, $\mu(\mathcal{X}_p) = \{i \in [n] : \langle p, i \rangle \in P_1\}$, and $\mu(\mathcal{Y}_p) = \{i \in [n] : \langle p, i \rangle \in P_2\}$ for all $p \in \mathcal{P}$. Therefore, the intended meaning of $Tr(n, \mathcal{X}, \mathcal{Y})$ is "the pair $\langle \mathcal{X}, \mathcal{Y} \rangle$ of placesets is (the preset and postset of) a transition if the size is n". We say that $\mathcal{N}(n)$ is an *instance* of $\mathcal{N}$.

In the following we use $\langle p, i \rangle$ and $p(i)$ as equivalent notations for the elements of $P_n = \mathcal{P} \times [n]$.

Example 1. We consider a version of the dining philosophers. Philosophers and forks are numbered 0, 1, ..., $n-1$. For every $i > 0$ the i-th philosopher first grabs the i-th and then the $(i \oplus_n 1)$-th fork, where $\oplus_n$ denotes addition modulo n. Philosopher 0 proceeds the other way round: she first grabs fork 1, and then fork 0. After eating all philosophers return their forks in one single step. We formalize this in the following parameterized net $\mathcal{N} = \langle \mathcal{P}, Tr \rangle$:

- $\mathcal{P} = \{\text{think}, \text{wait}, \text{eat}, \text{free}, \text{taken}\}$. Intuitively, $\{\text{think}(i), \text{wait}(i), \text{eat}(i)\}$ are the states of the i-th philosopher, and $\{\text{free}(i), \text{taken}(i)\}$ the states of the i-th fork.
- $Tr(n, \mathcal{X}, \mathcal{Y}) = \text{GrabFirst} \vee \text{GrabSecond} \vee \text{Release}$. We only present the formula for GrabFirst, the complete description can be found in the appendix.

$$\text{GrabFirst} := \begin{pmatrix} \exists x \,.\, 1 \leq x < n \wedge (\mathcal{X}_{\text{think}} = \mathcal{X}_{\text{free}} = \mathcal{Y}_{\text{wait}} = \mathcal{Y}_{\text{taken}} = \{x\}) \\ \wedge (\mathcal{X}_{\text{wait}} = \mathcal{X}_{\text{eat}} = \mathcal{X}_{\text{taken}} = \emptyset) \\ \wedge (\mathcal{Y}_{\text{think}} = \mathcal{Y}_{\text{eat}} = \mathcal{Y}_{\text{free}} = \emptyset) \end{pmatrix} \\ \vee \\ \begin{pmatrix} (\mathcal{X}_{\text{think}} = \mathcal{Y}_{\text{taken}} = \{0\}) \wedge (\mathcal{X}_{\text{free}} = \mathcal{Y}_{\text{wait}} = \{1\}) \\ \wedge (\mathcal{X}_{\text{wait}} = \mathcal{X}_{\text{eat}} = \mathcal{X}_{\text{taken}} = \emptyset) \\ \wedge (\mathcal{Y}_{\text{think}} = \mathcal{Y}_{\text{eat}} = \mathcal{Y}_{\text{free}} = \emptyset) \end{pmatrix}$$

Intuitively, the preset of GrabFirst is a philosopher in state think and her left (resp. right fork for philosopher 0) in state free; the postset puts the philosopher in state wait and the fork in state taken. The instance $\mathcal{N}(3)$ is shown in Fig. 1.

Parameterized Petri nets are parameterized nets with a WS1S-formula defining its initial markings:

Definition 2 (Parameterized Petri Nets). *A parameterized Petri net is a pair* $\langle \mathcal{N}, Initial \rangle$, *where* $\mathcal{N}$ *is a parameterized net, and* $Initial(n, \mathcal{M})$ *is a WS1S-formula over a first-order variable* n *and a placeset variable* $\mathcal{M}$.

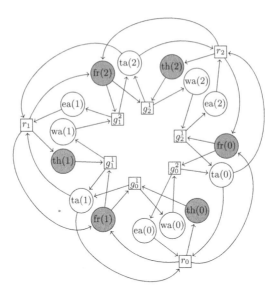

Fig. 1. $\mathcal{N}(3)$ for Example 1. Places which are colored green are initially marked w.r.t. *Initial*($\mathcal{X}$) from Example 2. Note the repeating structure for philosophers 1 and 2 while philosopher 0 grabs her forks in the opposite order. We abbreviate think(i) to th(i), and similarly with the other states.

A parameterized Petri net defines an infinite family of Petri nets. Loosely speaking, a Petri net $\langle N, M \rangle$ belongs to the family if N is an instance of $\mathcal{N}$, i.e., $N = \mathcal{N}(n)$ for some $n \geq 1$, and M is a 1-bounded marking of N satisfying *Initial*(n, M). For example, if $\mathcal{P} = \{p_1, p_2\}$, $n = 3$ and *Initial*($\{0, 1\}, \{0, 2\}$) holds, then the family contains a Petri net $\langle \mathcal{N}(3), M_3 \rangle$ such that M_3 is a 1-bounded marking with $\{M_3\} = \{p_1(0), p_1(1), p_2(0), p_2(2)\}$.

Example 2. The family of initial markings in which all philosophers think and all forks are free is modeled by:

$$Initial(\boldsymbol{n}, \mathcal{M}) := (\mathcal{M}_{\text{think}} = \mathcal{M}_{\text{free}} = [\boldsymbol{n}]) \wedge (\mathcal{M}_{\text{wait}} = \mathcal{M}_{\text{eat}} = \mathcal{M}_{\text{taken}} = \emptyset).$$

Example 3. Let us now model a simple version of the readers/writers system. A process can be idle, reading, or writing. An idle process can start to read if no other process is writing, and it can start to write if every other process is idle. We obtain the parameterized net $\mathcal{N} = \langle \mathcal{P}, Tr \rangle$, where

- $\mathcal{P} = \{\text{idle}, \text{rd}, \text{wr}, \text{not\_wr}\}$.
- $Tr(\boldsymbol{n}, \mathcal{X}, \mathcal{Y}) = \text{StartR} \vee \text{StopR} \vee \text{StartW} \vee \text{StopW}$. We give the formulae StartR and StartW, the other two being simpler.

$$\text{StartR} := \exists \boldsymbol{x} . \left(\begin{array}{c} (\mathcal{X}_{\text{idle}} = \{\boldsymbol{x}\} \wedge \mathcal{X}_{\text{not\_wr}} = \overline{\mathcal{X}_{\text{idle}}} \wedge (\mathcal{X}_{\text{rd}} = \mathcal{X}_{\text{wr}} = \emptyset)) \\ \wedge \mathcal{Y}_{\text{rd}} = \{\boldsymbol{x}\} \wedge \mathcal{Y}_{\text{not\_wr}} = \overline{\mathcal{X}_{\text{idle}}} \wedge (\mathcal{Y}_{\text{idle}} = \mathcal{Y}_{\text{wr}} = \emptyset) \end{array} \right)$$

$$\text{StartW} := \exists x \ . \ \begin{pmatrix} \mathcal{X}_{\text{idle}} = [n] \wedge \mathcal{X}_{\text{not\_wr}} = \{x\} \wedge (\mathcal{X}_{\text{rd}} = \mathcal{X}_{\text{wr}} = \emptyset) \\ \wedge \mathcal{Y}_{\text{idle}} = [n] \setminus \{x\} \wedge \mathcal{Y}_{\text{wr}} = \{x\} \wedge (\mathcal{Y}_{\text{rd}} = \mathcal{Y}_{\text{not\_wr}} = \emptyset) \end{pmatrix}$$

So the preset of a StartR transition is $\{\text{idle}(i), \text{not\_wr}(0), \ldots, \text{not\_wr}(n-1)\}$ for some i, and the postset is $\{\text{rd}(i), \text{not\_wr}(0), \ldots, \text{not\_wr}(n-1)\}$. The initial markings in which every process is initially idle are modeled by:

$$Initial(n, \mathcal{X}) := \mathcal{X}_{\text{idle}} = [n] \wedge \mathcal{X}_{\text{not\_wr}} = [n] \wedge (\mathcal{X}_{\text{rd}} = \mathcal{X}_{\text{wr}} = \emptyset)$$

Observe that in the dining philosophers transitions have presets and postsets of size 3, independently of the number of philosophers. On the contrary, in the readers and writers problems the transitions of $\mathcal{N}(n)$ have presets and postsets of size n. Intuitively, our formalism allows to model transitions involving all processes or, for example, all even processes. Observe also that in both cases the formula $Initial$ has exactly one model for every $n \geq 1$, but this is not required.

Proving Deadlock-Freedom for the Dining Philosophers. Let us now give a taste of what our paper achieves for Example 1. It is well known that this version of the dining philosophers is deadlock-free. However, finding a proof based on parameterized invariants of the systems is not so easy. Using the semi-automatic use of the approach we present we find the following five invariants to do so. The fully automatic analysis of this example gives ten properties of the system which collectively induce deadlock-freedom.

The first two invariants simply express that at every reachable marking M, and for every $0 \leq i \leq n-1$, the i-th philosopher is either thinking, waiting, or eating, and the i-th fork is either free or taken.

$$M(\text{think}(i)) + M(\text{wait}(i)) + M(\text{eat}(i)) = 1 \tag{1}$$

$$M(\text{free}(i)) + M(\text{taken}(i)) = 1. \tag{2}$$

The last three invariants provide the key insights. The last one holds for every $1 \leq i \leq n-2$:

$$M(\text{wait}(0)) + M(\text{eat}(0)) + M(\text{free}(1)) + M(\text{wait}(1)) + M(\text{eat}(1)) = 1 \tag{3}$$

$$M(\text{eat}(0)) + M(\text{free}(0)) + M(\text{eat}(n-1)) = 1 \tag{4}$$

$$M(\text{eat}(i)) + M(\text{eat}(i+1)) + M(\text{free}(i+1)) + M(\text{wait}(i+1)) = 1 \tag{5}$$

Let us sketch why (1)–(5) imply deadlock freedom. Let P_i denote the i-th philosopher and F_i the i-th fork. If P_0 is eating, then F_0 and F_1 are taken by (1)–(4), and there is no deadlock because P_0 can return them. The same holds if P_1 is eating by (1)–(3) and (5), or if any of $P_2, \ldots, P_{n-1}$ is eating by (1)–(2) and (5). If no philosopher eats, then by (1)–(3) and (5) either P_{i+1} is thinking and F_{i+1} is free for some $i \in \{1, \ldots, n-2\}$, or P_{i+1} is waiting for every $i \in \{1, \ldots, n-2\}$. In the first case P_{i+1} can grab F_{i+1}. In the second case P_{n-1} is waiting, and since F_0 is free by (1)–(2) and (4), it can grab F_0.

4   Checking 1-Boundedness

Our techniques work for parameterized Petri nets whose instances are 1-bounded. We present a technique that automatically checks 1-boundedness of all our examples. We say that a set of places Q of a Petri net $\langle N, M \rangle$, where $N = \langle P, T \rangle$, is

- *1-balanced* if for every transition $\langle P_1, P_2 \rangle \in T$ either $|P_1 \cap Q| = 1 = |P_2 \cap Q|$, or $|P_1 \cap Q| = 0 = |P_2 \cap Q|$, or $|P_1 \cap Q| \geq 2$.
- *1-bounded* at M if $M(Q) \leq 1$.

The following proposition is an immediate consequence of the definition:

Proposition 1. *If Q is a 1-balanced and 1-bounded set of places of $\langle N, M \rangle$, then $M'(Q) = M(Q)$ holds for every reachable marking M'.*

We abbreviate "1-bounded and 1-balanced set" to 1BB-set, and say that N is *covered* by 1BB-sets if every place belongs to some 1BB-set at initial marking M. By the proposition above, if N is covered by 1BB-sets at M, then $M'(p) \leq 1$ holds for every reachable marking M' and every place p, and so N is 1-bounded.

Given a parameterized Petri net $(\mathcal{N}, \textit{Initial})$, we can check if all instances are covered by 1BB-sets with the following formula:

$$1Bal(\boldsymbol{n}, \mathcal{X}) := \forall \mathcal{Y}, \mathcal{Z} : \textit{Tr}(\boldsymbol{n}, \mathcal{Y}, \mathcal{Z}) \rightarrow \quad (|\mathcal{X} \cap \mathcal{Y}| = 0 = |\mathcal{X} \cap \mathcal{Z}|) \vee$$
$$(|\mathcal{X} \cap \mathcal{Y}| = 1 = |\mathcal{X} \cap \mathcal{Z}|) \vee$$
$$(|\mathcal{X} \cap \mathcal{Y}| > 1)$$

$$1Bnd(\boldsymbol{n}, \mathcal{X}, \mathcal{M}) := |\mathcal{X} \cap \mathcal{M}| \leq 1$$
$$\textit{Cover} := \forall \boldsymbol{n}, \forall \mathcal{M} : \textit{Initial}(\boldsymbol{n}, \mathcal{M}) \rightarrow (\forall \boldsymbol{x} : \exists \mathcal{X} : \boldsymbol{x} \in \mathcal{X} \wedge$$
$$1Bal(\boldsymbol{n}, \mathcal{X}) \wedge$$
$$1Bnd(\boldsymbol{n}, \mathcal{X}, \mathcal{M}))$$

Observe that if Q is a 1BB-set then at every reachable marking exactly one of the places of Q is marked, with exactly one token. The sets of places corresponding to a philosopher, a fork, a reader, or a writer are 1BB-sets. Unsurprisingly, all our parameterized Petri net models are covered by 1BB-sets. Moreover, this can be automatically proved in a few seconds by checking the formula above. This gives us an automatic proof that all the Petri nets we consider are 1-bounded.

5   Checking Safety Properties

Let $\langle \mathcal{N}, \textit{Initial} \rangle$ be a parameterized Petri net, and let $\textit{Safe}(\boldsymbol{n}, \mathcal{M})$ be a WS1S-formula describing a set of "safe" markings of the instances of $\mathcal{N}$ (for example, "safe" could mean deadlock-free). It is easy to prove (using simulations of Turing machines by Petri nets like those of [23]) that the existence of some unsafe reachable marking in some instance of a given parameterized Petri net $\langle \mathcal{N}, \textit{Initial} \rangle$ is undecidable. In [14,15] we describe a semi-algorithm for the problem that derives from $\langle \mathcal{N}, \textit{Initial} \rangle$ a formula $\textit{PReach}(\boldsymbol{n}, \mathcal{M})$ describing a superset of the set of reachable markings of all instances, and checks that the formula

$$\textit{SafetyCheck} := \forall \boldsymbol{n} \forall \mathcal{M} : \textit{PReach}(\boldsymbol{n}, \mathcal{M}) \rightarrow \textit{Safe}(\boldsymbol{n}, \mathcal{M})$$

holds. We recall the main construction of [14,15], adapted and expanded.

1BB-Sets Again. Recall that if a marking M' of some instance $\langle N, M \rangle$ of a net $\langle \mathcal{N}, \mathit{Initial} \rangle$ is reachable from M, then $M'(Q) \leq 1$ holds for every 1BB-set of places Q of $\langle N, M \rangle$. So this latter property can be interpreted as a test for potential reachability: Only markings that pass the test can be reachable. We introduce a formula $\mathit{1BBTest}(\boldsymbol{n}, \mathcal{M}', \mathcal{M})$ expressing that $\mathcal{M}'$ passes the test with respect to $\mathcal{M}$ (i.e., $\mathcal{M}'$ might be reachable from $\mathcal{M}$).

$$\mathit{1BBTest}(\boldsymbol{n}, \mathcal{M}', \mathcal{M}) := \forall \mathcal{X} : \begin{pmatrix} \mathit{1Bal}(\boldsymbol{n}, \mathcal{X}) \\ \wedge \mathit{1Bnd}(\boldsymbol{n}, \mathcal{X}, \mathcal{M}) \end{pmatrix} \rightarrow \mathit{1Bnd}(\boldsymbol{n}, \mathcal{X}, \mathcal{M}')$$

Siphons and Traps. Let $\langle N, M \rangle$ be a Petri net with $N = \langle P, T \rangle$ and let $Q \subseteq P$ be a set of places. Q is a *trap* of N if ${}^\bullet Q \subseteq Q^\bullet$, and a *siphon* of N if $Q^\bullet \subseteq {}^\bullet Q$.

- If Q is a siphon and $M(Q) = 0$, then $M'(Q) = 0$ for all markings M' reachable from M.
- If Q is a trap and $M(Q) \geq 1$, then $M'(Q) \geq 1$ for all markings M' reachable from M.

If M' is reachable from M then it satisfies the following property: $M'(Q) \geq 1$ for every trap Q such that $M(Q) \geq 1$. A marking satisfying this property *passes the trap test* for $\langle N, M \rangle$. We construct a formula $\mathit{TrapTest}(\boldsymbol{n}, \mathcal{M})$ expressing that $\mathcal{M}$ passes the trap test for some instance of a parameterized Petri net. We first introduce a formula expressing that a set $\mathcal{X}$ of places is a trap.

$$\mathit{Trap}(\boldsymbol{n}, \mathcal{X}) := \forall \mathcal{Y}, \mathcal{Z} : (\mathit{Tr}(\boldsymbol{n}, \mathcal{Y}, \mathcal{Z}) \wedge \mathcal{X} \cap \mathcal{Y} \neq \emptyset) \rightarrow \mathcal{X} \cap \mathcal{Z} \neq \emptyset$$

Now we have:

$$\mathit{Marked}(\boldsymbol{n}, \mathcal{X}, \mathcal{M}) := \mathcal{X} \cap \mathcal{M} \neq \emptyset$$
$$\mathit{TrapTest}(\boldsymbol{n}, \mathcal{M}', \mathcal{M}) := \forall \mathcal{X} : \begin{pmatrix} \mathit{Trap}(\boldsymbol{n}, \mathcal{X}) \\ \wedge \mathit{Marked}(\boldsymbol{n}, \mathcal{X}, \mathcal{M}) \end{pmatrix} \rightarrow \mathit{Marked}(\boldsymbol{n}, \mathcal{X}, \mathcal{M}')$$

Similarly we obtain a formula for a siphon test:

$$\mathit{Empty}(\boldsymbol{n}, \mathcal{X}, \mathcal{M}) := \mathcal{X} \cap \mathcal{M} = \emptyset$$
$$\mathit{SiphTest}(\boldsymbol{n}, \mathcal{M}', \mathcal{M}) := \forall \mathcal{X} : \begin{pmatrix} \mathit{Siphon}(\boldsymbol{n}, \mathcal{X}) \\ \wedge \mathit{Empty}(\boldsymbol{n}, \mathcal{X}, \mathcal{M}) \end{pmatrix} \rightarrow \mathit{Empty}(\boldsymbol{n}, \mathcal{X}, \mathcal{M}')$$

We can now give the formula *PReach*:

$$\mathit{PReach}(\boldsymbol{n}, \mathcal{M}', \mathcal{M}) := \begin{pmatrix} \mathit{1BBTest}(\boldsymbol{n}, \mathcal{M}', \mathcal{M}) \\ \wedge \mathit{TrapTest}(\boldsymbol{n}, \mathcal{M}', \mathcal{M}) \\ \wedge \mathit{SiphTest}(\boldsymbol{n}, \mathcal{M}', \mathcal{M}) \end{pmatrix}$$
$$\mathit{PReach}(\boldsymbol{n}, \mathcal{M}') := \exists \mathcal{M} : \mathit{Initial}(\boldsymbol{n}, \mathcal{M}) \wedge \mathit{PReach}(\boldsymbol{n}, \mathcal{M}', \mathcal{M})$$

5.1   Automatic Computation of Parameterized Invariants

In [14] it was shown that many safety properties of parameterized Petri nets can be proved to hold *for all instances* by checking validity of the corresponding *PReach* formula. However, the technique does not return a set of invariants strong enough to prove the property. In this section we show how to overcome this problem. We design a CEGAR loop which, when successful, yields a finite set of *parameterized* invariants that imply the safety property being considered.

   We proceed as follows. In the first part of the section, we describe a CEGAR loop for the non-parameterized case. The input to the procedure is a parameterized Petri net $\langle \mathcal{N}, \mathit{Initial} \rangle$ and a number n such that all reachable markings of all instances $\mathcal{N}(1), \ldots, \mathcal{N}(n)$ are safe. The output is a set of invariants of $\mathcal{N}(1), \ldots, \mathcal{N}(n)$, derived from balanced sets, siphons, and traps, which are strong enough to prove safety. Since the set of all balanced placesets, siphons, and traps of these instances is finite, the procedure is guaranteed to terminate even if it computes one invariant at a time. Then we modify the loop by inserting an additional *parameterization procedure* that exploits the regularity of $\langle \mathcal{N}, \mathit{Initial} \rangle$. The procedure transforms a balanced placeset (siphon, trap) of a particular instance, say $\mathcal{N}(4)$, into a possibly infinite set of balanced placesets (siphons, traps) of all instances, encoded as the set of models of a WS1S-formula. This formula is a finite representation of the infinite set.

   For the sake of brevity, in the rest of the section we describe a CEGAR loop that only constructs traps. This allows us to avoid numerous repetitions of the phrase "balanced placesets, siphons, and traps". Since the structure of the loop is completely generic, this is purely a presentation issue without loss of generality[1].

A CEGAR Loop for the Non-parameterized Case. We need some preliminaries. Let $\mathcal{N} = \langle \mathcal{P}, \mathit{Tr} \rangle$ be a parameterized Petri net, and let $\mathcal{X}$ be a placeset variable. An *interpretation* of $\mathcal{X}$ is a pair $\mathbf{X} = \langle \ell, Q \rangle$, where $\ell \geq 1$ and Q is a set of places of $\mathcal{N}(\ell)$. We identify $\mathbf{X}$ and the tuple $\langle \mathbf{X}_p \rangle_{p \in \mathcal{P}}$, where $\mathbf{X}_p \subseteq [\ell]$, defined by $j \in \mathbf{X}_p$ iff $p(j) \in Q$. For example, if $\mathcal{P} = \{p, q, r\}$, $\ell = 1$, and $Q = \{p(0), p(1), q(1)\}$, then $\langle \mathbf{X}_p, \mathbf{X}_q, \mathbf{X}_r \rangle = \langle \{0, 1\}, \{1\}, \emptyset \rangle$. Given a formula $\phi(\ldots, \mathcal{X}, \ldots)$ and an interpretation $\mathbf{X} = (k, Q)$ of $\mathcal{X}$, we define the formula $\phi(\ldots, \mathbf{X}, \ldots)$ as follows:

$$x \in \mathbf{X}_p := \bigvee_{j \in \mathbf{X}_p} x = j$$

$$\mathcal{X} = \mathbf{X} := \boldsymbol{n} = k \wedge \bigwedge_{p \in \mathcal{P}} \forall \boldsymbol{x} \colon \boldsymbol{x} < \boldsymbol{n} \to (\boldsymbol{x} \in \mathcal{X}_p \leftrightarrow \boldsymbol{x} \in \mathbf{X}_p)$$

$$\phi(\ldots, \mathbf{X}, \ldots) := \forall \mathcal{X} \colon \mathcal{X} = \mathbf{X} \to \phi(\ldots, \mathcal{X}, \ldots)$$

   The CEGAR procedure maintains an (initially empty) set $\mathcal{T}$ of *indexed traps* of $\mathcal{N}(1), \mathcal{N}(2), \ldots, \mathcal{N}(n)$, where an indexed trap is a pair $\mathbf{T} = \langle i, Q \rangle$ such that

---

[1] The CEGAR loop for the non-parametric case could be formulated in SAT and solved using a SAT-solver. However, we formulate it in WS1S, since this allows us to give a uniform description of the non-parametric and the parametric cases.

$1 \leq i \leq n$ and Q is a trap of $\mathcal{N}(i)$. After every update of $\mathcal{T}$ the procedure constructs the formula $SafetyCheck_{\mathcal{T}}$, defined as follows:

$$TrapSet_{\mathcal{T}}(\boldsymbol{n}, \mathcal{X}) := \bigvee_{\mathbf{X} \in \mathcal{T}} \mathcal{X} = \mathbf{X}$$

$$PReach_{\mathcal{T}}(\boldsymbol{n}, \mathcal{M}', \mathcal{M}) := \forall \mathcal{X} : \left(\begin{array}{c} TrapSet(\boldsymbol{n}, \mathcal{X}) \\ \wedge Marked(\boldsymbol{n}, \mathcal{X}, \mathcal{M}) \end{array} \right) \rightarrow Marked(\boldsymbol{n}, \mathcal{X}, \mathcal{M}')$$

$$PReach_{\mathcal{T}}(\boldsymbol{n}, \mathcal{M}') := \exists \mathcal{M} : Initial(\boldsymbol{n}, \mathcal{M}) \wedge PReach_{\mathcal{T}}(\boldsymbol{n}, \mathcal{M}', \mathcal{M})$$

$$SafetyCheck_{\mathcal{T}} := \forall \boldsymbol{n} \forall \mathcal{M} : \boldsymbol{n} < n \wedge PReach_{\mathcal{T}}(\boldsymbol{n}, \mathcal{M}) \rightarrow Safe(\boldsymbol{n}, \mathcal{M})$$

Intuitively, $PReach_{\mathcal{T}}(\boldsymbol{n}, \mathcal{M}', \mathcal{M})$ states that according to the set $\mathcal{T}$ of (indexed) traps computed so far, $\mathcal{M}'$ could still be reachable from $\mathcal{M}$, because every trap of $\mathcal{T}$ marked at $\mathcal{M}$ is also marked at $\mathcal{M}'$. Therefore, if $SafetyCheck_{\mathcal{T}}$ holds then $\mathcal{T}$ is already strong enough to show that every reachable marking is safe.

If $\mathcal{T}$ is not strong enough, then the negation of $SafetyCheck_{\mathcal{T}}$ is satisfiable. The WS1S-checker returns a counterexample, i.e., a model $\mathbf{M} = \langle n, M \rangle$ of the formula $PReach_{\mathcal{T}}(\boldsymbol{n}, \mathcal{M}) \wedge \neg Safe(\boldsymbol{n}, \mathcal{M})$. Here M is a marking of $\mathcal{N}(n)$, potentially reachable from an initial marking but not safe. In this case we search for a witness trap $\mathcal{X}$ that is marked at every initial marking, but empty at $\mathbf{M}$, with the help of the formula

$$WTrap_{\mathbf{M}}(\boldsymbol{n}, \mathcal{X}) := \left(\begin{array}{c} Trap(\boldsymbol{n}, \mathcal{X}) \\ \wedge (\forall \mathcal{M} : Initial(\boldsymbol{n}, \mathcal{M}) \rightarrow Marked(\boldsymbol{n}, \mathcal{X}, \mathcal{M})) \\ \wedge Empty(\boldsymbol{n}, \mathcal{X}, \mathbf{M}) \end{array} \right)$$

If the formula is satisfiable, then the WS1S-checker returns a model $\mathbf{T} = (n, Q)$. The set Q is then a trap of $\mathcal{N}(n)$. We can now take $\mathcal{T} := \mathcal{T} \cup \{\mathbf{T}\}$, and iterate. Observe that after updating $\mathcal{T}$ the interpretation $\mathbf{M} = \langle n, M \rangle$ is no longer a model of $PReach_{\mathcal{T}}(\boldsymbol{n}, \mathcal{M}) \wedge \neg Safe(\boldsymbol{n}, \mathcal{M})$. Since $\mathcal{N}(1), \ldots, \mathcal{N}(n)$ only have finitely many traps, the procedure eventually terminates.

A CEGAR Approach for the Parameterized Case. In all nontrivial examples, proving safety of the infinitely many instances requires to compute infinitely many traps. Since the previous procedure only computes one trap per iteration, it does not terminate. The way to solve this problem is to insert a *parametrization step* that transforms the witness trap $\mathbf{T} = \langle k, Q \rangle$ into a formula $ParTrap_{\mathbf{T}}(\boldsymbol{n}, \mathcal{X})$ satisfying two properties: (1) all models of the formula are traps, and (2) $\mathbf{T}$ is a model. Since $ParTrap_{\mathbf{T}}(\boldsymbol{n}, \mathcal{X})$ can have infinitely many models, it constitutes a finite representation of an infinite set of traps. These models are also similar to each other and can be understood as capturing a single property of the system.

Example 4. Consider a parameterized net $\mathcal{N} = \langle \mathcal{P}, Tr \rangle$ exhibiting rotational symmetry: For every instance $\mathcal{N}(n)$, a pair (P_1, P_2) of sets is a transition of $\mathcal{N}(n)$ iff the pair $(P_1 \oplus_n 1, P_2 \oplus_n 1)$ is also a transition, where $P \oplus_n 1$ denotes the result of increasing all indices by 1 modulo n. Assume that $\mathcal{P} = \{p, q, r\}$ and

$\mathbf{T} = \langle 3, \{p(1), q(2)\}\rangle$, i.e., $\{p(1), q(2)\}$ is a trap of $\mathcal{N}(3)$. It is intuitively plausible (and we will later prove) that, due to the rotational symmetry, $\{p(i), q(i \oplus_m 1)\}$ is a trap of $\mathcal{N}(j)$ for every $m \geq 3$ and every $0 \leq i \leq m - 1$. We can then define the formula $ParTrap_{\mathbf{X}}(n, \mathcal{X})$ as:

$$ParTrap_{\mathbf{T}}(n, \mathcal{X}) := n \geq 3 \wedge \exists i : i < n$$

$$\wedge \forall x : x < n \rightarrow \left(\begin{array}{c} (x \in \mathcal{X}_p \leftrightarrow x = i) \\ \wedge (x \in \mathcal{X}_q \leftrightarrow x = i \oplus_n 1) \\ \wedge x \notin \mathcal{X}_r \end{array} \right).$$

Now, in order to describe the CEGAR procedure for the parameterized case we only need to *redefine* the formula $TrapSet_{\mathcal{T}}(n, \mathcal{X})$. Instead of the formula $TrapSet_{\mathcal{T}}(n, \mathcal{X}) := \bigvee_{\mathbf{T} \in \mathcal{T}} \mathcal{X} = \mathbf{T}$, which holds only when $\mathcal{X}$ is one of the finitely many traps in $\mathcal{T}$, we insert the parametrization procedure and define

$$TrapSet_{\mathcal{T}}(n, \mathcal{X}) := \bigvee_{\mathbf{T} \in \mathcal{T}} ParTrap_{\mathbf{T}}(n, \mathcal{X})$$

All the other formulas remain untouched. The question is how to obtain the formula $ParTrap_{\mathbf{T}}(n, \mathcal{X})$ from $\mathbf{T}$. We discuss this point in the rest of the section.

A Semi-automatic Approach. If we *guess* the formula $ParTrap_{\mathbf{T}}(n, \mathcal{X})$ we can use the WS1S-checker to automatically prove that the guess is correct. Indeed, it suffices to check that all models of $ParTrap_{\mathbf{T}}(n, \mathcal{X})$ are traps, which reduces to proving validity of the formula

$$\forall n \forall \mathcal{X} : ParTrap_{\mathbf{T}}(n, \mathcal{X}) \rightarrow Trap(n, \mathcal{X})$$

Let us see how this works in Example 1. Assume that the CEGAR procedure produces a trap $\mathbf{T} = \langle 3, \{p(1), q(2)\}\rangle$. The user finds it plausible that, due to the identical behavior of philosophers $1, 2, \ldots, n - 1$, the set $\{p(i), q(i \oplus 1)\}$ will be a trap of $\mathcal{N}(n)$ for every $n \geq 3$ and for every $1 \leq i \leq n - 2$ (i.e., the user excludes the case in which i or $i \oplus_n 1$ are equal to 0). So the user guesses a new formula

$$ParTrap_{\mathbf{T}}(n, \mathcal{X}) := n \geq 3 \wedge \exists i : (1 \leq i \leq n - 2) \wedge \forall x :$$
$$(x \in \mathcal{X}_p \leftrightarrow x = i) \wedge (x \in \mathcal{X}_q \leftrightarrow x = i \oplus_n 1) \wedge x \notin \mathcal{X}_r.$$

The user now automatically checks that all models of $ParTrap_{\mathbf{T}}(n, \mathcal{X})$ are traps. The formula can then be safely added to $TrapSet_{\mathcal{T}}(n, \mathcal{X})$ as a new disjunct.

An Automatic Approach for Specific Architectures. Parameterized Petri nets usually have a regular structure. For example, in the readers-writers problem all processes are indistinguishable, and in the philosophers problem, all right-handed processes behave in the same way. In the next sections we show how the structural properties of ring topologies and crowds (two common structure for parameterized systems) can be exploited to automatically compute the formula $ParTrap_{\mathbf{T}}(n, \mathcal{X})$ for each witness trap $\mathbf{T}$.

6   Trap Parametrization in Rings

Intuitively, a parameterized net $\mathcal{N}$ is a ring if for every transition of every instance $\mathcal{N}(n)$ there is an index $i \in [n]$ and sets $\mathcal{P}_L, \mathcal{P}_R, \mathcal{Q}_L, \mathcal{Q}_r \subseteq \mathcal{P}$ such that the preset of the transition is $(\mathcal{P}_L \times \{i\}) \cup (\mathcal{P}_R \times \{i \oplus_n 1\})$ and the post-set is $(\mathcal{Q}_L \times i) \cup (\mathcal{Q}_R \times i \oplus_n 1)$. In other words, every transition involves only two neighbor processes of the ring. In a fully symmetric ring all processes behave identically, while in a headed ring there is one distinguished process, as in Example 1. To ease presentation in this section we only consider fully symmetric rings. The extension to headed rings can be found in the appendix.

The informal statement "all processes behave identically" is captured by requiring the existence of a finite set of *transition patterns* $\langle \mathcal{P}_L, \mathcal{P}_R, \mathcal{Q}_L, \mathcal{Q}_R \rangle$ such that the transitions of $\mathcal{N}(n)$ are the result of "instantiating" each pattern with all pairs i and $i \oplus_n 1$ of consecutive indices.

Definition 3. *A parameterized net $\mathcal{N} = \langle \mathcal{P}, Tr \rangle$ is a* fully symmetric ring *if there is a finite set of transition patterns of the form $\langle \mathcal{P}_L, \mathcal{P}_R, \mathcal{Q}_L, \mathcal{Q}_R \rangle$, where $\mathcal{P}_L, \mathcal{P}_R, \mathcal{Q}_L, \mathcal{Q}_r \subseteq \mathcal{P}$, such that for every instance $\mathcal{N}(n)$ the following condition holds: $\langle P, Q \rangle$ is a transition of $\mathcal{N}(n)$ iff there is $i \in [n]$ and a pattern such that $P = \mathcal{P}_L \times \{i\} \cup \mathcal{P}_R \times \{i \oplus_n 1\}$ and $Q = \mathcal{Q}_L \times \{i\} \cup \mathcal{Q}_R \times \{i \oplus_n 1\}$.*

It is possible to decide if a given parameterized Petri net is a fully symmetric ring:

Proposition 2. *There is a formula of WS1S such that a parameterized net is a fully symmetric ring iff the formula holds.*

We need to distinguish between *global* and *local* traps of an instance. Loosely speaking, a global trap contains places of all processes, while a local trap does not. To understand why this is relevant, consider a fully symmetric ring $\mathcal{N} = \langle \mathcal{P}, Tr \rangle$ where $\mathcal{P} = \{p, q\}$ and the transitions of each instance $\mathcal{N}(n)$ are the pairs $\langle \{p(i), q(i \oplus_n 1)\}, \{p(i \oplus_n 1), q(i)\} \rangle$ for every $i \in [n]$. The sets $\{p(0), q(0)\}$ and $\{p(0), p(1), p(2), p(3)\}$ are both traps of $\mathcal{N}(4)$ (they are even 1-balanced sets). However, they are of different nature. Intuitively, in order to decide that $\{p(0), q(0)\}$ is a trap it is not necessary to inspect all of $\mathcal{N}(4)$, but only process 0 and its neighborhood. On the contrary, $\{p(0), \ldots, p(3)\}$ involves all the processes. This has consequences when parameterizing. Due to the symmetry of the ring, $\{p(i), q(i)\}$ is a trap of every instance $\mathcal{N}(n)$ for every $i \in [n]$. However, $\{p(i), p(i \oplus_n 1), \ldots, p(i \oplus_n 3)\}$ is *not* a trap of every instance for every $i \in [n]$, for example $\{p(0), \ldots, p(3)\}$ is not a trap of $\mathcal{N}(5)$. The correct parametrization is a different one, namely $\{p(0), p(1), \ldots, p(n-1)\}$. The difference between the two traps is captured by the following definition.

Definition 4. *Let $\mathcal{N} = \langle \mathcal{P}, Tr \rangle$ be a parameterized net. An indexed trap $\mathbf{T} = \langle n, Q \rangle$ of $\mathcal{N}$ is* global *if $Q \cap (\mathcal{P} \times \{i\}) \neq \emptyset$ for every $i \in [n]$, otherwise $\mathbf{T}$ is* local.

6.1   Parameterizing Local Traps

We first observe that local indexed traps can be "shifted" locally while maintaining their trap property.

Lemma 1. *Let $\mathcal{N} = \langle \mathcal{P}, Tr \rangle$ be a fully symmetric ring and let $\langle n, Q \rangle$ be a local indexed trap of $\mathcal{N}$. Then $\langle n, Q' \rangle$ with $Q' = \{\langle p, i \oplus_n 1 \rangle : \langle p, i \rangle \in Q\}$ is a local indexed trap of $\mathcal{N}$.*

Our second lemma states that for any indexed local traps $\langle n, Q \rangle$ with $Q \cap (\mathcal{P} \times \{n - 1\})$, the set Q remains a trap in any instance $\mathcal{N}(n')$ with $n \leq n'$.

Lemma 2. *Let $\mathcal{N}$ be a fully symmetric ring and $\langle n, Q \rangle$ a local indexed trap with $Q \cap (\mathcal{P} \times \{n - 1\}) = \emptyset$. Then $\langle n', Q \rangle$ is a local indexed trap for all $n' \geq n$.*

We can now show how to obtain a sound parameterization of a given indexed trap. The formula $ParTrap_{\mathbf{T}}(\mathcal{X})$ states that $\mathcal{X}$ is the result of "shifting" $\mathbf{T} = \langle n, Q \rangle$ in $\mathcal{N}(n')$ for some $n' \geq n$.

Theorem 1. *Let $\mathcal{N} = \langle \mathcal{P}, Tr \rangle$ be a fully symmetric ring and let $\langle n, Q \rangle$ be a local indexed trap of $\mathcal{N}(n)$ such that $Q \subseteq (\mathcal{P} \times I)$ for a minimal set $I \subset [n]$. Assume $I = \{i_0, \ldots, i_{k-1}\}$ with $0 \leq i_0 < i_1 < \ldots < i_{k-1} < n - 1$. Then every model of the formula*

$$ParTrap_{\mathbf{T}}(\boldsymbol{n}, \mathcal{X}) := n \leq \boldsymbol{n} \wedge \exists \boldsymbol{y} \colon \boldsymbol{y} < \boldsymbol{n} \wedge \bigwedge_{p \in \mathcal{P}} \forall \boldsymbol{x} \colon \boldsymbol{x} < \boldsymbol{n} \rightarrow$$

$$\left(\boldsymbol{x} \in \mathcal{X}_p \leftrightarrow \left(\begin{array}{c} \bigvee_{\langle i_0, p \rangle \in Q} \boldsymbol{x} = \boldsymbol{y} \\ \vee \bigvee_{j > 0, \langle i_j, p \rangle \in Q} \boldsymbol{x} = \boldsymbol{y} \oplus_{\boldsymbol{n}} (i_j - i_{j-1}) \end{array} \right) \right)$$

is an indexed trap of $\mathcal{N}$.

Remark 1. Since Theorem 1 requires $i_{k-1} < n-1$, it can only be applied to local traps $\langle n, Q \rangle$ such that $Q \cap (\mathcal{P} \times \{n - 1\}) = \emptyset$. However, for every local trap $\langle n, Q \rangle$ Lemma 1 allows us to find a local trap $\langle n, Q' \rangle$ satisfying $Q' \cap (\mathcal{P} \times \{n - 1\}) = \emptyset$, which we can then parameterize applying Theorem 1.

6.2   Parameterizing Global Traps

In contrast to local traps, global traps involve all indices $[n]$ of the instance $\mathcal{N}(n)$. Let $\langle n, Q \rangle$ be an indexed global trap. We denote with $Q[i]$ the set $P \subseteq \mathcal{P}$ such that $P \times \{i\} = Q \cap (\mathcal{P} \times \{i\})$; i.e., the set of places in Q at index i. Moreover, we say Q has period p if p is the smallest divisor of n such that for all $0 \leq j < p$ we have $Q[j] = Q[k \cdot p + j]$ for all $0 \leq k < \frac{n}{p}$. That is, Q is a repetition of the same p sets in a row. Since n is a period of Q we know that every Q has a period, which we denote p_Q. Recall the global trap $Q = \{p(0), p(1), p(2), p(3)\}$ from before. Then, $Q[0] = Q[1] = Q[2] = Q[3] = \{p\}$ and, consequently, $p_Q = 1$. Intuitively, we can repeat a period over and over again and still obtain a trap. So we can parameterize global traps by capturing the repetition of periodic behavior:

Theorem 2. *Let $\langle n, Q \rangle$ be an indexed global trap with $n \geq 2$. Then every model of the formula*

$$ParTrap_{\mathbf{T}}(\boldsymbol{n}, \mathcal{X}) := \exists \boldsymbol{P} : 0 \in \boldsymbol{P} \wedge \boldsymbol{n} \in \boldsymbol{P}$$

$$\wedge \; \forall \boldsymbol{x} : \boldsymbol{x} \leq \boldsymbol{n} \rightarrow \boldsymbol{x} \in \boldsymbol{P} \leftrightarrow \left(\begin{array}{c} \bigwedge\limits_{0 \leq k < p_Q} \boldsymbol{x} + k \notin \boldsymbol{P} \\ \wedge \; \boldsymbol{x} + p_Q \in \boldsymbol{P} \end{array} \right)$$

$$\wedge \; \forall \boldsymbol{x}_0, \dots, \boldsymbol{x}_{p_Q-1} : \left(\begin{array}{c} \bigwedge\limits_{0 < k \leq p_Q - 1} \boldsymbol{x}_{k-1} + 1 = \boldsymbol{x}_k \\ \wedge \; \boldsymbol{x}_{p_Q-1} < \boldsymbol{n} \wedge \boldsymbol{x}_0 \in \boldsymbol{P} \end{array} \right)$$

$$\rightarrow \bigwedge\limits_{0 \leq k < p_Q} \bigwedge\limits_{p \in Q[k]} \boldsymbol{x}_k \in \mathcal{X}_p \wedge \bigwedge\limits_{p \in \mathcal{P} \setminus Q[k]} \boldsymbol{x}_k \notin \mathcal{X}_p$$

is an indexed global trap.

7   Trap Parametrization in Barrier Crowds

Barrier crowds are parameterized systems in which communication happens by means of global steps in which each process makes a move. An initiator process decides to start a step, and all the other processes get a chance to veto it; if the step is not blocked (if all the processes accept it), all the processes, including the initiator, update their local state. Barrier crowds are slightly more general than broadcast protocols [25], which, loosely speaking, correspond to the special case in which no process makes use of the veto capability. Like broadcast protocols, barrier crowds can be used to model cache coherence protocols [18].

As for fully symmetric rings, transitions of the instances of a barrier crowd are generated from a finite set of "transition patterns". A transition pattern of a barrier crowd $\mathcal{N}$ is a pair $\langle \mathcal{I}, \mathbb{A} \rangle$, where $\mathcal{I} \in 2^{\mathcal{P}} \times 2^{\mathcal{P}}$ and $\mathbb{A} \subseteq 2^{\mathcal{P}} \times 2^{\mathcal{P}}$. Assume for example that each process can be in states p, q, r, and maintains a boolean variable with values $\{0, 1\}$. The corresponding parameterized net has $\mathcal{P} = \{p, q, r, 0, 1\}$ as set of places. Consider the transition pattern with $\mathcal{I} = \langle \{p, 0\}, \{q, 1\} \rangle$, and $\mathbb{A} = \{\langle \{p\}, \{p\} \rangle, \langle \{q, 0\}, \{r, 0\} \rangle, \langle \{q, 1\}, \{r, 0\} \rangle\}$. This pattern models that the initiator process, say process i, proposes a step that takes it from p to q, setting its variable to 1. Each other process reacts as follows, depending on its current state: if in p, it stays in p, leaving the variable unchanged; if in q, it moves to r, setting the variable to 0; if in r, it vetoes the step (because $\mathbb{A}$ does not offer a way to accept from state r).

Definition 5. *A parameterized Petri net $\mathcal{N} = \langle \mathcal{P}, Tr \rangle$ is a* barrier crowd *if there is a finite set of transition patterns of the form $\langle \mathcal{I}, \mathbb{A} \rangle$ such that for every instance $\mathcal{N}(n)$ the following condition holds: a pair $\langle P, Q \rangle$ is a transition of $\mathcal{N}(n)$ iff there exists a pattern $\langle \mathcal{I}, \mathbb{A} \rangle$ and $i \in [n]$ such that:*

- *$P \cap (\mathcal{P} \times \{i\}) = P_I \times \{i\}$ and $Q \cap (\mathcal{P} \times \{i\}) = Q_I \times \{i\}$, where $\mathcal{I} = \langle P_I, Q_I \rangle$.*

– *for every* $j \neq i$ *there is* $\langle P_A, Q_A \rangle \in \mathbb{A}$ *such that* $P \cap (\mathcal{P} \times \{j\}) = P_A \times \{j\}$
and $Q \cap (\mathcal{P} \times \{j\}) = Q_A \times \{j\}$.

Note that the number of transitions of $\mathcal{N}(n)$ grows quickly in n, even though the structure of the system remains simple, making parameterized verification particularly attractive.

In the rest of the section we present an automatic parametrization procedure for traps of barrier crowds. First we show that barrier crowds satisfy two important structural properties.

Given a set of places $P \subseteq \mathcal{P} \times [n]$ and a permutation $\pi \colon [n] \to [n]$, let $\pi(P)$ denote the set of places $\{p(\pi(i)) : p(i) \in P\}$. Given an index $0 \leq k < n$, let $drop_{k,n}(P)$ denote the set of places defined as follows: $p(i) \in drop_{k,n}(P)$ iff either $0 \leq i < k$ and $p(i) \in P$, or $k < i \leq n-1$ and $p(i+1) \in P$.

Definition 6. *Let $\mathcal{N}$ be a parameterized Petri net. A transition $\langle P_1, P_2 \rangle$ of $\mathcal{N}(n)$ is:*

– order invariant *if $\langle \pi(P_1), \pi(P_2) \rangle$ is also a transition of $\mathcal{N}(n)$ for every permutation $\pi \colon [n] \to [n]$.*
– homogeneous *if there is an index $0 \leq i < n$ such that for every $k \in [n] \setminus \{i\}$ the pair $\langle drop_{k,n}(P_1), drop_{k,n}(P_2) \rangle$ is a transition of $\mathcal{N}(n-1)$.*

$\mathcal{N}$ is homogeneous (order invariant) if all transitions of all instances $\mathcal{N}(n)$ is homogeneous (order invariant).

Intuitively, order invariance indicates that processes are indistinguishable. Homogeneity indicates that transitions in the large instances are not substantially different from the transitions in the smaller ones.

Proposition 3. *Barrier crowds are order invariant and homogeneous.*

7.1 Parameterizing Traps for Barrier Crowds

By order invariance, if Q is a trap of an instance, say $\mathcal{N}(n)$, then $\pi(Q)$ is also a trap for every permutation π. The set of all traps that can be obtained from Q by permutations can be described as a multiset $\mathcal{Q} \colon 2^{\mathcal{P}} \to [n]$. For example, assume $\mathcal{P} = \{p, q\}$, $n = 5$, and $Q = \{p(0), p(1), q(1), p(2), q(2), q(4)\}$. Then $\mathcal{Q}(\{p, q\}) = 2$ (because of indices 1 and 2), $\mathcal{Q}(\{p\}) = \mathcal{Q}(\{q\}) = 1$ (index 0 and 4, respectively), and $\mathcal{Q}(\emptyset) = 1$ (index 3). Any assignment of indices to the elements of $\mathcal{Q}$ results in a trap. We call $\mathcal{Q}$ the *trap family* of Q.

Proposition 4. *Let $\mathcal{N}$ be an order invariant and homogeneous parameterized Petri net, let Q be a trap of an instance $\mathcal{N}(n)$, and let $\mathcal{Q} \colon 2^{\mathcal{P}} \to [n]$ be the trap family of Q. We have:*

– *If $\mathcal{Q}(\emptyset) \geq 1$ and $\mathcal{Q}'$ is obtained from $\mathcal{Q}$ by increasing the multiplicity of $\emptyset$, then $\mathcal{Q}'$ is also a trap family of another instance of $\mathcal{N}$.*
– *For every $S \in 2^{\mathcal{P}}$, if $\mathcal{Q}(S) \geq 2$ and $\mathcal{Q}'$ is obtained from $\mathcal{Q}$ by increasing the multiplicity of S, then $\mathcal{Q}'$ is also a trap family of another instance of $\mathcal{N}$.*

Proposition 4 leads to a parameterization procedure for barrier crowds. Given a trap Q of some instance $\mathcal{N}(n)$ and its trap family $\mathcal{Q}$, consider all multisets obtained from $\mathcal{Q}$ by applying the operations of Proposition 4. We call this set of multisets the *extended trap family* of Q. Observe that $\mathcal{Q}$ represents a set of traps of $\mathcal{N}(n)$, while the extended family represents a set of traps across all instances $\mathcal{N}(n')$ with $n' \geq n$.

Give an indexed trap $\mathbf{T} = \langle n, Q \rangle$, we choose the formula $ParTrap_{\mathbf{T}}(\mathcal{X})$ so that its models correspond to the traps of the extended family of Q. For this, we capture the minimal required multiplicities of $\langle n, Q \rangle$ by quantifying for every $S \subseteq \mathcal{P}$ with $\mathcal{Q}(S) > 0$ indices $i_{S,1}, \ldots, i_{S,\mathcal{Q}(S)}$ for which precisely the places in S are marked. Making all indices introduced this way pairwise distinct ensures that any model of the formula at least covers the multiset $\mathcal{Q}$. Additionally, we can capture that the subset S of $\mathcal{P}$ which are marked in every other index are chosen such that Proposition 4 ensures that we still obtain a trap.

$$
ParTrap_{\mathbf{T}}(\boldsymbol{n}, \mathcal{X}) := \exists_{S \subseteq \mathcal{P}} i_{S,1}, \ldots, i_{S,\mathcal{Q}(S)} : \left\{ \left(\bigwedge_{(S,k) \neq (S',k')} (i_{S,k} \neq i_{S',k'}) \right) \right.
$$

$$
\wedge \forall j : j < \boldsymbol{n} \rightarrow \left[\left(\bigvee_{S \subseteq \mathcal{P}, k=1,\ldots,\mathcal{Q}(S)} \left(j = i_{S,k} \wedge \left(\begin{array}{c} \bigwedge_{p \in S} \boldsymbol{j} \in \mathcal{X}_p \\ \wedge \bigwedge_{p \in \mathcal{P} \setminus S} \boldsymbol{j} \notin \mathcal{X}_p \end{array} \right) \right) \right) \right.
$$

$$
\left. \vee \left(\left(\bigwedge_{S \subseteq \mathcal{P}, k=1,\ldots,\mathcal{Q}(S)} \boldsymbol{j} \neq i_{S,k} \right) \wedge \bigvee_{\substack{\emptyset \neq S \subseteq \mathcal{P}: \; \mathcal{Q}(S) \geq 2 \\ S = \emptyset: \; \mathcal{Q}(S) \geq 1}} \left(\begin{array}{c} \bigwedge_{p \in S} \boldsymbol{j} \in \mathcal{X}_p \\ \wedge \bigwedge_{p \in \mathcal{P} \setminus S} \boldsymbol{j} \notin \mathcal{X}_p \end{array} \right) \right) \right] \right\}.
$$

We immediately get:

Theorem 3. *Let $\mathcal{N} = \langle \mathcal{P}, Tr \rangle$ be a barrier crowd and let $\langle n, Q \rangle$ be a local indexed trap of $\mathcal{N}(n)$. Then every model of the formula $ParTrap_{\mathbf{T}}(\boldsymbol{n}, \mathcal{X})$ defined above is an indexed trap of $\mathcal{N}$.*

Remark 2. This theorem applies to all order invariant and homogeneous systems. It is easy to see that order invariance and homogeneity of a given parameterized net can be expressed in WS1S and verified automatically.

8   Experiments

We implemented the CEGAR loop and the parameterization techniques of Sects. 6 and 7 in our tool ostrich. ostrich heavily relies on MONA as a WS1S-solver. The results of our experiments are presented in Fig. 2. In the first two columns the table reports the topology and the name of the system to be verified. The array topology is a linear topology where agents can refer existentially or universally to agents with smaller or larger indices. Analogously to the other topologies we derive a sound parameterization technique for traps, 1-BB sets, and siphons. The rings are Dijsktra's token ring for mutual exclusion [33] and a model of the dining philosophers in which philosophers pick both forks simultaneously. For headed rings we consider Example 1 and a model of a message passing leader election algorithm. The array is Burns' mutual exclusion algorithm [38]. The crowds are Dijkstra's algorithm for mutual exclusion [20] and models of cache-coherence protocols taken from [18]. Note that we check inductiveness of the property; i.e., if it holds initially and there is no marking satisfying the property and the current abstraction and reaching in a single step a marking which violates the property. Additionally, we include in the specification of the parameterized Petri net a partition of the places $\mathcal{P}$ such that the places of every index in every instance form a 1BB-set. Collectively, this ensures that all examples are 1-bounded and yields invariants similar to (1), (2) for Example 1. Since `ostrich` does not compute but only checks these invariants we do not count them in Fig. 2 (leading to 3 semi-automatic invariants for Example 1 since we omit (1), and (2)). Moreover, these invariants already imply inductiveness of some safety properties; prominently deadlock-freedom for all considered cache-coherence protocols.

The third column gives the time `ostrich` needs to initialize the analysis; this includes verifying that the given parameterized Petri net is covered by 1BB-sets, and that it indeed has the given topology. The fourth column gives the property being checked. The specification of the cache coherence protocols consists of a number of consistency properties, specific for each protocol. The legend "consistency (x/y)" indicates that the specification consists of y properties, of which `ostrich` was able to automatically prove the inductiveness of x. Column 5 gives the time need to check the inductiveness the property (or, in the case of the cache-coherence protocols, either find a marking which satisfies all constraints imposed by 1BB-sets, traps or siphons, or prove the inductiveness of the properties together). Columns 6, 7, and 8 give the number of WS1S-formulas, each corresponding to a parameterized 1BB-sets, trap, or siphon that are computed by the CEGAR loop. Some of these WS1S-formulas have only one model, i.e., they correspond to a single trap, siphon, or 1BB-set of one instance. Such "artifacts" are needed when small instances (e.g., arrays of size 2) require ad-hoc proofs that cannot be parameterized. In these cases the "real" number of parametric invariants is the result of subtracting the number of artifacts from the total number. The last column reports the number of parameterized inductive invariants obtained by the semi-automatic CEGAR loop. There the user is presented a series of counter examples to the inductiveness of the property. The user can check for traps, siphons or 1BB-sets to disprove the counter example. If the user then provides an invariant which proves inductive it is used to refine the abstraction until no further counter example can be found. The response time of `ostrich` in this setting is immediate which provides a nice user experience. Dragon and MOESI are examples showing that the semi-automatic procedure can lead to proofs with fewer invariants. The last step of the automatic procedure is to remove invariants until no invariant can be removed without obtaining a counter example again.

For Example 1 `ostrich` automatically computes the following family of 1BB-sets (additionally to the invariants (1) and (2)): (For readability we omit some artifacts.)

$$2 \leq n \wedge \text{taken} = \text{think} = \emptyset \wedge \text{wait} = \text{eat} = \{0, 1\} \wedge \text{free} = \{1\}$$
$$3 \leq n \wedge \text{taken} = \text{wait} = \text{think} = \emptyset \wedge \text{eat} = \{n - 1, 0\} \wedge \text{free} = \{0\}$$
$$4 \leq n \wedge \text{taken} = \text{think} = \emptyset \wedge \text{free} = \text{wait} = \{n - 1\} \wedge \text{eat} = \{n - 2, n - 1\}$$
$$2 \leq n \wedge \exists i : 1 < i < n - 2 \wedge \left(\begin{array}{c} \text{taken} = \text{think} = \emptyset \wedge \text{free} = \text{wait} = \{i \oplus_n 1\} \\ \wedge \text{eat} = \{i, i \oplus_n 1\} \end{array} \right)$$

Topology	Example	Init. (ms)	Property	Check (ms)	1BB-sets	Traps	Siphons	Semi-automatic invariants
ring	Dijkstra ring	558	deadlock	40	1 (1)	0 (0)	0 (0)	2
			mutual exclusion	125	1 (1)	1 (1)	0 (0)	
headed ring	atomic phil.	409	deadlock	79	1 (1)	0 (0)	0 (0)	4
	lefty phil.	495	deadlock	294	7 (4)	0 (0)	0 (0)	3
	leader election	670	not 0 and $n - 1$ leader	965	1 (0)	0 (0)	2 (1)	1
			not two leaders	–	–	–	–	
array	Burns	501	deadlock	16	0 (0)	0 (0)	0 (0)	1
			mutual exclusion	379	0 (0)	8 (7)	0 (0)	
crowd	Dijkstra	1830	deadlock	88	2 (1)	0 (0)	0 (0)	3
			mutual exclusion	1866	0 (0)	3 (1)	0 (0)	
	Berkeley	544	deadlock	15	0 (0)	0 (0)	0 (0)	1
			consistency (1/3)	442	0 (0)	9 (1)	0 (0)	
	Dragon	673	deadlock	19	0 (0)	0 (0)	0 (0)	2
			consistency (7/7)	2015	25 (5)	11 (2)	0 (0)	
	Firefly	469	deadlock	15	0 (0)	0 (0)	0 (0)	1
			consistency (2/4)	617	0 (0)	14 (2)	0 (0)	
	Illinois	490	deadlock	15	0 (0)	0 (0)	0 (0)	1
			consistency (2/2)	184	0 (0)	5 (1)	0 (0)	
	MESI	407	deadlock	14	0 (0)	0 (0)	0 (0)	1
			consistency (2/2)	179	0 (0)	5 (1)	0 (0)	
	MOESI	439	deadlock	14	0 (0)	0 (0)	0 (0)	1
			consistency (7/7)	1496	0 (0)	35 (6)	0 (0)	
	Synapse	398	deadlock	14	0 (0)	0 (0)	0 (0)	1
			consistency (2/2)	18	0 (0)	0 (0)	0 (0)	

Fig. 2. Experimental results of `ostrich`. The complete data is available at [44].

9 Conclusion

We have refined the approach to parameterized verification of systems with regular architectures presented in [14]. Instead of encoding the complete verification question into large, monolithic WS1S-formula, our approach introduces a CEGAR loop which also outputs an explanation of why the property holds in the form of a typically small set of parameterized invariants (see Example 1). The explanation helps to uncover false positives, where the verification succeeds only because the system or the specification are incorrectly encoded in WS1S. It has also helped to find a subtle bug in the implementation of [14] which hid unnoticed in the complexity of the monolithic formula. Additionally, our incremental approach requires to check smaller WS1S-formulas, which often decreases the verification time (cp. the verification of Dijkstra's mutual exclusion algorithm [14] in 10 s to currently 2 s).

On the other hand, seeing the abstraction helps one understand the analyzed system. For example, we include in [44] a leader election algorithm for which the parameterization techniques of ostrich are too coarse to establish the general safety property of having always at most one leader. However, ostrich succeeds to prove the special case that not agents 0 and $n - 1$ can become leader at the same time. For this proof ostrich finds a family of siphons which hint to a general inductive invariant of the system. Using the semi-automatic mode of ostrich we can then verify this inductive invariant and, as a result of this, the general safety property.

Data Availability Statement and Acknowledgements. This work has received funding from the European Research Council(ERC) under the European Union's Horizon 2020 research and innovation programme under grant agreement No 787367 (PaVeS).

The tool ostrich and associated files are available at [44]. The current version is maintained at [30].

We thank the anonymous reviewers for their comments.

References

1. Abdulla, P.A., Cerans, K., Jonsson, B., Tsay, Y.: General decidability theorems for infinite-state systems. In: LICS, pp. 313–321. IEEE Computer Society (1996)
2. Abdulla, P.A., Delzanno, G., Henda, N.B., Rezine, A.: Regular model checking without transducers (on efficient verification of parameterized systems). In: Grumberg, O., Huth, M. (eds.) TACAS 2007. LNCS, vol. 4424, pp. 721–736. Springer, Heidelberg (2007). https://doi.org/10.1007/978-3-540-71209-1_56
3. Abdulla, P.A., Jonsson, B., Nilsson, M., Saksena, M.: A survey of regular model checking. In: Gardner, P., Yoshida, N. (eds.) CONCUR 2004. LNCS, vol. 3170, pp. 35–48. Springer, Heidelberg (2004). https://doi.org/10.1007/978-3-540-28644-8_3
4. Abdulla, P.A., Sistla, A.P., Talupur, M.: Model checking parameterized systems. Handbook of Model Checking, pp. 685–725. Springer, Cham (2018). https://doi.org/10.1007/978-3-319-10575-8_21
5. Apt, K.R., Kozen, D.C.: Limits for automatic verification of finite-state concurrent systems. Inf. Process. Lett. **22**(6), 307–309 (1986)
6. Athanasiou, K., Liu, P., Wahl, T.: Unbounded-thread program verification using thread-state equations. In: Olivetti, N., Tiwari, A. (eds.) IJCAR 2016. LNCS (LNAI), vol. 9706, pp. 516–531. Springer, Cham (2016). https://doi.org/10.1007/978-3-319-40229-1_35

7. Außerlechner, S., Jacobs, S., Khalimov, A.: Tight cutoffs for guarded protocols with fairness. In: Jobstmann, B., Leino, K.R.M. (eds.) VMCAI 2016. LNCS, vol. 9583, pp. 476–494. Springer, Heidelberg (2016). https://doi.org/10.1007/978-3-662-49122-5_23

8. Baukus, K., Bensalem, S., Lakhnech, Y., Stahl, K.: Abstracting WS1S systems to verify parameterized networks. In: Graf, S., Schwartzbach, M. (eds.) TACAS 2000. LNCS, vol. 1785, pp. 188–203. Springer, Heidelberg (2000). https://doi.org/10.1007/3-540-46419-0_14

9. Baukus, K., Lakhnech, Y., Stahl, K.: Parameterized verification of a cache coherence protocol: safety and liveness. In: Cortesi, A. (ed.) VMCAI 2002. LNCS, vol. 2294, pp. 317–330. Springer, Heidelberg (2002). https://doi.org/10.1007/3-540-47813-2_22

10. Bensalem, S., Bozga, M., Nguyen, T.-H., Sifakis, J.: D-finder: a tool for compositional deadlock detection and verification. In: Bouajjani, A., Maler, O. (eds.) CAV 2009. LNCS, vol. 5643, pp. 614–619. Springer, Heidelberg (2009). https://doi.org/10.1007/978-3-642-02658-4_45

11. Bloem, R., et al.: Decidability of parameterized verification. Synth. Lect. Distrib. Comput. Theory 6, 1–170 (2015)

12. Blondin, M., Esparza, J., Helfrich, M., Kučera, A., Meyer, P.J.: Checking qualitative liveness properties of replicated systems with stochastic scheduling. In: Lahiri, S.K., Wang, C. (eds.) CAV 2020. LNCS, vol. 12225, pp. 372–397. Springer, Cham (2020). https://doi.org/10.1007/978-3-030-53291-8_20

13. Blondin, M., Finkel, A., Haase, C., Haddad, S.: Approaching the coverability problem continuously. In: Chechik, M., Raskin, J.-F. (eds.) TACAS 2016. LNCS, vol. 9636, pp. 480–496. Springer, Heidelberg (2016). https://doi.org/10.1007/978-3-662-49674-9_28

14. Bozga, M., Esparza, J., Iosif, R., Sifakis, J., Welzel, C.: Structural invariants for the verification of systems with parameterized architectures. TACAS 2020. LNCS, vol. 12078, pp. 228–246. Springer, Cham (2020). https://doi.org/10.1007/978-3-030-45190-5_13

15. Bozga, M., Iosif, R., Sifakis, J.: Checking deadlock-freedom of parametric component-based systems. In: Vojnar, T., Zhang, L. (eds.) TACAS 2019. LNCS, vol. 11428, pp. 3–20. Springer, Cham (2019). https://doi.org/10.1007/978-3-030-17465-1_1

16. Browne, M., Clarke, E., Grumberg, O.: Reasoning about networks with many identical finite state processes. Inf. Comput. 81(1), 13–31 (1989)

17. Chen, Y., Hong, C., Lin, A.W., Rümmer, P.: Learning to prove safety over parameterised concurrent systems. In: FMCAD, pp. 76–83 (2017)

18. Delzanno, G.: Automatic verification of parameterized cache coherence protocols. In: CAV, pp. 53–68 (2000). https://doi.org/10.1007/10722167_8

19. Desel, J., Esparza, J.: Free Choice Petri Nets. Cambridge University Press, Cambridge (2005)

20. Dijkstra, E.W.: Cooperating sequential processes. In: Hansen, P.B. (ed.) The Origin of Concurrent Programming, pp. 65–138. Springer, New York (2002). https://doi.org/10.1007/978-1-4757-3472-0_2

21. Emerson, E.A., Kahlon, V.: Reducing model checking of the many to the few. In: McAllester, D. (ed.) CADE 2000. LNCS (LNAI), vol. 1831, pp. 236–254. Springer, Heidelberg (2000). https://doi.org/10.1007/10721959_19

22. Emerson, E.A., Namjoshi, K.S.: Reasoning about rings. In: POPL, pp. 85–94 (1995)

23. Esparza, J.: Decidability and complexity of petri net problems—an introduction. In: Reisig, W., Rozenberg, G. (eds.) ACPN 1996. LNCS, vol. 1491, pp. 374–428. Springer, Heidelberg (1998). https://doi.org/10.1007/3-540-65306-6_20

24. Esparza, J.: Parameterized verification of crowds of anonymous processes. In: Dependable Software Systems Engineering, pp. 59–71. IOS Press (2016)

25. Esparza, J., Finkel, A., Mayr, R.: On the verification of broadcast protocols. In: LICS, pp. 352–359. IEEE Computer Society (1999)

26. Esparza, J., Ledesma-Garza, R., Majumdar, R., Meyer, P., Niksic, F.: An SMT-based approach to coverability analysis. In: Biere, A., Bloem, R. (eds.) CAV 2014. LNCS, vol. 8559, pp. 603–619. Springer, Cham (2014). https://doi.org/10.1007/978-3-319-08867-9_40

27. Esparza, J., Melzer, S.: Verification of safety properties using integer programming: beyond the state equation. Formal Methods Syst. Des. **16**(2), 159–189 (2000)

28. Esparza, J., Meyer, P.J.: An SMT-based approach to fair termination analysis. In: FMCAD, pp. 49–56. IEEE (2015)

29. Esparza, J., Raskin, M., Welzel, C.: Computing parameterized invariants of parameterized petri nets (2021). https://arxiv.org/abs/2103.10280

30. Esparza, J., Raskin, M., Welzel, C.: Computing parameterized invariants of parameterized petri nets (2021). https://gitlab.lrz.de/i7/ostrich

31. Finkel, A., Haddad, S., Khmelnitsky, I.: Minimal coverability tree construction made complete and efficient. FoSSaCS 2020. LNCS, vol. 12077, pp. 237–256. Springer, Cham (2020). https://doi.org/10.1007/978-3-030-45231-5_13

32. Finkel, A., Schnoebelen, P.: Well-structured transition systems everywhere!. Theor. Comput. Sci. **256**(1–2), 63–92 (2001)

33. Fribourg, L., Olsén, H.: Reachability sets of parameterized rings as regular languages. Electr. Notes Theor. Comput. Sci. **9**, 40 (1997). https://doi.org/10.1016/S1571-0661(05)80427-X

34. Geffroy, T., Leroux, J., Sutre, G.: Occam's razor applied to the petri net coverability problem. Theor. Comput. Sci. **750**, 38–52 (2018)

35. German, S.M., Sistla, A.P.: Reasoning about systems with many processes. J. ACM **39**(3), 675–735 (1992)

36. Henriksen, J.G., et al.: Mona: monadic second-order logic in practice. In: Brinksma, E., Cleaveland, W.R., Larsen, K.G., Margaria, T., Steffen, B. (eds.) TACAS 1995. LNCS, vol. 1019, pp. 89–110. Springer, Heidelberg (1995). https://doi.org/10.1007/3-540-60630-0_5

37. Jacobs, S., Sakr, M.: Analyzing guarded protocols: better cutoffs, more systems, more expressivity. VMCAI 2018. LNCS, vol. 10747, pp. 247–268. Springer, Cham (2018). https://doi.org/10.1007/978-3-319-73721-8_12

38. Jensen, H.E., Lynch, N.A.: A proof of burns N-process mutual exclusion algorithm using abstraction. In: Steffen, B. (ed.) TACAS 1998. LNCS, vol. 1384, pp. 409–423. Springer, Heidelberg (1998). https://doi.org/10.1007/BFb0054186

39. Kesten, Y., Maler, O., Marcus, M., Pnueli, A., Shahar, E.: Symbolic model checking with rich assertional languages. Theor. Comput. Sci **256**(1), 93–112 (2001)

40. Murata, T.: Petri nets: properties, analysis and applications. Proc. IEEE **77**(4), 541–580 (1989)

41. Reisig, W.: Understanding Petri Nets - Modeling Techniques, Analysis Methods, Case Studies. Springer, Heidelberg (2013). https://doi.org/10.1007/978-3-642-33278-4

42. Reynier, P.-A., Servais, F.: On the computation of the minimal coverability set of petri nets. In: Filiot, E., Jungers, R., Potapov, I. (eds.) RP 2019. LNCS, vol. 11674, pp. 164–177. Springer, Cham (2019). https://doi.org/10.1007/978-3-030-30806-3_13
43. The MONA Project: MONA. https://www.bricks.dk/mona
44. Welzel, C., Esparza, J., Raskin, M.: Ostrich (2020). https://doi.org/10.5281/zenodo.4499091
45. Wimmel, H., Wolf, K.: Applying CEGAR to the Petri net state equation. Log. Methods Comput. Sci 8(3), (2012)

On the Combination of Polyhedral Abstraction and SMT-Based Model Checking for Petri Nets

Nicolas Amat$^{(\boxtimes)}$ (iD), Bernard Berthomieu (iD), and Silvano Dal Zilio (iD)

LAAS-CNRS, Université de Toulouse, CNRS, Toulouse, France
namat@laas.fr

Abstract. We define a method for taking advantage of net reductions in combination with a SMT-based model checker. We prove the correctness of this method using a new notion of equivalence between nets that we call polyhedral abstraction. Our approach has been implemented in a tool, named SMPT, that provides two main procedures: Bounded Model Checking (BMC) and Property Directed Reachability (PDR). Each procedure has been adapted in order to use reductions and to work with arbitrary Petri nets. We tested SMPT on a large collection of queries used during the 2020 edition of the Model Checking Contest. Our experimental results show that our approach works well, even when we only have a moderate amount of reductions.

1 Introduction

A significant focus in model checking research is finding algorithmic solutions to avoid the "state explosion problem", that is finding ways to analyse models that are out of reach from current methods. To overcome this problem, it is often useful to rely on symbolic representation of the state space (like with decision diagrams) or on an abstraction of the problem, for instance with the use of logical approaches like SAT solving. We can also benefit from optimizations related to the underlying model. When analysing Petri nets, for instance, a valuable technique relies on the transformation and decomposition of nets, a method pioneered by Berthelot [5] and known as *structural reduction*.

We recently proposed a new abstraction technique based on reductions [6,7]. The idea is to compute reductions of the form (N, E, N'), where: N is an initial net (that we want to analyse); N' is a residual net (hopefully simpler than N); and E is a system of linear equations. The idea is to preserve enough information in E so that we can rebuild the reachable markings of N knowing only the ones of N'. In a nutshell, we capture and abstract the effect of reductions using a set of linear constraints between the places of N and N'.

In this paper, we show that this approach works well when combined with SMT-based verification. In particular, it provides an elegant way to integrate reductions into known verification procedures. To support this statement, we

© Springer Nature Switzerland AG 2021
D. Buchs and J. Carmona (Eds.): PETRI NETS 2021, LNCS 12734, pp. 164–185, 2021.
https://doi.org/10.1007/978-3-030-76983-3_9

provide a full theoretical framework based on the definition of a new equivalence-relation (Sect. 3) and show how to use it for checking safety and invariant properties on nets (Sect. 4).

We have previously applied this technique in a symbolic model checker, called Tedd, that uses Set Decision Diagrams [33] in order to generate an abstract representation for the state space of a net N. In practice, we can often reduce a Petri net N with n places (from a high dimensional space) into a residual net N' with far fewer places, say n' (in a lower-dimensional space). Hence, with our approach, we can represent the state space of N as the "inverse image", by the linear system E, of a subset of vectors of dimension n'. This technique can result in a very compact representation of the state space. We observed this effect during the recent editions of the Model Checking Contest (MCC) [2], where Tedd won the competition for the *State Space* category. In this paper, we show that we can benefit from the same "dimensionality reduction" effect when using automatic deduction procedures. Actually, since we are working with (possibly unbounded) vectors of integers, we need to consider SMT instead of SAT solvers. We show that it is enough to use solvers for the theory of Quantifier-Free formulas on Linear Integer Arithmetic, what is known as QF-LIA in SMT-LIB [4].

To adapt our approach with the theory of SMT solving, we define an abstraction based on Boolean combinations of linear constraints between integer variables (representing the marking of places). This results in a new relation $N \rhd_E N'$, which is the counterpart of the tuple (N, E, N') in a SMT setting. We named this relation a *polyhedral abstraction* in reference to "polyhedral models" used in program optimization and static analysis [8, 20]. (Like in these works, we propose an algebraic representation of the relation between a model and its state space based on the sets of solutions to systems of linear equations.) One of our main results is that, given a relation $N \rhd_E N'$, we can derive a formula $\tilde{E}$ such that F is an invariant for N if and only if $\tilde{E} \wedge F$ is an invariant for the net N'. Since the residual net may be much simpler than the initial one, we expect that checking the invariant $\tilde{E} \wedge F$ on N' is more efficient than checking F on N.

Our approach has been implemented and computing experiments show that reductions are effective on a large benchmark of queries. We provide a prototype tool, called SMPT, that includes an adaptation of two procedures, Bounded Model Checking (BMC) [9] and Property Directed Reachability (PDR) [13,14]. Each of these methods has been adapted in order to use reductions and to work with arbitrary Petri nets. We tested SMPT on a large collection of queries (13 265 test cases) used during the 2020 edition of the Model Checking Contest. Our experimental results show that our approach works well, even when we only have a moderate amount of reductions.

Outline and Contributions. This paper summarises the key ideas and results of [1], to which we refer the reader for full details. The paper is organized as follows. In Sect. 3, we define our notion of polyhedral abstraction and prove several of its properties. This definition relies on a presentation of Petri net semantics that emphasizes the relationship with the QF-LIA theory (Sect. 2).

We use these results in Sect. 4 to describe our adaptation of general SMT-based algorithms with reductions and prove their correctness. In Sect. 5 we describe our adaptation of BMC and PDR with reductions. Before concluding, we report on experimental results on an extensive collection of nets and queries.

2   Petri Nets and Linear Arithmetic Constraints

A *Petri net* N is a tuple $(P, T, \mathbf{pre}, \mathbf{post})$ where $P = \{p_1, \ldots, p_n\}$ is a finite set of places, $T = \{t_1, \ldots, t_k\}$ is a finite set of transitions (disjoint from P), and $\mathbf{pre}: T \rightarrow (P \rightarrow \mathbb{N})$ and $\mathbf{post}: T \rightarrow (P \rightarrow \mathbb{N})$ are the pre- and post-condition functions (also called the flow functions of N). A state m of a net, also called a *marking*, is a mapping $m : P \rightarrow \mathbb{N}$ which assigns a number of *tokens*, $m(p)$, to each place p in P. A marked net (N, m_0) is a pair composed from a net and an initial marking m_0. In the following, we will often consider that each transition is associated with a label (a symbol taken from an alphabet Σ). In this case, we assume that a net is associated with a labeling function $l : T \rightarrow \Sigma \cup \{\tau\}$, where τ is a special symbol for the silent action name. Every net has a default labeling function l_N such that $\Sigma = T$ and $l_N(t) = t$ for every transition $t \in T$.

A transition $t \in T$ is *enabled* at marking $m \in \mathbb{N}^P$ when $m(p) \geq \mathbf{pre}(t, p)$ for all places p in P. (We can also simply write $m \geq \mathbf{pre}(t)$, where $\geq$ stands for the component-wise comparison of markings.) A marking $m' \in \mathbb{N}^P$ is reachable from a marking $m \in \mathbb{N}^P$ by firing transition t, denoted $m \xrightarrow{t} m'$, if: (1) transition t is enabled at m; and (2) $m' = m - \mathbf{pre}(t) + \mathbf{post}(t)$. By extension, we say that a *firing sequence* $\sigma = t_1 \ldots t_n \in T^*$ can be fired from m, denoted $m \xRightarrow{\sigma} m'$, if there exist markings $m_0, \ldots, m_n$ such that $m = m_0$, $m' = m_n$ and $m_i \xrightarrow{t_{i+1}} m_{i+1}$ for all $i < n$.

We denote $R(N, m)$ the set of markings reachable from m in N. A marking m is k-bounded when each place has at most k tokens; property $\bigwedge_{p \in P} m(p) \leq k$ is true. Likewise, a marked Petri net (N, m_0) is bounded when there is k such that all reachable markings are k-bounded. A net is *safe* when it is 1-bounded. In our work, we consider *generalized* Petri nets (in which net arcs may have weights larger than 1) and we do not restrict ourselves to bounded nets.

We can extend the notion of labels to sequences of transitions in a straightforward way. Given a relabeling function, l, we can extend it into a function from $T^* \rightarrow \Sigma^*$ such that $l(\epsilon) = \epsilon$, $l(\tau) = \epsilon$ and $l(\sigma t) = l(\sigma) l(t)$. Given a sequence of labels σ in Σ^*, we write $(N, m) \xRightarrow{\sigma} (N, m')$ when there is a firing sequence ϱ in T^* such that $(N, m) \xRightarrow{\varrho} (N, m')$ and $\sigma = l(\varrho)$. We say in this case that σ is an *observable sequence* of the marked net (N, m).

We use the standard graphical notation for nets, where places are depicted as circles and transitions as squares. With the net displayed in Fig. 1 (left), the initial marking is $m_1 \triangleq p_0*5 \ p_6*4$ (only 5 and 4 tokens in places p_0 and p_6). We have $m_1 \xRightarrow{\sigma} m_1'$ with $\sigma \triangleq t_0 t_0 t_1 t_1 t_2 t_3 t_4$ and $m_1' \triangleq p_0*3 \ p_2*1 \ p_3*1 \ p_6*3$; and therefore $m_1 \xRightarrow{a\,a\,b\,c} m_1'$.

We can define many properties on the markings of a net N using Boolean combinations of linear constraints with integer variables. Assume that we have

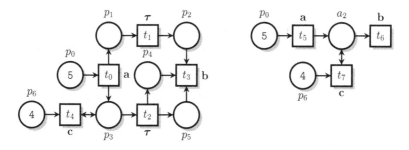

Fig. 1. An example of Petri net, M_1 (left), and one of its polyhedral abstraction, M_2 (right), with $E_M \triangleq (p_5 = p_4) \wedge (a_1 = p_1 + p_2) \wedge (a_2 = p_3 + p_4) \wedge (a_1 = a_2)$.

a marked net (N, m_0) with set of places $P = \{p_1, \ldots, p_n\}$. We can associate a marking m over P to the formula $\underline{m}(x_1, \ldots, x_n)$, below. In this context, an equation $x_i = k$ means that there must be k tokens in place p_i. Formula $\underline{m}$ is obviously a conjunction of literals, what is called a *cube* in [13].

$$\underline{m}(x_1, \ldots, x_n) \triangleq (x_1 = m(p_1)) \wedge \cdots \wedge (x_k = m(p_k)) \tag{1}$$

In the remainder, we use the notation $\phi(\vec{x})$ for the declaration of a formula ϕ with variables in $\vec{x}$, instead of the more cumbersome notation $\phi(x_1, \ldots, x_n)$. We also simply use $\phi(\vec{v})$ instead of $\phi\{x_1 \leftarrow v_1\} \ldots \{x_n \leftarrow v_n\}$, for the substitution of $\vec{x}$ with $\vec{v}$ in ϕ. We should often use place names as variables (or parameters) and use $\vec{p}$ for the vector $(p_1, \ldots, p_n)$. We also often use $\underline{m}$ instead of $\underline{m}(\vec{p})$.

Definition 1 (Models of a Formula). *We say that a marking m is a model of (or m satisfies) property ϕ, denoted $m \models \phi$, when formula $\phi(\vec{x}) \wedge \underline{m}(\vec{x})$ is satisfiable. In this case ϕ may use variables that are not necessarily in P.*

We can use this approach to reframe many properties on Petri nets. For instance the notion of safe markings, described previously: a marking m is safe when $m \models \text{SAFE}_1(\vec{x})$, where $\text{SAFE}_k(\vec{x}) \triangleq \bigwedge_{i \in 1..n}(x_i \leq k)$.

Likewise, the property that transition t is enabled corresponds to formula $\text{ENBL}_t(\vec{x}) \triangleq \bigwedge_{i \in 1..n}(x_i \geq \mathbf{pre}(t, p_i))$, in the sense that t is enabled at m when $m \models \text{ENBL}_t(\vec{x})$. Another example is the definition of *deadlocks*, which are characterized by formula $\text{DEAD}(\vec{x}) \triangleq \bigwedge_{t \in T} \neg\text{ENBL}_t(\vec{x})$. We give other examples in Sect. 4, when we encode the transition relation of a Petri net using formulas.

In our work, we focus on the verification of *safety* properties on the reachable markings of a marked net (N, m_0). Examples of properties that we want to check include: checking if some transition t is enabled (commonly known as *quasi-liveness*); checking if there is a deadlock; checking whether some invariant between place markings is true; ...

Definition 2 (Invariant and Reachable Properties). *Property ϕ is an invariant on (N, m_0) if and only if we have $m \models \phi$ for all $m \in R(N, m_0)$. We say that ϕ is reachable when there exists $m \in R(N, m_0)$ such that $m \models \phi$.*

In our experiments, we consider the two main kinds of *reachability formulas* used in the MCC: AG ϕ (true only when ϕ is an invariant), and EF ϕ (true when ϕ is reachable), where ϕ is a Boolean combination of atomic properties (it has no modalities).

3   Polyhedral Abstraction and E-Equivalence

We define a new notion, called E-*abstraction equivalence*, that is used to state a correspondence between the set of reachable markings of two Petri nets "modulo" some system of linear equations, E. Basically, we have that (N_1, m_1) is E-equivalent to (N_2, m_2) when, for every sequence $m_2 \overset{\sigma_2}{\Longrightarrow} m_2'$ in N_2, there must exist a sequence $m_1 \overset{\sigma_1}{\Longrightarrow} m_1'$ in N_1 such that $E \wedge \underline{m_1'} \wedge \underline{m_2'}$ is satisfiable (and reciprocally). Therefore, knowing E, we can compute the reachable markings of N_1 from those of N_2, and vice versa. We also ask for the observable sequences, σ_1 and σ_2 in this case, to be equal. As a result, we will prove that our equivalence is also a congruence.

We can illustrate these notions using the two nets M_1, M_2 in Fig. 1, we have that $m_1' \triangleq p_0{*}3 \ p_2{*}1 \ p_3{*}1 \ p_6{*}3$ is reachable in M_1 and $E_M \wedge \underline{m_1'}$ entails $\underline{m_1'} \wedge (a_1 = 1) \wedge (a_2 = 1)$, which means that marking $m_2' \triangleq a_2{*}1 \ p_0{*}3 \ p_6{*}3$ is reachable in M_2. Conversely, we have several markings (exactly 4) in M_1 that corresponds to the constraint $E_M \wedge \underline{m_2'} \equiv (p_5 = p_4) \wedge (p_1 + p_2 = 1) \wedge (p_3 + p_4 = 1) \wedge \underline{m_2'}$. All these markings are reachable in M_1 using the same observable sequence **a a b c**. More generally, each marking m_2' of N_2 can be associated to a convex set of markings of N_1, defined as the set of positive integer solutions of $E \wedge \underline{m_2'}$. Moreover, these sets form a partition of $R(N_1, m_1)$. This motivates our choice of calling this relation a *polyhedral abstraction*.

While our approach does not dictate a particular method for finding pairs of equivalent nets, we rely on an automatic approach based on the use of *structural net reductions*. When the net N_1 can be reduced, we will obtain a resulting net (N_2) and a condition (E) such that N_2 is a polyhedral abstraction of N_1. In this case, E will always be expressed as a conjunction of equality constraints between linear combinations of integer variables (the marking of places). This is why we should often use the term *reduction equations* when referring to E. Our goal is to transform any reachability problem on the net N_1 into a reachability problem on the (reduced) net N_2, which is typically much easier to check.

Solvable Systems and E-Equivalence. Before defining our equivalence more formally, we need to introduce some constraints on the condition, E, used to correlate the markings of two different nets. We say that a pair of markings (m_1, m_2) are *compatible* (over respective sets of places P_1 and P_2) when they have equal marking on their shared places, meaning $m_1(p) = m_2(p)$ for all p in $P_1 \cap P_2$. This is a necessary and sufficient condition for formula $\underline{m_1} \wedge \underline{m_2}$ to be satisfiable. When this is the case, we denote $m_1 \uplus m_2$ the unique marking in $(P_1 \cup P_2)$ such that $(m_1 \uplus m_2)(p) = m_1(p)$ if $p \in P_1$ and $(m_1 \uplus m_2)(p) = m_2(p)$ otherwise. Hence, with our conventions, $\underline{m_1 \uplus m_2} \Leftrightarrow \underline{m_1} \wedge \underline{m_2}$.

In the following we ask that condition E be *solvable for* N_1, N_2, meaning that for all reachable marking m_1 in N_1 there must exist at least one marking m_2 of N_2, compatible with m_1, such that $m_1 \uplus m_2 \models E$ (see condition A2). While this property is not essential for most of our results, it simplifies our presentation and it will always be true for the reduction equations generated with our method. On the other hand, we do not prohibit to use variables in E that are not in $P_1 \cup P_2$. Actually, such a situation will often occur in practice, when we start to chain several reductions.

Definition 3 (E-Abstraction Equivalence). *Assume N_1, N_2 are two Petri nets with respective sets of places P_1, P_2 and labeling functions l_1, l_2, over the same alphabet Σ. We say that the marked net (N_2, m_2) is an E-abstraction of (N_1, m_1), denoted $(N_1, m_1) \sqsupseteq_E (N_2, m_2)$, if and only if:*

(A1) *the initial markings are compatible with E, meaning $m_1 \uplus m_2 \models E$.*
(A2) *for all observation sequences $\sigma \in \Sigma^\star$ such that $(N_1, m_1) \xRightarrow{\sigma} (N_1, m_1')$ then there is at least one marking $m_2' \in R(N_2, m_2)$ such that $m_1' \uplus m_2' \models E$ (we say E solvable), and for all markings m_2' over P_2, we have that $m_1' \uplus m_2' \models E$ implies $(N_2, m_2) \xRightarrow{\sigma} (N_2, m_2')$.*

We say that (N_1, m_1) is E-equivalent to (N_2, m_2), denoted $(N_1, m_1) \triangleright_E (N_2, m_2)$, when we have both $(N_1, m_1) \sqsupseteq_E (N_2, m_2)$ and $(N_2, m_2) \sqsupseteq_E (N_1, m_1)$.

Notice that condition (A2) is defined only for sequences starting from the initial marking of N_1. Hence the relation is usually not true on every pair of matching markings; it is not a bisimulation.

By definition, relation $\triangleright_E$ is symmetric. We deliberately use a "comparison symbol" for our equivalence, $\triangleright$, in order to stress the fact that N_2 should be a reduced version of N_1. In particular, we expect that $|P_2| \leq |P_1|$.

Basic Properties of Polyhedral Abstraction. We prove that we can use E-equivalence to check the reachable markings of N_1 simply by looking at the reachable markings of N_2. We give a first property that is useful in the context of bounded model checking, when we try to find a counter-example to a property by looking at firing sequences with increasing length. Our second property is useful for checking invariants, and is at the basis of our implementation of the PDR method for Petri nets.

Lemma 1 (Bounded Model Checking). *Assume $(N_1, m_1) \triangleright_E (N_2, m_2)$. Then for all m_1' in $R(N_1, m_1)$ there is m_2' in $R(N_2, m_2)$ such that $m_1' \uplus m_2' \models E$.*

Proof. Since m_1' is reachable, there must be a firing sequence σ_1 in N_1 such that $(N_1, m_1) \xRightarrow{\sigma_1} (N_1, m_1')$. By condition (A2), there must be some marking m_2' over P_2, compatible with m_1', such that $m_1' \uplus m_2' \models E$ and $(N_2, m_2) \xRightarrow{\sigma_2} (N_2, m_2')$ (for some firing sequence σ_2). Therefore we have m_2' reachable in N_2 such that $m_1' \uplus m_2' \models E$. □

Lemma 1 can be used to find a counter-example m_1', to some property F in N_1, just by looking at the reachable markings of N_2. Indeed, it is enough to find a marking m_2' reachable in N_2 such that $m_2' \models E \wedge \neg F$. This is the result we use in our implementation of the BMC method.

Our second property can be used to prove that every reachable marking of N_2 can be traced back to at least one marking of N_1 using the reduction equations. (While this mapping is surjective, it is not a function, since a state in N_1 could be associated with multiple states in N_2.)

Lemma 2 (Invariance Checking). *Assume* $(N_1, m_1) \rhd_E (N_2, m_2)$. *Then for all pairs of markings* m_1', m_2' *over* N_1, N_2 *such that* $m_1' \uplus m_2' \models E$ *and* $m_2' \in R(N_2, m_2)$ *it is the case that* $m_1' \in R(N_1, m_1)$.

Proof. Take m_1', m_2' a pair of markings in N_1, N_2 such that $m_1' \uplus m_2' \models E$ and $m_2' \in R(N_2, m_2)$. Hence there is a firing sequence σ_2 such that $(N_2, m_2) \overset{\sigma_2}{\Longrightarrow} (N_2, m_2')$. By condition (A2), since $m_1' \uplus m_2' \models E$, there must be a firing sequence in N_1, say σ_1, such that $(N_1, m_1) \overset{\sigma_1}{\Longrightarrow} (N_1, m_1')$. Hence $m_1' \in R(N_1, m_1)$. $\qquad\square$

Using Lemma 2, we can easily extract an invariant on N_1 from an invariant on N_2. Basically, if property $E \wedge F$ is an invariant on N_2 (where F is a formula whose variables are in P_1) then we can prove that F is an invariant on N_1. This property (the *invariant conservation* theorem of Sect. 4) ensures the soundness of the model checking technique implemented in our tool.

Next we prove that polyhedral abstractions are closed by synchronous composition, relabeling, and chaining. Before defining these operations, we start by describing sufficient conditions in order to safely compose equivalence relations. The goal here is to avoid inconsistencies that could emerge if we inadvertently reuse the same variable in different reduction equations.

The *fresh variables* in an equivalence statement $EQ : (N_1, m_1) \rhd_E (N_2, m_2)$ are the variables occurring in E but not in $P_1 \cup P_2$. (These variables can be safely "alpha-converted" in E without changing any of our results.) We say that a net N_3 is *compatible* with respect to EQ when $(P_1 \cup P_2) \cap P_3 = \emptyset$ and there are no fresh variables of EQ that are also places in P_3. Likewise we say that the equivalence statement $EQ' : (N_2, m_2) \rhd_{E'} (N_3, m_3)$ is *compatible* with EQ when $P_1 \cap P_3 \subseteq P_2$ and the fresh variables of EQ and EQ' are disjoint.

In this section we rely on the classical synchronous product operation between labeled Petri nets [28]. Let N_1 and N_2 be two labeled Petri nets with respective sets of places P_1, P_2 and with labeling functions l_1 and l_2 on the respective alphabets Σ_1 and Σ_2. We can assume, without loss of generality, that the sets P_1 and P_2 are disjoint. We denote $N_1 \| N_2$ the *synchronous product* between N_1 and N_2. Since the places in N_1 and N_2 are disjoint, we can always see a marking m in $N_1 \| N_2$ as the disjoint union of two markings m_1, m_2 from N_1, N_2. In this case we simply write $m = m_1 \| m_2$. More generally, we extend this product operation to marked nets and write $(N_1, m_1) \| (N_2, m_2)$ for the marked net $(N_1 \| N_2, m_1 \| m_2)$.

Another standard operation on labeled Petri net is *relabeling*, denoted as $N[a/b]$, that apply a substitution to the labeling function of a net. Assume l is the labeling function over the alphabet Σ. We denote $l[a/b]$ the labeling function on

$(\Sigma \setminus \{a\}) \cup \{b\}$ such that $l[a/b](t) = b$ when $l(t) = a$ and $l[a/b](t) = l(t)$ otherwise. Then $N[a/b]$ is the same as net N but equipped with labeling function $l[a/b]$. Relabeling has no effect on the marking of a net. The relabeling law is true even in the case where b is the silent action τ. In this case we say that we *hide* action a from the net.

Theorem 1 (E-Equivalence is a Congruence). *Assume we have two compatible equivalence statements* $(N_1, m_1) \rhd_E (N_2, m_2)$ *and* $(N_2, m_2) \rhd_{E'}$ (N_3, m_3), *and that* M *is compatible with respect to these equivalences, then:*

- $(N_1, m_1) \| (M, m) \rhd_E (N_2, m_2) \| (M, m)$.
- $(N_1, m_1) \rhd_{E,E'} (N_3, m_3)$.
- $(N_1[a/b], m_1) \rhd_E (N_2[a/b], m_2)$ *and* $(N_1[a/\tau], m_1) \rhd_E (N_2[a/\tau], m_2)$.

Proof (sketch). The result for the first composition law derives from the fact that we can always compute a unique pair of firing sequences for N_1, M from a firing sequence of $N_1 \| M$. The proof for the other laws are similar, see [1]. □

The composition laws stated in Theorem 1 are useful to build larger equivalences from simpler axioms (reductions rules). We show some examples of reductions in the next paragraph and how they occur in the example of Fig. 1.

Deriving E-Equivalences Using Reductions. We can compute net reductions by reusing a tool, called Reduce, that was developed in our previous work [7]. The tool takes a marked Petri net as input and returns a reduced net and a sequence of linear equations. For example, given the net M_1 of Fig. 1, Reduce returns net M_2 and equations $(p_5 = p_4), (a_1 = p_1 + p_2), (a_2 = p_3 + p_4)$, and $(a_1 = a_2)$, that corresponds to formula E_M in Fig. 1. The tool works by applying successive reduction rules, in a compositional way. We give an example of such rule in Fig. 2 (above), which states that we can fuse places inside a "deterministic sequence" of transitions. This is one of the many *agglomeration* rules defined in [7] and also one of the original rules found in [5].

It is possible to prove that each reduction step computed by Reduce, from a net (M_i, m_i) to (M_{i+1}, m_{i+1}) with equations E_i, is such that $(M_i, m_i) \rhd_{E_i}$ (M_{i+1}, m_{i+1}). Therefore, by Theorem 1, the results computed by Reduce always translate into valid polyhedral abstractions.

We can look at our running example to explain the inner working of Reduce. It is always safe to remove a *redundant place*, e.g. a place with the same **pre** and **post** relations than another one. This is the case with places p_4, p_5 (see Fig. 2). Redundant places can sometimes be found by looking at the structure of the net, but our tool can also find more elaborate occurrences of redundant places by solving an integer linear programming problem [30].

After the removal of p_5, we are left with a residual net similar to the one in the second equivalence of Fig. 2. In this case, we can use our agglomeration rules to simplify places p_1 and p_3. Similar situations, where we can aggregate several places together, can be found by searching patterns in the net. After this step (introducing two new places a_1 and a_2), we find a new opportunity

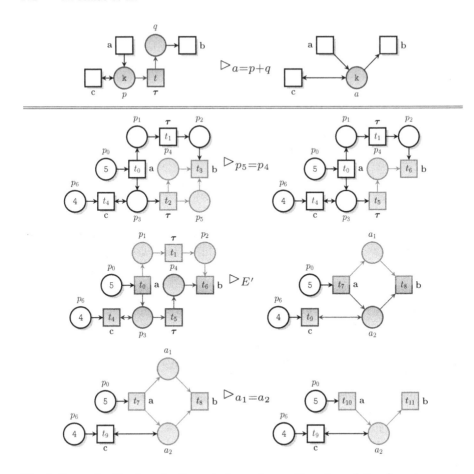

Fig. 2. Example of basic reduction rule for agglomerating places (above), and sequence of three reductions (below) leading from the net M_1 to M_2 from Fig. 1, with $E' \triangleq (a_1 = p_1 + p_2) \wedge (a_2 = p_3 + p_4)$.

to reduce a pair of redundant places, (a_1, a_2). Besides these two main kinds of reduction rules (redundancy and agglomeration), the Reduce tool can also identify other opportunities for reductions. For instance specific structural or behavioural restrictions, such as nets that are marked graphs or other cases where the set of reachable markings is exactly defined by the solutions of the state equation [24].

In conclusion, we can use Reduce to compute polyhedral abstractions automatically. In the other direction, we can use our notion of equivalence to prove the correctness of new reduction patterns that could be added in the tool. While it is not always possible to reduce the complexity of a net using this approach, we observed in our experiments (Sect. 6) that, on a benchmark suite that includes almost 1 000 instances of nets, about half of them can be reduced by a factor of more than 30%.

4   SMT-Based Model Checking Using Abstractions

We introduce a general method for combining polyhedral abstraction with SMT-based model checking procedures. Assume we have $(N_1, m_1) \rhd_E (N_2, m_2)$, where the nets N_1, N_2 have sets of places P_1, P_2 respectively. In the following, we use $\vec{p_1} \triangleq (p_1^1, \ldots, p_k^1)$ and $\vec{p_2} \triangleq (p_1^2, \ldots, p_l^2)$ for the places in P_1 and P_2. We also consider (disjoint) sequences of variables, $\vec{x}$ and $\vec{y}$, ranging over (the places of) N_1 and N_2. With these notations, we denote $\tilde{E}(\vec{x}, \vec{y})$ the formula obtained from E where place names in N_1 are replaced with variables in $\vec{x}$, and place names in N_2 are replaced with variables in $\vec{y}$. When we have the same place in both nets, say $p_i^1 = p_j^2$, we also add the constraint $(x_i = y_j)$ to $\tilde{E}$ in order to avoid shadowing variables. (Remark that $\tilde{E}(\vec{p_1}, \vec{p_2})$ is equivalent to E, since equalities $x_i = y_j$ become tautologies in this case.)

$$\tilde{E}(\vec{x}, \vec{y}) \triangleq E\{\vec{p_1} \leftarrow \vec{x}\}\{\vec{p_2} \leftarrow \vec{y}\} \wedge \bigwedge_{\{(i,j)|p_i^1=p_j^2\}} (x_i = y_j) \tag{2}$$

Given a formula F, we denote $fv(F)$ the set of free variables contained in it. Assume F_1 is a property that we want to study on N_1, without loss of generality we can enforce the condition $(fv(F_1) \setminus P_1) \cap (fv(E) \setminus P_1) = \emptyset$ (meaning we can always rename the variables in F_1 and E that are not places in N_1). This condition ensures that the studied property on the initial net does not contain any new variable introduced during the reduction.

Definition 4 (E-Transform Formula). *Assume $(N_1, m_1) \rhd_E (N_2, m_2)$ and take F_1 a property with variables in P_1 such that $(fv(F_1) \setminus P_1) \cap (fv(E) \setminus P_1) = \emptyset$. Formula $F_2(\vec{y}) \triangleq \tilde{E}(\vec{x}, \vec{y}) \wedge F_1(\vec{x})$ is the E-transform of F_1.*

The following property states that, to check an invariant F_1 on the reachable markings of N_1, it is enough to check the corresponding E-transform formula F_2 on the reachable markings of N_2.

Theorem 2 (Invariant Conservation). *Assume $(N_1, m_1) \rhd_E (N_2, m_2)$ and that $F_2(\vec{y})$ is the E-transform of formula F_1 on N_1. Then F_1 is an invariant on N_1 if and only if $F_2(\vec{p_2})$ is an invariant on N_2.*

Proof. Assume $(N_1, m_1) \rhd_E (N_2, m_2)$ and property F_1 is an invariant on N_1. Consider m_2' a reachable marking in N_2. By definition of E-abstraction, we have at least one reachable marking m_1' in N_1 such that $m_1' \uplus m_2' \models E$. Since F_1 is an invariant on N_1 we have $m_1' \models F_1$. The condition $m_1' \uplus m_2' \models E$ is equivalent to $\underline{m_1' \wedge m_2' \wedge E}$ satisfiable. By definition we have $\tilde{E}(\vec{p_1}, \vec{p_2}) \equiv E$, which implies $\underline{m_1' \wedge m_2' \wedge \tilde{E}(\vec{p_1}, \vec{p_2}) \wedge F_1(\vec{p_1})}$ satisfiable, since the only variables that are both in F_1 and E must also be in N_1. Hence, m_2' satisfies the E-transform formula of F_1. The proof is similar in the other direction. $\qquad\square$

Since F_1 invariant on N_1 is equivalent to $\neg F_1$ not reachable, we can directly infer an equivalent conservation theorem for reachability: to find a model of F_1 in N_1, it is enough to find a model for $F_1(\vec{p_1}) \wedge \tilde{E}(\vec{p_1}, \vec{p_2})$ in N_2.

Theorem 3 (Reachability Conservation). *Assume* $(N_1, m_1) \rhd_E (N_2, m_2)$ *and that* $F_2(\vec{y})$ *is the E-transform of formula* F_1 *on* N_1. *Then formula* F_1 *is reachable in* N_1 *if and only if* $F_2(\vec{p_2})$ *is reachable in* N_2.

5   BMC and PDR Implementation

We developed a prototype model checker that takes advantage of net reductions. The tool offers two main analysis methods that have been developed for generalized Petri nets. (No specific optimizations are applied when we know the net is safe.) These options correspond to the implementation of the BMC and PDR methods, that we sketch below.

Bounded Model Checking (BMC) is an iterative method for exploring the state space of finite-state systems by unrolling their transitions [9]. The method was originally based on an encoding of transition systems into (a family of) propositional logic formulas and the use of SAT solvers to check these formulas for satisfiability [17]. More recently, this approach was extended to more expressive models, and richer theories, using SMT solvers [3].

In BMC, we try to find a reachable marking m that is a model for a given formula F, that usually models a set of "feared events". The algorithm starts by computing a formula, say ϕ_0, representing the initial marking and checking whether $\phi_0 \wedge F$ is satisfiable (meaning F is initially true). If the formula is *UNSAT*, we compute a formula ϕ_1 representing all the markings reachable in one step, or less, from the initial marking and check $\phi_1 \wedge F$. This way, we compute a sequence of formulas $(\phi_i)_{i \in \mathbb{N}}$ until either $\phi_i \wedge F$ is *SAT* (in which case a counter-example is found) or we have $\phi_{i+1} \Rightarrow \phi_i$ (in which case we reach a fixed point and no counter-example exists). The BMC method is not complete since it is not possible, in general, to bound the number of iterations needed to give an answer. Also, when the net is unbounded, we may very well have an infinite sequence of formulas $\phi_0 \subsetneq \phi_1 \subsetneq \ldots$ However, in practice, this method can be very efficient to find a counter-example when it exists.

The crux of the method is to compute formulas ϕ_i that represent the set of markings reachable using firing sequences of length at most i. We show how we can build such formulas incrementally. We assume that we have a marked net (N, m_0) with places $P = \{p_1, \ldots, p_n\}$ and transitions $T = \{t_1, \ldots, t_k\}$. In the remainder of this section, we build formulas that express constraints between markings m and m' such that $m \to m'$ in N. Hence we define formulas with $2n$ variables. We use the notation $\psi(\vec{x}, \vec{x}')$ as a shorthand for $\psi(x_1, \ldots, x_n, x_1', \ldots, x_n')$.

We already defined (Sect. 2) a helper formula, or *operator*, $\text{ENBL}_t(\vec{x})$ such that $\text{ENBL}_t(\vec{x}) \wedge \underline{m}(\vec{x})$ is true when t is enabled at m. We can define, in the same way, an operator Δ_t that describes the evolution of a marking after transition t fires, see (3) below. It can be used to define another helper formula, $t(\vec{x}, \vec{x}')$, such that $(\underline{m}(\vec{x}) \wedge t(\vec{x}, \vec{x}') \wedge \underline{m}'(\vec{x}'))$ entails that $m \xrightarrow{t} m'$, when t is enabled at m, or $m = m'$ otherwise. With all these notations, we can define $\text{T}(\vec{x}, \vec{x}')$ as the disjunction of all the transition formulas $t(\vec{x}, \vec{x}')$. By construction, formula $T(m, m') \triangleq \underline{m}(\vec{x}) \wedge T(\vec{x}, \vec{x}') \wedge \underline{m}'(\vec{x}')$ is true when $m \rightarrow m'$, or when $m = m'$.

$$\Delta_t(\vec{x}, \vec{x}') \triangleq \bigwedge_{i \in 1..n} (x_i' = x_i + \mathbf{post}(t, p_i) - \mathbf{pre}(t, p_i)) \tag{3}$$

$$EQ(\vec{x}, \vec{x}') \triangleq \bigwedge_{i \in 1..n} x_i = x_i' \tag{4}$$

$$t(\vec{x}, \vec{x}') \triangleq (ENBL_t(\vec{x}) \Rightarrow \Delta_t(\vec{x}, \vec{x}')) \wedge (\neg ENBL_t(\vec{x}) \Rightarrow EQ(\vec{x}, \vec{x}')) \tag{5}$$

$$T(\vec{x}, \vec{x}') \triangleq EQ(\vec{x}, \vec{x}') \vee \bigvee_{t \in T} (ENBL_t(\vec{x}) \wedge \Delta_t(\vec{x}, \vec{x}')) \tag{6}$$

Formula ϕ_i is the result of connecting i successive occurrences of formulas of the form $T(\vec{x}_j, \vec{x}_{j+1})$. We define the formulas inductively, with a base case (ϕ_0) which states that only m_0 is reachable initially. To define the ϕ_i's, we assume that we have a collection of (pairwise disjoint) sequences of variables, $(\vec{x}_i)_{i \in \mathbb{N}}$.

$$\phi_0(N, m_0) \triangleq \underline{m_0}(\vec{x}_0) \qquad \phi_{i+1}(N, m_0) \triangleq \phi_i(N, m_0) \wedge T(\vec{x}_i, \vec{x}_{i+1})$$

We can prove that this family of BMC formulas provide a way to check reachability properties, meaning that formula F is reachable in (N, m_0) if and only if there exists $i \geq 0$ such that $F(\vec{x}_i) \wedge \phi_i(N, m_0)$ is satisfiable. The approach we describe here is well-known (see for instance [9]). It is also quite simplified. Actual model checkers that rely on BMC apply several optimizations techniques, such as compositional reasoning; acceleration methods; or the use of invariants on the underlying model to add extra constraints. We do not consider such optimizations here, on purpose, since our motivation is to study the impact of polyhedral abstractions. We believe that our use of reductions is orthogonal and does not overlap with many of these optimizations, in the sense that we do not preclude them, and that the performance gain we observe with reductions could not be obtained with these optimizations.

Assume we have $(N_1, m_1) \rhd_E (N_2, m_2)$. We denote T_1, T_2 the equivalent of formula T, above, for the nets N_1, N_2 respectively. We also use $\vec{x}, \vec{y}$ for sequences of variables ranging over (the places of) N_1 and N_2 respectively. We should use $\phi(N_1, m_1)$ for the family of formulas built using operator T_1 and variables $\vec{x}_0, \vec{x}_1, \dots$ and similarly for $\phi(N_2, m_2)$, where we use T_2 and variables of the form $\vec{y}$. The following property states that, to find a model of F in the reachable markings of N_1, it is enough to find a model for its E-transform in N_2.

Theorem 4 (BMC with E-Transform). *Assume $(N_1, m_1) \rhd_E (N_2, m_2)$ and that $F_2(\vec{y})$ is the E-transform of $F_1(\vec{x})$. Formula $F_1(\vec{x})$ is reachable in N_1 if and only if there exists $j \geq 0$ such that $F_2(\vec{y}_j) \wedge \phi_j(N_2, m_2)$ is satisfiable.*

Proof (sketch). We start by proving that F_1 reachable in N_1 is equivalent to $F_1 \wedge \phi_i(N_1, m_1)$ satisfiable for some $i \geq 0$. The proof is by induction on the value of i and uses the fact that $T_1(m, m')$ entails $m \Rightarrow m'$ in N_1. As a result, we can prove the existence of a firing sequence $m_1 \overset{\sigma}{\Rightarrow} m_1'$, of length at most i, such that $m_1' \models F_1$. The result follows by our *conservation of reachability* property (Theorem 3), F_1 reachable in N_1 means F_2 reachable in N_2. Therefore F_1 is reachable iff there is $j \geq 0$ such that $F_2(\vec{y}_j) \wedge \phi_j(N_2, m_2)$ is satisfiable.   □

We can give a stronger result, comparing the value of i and j, when the reductions used in proving the E-abstraction equivalence never introduce new transitions. This is the case, for example, with the reductions computed using the Reduce tool. Indeed, in this case, we can show that we may find a witness of length i in N_1 (a firing sequence of length i showing that F_1 is reachable in N_1) when we find a witness of length $j \leq i$ in N_2. This is because, in this case, reductions may compact a sequence of several transitions into a single one or, at worst, not change it. Take the example of the agglomeration rule in Fig. 2. Therefore BMC benefits from reductions in two ways. First because we can reduce the size of formulas ϕ (which are proportional to the size of the net), but also because we can accelerate transition unrolling in the reduced net.

Property Directed Reachability (PDR). While BMC is the right choice when we try to find counter-examples, it usually performs poorly when we want to check an invariant property, AG F. There are techniques that are better suited to prove *inductive invariants* in a transition system; that is a property that is true initially and stays true after firing any transition.

In order to check invariants with SMPT, we have implemented a method called PDR [13,14] (also known as IC3), which incrementally generates clauses that are inductive "relative to stepwise approximate reachability information". PDR is a combination of induction, over-approximation, and SAT solving. For SMPT, we developed a similar method that uses SMT solving, to deal with markings and transitions, and that can take advantage of polyhedral abstractions.

We use the same notations as with BMC. The PDR method requires to define a set of *safe states*, described as the models of some property F. It also requires a set of initial states, I. In our case $I \triangleq m_0(\vec{x})$. The procedure is complete for finite transition systems, for instance with bounded Petri nets. We can also prove termination in the general case when property $\neg F$ is *monotonic*, meaning that $m \models \neg F$ implies that $m' \models \neg F$ for all markings m' that covers m (that is when $m' \geq m$, component-wise). An intuition is that it is enough, in this case, to check the property on the minimal coverability set of the net, which is always finite (see e.g. [21]).

A formula F is *inductive* [14] when $I \Rightarrow F$ and $F(\vec{x}) \wedge \mathrm{T}(\vec{x}, \vec{x}') \Rightarrow F(\vec{x}')$ hold. It is *inductive relative* to formula G if both $I \Rightarrow F$ and $G(\vec{x}) \wedge F(\vec{x}) \wedge \mathrm{T}(\vec{x}, \vec{x}') \Rightarrow F(\vec{x}')$ hold. With PDR we compute *Over Approximated Reachability Sequences* (OARS), meaning sequences of formulas $(F_0, \ldots, F_{k+1})$, with variables in $\vec{x}$, that are monotonic: $F_0 = I$, $F_i \Rightarrow F_{i+1}$ for all $i \in 0..k$, and $F_{k+1} \Rightarrow F$; and satisfies consecution: $F_i(\vec{x}) \wedge \mathrm{T}(\vec{x}, \vec{x}') \Rightarrow F_{i+1}(\vec{x}')$ for all $i \leq k + 1$. The formulas F_i change at each iteration of the procedure (each time we increase k). The procedure stops when we find an index i such that $F_i = F_{i+1}$. In this case we know that F is an invariant. We can also stop during the iteration if we find a counter-example.

Our implementation follows closely the algorithm for IC3 described in [14]. We only give a brief sketch of the OARS construction. Each of the F_i is computed as a formula in CNF (the conjunction of a set of clauses $CL(F_i)$) such that $CL(F_{i+1}) \subseteq CL(F_i)$. Intuitively, each clause is built from a *witness*, a marking such that $F_i(\vec{x}) \wedge \mathrm{T}(\vec{x}, \vec{x}') \wedge (\neg F)(\vec{x}')$ is satisfiable. The procedure iterates through possible witnesses, say m, and pushes the clause $\neg \underline{m}(\vec{x})$ to the formulas F_k with $k < i$. Actually, we push a *minimal inductive cube* (MIC), c, such that $c \Rightarrow \neg \underline{m}$ and c is inductive relative to F_k. To overcome the problem with a potential infinite number of witnesses, we define the formula $\hat{m}(\vec{x}) \triangleq \bigwedge_{i \in 1..n}(x_i \geq m(p_i))$ that is valid for every marking that covers m; in the sense that $m' \models \hat{m}$ only when $m' \geq m$. By virtue of the monotonicity of the flow function of Petri nets, when $\neg F$ is monotonic and m is a witness, we know that all models of $\hat{m}$ are also witnesses. Hence we can improve the method by generating minimal inductive clauses from $\neg \hat{m}(\vec{x})$ instead of $\neg \underline{m}(\vec{x})$. Another benefit of this choice is that $\hat{m}$ is a conjunction of inequalities of the form $(x_j \geq k_i)$, which greatly simplifies the computation of the MIC. When F is anti-monotonic ($\neg F$ is monotonic), we can prove the completeness of the procedure using an adaptation of Dickson's lemma, which states that we cannot find an infinite decreasing chain of witnesses (but the number of possible witness may be extremely large).

Assume we have $(N_1, m_1) \rhd_E (N_2, m_2)$ and that $G_2(\vec{y})$ is the E-transform of formula $G_1(\vec{x})$ on N_1. We also assume that G_1 and G_2 are monotonic, in order to ensure the termination of the PDR procedure. (We can prove that $\tilde{E}$ is monotonic for systems E computed with the Reduced tool when the initial net does not use inhibitor arcs.) To check that formula G_1 is an invariant on N_1, it is enough [13] to incrementally build OARS $(F_0, \ldots, F_{k+1})$ on N_1 until $F_i = F_{i+1}$ for some index $i \in 0..k$. In this context, $F_0 = \underline{m_1}$ and $F_{k+1} \Rightarrow G_1$. In a similar way than with our extension of BMC with reductions, a corollary of our *invariant conservation* theorem (Theorem 2) is that, to check that G_1 is an invariant on N_1, it is enough to build OARS $(F_0', \ldots, F_{l+1}')$ on N_2 where $F_0' = \underline{m_2}$ and $F_{l+1}' \Rightarrow G_2$.

Theorem 5 (PDR with E-Transform). *Assume $(N_1, m_1) \rhd_E (N_2, m_2)$ and that $G_2(\vec{y})$ is the E-transform of $G_1(\vec{x})$, both monotonic formulas. Formula G_1 is an invariant on N_1 if and only if there exists $i \geq 0$ such that $F_i' = F_{i+1}'$ in the OARS built from net N_2 and formula G_2.*

Combination of BMC and PDR. In the next section, we report on the results obtained with our implementation of BMC and PDR (with and without reductions), on an independent and comprehensive set of benchmarks.

With PDR, we restrict ourselves to the proof of liveness properties, EF ϕ where ϕ is monotonic (or equivalently, invariants AG ϕ with ϕ anti-monotonic). In practice, we do not check if ϕ is monotonic using our "semantical" definition. Instead, our implementation uses a syntactical restriction that is a sufficient condition for monotonicity. This is the case, for example, when testing the quasi-liveness of a set of transitions. On the other hand, deadlock is not monotonic. In such cases, we can only rely on the BMC procedure, which may not terminate if the net has no deadlocks. Hence, our best-case scenario is when we check a monotonic property (or if a model for the property exists). In our benchmarks, we find that almost 30% of all the properties are monotonic.

We have plans to improve our PDR procedure to increase the set of properties that can be handled. In particular, we know how to do better when the net is k-bounded (and we know the value of k). We also have several proposals to improve the computation of a good witness, and its MIC, in the general case. We should explore all these ideas in a future work.

6   Experimental Results

We have implemented the approach described in Sect. 5 into a new tool, called SMPT (for Satisfiability Modulo P/T Nets). The tool is open-source, under the GPLv3 license, and is freely available on GitHub (https://github.com/nicolasAmat/SMPT/). In this section, we report on some experimental results obtained with SMPT on an extensive benchmark of models and formulas provided by the Model Checking Contest (MCC) [2,23].

SMPT serves as a front-end to generic SMT solvers, such as z3 [10,29]. The tool can output sets of constraints using the SMT-LIB format [4] and pipe them to a z3 process through the standard input. We have implemented our tool with the goal to be as interoperable as possible, but we have not conducted experiments with other solvers yet. SMPT takes as inputs Petri nets defined using the .net format of the TINA toolbox. For formulas, we accept properties defined with the XML syntax used in the MCC competition. The tool does not compute net reductions directly but relies on the tool Reduce, that we described at the end of Sect. 3.

Benchmarks and Distribution of Reduction Ratios. Our benchmark suite is built from a collection of 102 models used in the MCC competition. Most of the models are parametrized, and therefore there can be several different *instances* for the same model. There are about 1 000 different instances of Petri nets whose size vary widely, from 9 to 50 000 places, and from 7 to 200 000 transitions. Most nets are ordinary, but a significant number of them use weighted arcs. Overall, the collection provides a large number of examples with various structural and behavioral characteristics, covering a large variety of use cases.

Since our approach relies on the use of net reductions, it is natural to wonder if reductions occur in practice. To answer this question, we computed the reduction ratio (r), obtained using Reduce, as a quotient between the number of places before (p_{init}) and after (p_{red}) reduction: $r = (p_{\text{init}} - p_{\text{red}})/p_{\text{init}}$. We display the results for the whole collection of instances in Fig. 3, sorted in descending order. A ratio of 100% ($r = 1$) means that the net is *fully reduced*; the resulting net has only one (empty) marking. We see that there is a surprisingly high number of models that are totally reducible with our approach (about 20% of the total number), with approximately half of the instances that can be reduced by a ratio of 30% or more.

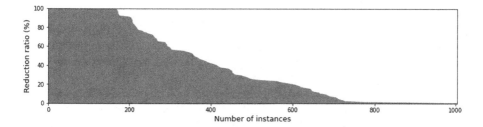

Fig. 3. Distribution of reduction ratios over the instances in the MCC

For each edition of the MCC, a collection of about 30 random reachability properties are generated for each instance. We evaluated the performance of SMPT using the formulas of the MCC2020, on a selection of 426 Petri nets taken from instances with a reduction ratio greater than 1%. (To avoid any bias introduced by models with a large number of instances, we selected at most 5 instances with a similar reduction ratio from each model).

A pair of an instance and a formula is called a *test case*. For each test case, we check the formulas with and without the help of reductions (using both the BMC and PDR methods in parallel) and with a fixed timeout of 120 s. This adds up to a total of 13 265 *test cases* which required the equivalent of 447 h of CPU time.

Impact on the Number of Solvable Queries. We report our results in the table below. We compared our results with the ones provided by an *oracle* [31], which gives the expected answer (as computed by a majority of tools, using different techniques, during the MCC competition). We achieve 100% reliability on the benchmark; meaning we always give the answer predicted by the oracle.

We give the number of computed results for four different categories of test cases: *Full* contains only the fully reducible instances (the best possible case with our approach); while *Low/Good/High* correspond to instances with a low/moderate/high level of reduction. We chose the limits for these categories in order to obtain samples with comparable sizes. We also have a general category, *All*, for the complete set of benchmarks.

	Reduction Ratio (r)	# Test cases	Results (BMC/PDR)	
			With reductions	Without
All	$r \in {]}0,1]$	13 265	6 986	3 555 (3 261/294)
Low	$r \in {]}0, 0.25[$	4 586	1 662 (1 532/130)	1 350 (1 247/103)
Good	$r \in [0.25, 0.5[$	2 823	1 176 (1 084/92)	704 (631/73)
High	$r \in [0.5, 1[$	3 298	1 591 (1 412/179)	511 (457/54)
Full	$r = 1$	2 558	2 557	990 (926/64)

We observe that we are able to compute almost twice as many results when we use reductions than without. This gain is greater on the *High* ($\times$3.1) than on the *Good* (×1.7) instances. Nonetheless, the fact that the number of additional queries solved using reductions is still substantial, even for a reduction ratio under 50%, indicates that our approach can benefit from all the reductions we can find in a model (and that our results are not skewed by the large number of fully reducible instances).

In the special case of *fully reducible* nets, checking a query amounts to solving a linear system on the initial marking of the reduced net. There are no iterations. Moreover this is the same system for both the BMC and PDR procedures. For this category, we are able to compute a result for all but one of the queries (that could be computed using a timeout of 180 s). Most of these queries can be solved in less than a few seconds.

When the distinction makes sense, we also report the number of cases solved using BMC/PDR. (As said previously, the two procedures coincide in category *Full*, with reductions.) We observe that the contribution of PDR is poor. This can be explained by several factors. First, we restricted our implementation of PDR to monotonic formulas (which represents 30% of all properties). Among these, PDR is useful only when we have an invariant that is true (meaning BMC will certainly not terminate). On the other hand, PDR is able to give answers on the most complex cases. Indeed, it is much more difficult to prove an invariant than to find a counter-example (and we have other means to try and find counter-examples, like simulation for instance). This is why we intend to improve the performances and the "expressiveness" of our PDR implementation. Another factor, already observed in [32], is the existence of a bias in the MCC benchmark: in more than 60% of the cases, the result follows from finding a counter-example (meaning an invariant that is false or a reachability property that is true).

Impact on Computation Time. To better understand the impact of reductions on the computation time, we compare the computation time, with or without reductions, for each test case. These results do not take into account the time spent for reducing each instance. This time is negligible when compared to each test, usually in the order of 1 s. Also, we only need to reduce the net once when checking the 30 properties for the same instance.

We display our results in Fig. 4, where we give four scatter plots comparing the computation time "with" (*y*-axis) and "without" reductions (*x*-axis), for the *Low*, *Good*, *High* and *Full* categories of instances. Each chart uses a logarithmic scale. We also display a histogram, for each axis on the charts, that gives the density of points for a given duration. To avoid overplotting, we removed all the "trivial" properties (the bottom left part of the chart), that can be computed with and without reduction in less than 10 ms. These "trivial" queries (507 in total) correspond to instances with a small state space or to situations where a counter-example can be found very quickly.

We observe that almost all the data points are below the diagonal, meaning reductions accelerate the computation, with many test cases exhibiting speed-ups larger than ×100. We have added two light-coloured, dashed lines to materialize data points with speed-ups larger than ×10 and ×100 respectively.

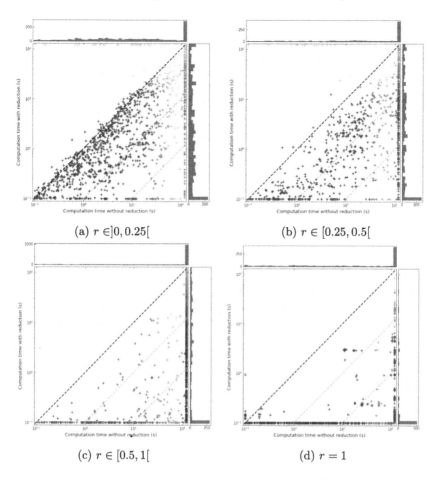

Fig. 4. Comparing computation time, "with" (y-axis) and "without" (x-axis) reductions for categories *Low* (a), *Good* (b), *High* (c) and *Full* (d).

On our 13 265 test cases, we timeout with reductions but compute a result without on only 51 cases (0.4%). These exceptions can be explained by border cases where the order in which transitions are processed has a sizeable impact.

Another interesting point is the ratio of properties that can be computed only using reductions. This is best viewed when looking at the histogram values. A vast majority of the points in the charts are either on the right border (computation without reductions timeout) or on the x-axis (they can be computed in less than 10 ms using reductions).

7   Related Work and Conclusion

We propose a new method to combine structural reductions with SMT solving in order to check invariants on arbitrary Petri nets. While this idea is not original, the framework we developed is new. Our main innovation resides in the use of a principled approach, where we can trace back reachable markings (between an initial net and its residual) by means of a conjunction of linear equalities (the formula $\tilde{E}$). Basically, we show that we can adapt a SMT-based procedure for checking a property on a net (that relies on computing a family of formulas of the form $(\phi_i)_{i \in I}$) into a procedure that relies on a reduced version of the net and formulas of the form $(\phi_i \wedge \tilde{E})_{i \in J}$.

As a proof of concept, we apply our approach to two basic implementations of the BMC and PDR procedures. Our empirical evaluation shows promising results. For example, we observe that we are able to compute twice as many results using reductions than without. We believe that our approach can be adapted to more decision procedures and could easily accommodate various types of optimizations.

Related Work. Our main theoretical results (the conservation theorems of Sect. 4) can be interpreted as examples of *reduction theorems* [26,27], that allow to deduce properties of an initial model (N) from properties of a simpler, coarser-grained version (N^R). While these works are related, they mainly focus on reductions where one can group a sequence of transitions into a single, atomic action. Hence, in our context, they correspond to a restricted class of reductions, similar to a subset of the agglomeration rules used in [7].

We can also mention approaches where the system is simplified with respect to a given property, for instance by eliminating parts that cannot contribute to its truth value, like with the slicing or *Cone of Influence* abstractions [16] used in some model checkers. Finding such "parts" (places and transitions) in a Petri net is not always easy, especially when the formula involves many places. This is not a problem with our approach, since we can always abstract away a place, as long as its effect is preserved in the E-transform formula.

In practice, we derive polyhedral abstractions using *structural reductions*, a concept introduced by Berthelot in [5]. In our work, we are interested in reductions that preserves the reachable states. This is in contrast with most works about reductions, where more powerful transformations can be applied when we focus on specific properties, such as the absence of deadlocks. Several tools use reductions for checking reachability properties. TAPAAL [11], for instance, is an explicit-state model checker that combines Partial-Order Reduction techniques and structural reductions and can check property on Petri nets with weighted arcs and inhibitor arcs.

A more relevant example is ITS Tools [32], which combines several techniques, including structural reductions and the use of SAT and SMT solvers. This tool relies on efficient methods for finding counter-examples—with the goal to invalidate an invariant—based on the collaboration between pseudo-random exploration techniques; hints computed by an SMT engine; and reductions that may simplify atoms in the property or places and transitions in the net. It also describes a semi-decision procedure, based on an over-approximation of the state space, that may detect when an invariant holds (by ruling out infeasible behaviours). This leads to a very efficient tool, able to compute a result for most of the queries in our benchmark, when we solve only 52% of our test cases. Nonetheless, we are able to solve 46 queries with SMPT (with a timeout of 120 s) that are not in the oracle results collected from ITS Tools [31].

It has to be kept in mind, though, that our goal is to study the impact of polyhedral abstractions, in isolation from other techniques. However, the methods described in [32] provide many ideas for improving our approach, such as: using linear arithmetic over reals—which is more tractable than integer arithmetic—to over-approximate the state space of a net; adding extra constraints to strengthen invariants (for instance using the state equation or constraints derived from traps); dividing up a formula into smaller sub-parts, and checking them incrementally or separately; ... But the main lesson to be learned is that there is a need for a complete decision procedure devoted to the proof of satisfiable invariants, which further our interest in improving our implementation of PDR.

Indeed, a byproduct of our work is to provide a partial implementation of PDR that is correct and complete when the property is monotonic (see Sect. 4), even in the case of nets that are not bounded. Our current solution can be understood as a restriction to the case of "coverability properties", which seems to be the current state-of-the-art with Petri nets; see for example [19] or the extension of PDR to "well-structured transition systems" [25]. We can also mention the works on inductive procedures for infinite-state and/or parametrized systems, such as the verification methods used in Cubicle [18], or in [15,22].

Future Work. We propose a new method that adapts our approach—initially developed for model checking with decision diagrams [6,7]—for use with SMT solvers. We plan to continue in this direction, trying new verification methods and tackling properties more complex than reachability. For example, we already have plans [1] to apply our notion of polyhedral abstraction to the concurrent places problem [12].

There is also ample room for improving our tool. We already mentioned some ideas for enhancements that we could borrow from ITS Tools, but we also plan to specialize our verification procedures in some specific cases, for example when we know that a net is 1-safe. A first step should be to compare our performances with other tools in more details. This is what motivate our participation to the next edition of the MCC, with SMPT alone in the reachability examinations, even though it is common knowledge that winning tools need to combine several different techniques.

Finally, the most promising part of our work is to improve our adaptation of PDR, which raises several interesting problems. We have several ideas on how to improve our adaptation of PDR, and the computation of the Minimal Inductive Cube (MIC), while retaining completeness in the case of bounded nets. This will be the subject of a future work.

References

1. Amat, N.: A New Approach for the Symbolic Model Checking of Petri nets. Master's thesis, University of Grenoble (2020)

2. Kordon, F., et al.: MCC'2017 – the seventh model checking contest. In: Koutny, M., Kristensen, L.M., Penczek, W. (eds.) Transactions on Petri Nets and Other Models of Concurrency XIII. LNCS, vol. 11090, pp. 181–209. Springer, Heidelberg (2018). https://doi.org/10.1007/978-3-662-58381-4_9

3. Armando, A., Mantovani, J., Platania, L.: Bounded model checking of software using SMT solvers instead of SAT solvers. In: Valmari, A. (ed.) SPIN 2006. LNCS, vol. 3925, pp. 146–162. Springer, Heidelberg (2006). https://doi.org/10.1007/11691617_9

4. Barrett, C., Fontaine, P., Tinelli, C.: The SMT-LIB Standard: Version 2.6. Technical report, Department of Computer Science, The University of Iowa (2017). http://www.smt-lib.org/

5. Berthelot, G.: Transformations and decompositions of nets. In: Brauer, W., Reisig, W., Rozenberg, G. (eds.) ACPN 1986, Part I. LNCS, vol. 254, pp. 359–376. Springer, Heidelberg (1987). https://doi.org/10.1007/978-3-540-47919-2_13

6. Berthomieu, B., Le Botlan, D., Dal Zilio, S.: Petri net reductions for counting markings. In: Gallardo, M.M., Merino, P. (eds.) SPIN 2018. LNCS, vol. 10869, pp. 65–84. Springer, Cham (2018). https://doi.org/10.1007/978-3-319-94111-0_4

7. Berthomieu, B., Le Botlan, D., Dal Zilio, S.: Counting Petri net markings from reduction equations. Int. J. Softw. Tools Technol. Transfer (2019). https://doi.org/10.1007/s10009-019-00519-1

8. Besson, F., Jensen, T., Talpin, J.-P.: Polyhedral analysis for synchronous languages. In: Cortesi, A., Filé, G. (eds.) SAS 1999. LNCS, vol. 1694, pp. 51–68. Springer, Heidelberg (1999). https://doi.org/10.1007/3-540-48294-6_4

9. Biere, A., Cimatti, A., Clarke, E., Zhu, Y.: Symbolic model checking without BDDs. In: Cleaveland, W.R. (ed.) TACAS 1999. LNCS, vol. 1579, pp. 193–207. Springer, Heidelberg (1999). https://doi.org/10.1007/3-540-49059-0_14

10. Bjørner, N.: The z3 theorem prover (2020). https://github.com/Z3Prover/z3/

11. Bønneland, F.M., Dyhr, J., Jensen, P.G., Johannsen, M., Srba, J.: Stubborn versus structural reductions for petri nets. J. Logic. Algebraic Methods Program. **102**, 46–63 (2019). https://doi.org/10.1016/j.jlamp.2018.09.002

12. Bouvier, P., Garavel, H., Ponce-de-León, H.: Automatic decomposition of petri nets into automata networks – a synthetic account. In: Janicki, R., Sidorova, N., Chatain, T. (eds.) PETRI NETS 2020. LNCS, vol. 12152, pp. 3–23. Springer, Cham (2020). https://doi.org/10.1007/978-3-030-51831-8_1

13. Bradley, A.R.: SAT-based model checking without unrolling. In: Jhala, R., Schmidt, D. (eds.) VMCAI 2011. LNCS, vol. 6538, pp. 70–87. Springer, Heidelberg (2011). https://doi.org/10.1007/978-3-642-18275-4_7

14. Bradley, A.R.: Understanding IC3. In: Cimatti, A., Sebastiani, R. (eds.) SAT 2012. LNCS, vol. 7317, pp. 1–14. Springer, Heidelberg (2012). https://doi.org/10.1007/978-3-642-31612-8_1

15. Cimatti, A., Griggio, A., Mover, S., Tonetta, S.: Infinite-state invariant checking with IC3 and predicate abstraction. Formal Methods Syst. Des. **49**(3), 190–218 (2016). https://doi.org/10.1007/s10703-016-0257-4

16. Clarke, E.M., Grumberg, O., Peled, D.: Model Checking. MIT Press, Cambridge (1999)

17. Clarke, E., Biere, A., Raimi, R., Zhu, Y.: Bounded model checking using satisfiability solving. Formal Methods Syst. Des. **19**(1), 7–34 (2001). https://doi.org/10.1023/A:1011276507260

18. Conchon, S., Goel, A., Krstić, S., Mebsout, A., Zaïdi, F.: Cubicle: a parallel SMT-based model checker for parameterized systems. In: Madhusudan, P., Seshia, S.A. (eds.) CAV 2012. LNCS, vol. 7358, pp. 718–724. Springer, Heidelberg (2012). https://doi.org/10.1007/978-3-642-31424-7_55

19. Esparza, J., Ledesma-Garza, R., Majumdar, R., Meyer, P., Niksic, F.: An SMT-based approach to coverability analysis. In: Biere, A., Bloem, R. (eds.) CAV 2014. LNCS, vol. 8559, pp. 603–619. Springer, Cham (2014). https://doi.org/10.1007/978-3-319-08867-9_40

20. Feautrier, P.: Automatic parallelization in the polytope model. In: Perrin, G.-R., Darte, A. (eds.) The Data Parallel Programming Model. LNCS, vol. 1132, pp. 79–103. Springer, Heidelberg (1996). https://doi.org/10.1007/3-540-61736-1_44

21. Finkel, A.: The minimal coverability graph for Petri nets. In: Rozenberg, G. (ed.) ICATPN 1991. LNCS, vol. 674, pp. 210–243. Springer, Heidelberg (1993). https://doi.org/10.1007/3-540-56689-9_45

22. Gurfinkel, A., Shoham, S., Meshman, Y.: SMT-based verification of parameterized systems. In: International Symposium on Foundations of Software Engineering. ACM (2016). https://doi.org/10.1145/2950290.2950330

23. Hillah, L.M., Kordon, F.: Petri nets repository: a tool to benchmark and debug petri net tools. In: van der Aalst, W., Best, E. (eds.) PETRI NETS 2017. LNCS, vol. 10258, pp. 125–135. Springer, Cham (2017). https://doi.org/10.1007/978-3-319-57861-3_9

24. Hujsa, T., Berthomieu, B., Dal Zilio, S., Le Botlan, D.: Checking marking reachability with the state equation in petri net subclasses (2020)

25. Kloos, J., Majumdar, R., Niksic, F., Piskac, R.: Incremental, inductive coverability. In: Sharygina, N., Veith, H. (eds.) CAV 2013. LNCS, vol. 8044, pp. 158–173. Springer, Heidelberg (2013). https://doi.org/10.1007/978-3-642-39799-8_10

26. Cohen, E., Lamport, L.: Reduction in TLA. In: Sangiorgi, D., de Simone, R. (eds.) CONCUR 1998. LNCS, vol. 1466, pp. 317–331. Springer, Heidelberg (1998). https://doi.org/10.1007/BFb0055631

27. Lipton, R.J.: Reduction: a method of proving properties of parallel programs. Commun. ACM **18**(12), 717–721 (1975). https://doi.org/10.1145/361227.361234

28. Lloret, J.C., Azéma, P., Vernadat, F.: Compositional design and verification of communication protocols, using labelled petri nets. In: Clarke, E.M., Kurshan, R.P. (eds.) CAV 1990. LNCS, vol. 531, pp. 96–105. Springer, Heidelberg (1991). https://doi.org/10.1007/BFb0023723

29. de Moura, L., Bjørner, N.: Z3: an efficient SMT solver. In: Ramakrishnan, C.R., Rehof, J. (eds.) TACAS 2008. LNCS, vol. 4963, pp. 337–340. Springer, Heidelberg (2008). https://doi.org/10.1007/978-3-540-78800-3_24

30. Silva, M., Terue, E., Colom, J.M.: Linear algebraic and linear programming techniques for the analysis of place/transition net systems. In: Reisig, W., Rozenberg, G. (eds.) ACPN 1996. LNCS, vol. 1491, pp. 309–373. Springer, Heidelberg (1998). https://doi.org/10.1007/3-540-65306-6_19

31. Thierry-Mieg, Y.: Oracle for the MCC 2020 edition (2020). https://github.com/yanntm/pnmcc-models-2020

32. Thierry-Mieg, Y.: Structural reductions revisited. In: Janicki, R., Sidorova, N., Chatain, T. (eds.) PETRI NETS 2020. LNCS, vol. 12152, pp. 303–323. Springer, Cham (2020). https://doi.org/10.1007/978-3-030-51831-8_15

33. Thierry-Mieg, Y., Poitrenaud, D., Hamez, A., Kordon, F.: Hierarchical set decision diagrams and regular models. In: Kowalewski, S., Philippou, A. (eds.) TACAS 2009. LNCS, vol. 5505, pp. 1–15. Springer, Heidelberg (2009). https://doi.org/10.1007/978-3-642-00768-2_1

Skeleton Abstraction for Universal Temporal Properties

Sophie Wallner and Karsten Wolf[(✉)]

Universität Rostock, Rostock, Germany
{sophie.wallner,karsten.wolf}@uni-rostock.de

Abstract. Uniform coloured Petri nets can be abstracted to their *skeleton*, the place/transition net that simply turns the coloured tokens into black tokens. A coloured net and its skeleton are related by a *net morphism* [Des91,PGE98]. For the application of the skeleton as an abstraction method in the model checking process, we need to establish a *simulation relation* [Mil89] between the state spaces of the two nets. Then, universal temporal properties (properties of the $ACTL^*$ logic) are preserved. The abstraction relation induced by a net morphism is not necessarily a simulation relation, due to a subtle issue related to deadlocks [Fin92]. We discuss several situations where the abstraction relation induced by a net morphism is as well a simulation relation, thus preserving $ACTL^*$ properties. We further propose a partition refinement algorithm for folding a place/transition net into a coloured net. This way, skeleton abstraction becomes available for models given as place/transition nets. Experiments demonstrate the capabilities of the proposed technology. Using skeleton abstraction, we are capable of solving problems that have not been solved before in the Model Checking Contest [KGH+19].

1 Introduction

In the model checking process for coloured Petri nets, one of the biggest issues is the state explosion problem, which makes the verification of a property impossible, as the state space is getting too big to handle. A way to deal with these big systems, is the wellknown technique of abstraction. Given a coloured Petri net C, we can form its *skeleton* S, which has the structure of C and simply decolours its components and tokens. This skeleton is an abstraction of the coloured net, its behaviour includes the behaviour of C. To use this abstraction technique in the model checking process, we need to guarantee, that properties are preserved through this abstraction, i.e. that the validity of property in S indicates the validity of the property in C. Unfortunately, this is not the case for every coloured net. The issue is that some deadlocks of C are not preserved in S, such that the additional behaviour of S changes the validity of the property. Deadlocks in a coloured net can have two different causes. First, they can be caused by an insufficient number of tokens in the preset of a transition. These deadlocks are preserved in the skeleton, as the number of tokes will neither be sufficient in the skeleton. Second, they can be caused by a wrong colour set of

© Springer Nature Switzerland AG 2021
D. Buchs and J. Carmona (Eds.): PETRI NETS 2021, LNCS 12734, pp. 186–207, 2021.
https://doi.org/10.1007/978-3-030-76983-3_10

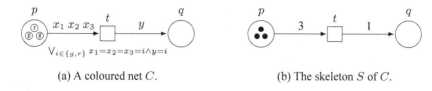

(a) A coloured net C.   (b) The skeleton S of C.

Fig. 1. A coloured net C with a deadlock not preserved in its skeleton S.

tokens, as the number of tokens in the preset of a transition is sufficient, but the colour distribution of the tokens violates the guard of the transition. This type of deadlocks is usually not preserved in the skeleton, as the skeleton does not distinguish colors at all. Consider the following example:

Example 1. Let C be a coloured Petri net, for which we can build the skeleton S, just by removing the colour sets of the places, the guard of the transition and making the tokens all indistinguishable. The two nets are pictured in Fig. 1. We consider the $ACTL^*$ formula $\varphi : \mathbf{AF}p \leq 1$. The colour sets of C are $\chi(p) = \chi(q) = \{g, r\}$. The guard of the only transition t is $\gamma(t) : \bigvee_{i \in \{g,r\}} x_1 = x_2 = x_3 = i \wedge y = i$, which means that t requires three tokens of the same colour and produces one token of this colour. In the given marking of C, t is not enabled, so this marking is a deadlock. The corresponding marking of S is not a deadlock, as the number of tokens is sufficient and t is activated. Firing t in S leads to the marking, where all tokens are removed from p, so φ is true for S. Meanwhile, φ is not true for C. Transferring the validity of φ from S to C will draw a wrong conclusion.

This work will give a detailed analysis of situations, where the skeleton as an abstraction technique is soundly applicable in the model checking process for coloured Petri nets. It is structured as follows: The next Sect. 2 will give an overview of the application of skeleton nets in other contexts, Sect. 3 provides necessary basic definitions. After that, Sect. 4 will introduce the different concepts of relations, which can hold between reachability graphs of Petri nets. Section 5 will set the focus on the simulation relation between reachability graphs as the core concept for keeping validity through abstraction. We there present an survey, in which cases, the skeleton is a valid abstraction method for a coloured Petri net, distinguishing different classes of nets and types of formulas. With the folding algorithm in Sect. 6, we extend the scope of application of the skeleton abstraction to place/transition nets. The experimental results in Sect. 7 underline the powerfullness of this abstraction method.

2   Related Work

The idea of a skeleton-based analysis of a Petri net is subject of [Vau87]. Based on this, [Fin92] examines the role of deadlocks within this topic more precisely. The results are also applied in other contexts. In [RB01] extended Pr/T-Nets are used as a modeling formalism for embedded real-time systems due to the multitude of analysis methods for Petri nets. With the skeleton of a Pr/T-Net, properties like reachability of states or deadlock freeness can be examined. [Lil95] transfers Findlow's results [Fin92] to algebraic nets. They are used as an application for a folding construction, which is described there. Findlow's observations on deadlock-preserving skeletons are further used in [SMS99] for a skeleton-based analysis of G-Nets, an object-based Petri net formalism. The preservation of predicates in temporal logic under morphisms has also already been discussed. [PGE98] describes a rule-based modification of algebraic high-level nets extended with morphisms such that safety properties described in temporal logic are preserved. This provides a technique which allows to transfer safety properties between the source and the target net.

3   Basic Definitions

First, we present definitions for place/transition nets.

Definition 1 (Place/Transition Net). *A place/transition net (P/T net) is a tuple $N = [P, T, F, W, m_0]$, where P is a finite set of* places *and T is a finite set of* transitions *with $P \cap T = \emptyset$. The arcs $F \subseteq (P \times T) \cup (T \times P)$ of the net are labeled by a* weight function *$W : F \to \mathbb{N}$, with $W(x, y) = 0$ iff $(x, y) \notin F$. m_0 is the* initial marking *of the net, whereas $m : P \to \mathbb{N}$ describes a marking.*

The behavior of a P/T net is defined by the transition rule.

Definition 2 (Transition rule of a P/T net). *Let $N = [P, T, F, W, m_0]$ be a P/T net. Transition $t \in T$ is enabled in marking m if $\forall p \in P : W(p, t) \leq m(p)$. Firing Transition t leads from marking m to marking m' (denoted as $m \xrightarrow{t} m'$) in N, if t is enabled in m and $\forall p : m'(p) = m(p) - W(p, t) + W(t, p)$.*

A marking m' is *reachable* from a marking m (denoted as $m \xrightarrow{*} m'$), if there is a firing sequence $t_1 t_2 \ldots t_n \in T^*$, such that $m \xrightarrow{t_1} m_1 \xrightarrow{t_2} \ldots \xrightarrow{t_n} m'$. We extend the notation of reachability to firing sequences $\omega \in T^*$ and we call $RS(N) = \{m \mid \exists \omega \in T^* : m_0 \xrightarrow{\omega} m\}$ the *reachability set*, which contains all of N's reachable markings. Using the transition rule, a Petri net induces a labeled transition system, called the *reachability graph*.

Definition 3 (Labeled Transition System, Reachability Graph). *A transition system $TS = [Q, q_0, R, A]$ is a labeled, directed graph, where Q is the set of states, $q_0 \in Q$ is the initial state and a transition relation $R \subseteq Q \times A \times Q$ with some set of actions A. The reachability graph $R_N(m_0)$ of a Petri net N is a transition system, where the set of states is $RS(N)$, m_0 serves as the initial state and $(m, t, m') \in R$ iff $m \xrightarrow{t} m'$.*

Furthermore, we introduce a simple notion for coloured Petri nets with finite colour domains.

Definition 4 (Coloured Petri net). *A coloured Petri net* $C = [P_C, T_C, F_C, W_C, \chi, \gamma, m_{0C}]$ *consists of a finite set* P_C *of places, a finite set* T_C *of transitions where* $P_C \cap T_C = \emptyset$ *and a set of arcs* $F_C \subseteq (P_C \times T_C) \cup (T_C \times P_C)$. *The weight function* W_C *assigns a finite set of variables to each element of* F_C. *If* $(x,y) \notin F_C$, *we assume* $W_C(x,y) = \emptyset$. *The colouring function* χ *assigns a finite set* $\chi(p)$ *of colours to each place* $p \in P_C$, *called colour domain of* p. *The guard function* γ *assigns a boolean predicate* $\gamma(t)$ *to each transition* $t \in T_C$, *which ranges over the variables of* $W_C(p,t) \cup W_C(t,p)$ *for all* $p \in P_C$. *The initial marking* m_0 *is a multiset over* $\chi(p)$ *for every* $p \in P_C$. *The number of tokens of colour* c *on place* p *in marking* m *is described as* $m(p)(c)$.

For a Transition $t \in T_C$, we define a *firing mode* of t as a mapping $g : \bigcup_{p \in P_C} (W_C(p,t) \cup W_C(t,p)) \rightarrow \bigcup_{p \in P_C} \chi(p)$, which assigns a colour from $\chi(p)$ for every place $p \in P_C$ and for each variable $x \in W_C(p,t) \cup W_C(t,p)$. A firing mode g of a transition t satisfies the guard $\gamma(t)$, denoted as $g \models \gamma(t)$, if the assignment of colours to variables is a model of the guard. Usually, definitions of coloured nets permit a richer syntax for arc weights. However, any complex arc inscription w can be replaced by a simple variable x and by adding $x = w$ to the guard of the involved transition. Thus, simplicity of our definition does not undermine expressivity. For a coloured net, its unfolding can be defined.

Definition 5 (Unfolding). *Let* $C = [P_C, T_C, F_C, W_C, \chi, \gamma, m_{0C}]$ *be a coloured Petri net. A P/T net* $U = [P_U, T_U, F_U, W_U, m_{0U}]$ *is the* unfolding *of* C *if*

- $P_U = \{[p,c] \mid p \in P_C, c \in \chi(p)\}$
- $T_U = \{[t,g] \mid t \in T_C, g \models \gamma(t)\}$
- $([p,c], [t,g]) \in F_U$, *iff* $(p,t) \in F_C$ *and* $c \in g(W_C(p,t))$
- $([t,g], [p,c]) \in F_U$, *iff* $(t,p) \in F_C$ *and* $c \in g(W_C(t,p))$
- $W_U([p,c], [t,g]) = \mathrm{card}(\{x \mid x \in W_C(p,t), g(x) = c\})$
- $W_U([t,g], [p,c]) = \mathrm{card}(\{x \mid x \in W_C(t,p), g(x) = c\})$
- $m_{0U}([p,c]) = m_{0C}(p)(c)$.

In the sequel, refer to the transition system defined by a coloured net C as the transition system of its unfolding U, as they are isomorphic [JK09]. Coloured nets as defined above are *uniform*. This means that the number of tokens consumed or produced on any arc are independent of the particular firing mode, i.e., always $\mathrm{card}(W(x,y))$ tokens. There exist non-uniform variants of coloured nets. They use variables that take multisets over $\chi(p)$ as values of their variables. They are, however, out of the scope of this article since the core artifact studied in this paper, the skeleton, is not applicable to non-uniform nets. For a uniform net, we can assign a second P/T net, its skeleton.

Definition 6 (Skeleton). *Let* $C = [P_C, T_C, F_C, W_C, \chi, \gamma, m_{0C}]$ *be a coloured net. Its skeleton* $S = [P_S, T_S, F_S, W_S, m_{0S}]$ *is a P/T net where*

- $P_S = P_C$, $T_S = T_C$, $F_S = F_C$
- *for all* $x,y \in P \cup T : W_S(x,y) = \mathrm{card}(W_C(x,y))$
- *for all* $p \in P : m_{0S}(p) = \sum_{c \in \chi(p)} m_0(p)(c)$.

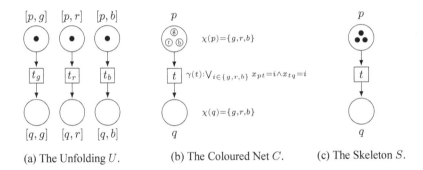

(a) The Unfolding U.   (b) The Coloured Net C.   (c) The Skeleton S.

Fig. 2. A coloured Petri net C, its unfolding U and its skeleton net S.

The following example will help to understand the concepts of the unfolding and the skeleton of a coloured net.

Example 2. Let C be the given coloured Petri net, as depicted in Fig. 2. Place p and q have the colour domain $\chi(p) = \chi(q) = \{g, r, b\}$. Unfolding C leads to the corresponding places $[p, g], [p, r], [p, b]$ resp. $[q, g], [q, r], [q, b]$. For every firing mode, which satisfies the guard of transition t, we introduce one transition in the unfolding, so the unfolding has three transitions t_g, t_r, t_b. Building the skeleton makes all tokens on p indistinguishable and removes the colour sets of p and q. The transition t in the skeleton has no guard and is simply acitvated, if there is a sufficient number of tokens on p.

In the sequel, unless stated otherwise, let C be an arbitrary but fixed coloured net, U its unfolding, and S its skeleton. U and S are related by a net morphism.

Definition 7 (Net Morphism [Des91]). *Let $N_1 = [P_1, T_1, F_1, W_1, m_{01}]$ and $N_2 = [P_2, T_2, F_2, W_2, m_{02}]$ be arbitary P/T nets. A net morphism from N_1 to N_2 is a mapping $\mu : (P_1 \cup T_1) \rightarrow (P_2 \cup T_2)$ such that $\mu(P_1) \subseteq P_2$, $\mu(T_1) \subseteq T_2$ and $\forall x, y \in P_1 \cup T_1 : W(\mu(x), \mu(y)) = W(x, y)$. For the initial markings, it holds that $\forall p_2 \in P_2 : m_{02}(p_2) = \sum_{p_1 \in P_1 : (p_1, p_2) \in \mu} m_{01}(p_1)$.*

A net morphism can be extended to a mapping from markings of N_1 to markings of N_2 by setting $m_2(p_2) = \sum_{p_1 \in P_1 : (p_1, p_2) \in \mu} m_1(p_1)$ for all $p_2 \in P_2$, where $m_1 \in RS(N_1)$ and $m_2 \in RS(N_2)$. A net morphism preserves the reachability between the related nets.

Proposition 1 (Net Morphism preserves reachability [Pin11]). *Let N_1, N_2 be two P/T nets, related by a net morphism μ. The transition $m \xrightarrow{t} m'$ in N_1 implies the transition $\mu(m) \xrightarrow{\mu(t)} \mu(m')$ in N_2.*

It is easy to see that U and S are related by a net morphism.

Proposition 2 (Net morphism from unfolding to skeleton [Des91, PGE98]). *Let C be a coloured Petri net, U its unfolding and S its skeleton. The mapping $\mu : (P_U \cup T_U) \rightarrow (P_s \cup T_s)$ is a net morphism from U to S, where*

- $\forall [p, c] \in P_U : \mu([p, c]) = p \in P_s$
- $\forall [t, g] \in T_U : \mu([t, g]) = t \in T_s$

The net morphism μ between U and S can as well be extended to the markings of U and S, such as $m_s(p) = \sum_{[p,c] \in P_U : ([p,c],p) \in \mu} : m_U([p, c])$ for all $p \in P_s$, where $m_U \in RS(U)$ and $m_s \in RS(S)$.

We continue with the introduction of the syntax and semantics of the temporal logic CTL^*. The foundation for this logic are atomic propositions, properties which is either true or false. CTL^* distinguishes state formulas and path formulas.

Definition 8 (Syntax of CTL^*). *The temporal logic CTL^* is inductively defined as follows:*

- *every atomic proposition is a state formula*
- *if φ and ψ are state formulas, so are $(\varphi \wedge \psi)$, $(\varphi \vee \psi)$, and $\neg \varphi$*
- *every state formula is a path formula*
- *if φ and ψ are path formulas, so are $(\varphi \wedge \psi)$, $(\varphi \vee \psi)$, $\neg \varphi$, $\mathbf{X} \varphi$, $\mathbf{F} \varphi$, $\mathbf{G} \varphi$, $(\varphi \mathbf{U} \psi)$, $(\varphi \mathbf{W} \psi)$, and $(\varphi \mathbf{R} \psi)$*
- *if φ is a path formula then $\mathbf{E} \varphi$ and $\mathbf{A} \varphi$ are state formulas.*

The semantics of CTL^* relies on the concept of paths in the considered system, given as a transition system.

Definition 9 (Path,Suffix). *Let $TS = [Q, q_0, R, A]$ be a transition system. A finite path starting in state q_0 is a sequence $\pi = q_0 \ldots q_n$ of states where $\forall i \in \{0, \ldots, n-1\} : (q_i, a, q_{i+1}) \in R$. An infinite path starting in q_0 is an infinite sequence $\pi = q_0 q_1 \ldots$ where $\forall i \in \mathbb{N} : (q_i, a, q_{i+1}) \in R$. A path is a finite or infinite path. A path is maximal, if it is infinite, or is a finite path $q_1 \ldots q_n$ where q_n is a deadlock, i.e., a state where, for all $q \in Q$, $(q_n, a, q) \notin R$. As a Suffix of a path π we define π_i as the part of π, starting in q_i.*

The semantics of CTL^* is defined on infinite paths, as we find them in *Kripke structures*. A Kripke structure is a transition system $K = [Q, q_0, R, A, L]$ where R is total, i.e., every state has at least one successor state. Thus the maximal paths are always infinite here. Aditionally, Kripke structures have a labelling function $L : Q \to 2^{AP}$, which assigns the set of atomic propositions to every state, which are true in this state. Every transition system can canonically be transformed into a Kripke structure by adding a *silent transition action* (q_d, τ, q_d) to R for each deadlock state q_d of the system, which does not have a successor state. The semantics of CTL^* is defined by two satisfaction relations, both denoted with $\models$, that relate markings and state formulas resp. infinite paths and path formulas according to the following rules.

Definition 10 (Semantics of CTL^*). *Let $K = [Q, q_0, R, A, L]$ be a Kripke structure. Let $q \in Q$ be a state and $\pi = q_0 q_1 \ldots$ an infinite path of the system. The satisfaction of a CTL^* formula is defined:*

- *For an atomic proposition φ: let $q \models \varphi$ corresponding to Definition 14.*
- *For a state formula φ: $\pi \models \varphi$, if $q_0 \models \varphi$.*
- *Boolean connectors:*
  - $q \models \neg\varphi$, if $q \not\models \varphi$; $\pi \models \neg\varphi$ if $q_0 \not\models \varphi$
  - $q \models (\varphi \wedge \psi)$, if $q \models \varphi$ and $q \models \psi$; $\pi \models (\varphi \wedge \psi)$ if $\pi \models \varphi$ and $\pi \models \psi$.
- *Temporal operators:*
  - $\pi \models \mathbf{X}\varphi$, if $\pi_1 \models \varphi$
  - $\pi \models (\varphi \mathbf{U} \psi)$, if $\exists i \geq 0 : \pi_i \models \psi$ and $\forall 0 \leq j < i : \pi_j \models \varphi$.
- *Path quantifier: $q_1 \models \mathbf{E}\varphi$, if $\exists \pi : \pi \models \varphi$.*

Let further $\varphi \vee \psi$ be equivalent to $\neg(\neg\varphi \wedge \neg\psi)$, $\mathbf{F}\varphi$ to true $\mathbf{U}\varphi$, $\mathbf{G}\varphi$ to $\neg\mathbf{F}\neg\varphi$, $\varphi \mathbf{R}\psi$ to $\neg(\neg\varphi \mathbf{U} \neg\psi)$, $\varphi \mathbf{W}\psi$ to $\mathbf{G}\varphi \vee (\varphi \mathbf{U} \psi)$ and $\mathbf{A}\varphi$ to $\neg\mathbf{E}\neg\varphi$.

A Kripke structure satisfies a state formula if its initial states does. It satisfies a path formula if all paths starting in the initial state do. For CTL^*, several fragments are frequently studied.

Definition 11 (Fragments of CTL^*). CTL^* *formula φ is in*

- *LTL if φ does neither contain $\mathbf{E}$ nor $\mathbf{A}$*
- *ACTL* if φ does neither contain $\mathbf{E}$ nor $\neg$*
- *CTL if every occurrence of $\mathbf{X}, \mathbf{U}, \mathbf{F}, \mathbf{G}, \mathbf{R}$ is immediately preceded by an occurrence of $\mathbf{A}$ or $\mathbf{E}$*
- *ACTL if φ is in ACTL* and CTL*
- *for any fragment F, φ is in the fragment F_X if φ is in F and does not contain $\mathbf{X}$.*

Since CTL and LTL contain, for all their operators, the dual operator w.r.t negation, we can push negations to the bottom of formulas. Consequently, LTL is indeed a subset of $ACTL^*$.

4   Relations Between Reachability Graphs

For describing relations between reachability graphs, we use the concepts of abstraction relation and simulation relation, defined for Kripke structures.

Definition 12 (Abstraction Relation [Mil89]). *Let $K = [Q, q_0, R, A, L]$ and $\hat{K} = [\hat{Q}, \hat{q}_0, \hat{R}, \hat{A}, \hat{L}]$ be Kripke structures. An abstraction relation exists between K and $\hat{K}$, if there is a surjective abstraction function $\sigma : Q \rightarrow \hat{Q}$, for which it holds that for every $\hat{q} \in \hat{Q}$ and $\forall a \in AP : \hat{q} \models a \Leftrightarrow \forall q \in Q$ with $(q, \hat{q}) \in \sigma : q \models a$.*

If such an abstraction relation exists between K and $\hat{K}$, we say that $\hat{K}$ (abstract system) abstracts K (concrete system). A particular type of abstraction relation is the simulation relation.

Definition 13 (Simulation Relation [GL94]). *An abstraction relation between $K = [Q, q_0, R, A, L]$ and $\hat{K} = [\hat{Q}, \hat{q}_0, \hat{R}, \hat{A}, \hat{L}]$ with the abstraction function $\sigma : Q \rightarrow \hat{Q}$ is a simulation relation, if $\forall q, q_1 \in Q : q \xrightarrow{*} q_1$ and $(q, \hat{q}) \in \sigma \rightarrow \exists \hat{q}_1 \in \hat{Q} : \hat{q} \xrightarrow{*} \hat{q}_1$ for $\hat{q} \in \hat{Q}$ and $(q_1, \hat{q}_1) \in \sigma$.*

If a simulation relation exists between K and $\hat{K}$, we say that $\hat{K}$ simulates K. $ACTL^*$ properties are preserved through a simulation relation.

Proposition 3 (Simulation Relation preserves $ACTL^*$ [CES86]). *Let* $K = [Q, q_0, R, A, L]$ *and* $\hat{K} = [\hat{Q}, \hat{q}_0, \hat{R}, \hat{A}, \hat{L}]$ *be Kripke structures. If there is a simulation relation between K and $\hat{K}$, for every $ACTL^*$ formula φ, it holds that* $\hat{K} \models \varphi \Rightarrow K \models \varphi$.

As deadlocks may occur in Petri nets, a reachability graph is not necessarily a Kripke structure. To make the concepts of abstraction and simulation formally applicable to Petri nets, we need to transform the reachability graphs into Kripke structures, as described above. Thus, for every deadlock marking m_d in a reachability graph, we add a self-loop (m_d, τ, m_d) with a silent transition τ to R. From now on, consider the reachability graphs of Petri nets as Kripke structures, arised out of this transformation. For an abstraction relation, atomic propositions are essential, so we first specify atomic propositions in the context of Petri nets.

Definition 14 (Atomic proposition). *Let N be a Petri net. An* atomic proposition *is one of the constants* true *and* false *or an expression* $k_1 p_1 + \ldots k_n p_n \leq k$, *for some $n \in \mathbb{N}$ with $k_1, \ldots, k_n, k \in \mathbb{Z}$, and $p_1, \ldots, p_n \in P$, where P is the set of places of N. A marking m of a P/T net satisfies the proposition $k_1 p_1 + \cdots + k_n p_n \leq k$, iff the term $\sum_{i=1}^{n} k_i \cdot m(p_i)$ evaluates to a number less or equal to k. A marking m of coloured net satisfies proposition $k_1 p_1 + \cdots + k_n p_n \leq k$, iff the term $\sum_{i=1}^{n} k_i \cdot \sum_{c \in \chi(p_i)} m(p_i)(c)$ evaluates to a number less or equal to k. For both, $m \models a$ denote the fact that m satisfies atomic proposition a.*

The net morphism $\mu : (P_U \cup T_U) \to (P_S \cup T_S)$ between the unfolding U of a coloured net C and its skeleton S induces an abstraction relation between their reachability graphs. To show this, we need to specify the unfolding of atomic propositions of coloured nets. As S resp. C normally have another set of places as U, the equisatisfiability between the concrete and the abstract states, required in Definition 12, is not trivial.

Definition 15 (Unfolding of Atomic Propositions). *Let $\mu : (P_U \cup T_U) \to (P_S \cup T_S)$ be the net morphism between U and S. Let $a_c \in AP_C$ an atomic proposition of a coloured net C. Proposition a_c can be unfolded to an atomic proposition $a_U \in AP_U$ by substituting every occurrence of any place $p \in P_C$ by $\sum_{[p,c] \in P_U : \mu([p,c]) = p} [p, c]$ for $[p, c] \in P_U$.*

An atomic proposition a_C and its unfolding a_U are equisatisfiable. To make the unfolding of atomic propositions more clear, consider Example 1 and the atomic proposition $p \leq 3$. As the colour domain of p is $\chi(p) = \{g, r, b\}$ and p would be unfolded to the places $[p, g], [p, r]$, and $[p, b]$, we unfold the atomic proposition to $[p, g] + [p, r] + [p, b] \leq 3$.

With this definition we can build an abstraction relation between a the unfolding of a coloured net and its skeleton.

Proposition 4 (Abstraction Relation between U and S). *Let U and S be related by the net morphism $\mu : (P_U \cup T_U) \to (P_s \cup T_s)$ from Proposition 2. The extension of μ on the markings of U and S yields to a surjective abstraction function σ with $(m_U, m_s) \in \sigma$ for $(m_U, m_s) \in \mu$. Therefore, an abstraction relation between the markings of U and S exists.*

It is worth mentioning, that markings here include reachable and non-reachable markings.

Proof. Let $a_U \in AP_U$, $a_C \in AP_C$ and $a_s \in AP_s$ be atomic propositions. The relation σ is an abstraction relation indeed, if for a marking m_s of S, $m_s \models a_s \Leftrightarrow \forall m_U \models a_U$ with $(m_U, m_s) \in \sigma$. If $m_s \models a_s$, then $\sum_{i=1}^{n} k_i \cdot m_s(p_i) \leq k$. For every corresponding marking m_C of C, it holds that $m_C \models a_C$, as $m_s(p_i) = \sum_{c \in \chi(p_i)} m_C(p_i)(c)$ for every $i \in \{1, \ldots, n\}$ and so, $\sum_{i=1}^{n} k_i \cdot \sum_{c \in \chi(p_i)} m_C(p_i)(c) \leq k$. Notice, that m_C may be unreachable. As the corresponding markings of the unfoldings are equisatisfiable, for every m_U of U, it holds that $m_U \models a_U$. Reversed, it must hold that if for a marking m_s with $m_s \not\models a_s$, there is a marking m_U with $(m_U, m_s) \in \sigma$, for which it holds that $m_U \not\models a_U$. Let $\sum_{i}^{n} k_i \cdot m_s(p_i) > k$. For the marking m_C also holds that $m_C \not\models a_C$. This m_C might be unreachable again. We can see, that for m_U, $m_U \not\models a_U$ as well. $\qquad\square$

The existence of an abstraction relation is not sufficient for transferring the validity results on the markings of S to U. We need in fact a simulation relation. A simulation σ requires the preservation of the transitions between the simulating systems, so it should hold that $\forall m_U, m_{U1} \in RS(U) : m_U \xrightarrow{*} m_{U1}$ and $(m_U, m_s) \in \sigma \Rightarrow \exists m_{s1} \in RS(S) : m_s \xrightarrow{*} m_{s1}$ and $(m_{U1}, m_{s1}) \in \sigma$ for $m_s \in RS(S)$. As the coloured net may have deadlocks, which is not preserved in the skeleton as shown in the opening example, there may be silents transitions at the deadlock states of U, which are not preserved in the skeleton S. Let $m_U \in RS(U)$ be a deadlock of U not preserved in S, so $m_U \xrightarrow{\tau} m_U$. Let $m_s \in RS(S)$ be the corresponding marking of S with $(m_U, m_s) \in \sigma$. Since m_s is not a deadlock, there is no silent transition added for m_s and consequentially, there is no marking $m_{s1} \in RS(S)$ with $m_s \xrightarrow{*} m_{s1}$ and $(m_U, m_{s1}) \in \sigma$, as m_U, m_{s1} do not fulfill atomic propositions equally.

5   Simulation Relation Between Reachability Graphs

In this section, we discuss the existence of a simulation relation between the unfolding and the skeleton under various conditions. As mentioned above, deadlocks that are not preserved in the skeleton, may cause problems. We therefore distinguish coloured nets, where

a) no deadlocks occur at all (Sect. 5.1),
b) all deadlocks are preserved in the skeleton (Sect. 5.2),
c) deadlocks are not always preserved. (Sect. 5.3)

The kind of the $ACTL^*$ formula is significant as well. $ACTL^*$ safety properties permit the use of the skeleton approach even if deadlocks are not preserved, as shown in Sect. 5.4.

5.1   Deadlock-Free Nets

When the net C resp. U has no deadlocks, the net morphism directly leads to a simulation relation between the markings of U and S. There is no need to add silent transitions in U that are not preserved in S.

Proposition 5. *Let C be a coloured net without deadlocks, U its unfolding and S its skeleton. The net morphism $\mu : (P_U \cup T_U) \rightarrow (P_s \cup T_s)$ from Proposition 2 induces a simulation relation between the markings of U and S.*

Proof. The reachability graph $R_U(m_0)$ is a Kripke structure, without adding silent transitions. Let $m_U, m_{U_1} \in RS(U)$ and $m_s \in RS(S)$ be the corresponding marking of m_U with $(m_U, m_s) \in \mu$. The markings m_U and m_s are then related by the abstraction relation σ from Proposition 4: $(m_U, m_s) \in \sigma$. Because net morphisms preserve reachability, for $t_U \in T_U$, if it holds that if $m_U \xrightarrow{t_U} m_{U_1}$ in $R_U(m_0)$, then there is a marking $m_{s_1} \in RS(S)$ with $m_s \xrightarrow{\mu(t)} m_{s_1}$ in $R_s(m_0)$, for which $(m_{U_1}, m_{s_1}) \in \mu$. Consequently, for all markings $m_U, m_{U_1} \in RS(U)$ with $m_U \xrightarrow{*} m_{U_1}$ and $(m_U, m_s) \in \sigma$, there is a marking $m_{s_1} \in RS(S)$ with $m_s \xrightarrow{*} m_{s_1}$ in $R_s(m_0)$ and $(m_{U_1}, m_{s_1}) \in \sigma$.   □

Thus, according to Proposition 3, $ACTL^*$ properties are preserved. If we can guarantee, that the considered net is deadlock-free, the skeleton abstraction can be used for transferring positive results of an $ACTL^*$ verification in S to U.

5.2   Deadlock Preservation

We now consider the case where a Petri nets has deadlocks. The reachability graph of this net is not readily a Kripke structure, hence all deadlock states were extended with a self loop with a silent action. In [Fin92], a necessary and sufficient criterium is formulated, defining a class of coloured Petri nets, which have a *deadlock-preserving skeleton*. This means that every dead marking of the coloured net has a dead skeletal image, thus no deadlock of the coloured net is invisible in the skeleton and it is possible to detect all deadlocks just by the skeletal analysis. With this criterium, we can show that the simulation relation between a coloured net and its skeleton is preserved for this subclass of coloured nets. As we assume that the coloured net $C = [P_C, T_C, F_C, W_C, \chi, \gamma, m_{0c}]$ is uniform, the number of input tokens, a transition $t \in T_c$ needs from each place p_i for $i \in \{1, \ldots n\}$ with $n = |P_C|$ is unambiguous, from now on denoted as $f_i(t)$, where $f_i(t) = |W_c(p_i, t)|$. These numbers form an input vector $f : T \rightarrow \mathbb{N}^n$ for every transition $t \in T_c : f(t) = (f_1(t), f_2(t), \ldots, f_n(t))$. Building on that, we can determine a preorder of the transitions of C, such that $\forall t, t' \in T_c : t \leq t'$, iff $f(t) \leq f(t')$, which leads to an equivalence relation $\sim$ on T_c, such that $\forall t, t' \in T_c : t \sim t'$, iff $f(t) = f(t')$. Transitions with an identical input are aggregated in one equivalence class. Let $T_C/\sim$ be the set of equivalence classes of T_c. The preorder of the transitions induces a partial order $(T_C/\sim, \leq)$ on $T_C/\sim$, such that $\forall [t], [t'] \in T_c/\sim : [t] \leq [t']$, iff $t \leq t'$.

Definition 16 (Full Transition Class). *An equivalence class $[t] \in T_c/\sim$ is full, if for every marking m_c of C with $|m_c(p_i)| = f_i(t)$ for all $i \in \{1, \ldots, n\}$, there is a transition $t \in [t]$ that is enabled in m_c.*

In other words, $[t]$ is full, if any collection of bags matching the input size requirements of $[t]$ also matches the input colour distribution requirements of one $t \in [t]$. This leads to the following proposition:

Proposition 6 (Deadlock-Preserving Skeleton [Fin92]**).** *Let C be a uniform, coloured Petri net. Iff every minimal transition class of C in $(T_c/\sim, \leq)$ is full, then C has a deadlock-preserving skeleton.*

This is a necessary and sufficient condition. In this case, the net morphism μ : $(P_U \cup T_U) \to (P_s \cup T_s)$ between C and S induces a simulation relation. If a coloured net has a deadlock-preserving skeleton, for every added silent transition at a dead marking $m_U \in RS(U)$ of U, a silent transition is added as well in the dead skeletal image $m_s \in RS(S)$ with $(m_U, m_s) \in \mu$. With regard to Proposition 3, we can verify the $ACTL^*$ properties only in S without risking wrong conclusions about the behavior of C.

Next, we shall discuss how to algorithmically check whether a transition class is full. A brute-force enumeration of all firing modes may be very inefficient, as already observed in [SRL+20]. It would in particular prevent the application of the skeleton approach to coloured nets that have an unfolding too large to be constructed. Following the approach of [SRL+20], we rather create an automaton, or an interval decision diagram (IDD) that accepts precisely those assignments to the variables of a transition t that satisfy the guard of the t. The construction of that automaton (IDD) is outlined in [SRL+20] and proceeds by recursively translating the operators used in the guard into operations on the automaton (IDD). The resulting automaton is a compact representation of all valid firing modes of t. By a simple projection of this automaton to the variables occurring in the incoming arcs of t, we obtain an automaton that accepts all tuples of colours that t may consume in any of its firing modes. Using the automata construction for language union, we can then determine the set of tuples that can be consumed by any transition in the class of t. The class is full if and only if the resulting automaton accepts the set of all tuples of colours of the input places of t (which is the same for all members of the class). Our experiments revealed that we are able to decide the full transition class criterion for coloured nets where, so far, no participant in the model checking contest could create its unfolding. We managed to get some verification results for those nets using the skeleton approach.

5.3   Inject Deadlocks to Skeleton

The main focus of this section are nets with deadlocks, but without deadlock-preserving skeleton. Here, the net morphism does not induce a simulation relation, so the $ACTL^*$ results cannot be transferred from the skeleton to the coloured net. We present an approach to modify the skeleton net such that every deadlock of the unfolding occurs in the new skeleton, but potentially with some delay. In this case we cannot guarantee that every dead marking has a dead skeletal image, but we can at least guarantee that for a dead marking, the corresponding skeletal deadlock occurs after a finite number of actions.

Definition 17 (Modified Skeleton Net). *Let* $C = [P_c, T_c, F_c, W_c, \chi, \gamma, m_{0c}]$ *be a uniform coloured net. The* modified skeleton S' *can be constructed from the skeleton* S *as, for every preset place* $p \in P_c$ *of a non-full minimal transition class* $[t]$*, a complement place* $\bar{p}$ *and a recipient transition* t_r *with* $\bullet t_r = \{p\}$ *and* $t_r \bullet = \{\bar{p}\}$ *are introduced with* $W(p, t_r) = W(t_r, \bar{p}) = 1$*. Apart from that,* S' *and* S *are identical.*

The modified skeleton has another behaviour than the original skeleton. Every recipient transition t_r can successively empty its preset place p and stores the tokens on the complement place $\bar{p}$. These actions can be considered as silent actions of S'. Once a token is stored on $\bar{p}$, it cannot leave this place anymore. So, after a finite number of actions of the recipient transitions, the preset of $[t]$ is empty and the transitions in $[t]$ cannot fire anymore. The deadlock of U occurs in S' after a finite number of silent actions of the recipient transitions. Between U and S' a stuttering simulation holds, which is a weakened version of a simulation relation.

Definition 18 (Stuttering Simulation [PSGK00]**).** *Let* $K = [Q, q_0, R, A, L]$ *and* $\hat{K} = [\hat{Q}, \hat{q}_0, \hat{R}, \hat{A}, \hat{L}]$ *be Kripke structures and* a *be an atomic proposition. A mapping* $\sigma_s : Q \to \hat{Q}$ *is a* stuttering simulation relation *if the following conditions hold:*

- $(q_0, \hat{q}_0) \in \sigma_s$
- $(q, \hat{q}) \in \sigma_s \Rightarrow q \models a \Leftrightarrow \hat{q} \models a$ *and for every path* $\pi = q_0 q_1 q_2 \ldots$ *of* K*, there is a path* $\hat{\pi} = \hat{q}_0 \hat{q}_1 \hat{q}_2 \ldots$ *of* $\hat{K}$*, such that we can find the partitions* $B_0, B_1, B_2, \ldots$ *for* π *resp.* $\hat{B}_0, \hat{B}_1, \hat{B}_2, \ldots$ *for* $\hat{\pi}$ *for which holds that:*
  - $\forall i \geq 0 : B_i, \hat{B}_i$ *are not empty and finite*
  - *every state of* $\hat{B}_i$ *is related with every state of* B_i *by* σ_s*.*

If two systems are related by a stuttering simulation, the behaviour of the concrete system K is simulated by the abstract system $\hat{K}$, but $\hat{K}$ can run internal silent actions while simulating. Between the unfolding and the modified skeleton, we can observe this stuttering simulation. To prove this, we first need to establish a relation between the markings of U and S'. Therefore, we create a relation between the markings of S and the markings of S'. A Marking m_s of S and a marking $m_{s'}$ of S' are related, if

- $m_s(p) = m_{s'}(p) + m_{s'}(\bar{p})$ for $p \in \bullet[t]$, where $[t]$ is a non-full minimal transition class
- $m_s(p) = m_{s'}(p)$ otherwise.

The relation between a marking m_U and a marking $m_{s'}$ can then be established by composing the abstraction relation from m_U to m_s and with the one just defined. Thus, the relation between the markings of U and S' is an abstraction relation. The silent actions of the recipient transitions move the tokens of the preset places to their complementary places. No matter if they have moved one or all tokens, the sum over the places p and $\overline{p}$ is always invariant.

Proposition 7 (Stuttering Simulation between U and S'). *Let C be a uniform coloured net, U its unfolding and S' its modified skeleton. Between the markings of U and S', a stuttering simulation σ_s holds.*

Proof. The definition of the marking guarantees, that an abstraction relation exists between the markings of U and S'. States, which are related by the σ_s, fullfil atomic propositions equally. Between the initial markings m_{0U} and $m_{0s'}$ the stuttering simulation holds. Now consider the path $\pi_U = m_{1U}m_{2U}\dots$ of U and the corresponding path $\pi_{s'} = m_{1s'}m_{2s'}\dots$ of S', where $(m_{iU}, m_{is'}) \in \mu$ for all i. The the partitioning of π_U and the corresponding path $\pi_{s'}$ in S' is obtained as follows: For a marking m_{iU} of path π_U, which is not a deadlock, the corresponding part of $\pi_{s'}$ is simply $m_{is'}$ with $(m_{iU}, m_{is'}) \in \sigma_s$. The partitioning of the paths for this parts is trivial: $B_{iU} = \{m_{iU}\}$ resp. $B_{is'} = \{m_{is'}\}$. Let now be m_{iU} a deadlock, which is only followed by the self-loop-τ-actions. The corresponding marking $m_{is'}$ is not necessarily a deadlock. Firing the recipient transitions in $m_{is'}$ yields to a sequence τ^* which ends in a deadlock marking $m_{ds'}$, where only the self-loop-τ-action is possible as well. For partitioning, B_{iU} contains only the deadlock state m_{iU} of U. $B_{is'}$ contains the states $m_{is'}, m_{i+1s'}, m_{i+2s'}, \dots, m_{ds'}$, where $m_{i+1s'}, m_{i+2s'}, \dots$ are the markings, reached by actions of the recipient transitions and $m_{ds'}$ is the delayed deadlock marking. All states in $B_{is'}$ have the same validity of atomic propositions and so they can be related with m_{iU} by σ_s. So, between U and S' a stuttering simulation holds. $\square$

A stuttering simulation preserves $ACTL_X^*$ properties.

Proposition 8 (Stuttering simulation preserves $ACTL_X^*$ [PSGK00]). *Let $K = [Q, q_0, R, A, L]$ and $\hat{K} = [\hat{Q}, \hat{q}_0, \hat{R}, \hat{A}, \hat{L}]$ be Kripke structures, which are related by a stuttering simulation. Then $K \models \varphi \Rightarrow \hat{K} \models \varphi$ for any $ACTL_X^*$ formula φ.*

$ACTL_X^*$ formulas permit claims about the overall behaviour of the system, except for referring next states. The silent actions in the abstract system can generate new next states, which replace the actual simulating next state. Because of this, assumptions on next states can be falsified, which explains the restriction to $ACTL_X^*$. Nevertheless, the validity of at least a subset of $ACTL^*$ formulas can be transferred from the modified skeleton to the unfolding.

5.4   Safety Properties

In the context of net morphisms, safety properties make an exception with regard of their validation.

Definition 19 (Safety Property [KGG99]). *An $ACTL^*$ property is a* safety property, *if only the temporal operators* $\mathbf{W}, \mathbf{X}$ *and the path quantifier* $\mathbf{A}$ *occur.*

We claim that a safety property φ is preserved by a net morphism even if that morphism does not induce a simulation relation. In the context of the skeleton abstraction, the abstraction relation between the markings of U and S is sufficient for the preservation of $ACTL^*$ safety formulas. This fact was already informally mentioned in [PGE98]. However, that paper did not precisely define the class of properties and did not prove the claim.

Proposition 9 (Net Morphisms preserve $ACTL^*$ Safety Properties). *Let C be a coloured net, U its unfolding and S its skeleton. Let $\mu : (P_U \cup T_U) \to (P_s \cup T_s)$ a net morphism, φ_s an $ACTL^*$ safety property and φ_U its unfolding after Definition 15. Let m be a marking of U. Then it holds that: $\mu(m) \models \varphi_s \Rightarrow m \models \varphi_U$.*

Proof. We prove the contraposition $m \not\models \varphi_U \Rightarrow \mu(m) \not\models \varphi_s$ by induction on the structure of φ_C.

Base: If $m \not\models \varphi_U$, then $\mu(m) \not\models \varphi_s$, corresponding to Definition 12.

Step: We therefore distinguish between the possible structures of φ_U and φ_s:

1. $\varphi_U = \psi_U \wedge \xi_U$ resp. $\varphi_U = \psi_U \vee \xi_U$: the induction hypothesis can directly be applied to ψ_U and ξ_U.

2. $\varphi_U = \mathbf{A}\psi_U$: If $m \not\models \varphi_U$, there is a path $\pi = m\, m_1 m_2 \ldots$ with $\pi \not\models \psi_U$. Because reachability is preserved, there is a path $\mu(\pi) = \mu(m)\mu(m_1)\mu(m_2) \ldots$ with $\mu(\pi) \not\models \psi_s$. So, $\mu(m) \not\models \mathbf{A}\psi_s$ resp. $\mu(m) \not\models \varphi_s$.

3. $\varphi_U = \mathbf{A}\mathbf{X}\psi_U$: If $m \not\models \varphi_U$, there is a path $\pi = m\, m_1 m_2 \ldots$ where $m_1 \not\models \psi_U$. For the skeleton, there is a path $\mu(\pi) = \mu(m)\mu(m_1)\mu(m_2) \ldots$ where $\mu(m_1) \not\models \psi_s$. So, $\mu(m) \not\models \mathbf{A}\mathbf{X}\psi_s$ resp. $\mu(m) \not\models \varphi_s$.

4. $\varphi_U = \mathbf{A}\psi_U \mathbf{W}\xi_U$, which is the disjunction between

    a) $\varphi_U = \mathbf{A}\mathbf{G}\psi_U$: If $m \not\models \mathbf{A}\mathbf{G}\psi_U$, there is a path $\pi = m\, m_1 m_2 \ldots$ with a marking $m_i \not\models \psi_U$. Hence, $\pi \not\models \mathbf{G}\psi_U$. Again, the preservation of reachability leads to a path $\mu(\pi) = \mu(m)\mu(m_1)\mu(m_2) \ldots$ with a marking $\mu(m_i) \not\models \psi_s$. So, $\mu(\pi) \not\models \mathbf{G}\psi_s$ and thus $\mu(m) \not\models \mathbf{A}\mathbf{G}\psi_s$;

    b) $\varphi_U = \mathbf{A}\psi_U \mathbf{U}\xi_U$: If $m \not\models \mathbf{A}\psi_U \mathbf{U}\xi_U$ there is a path $\pi = m\, m_1 m_2 \ldots$ with $\pi \not\models (\psi_U \mathbf{U}\xi_U)$. This is possible in two different ways: On the one hand, for all $i \geq 0 : m_i \not\models \xi_U$ can hold, hence $\pi \not\models \mathbf{G}\xi_U$. This can be treated analogously to case 4.a). On the other hand, there might be a $m_i \models \xi_U$, but there is also a m_j with $j < i$ and $m_j \not\models \psi_U$. Then, there is a path $\mu(\pi) = \mu(m)\mu(m_1)\mu(m_2) \ldots$ with $\mu(m_i)$ and $\mu(m_j)$ with $\mu(m_i) \models \xi_s$ and $\mu(m_j) \not\models \psi_s$ as well. Hence, it holds $\mu(m) \not\models \mathbf{A}\psi_s \mathbf{U}\xi_s$.

    In both cases, $\mu(m) \not\models \varphi_s$.

The invalidity of φ_U can always be proven with a finite counterexample path. Deadlocks may just occur in the last marking of this path. Let m_i, the last marking of the counterexample were we can see the invalidity of φ_U, be a deadlock. Because we consider Kripke structures, every path of the system is infinite. The counterexample path π is therefore continued to an infinite path $\pi = m\,m_1 m_2 \ldots m_i m_i m_i \ldots$ by repeating the deadlock state m_i. This repetition does not change the finiteness of the counterexample. If the deadlock m_i is preserved in the skeleton, this leads to an corresponding path $\mu(\pi)$ with repetitions as well: $\mu(\pi) = \mu(m)\mu(m_1)\mu(m_2)\ldots\mu(m_i)\mu(m_i)\mu(m_i)\ldots$. The invalidity of φ_s remains unchanged. If the deadlock is not preserved, the path $\mu(\pi)$ has another sequel: $\mu(\pi) = \mu(m)\mu(m_1)\mu(m_2)\ldots\mu(mi)\mu(m_{i+1})\mu(m_{i+2})\ldots$ with $\mu(m_{i+1}) \neq \mu(m_i)$. The counterexample is transferred exactly up to and including m_i, the markings $\mu(m_{i+1})\mu(m_{i+2})\ldots$ do not change the invalidity of φ_s.

$\square$

6   Folding Place/Transition Nets

With the results presented so far, the skeleton abstraction is only available for systems modeled as coloured Petri nets. In this section, we extend the applicability to nets that are originally modeled as P/T nets. There exist translations from various high level system descriptions directly into P/T nets that could as well have been translated into coloured nets. We propose an efficient procedure to fold a P/T net N into a coloured net C_N, for which we then can build the skeleton S_N. The idea of folding a P/T net into a coloured net is as old as coloured nets as such. To our best knowledge, however, the efficiency of an actual implementation has not been observed so far. Our approach is based on partition refinement. The goal here is to partition a set $\mathcal{M}$ into a partition M of disjunct subsets $M_1, \ldots, M_n$. First, M contains only one subset, which is $\mathcal{M}$. The Partition is then refined by the application of a split function.

Definition 20. *Let* $M = \{M_1, \ldots, M_n\}$ *be a partition of the set* $\mathcal{M}$ *and* $f : \mathcal{M} \to \mathbb{Z}$ *be a split function. The application of* f *on the partition* M *is defined as:* $\mathrm{split}(M, f) = \{\{x \mid x \in M_i, f(x) = j\}\}$ *for* $1 \le i \le n, j \in \mathbb{Z}$.

Informally, we separate elements, where f yields different values. This leads to a new partition of $\mathcal{M}$. For two subsets M_i, M_j, it should hold that $M_i \neq \emptyset$, $M_i \cap M_j = \emptyset$ and also $M_1 \cup \cdots \cup M_n = \mathcal{M}$. For implementing a split operation, we assume an array where every element of $\mathcal{M}$ appears exactly once. For every class in the partition, there is a pair of indices i and j such that the elements of the class are the array entries between i and j. For a split operation, we separately sort the elements of each class and then introduce new classes where adjacent elements have different f-values. Given a P/T net $N = [P, T, F, W, m_0]$, the initial set, which should be partitioned is $\mathcal{M} = P \cup T$. The coarsest partition fitting all requirements is $M = \{P, T\}$. We refine this partition such that, ultimately, every class of places of the given net serves as a place of the resulting coloured net C_N while every class of transitions of the given net serves as a transition. We have to take care that we obey the restrictions of uniformity, that building a skeleton is possible. The folding happens with regard to an $ACTL^*$ formula φ. Let AP_φ denote the set of atomic propositions occuring in φ. The procedure for folding a P/T net into a coloured net is described in Fig. 3. To make this algorithm more understandable, we demonstrate it with an example.

Input: $P \cup T$ of a Petri net $N = [P, T, F, W, m_0]$
Output: Partition M of $P \cup T$ resp. $C_N = [P_{CN}, T_{CN}, F_{CN}, W_{CN}, \chi, \gamma, m_{0CN}]$
Split P and T;
Split class $M^* \in M$ according to equivalence $\sim$ where $x, y \in P \cup T : x \sim y$, iff $|\bullet x| = |\bullet y|$;
Split class $M^* \in M$ according to equivalence $\sim$ where $x, y \in P \cup T : x \sim y$, iff $|x \bullet| = |y \bullet|$;
Split each place class $M^* \in M$ according to equivalence $\sim$ regarding every $(k_1 p_1 + \cdots + k_n p_n <= k) \in AP_\varphi$ where $p_i \sim p_j$, iff $k_i = k_j$ for $p_i, p_j \in M^*$ and $i, j \in \{1, \ldots, n\}$;
repeat
    for all classes $M^* \in M$ **do**
        for all weights $w^* \in W$ **do**
            Split M^* according to equivalence $\sim$ where $x, y \in P \cup T : x \sim y$
            iff $\mathrm{card}(\{z | z \in M^*, W(x, z) = w^*\}) = \mathrm{card}(\{z | z \in M^*, W(y, z = w^*\})$;
            Split M^* according to equivalence $\sim$ where $x, y \in P \cup T : x \sim y$
            iff $\mathrm{card}(\{z | z \in M^*, W(z, x) = w^*\}) = \mathrm{card}(\{z | z \in M^*, W(z, y) = w^*\})$;
        end for
    end for
until nothing changes anymore
P_{CN} = place classes, T_{CN} = transition classes, $(M^*, M^{*\prime}) \in F_{CN}$, $W_{CN}(M^*, M^{*\prime}) = \{v_0, \ldots, v_k\}$, iff $\exists x \in M^*, \exists y \in M^{*\prime} : (x, y) \in F, W(x, y) = k$ for $k \in \mathbb{N}$, $M^*, M^{*\prime} \in M$;
$m_{0CN}(M^*) = \sum_{p \in M^*} m_0(p)$, $\chi(M^*)$ = elements of M^* for every place class $M^* \in M$;
$\gamma(M^*)$ = disjunction of all firing modes of M^* for every transition class $M^* \in M$;

Fig. 3. Algorithm for folding a P/T net into a coloured net.

Example 3. We consider a P/T net N, which shows the dilemma of five dining philosophers. The P/T net is structured as follows: For every $i \in \{0, \ldots, 4\}$ there is a place th_i (philosopher i is thinking), a place hl_i (has left fork), hr_i (has right fork), ea_i (philosopher i is eating) and fo_i (fork i is on the table). There are the transitions tl_i (take left fork) that consume tokens from th_i and fo_i, and produce on hl_i, transitions tr_i (take right fork) that consume from hl_i and $fo_{i+1 \bmod 5}$ and produce on ea_i, transitions rl_i (release left fork) that consume from ea_i and produce on hr_i and fo_i, and, finally, transitions rr_i (release right fork) that consume from hr_i and produce on $fo_{i+1 \bmod 5}$ and th_i. Places th_i and fo_i are initially marked, and all arc weights are 1. For better readability, x_i describes the set $x_0, \ldots, x_4$ for every node $x \in P \cup T$ of the net. The folding is regarding the $ACTL^*$ formula $\varphi : \mathbf{AG} \neg (\sum_i hr.i = 4) \wedge (\sum_i hl.i = 1)$ for $i \in \{0, \ldots, 4\}$. Initially the coarsest partition distinguishes between places and transitions: $M = \{\{th_i, ea_i, fo_i, hl_i, hr_i\}, \{tr_i, tl_i, rl_i, rr_i\}\}$. Then, the sets are split according to the *number of incoming* and *outgoing arcs*. This leads to partition $M = \{\{th_i, ea_i, hl_i, hr_i\}, \{fo_i\}, \{tr_i, tl_i\}, \{rl_i, rr_i\}\}$. Afterwards, the nodes are partitioned with respect to the *cardinality of the incoming* and *outgoing arcs*. For all occurring weights, a node is mapped to its number of incoming resp. outgoing arcs with this weight. In the example, these two steps do not lead to another partition, as all weights are 1. The *atomic propositions* of φ give additional restrictions, as the elements of the subsets finally should satisfy those propositions equally. So, for the atomic propositions $\sum_{i=1}^{4} hr.i = 4$ and $\sum_{i=1}^{4} hl.i = 1$, every place is mapped to its coefficient in the corresponding proposition. Transitions are not affected here, so $M = \{\{th_i, ea_i\}, \{hl_i\}, \{hr_i\}, \{fo_i\}, \{tr_i, tl_i\}, \{rl_i, rr_i\}\}$. For obtaining *uniformity*, we proceed as follows until nothing changes: Pick a class M^* and an arc weight $w*$, and split the partition corresponding to the following mappings:

- for all $x \in P \cup T$, map node x to $\mathrm{card}(\{y | y \in M^*, W(x, y) = w^*\})$;
- for all $x \in P \cup T$, map node x to $\mathrm{card}(\{y | y \in M^*, W(y, x) = w^*\})$.

Picking the class that contains all hl places, the two transition classes are split into the three classes of all tl transitions (have a hl place as post-place), of all tr transitions (have a hl place as pre-place), and all remaining transitions (are not connected to hl). Picking the class of tl transitions, we separate the th places (outgoing are to a tl transition) from the hr and ea places. With the th place class, the rl transitions (not connected to any th place) are separated from the rr transitions (arc to some th place). Finally, the rl transitions separate the ea places from the hr places. The partition is $M = \{\{th_i\}, \{ea_i\}, \{hl_i\}, \{hr_i\}, \{fo_i\}, \{tr_i\}, \{tl_i\}, \{rl_i\}, \{rr_i\}\}$. Finally, every place class M_i is turned into a place p_{M_i} with $\sum_{p \in M_i} m(p)$ tokens and the colour domain $\chi(p_{M_i}) = M_i$, and every transition class M_j into a transition t_{M_j}. We obtain the places th, ea, fo, hl, hr and the transitions tr, tl, rl, rr. Let there be an arc from p_{M_i} to t_{M_j}, if there exists some $p \in M_i$ and $t \in M_j$ with $(p, t) \in F$. Arcs from transitions to places are formed analogously. An Arc (p_{M_i}, t_{M_j}) ist assigned with the variables $x_1, \ldots, x_w$ where $w = W(p, t)$ for $p \in M_i$ and $t \in M_j$. In the end, we need to formulate the guard of the transitions, which needs to ensure, that the coloured transition only fires if the right coloured tokens lay on the pre-places. We therefore build the disjunction of the input requirements of the transitions according the arcs and weights of N. The guard of transition tl is $\gamma(i) = \bigvee_i x_{thtl} = th.i \wedge x_{fotl} = fo.i$, where x_{thtl}, x_{fotl} are the variables of the corresponding arcs, for instance. Figure 4 presents the coloured net, which results of the described folding. The coloured net can subsequently be decoloured to a skeleton.

The resulting coloured net does not necessarily have full minimal transition classes (cf. Sect. 5.2), thus it does not have a deadlock-preserving skeleton. Due to the process for deriving the folded coloured net, the guards do not permit the approach outlined there for checking whether or not a transition class is full. We may, however, approach that criterion differently. The idea is to check the criterion right after the folding procedure, just as we have the final partitioning of the P/T nodes. If we cannot prove the deadlock preservation at this point, we can abort the skeletal analysis of this net, as we don't expect useful results.

Proposition 10 (*Deadlock Preservation for P/T Nets*). *Let N be a P/T net and C_N its folding. Let $[t] = \{t_1, t_2, \ldots, t_k\}$ be a minimal transition class of C_N, where each transition t_j has s_j firing modes for $j \in \{1, \ldots, k\}$. Let $p_1, p_2, \ldots, p_\ell$ be the pre-places of $[t]$, with the colour domains $\chi(p_i)$ for $i \in \{1, \ldots, \ell\}$. Every*

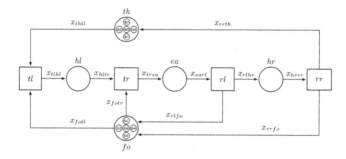

Fig. 4. A coloured net version of the five dining philosophers.

pre-place p_i is connected to every transition t_j of $[t]$ by an arc with the weight w_{ij}. The folding C_N resp. the underlying P/T net N has a deadlock-preserving skeleton, if $\prod_{i=1}^{\ell} \binom{|\chi(p_i)|}{w_{ij}} = \sum_{j=1}^{k} s_j$ for every minimal transition class of C_N.

Proof. The folding C_N has a deadlock-preserving skeleton, if all of its minimal transition classes are full. A minimal transition class $[t] = \{t_1, t_2, \ldots, t_k\}$ is full, if for every marking m_{C_N} with $|m_{C_N}(p_i)| = f_i(t)$ for $i \in \{1, \ldots, \ell\}$, there is one transition in $[t]$, for which the marking is a firing mode. This is expressed by the equation $\prod_{i=1}^{\ell} \binom{|\chi(p_i)|}{w_{ij}} = \sum_{j=1}^{k} s_j$. The binomial coefficient $\binom{|\chi(p_i)|}{w_{ij}}$ gives the number of sufficient tokensubsets of $\chi(p_i)$ for one pre-place p_i of $[t]$, where $i \in \{1, \ldots, \ell\}$ and $j \in \{1, \ldots, k\}$. Multiplying these numbers for every pre-place p_i for $i \in \{1, \ldots, \ell\}$, leads to the total number of sufficient combinations of tokens, i.e., the number of possible markings m_{C_N} with sufficient input requirements $|m(p_i)| = f_i(t)$ with $i \in \{1, \ldots, \ell\}$ for $[t]$. Each of the transitions t_j in $[t]$ for $j \in \{1, \ldots, k\}$ has s_j firing modes, thus in $[t]$, we have $\sum_{j=1}^{k} s_j$ firing modes overall. If $\prod_{i=1}^{\ell} \binom{|\chi(p_i)|}{w_{ij}} = \sum_{j=1}^{k} s_j$, for every sufficient combination of the coloured tokens, there is a firing mode of one transition in $[t]$. If $\prod_{i=1}^{\ell} \binom{|\chi(p_i)|}{w_{ij}} = \sum_{j=1}^{k} s_j$, the minimal transition class $[t]$ is full, thus if the equation holds for every minimal transition class, C_N has a deadlock-preserving skeleton. Transferring this to N, for every combination of tokens of the P/T pre-places (represented by $\chi(p_i)$), there is one P/T transition related to $[t]$ enabled (representend by s_j). So, if a marking of N fits with regard to the cardinality, there must be one activated P/T transition if the equation holds. Then, N is has a deadlock-preserving skeleton. It is important to mention, that the firing modes s_j need to be all different from each other, resp. all of the P/T transitions need to have different presets. Otherwise the equality of combinations and firing modes will not hold, although every combination activates a transition. □

This equation is sufficient for the fullness of $[t]$, but it is not necessary. If $\prod_{i=1}^{l} \binom{|\chi(p_i)|}{w_{ij}} > \sum_{j=1}^{k} s_j$, which means there is a combination of tokens which does not activate a transition, these too many combinations might be unreachable, thus are not in need of an activated transition. If the equation holds for every minimal transition class we know that C_N will have a deadlock-preserving skeleton and the method of skeletal abstraction can be applied to N and all its $ACTL^*$ formulas.

7   Experimental Results

We conducted our experiments on the benchmark provided by the Model Checking Contest (MCC) 2019 [KGH+19]. On that page, the reader may find a detailed specification of the machine "tajo" that was used to execute the experiments. The benchmark comprises 1018 nets (193 coloured nets and 825 P/T nets). For the majority of colored nets, their unfolding is among the P/T nets of the benchmark, too. We covered the three categories Reachability, CTL, and LTL where the skeleton approach makes sense. For every net and category, there are 16 formulas with place-based atomic propositions and 16 formulas with transition-based propositions. That makes a total of 97,728 formulas. If a P/T net is the unfolding of a coloured net, some but not all formulas of the P/T net accord with formulas used for the coloured net.

In every single run, we allowed 4 cores, 1800 s, and 16 MB of RAM for the verification of a group of 16 formulas. The runs used the full portfolio [Wol20] of verification methods available in our tool LoLA [Wol18], now including the skeleton approach. For the skeleton, we applied the same search based model checking routines as for the unfolded net, and the state equation approach [WW12]. In our approach, the skeleton is directly derived from the PNML description of a coloured net, so the skeleton related verification tasks start before the unfolding of the net is generated in parallel (and only then the remaining verification routines are launched). If the input is a P/T net, we first launch the verification tasks for the given net, before trying to fold the net (in parallel to the already running routines). We launch skeleton related tasks only if the size of the skeleton is less than one third of the size of the given P/T net. This way, we avoid spending resources in situations where the skeleton is too close to the given net. Since folding depends on the formula, we have to execute up to 16 individual folding procedures per run. We do not fold a net if the formula is trivial (i.e., does not contain temporal operators). Trivial formulas are mostly the result of sophisticated application of logical tautologies and preprocessing based on linear programming [BDJ+18]. We also stop the folding procedure as soon as some other portfolio member has determined the value of the formula.

For the 79,200 formulas for P/T nets, 66,546 skeletons were created. Of the remaining 12,654 formulas, 11,086 contain no temporal operators, so no folding was launched. For the remaining 1,568 formulas, some other portfolio member may have delivered a result before folding completed. Folding took at most 287 s, with an average of half a second. Generation of the skeleton for a coloured net takes no time at all as it appears as an intermediate step of the unfolding process. Generating the skeleton naturally succeeded for all coloured nets. It also succeeded whenever both net and formula were derived from a coloured net. There are few other cases where the skeleton could be generated. Although more nets have a regular, foldable structure, formulas, if not derived from coloured nets, are generated randomly, so they tend to break symmetry more frequently than in practical situations. On the other hand, the formula syntax for coloured nets in the MCC does not permit references to individual colours or firing modes, so the skeleton approach is applicable more frequently than in practice. Since there are more P/T nets than coloured nets in the contest, results obtained for the MCC benchmark should be a lower bound for the performance to be observed in practice.

The 66,546 skeletons include those that are considered too large to make a difference compared to the given net. After ruling them out, 34,906 formulas have useful skeletons, including coloured and P/T nets. In 15,315 cases, we launched the skeleton related tasks. In the remaining 19,591 cases, the formula (nor its negation) are not in $ACTL^*$, or none of the criteria discussed in the paper would certify preservation of the formula. We need to mention here that deadlock injection has not been implemented so far.

Of the 15,315 formulas where we launched the skeleton related tasks, they were the first (among the whole portfolio) to deliver results in 3168 cases. The remaining 12,147 formulas include those where some other portfolio member responded earlier, or where the skeleton approach evaluated its $ACTL^*$ query to false (so the value is not inherited by the unfolded net).

Among the considered 97,728 formulas considered, there have been 7,768 formulas none of the participants in the MCC 2019 could solve. With the skeleton approach, we have now been able to solve 226 of these particularly involved problems. These include but are not restricted to nets that have a prohibitively large unfolding.

Given that we run the skeleton approach as part of a powerful portfolio, with LoLA being a competitive participant in the MCC, we may conclude that the skeleton approach nicely complements the existing portfolio.

8   Conclusion

With our contribution, we turned the skeleton approach into an executable and useful member of a verification portfolio. We investigated the gap between the concepts of net morphisms and simulation relations and proposed algorithms for checking the required criteria. Through folding, we extended the approach to P/T nets. Experiments underpin the usefulness of the approach.

Future work may include the implementation of deadlock injection. Furthermore, we may enhance the approach to the full transition classes. First, we may try to use place invariants to rule out certain token distributions in the pre-set of transition classes thus being able to certify more transition classes as full. Second, we may try to split places and transitions in the skeleton, turning non-full transition classes into full ones.

References

[BDJ+18] Bønneland, F., Dyhr, J., Jensen, P.G., Johannsen, M., Srba, J.: Simplification of CTL formulae for efficient model checking of Petri Nets. In: Khomenko, V., Roux, O.H. (eds.) PETRI NETS 2018. LNCS, vol. 10877, pp. 143–163. Springer, Cham (2018). https://doi.org/10.1007/978-3-319-91268-4_8

[CES86] Clarke, E.M., Emerson, E.A., Sistla, A.P.: Automatic verification of finite-state concurrent systems using temporal logic specifications. ACM Trans. Program. Lang. Syst. **8**(2), 244–263 (1986)

[Des91] Desel, J.: On abstractions of nets. In: Rozenberg, G. (ed.) ICATPN 1990. LNCS, vol. 524, pp. 78–92. Springer, Heidelberg (1991). https://doi.org/10.1007/BFb0019970

[Fin92] Findlow, G.: Obtaining deadlock-preserving skeletons for coloured nets. In: Jensen, K. (ed.) ICATPN 1992. LNCS, vol. 616, pp. 173–192. Springer, Heidelberg (1992). https://doi.org/10.1007/3-540-55676-1_10

[GL94] Grumberg, O., Long, D.E.: Model checking and modular verification. ACM Trans. Program. Lang. Syst. **16**(3), 843–871 (1994)

[JK09] Jensen, K., Kristensen, L.M.: Coloured Petri Nets. Modelling and Validation of Concurrent Systems. Springer, Heidelberg (2009). https://doi.org/10.1007/b95112

[KGG99] Katz, S., Grumberg, O., Geist, D.: "Have i written enough properties?" - a method of comparison between specification and implementation. In: Pierre, L., Kropf, T. (eds.) CHARME 1999. LNCS, vol. 1703, pp. 280–297. Springer, Heidelberg (1999). https://doi.org/10.1007/3-540-48153-2_21

[KGH+19] Kordon, F., et al.: Complete Results for the 2019 Edition of the Model Checking Contest (2019). http://mcc.lip6.fr/2019/results.php

[Lil95] Lilius, J.: On the Folding of Algebraic Nets. Helsinki University of Technology (1995)

[Mil89] Milner, R.: Communication and Concurrency. Prentice Hall International Series in Computer Science. Prentice Hall, New York (1989)

[PGE98] Padbergx, J., Gajewsky, M., Ermel, C.: Rule-based refinement of high-level nets preserving safety properties. In: Proceedings of the FASE, vol. 1382, pp. 221–238 (1998)

[Pin11] Pinna, G.M.: How much is worth to remember? A taxonomy based on Petri Nets unfoldings. In: Kristensen, L.M., Petrucci, L. (eds.) PETRI NETS 2011. LNCS, vol. 6709, pp. 109–128. Springer, Heidelberg (2011). https://doi.org/10.1007/978-3-642-21834-7_7

[PSGK00] Penczek, W., Szreter, M., Gerth, R., Kuiper, R.: Improving partial order reductions for universal branching time properties. Fundamenta Informaticae **43**(14), 245–267 (2000)

[RB01] Rust, C., Tacken, J., Böke, C.: Pr/T-Net based seamless design of embedded real-time systems. In: Colom, J.-M., Koutny, M. (eds.) ICATPN 2001. LNCS, vol. 2075, pp. 343–362. Springer, Heidelberg (2001). https://doi.org/10.1007/3-540-45740-2_20

[SMS99] Sliva, V.P., Murataxx, T., Shatz, S.M.: Protocol specification design using an object-based Petri Net formalism. Int. J. Softw. Eng. Knowl. Eng. **09**(01), 97–125 (1999)

[SRL+20] Schwarick, M., Rohr, C., Liu, F., Assaf, G., Chodak, J., Heiner, M.: Effi-
          cient unfolding of coloured Petri Nets using interval decision diagrams. In:
          Janicki, R., Sidorova, N., Chatain, T. (eds.) PETRI NETS 2020. LNCS,
          vol. 12152, pp. 324–344. Springer, Cham (2020). https://doi.org/10.1007/
          978-3-030-51831-8_16

[Vau87]   Vautherin, J.: Parallel systems specifications with coloured Petri nets and
          algebraic specifications. In: Rozenberg, G. (ed.) APN 1986. LNCS, vol. 266,
          pp. 293–308. Springer, Heidelberg (1987). https://doi.org/10.1007/3-540-
          18086-9_31

[Wol18]   Wolf, K.: Petri Net model checking with LoLA 2. In: Khomenko, V., Roux,
          O.H. (eds.) PETRI NETS 2018. LNCS, vol. 10877, pp. 351–362. Springer,
          Cham (2018). https://doi.org/10.1007/978-3-319-91268-4_18

[Wol20]   Wolf, K.: Portfolio management in explicit model checking. In: Proceedings
          of the PNSE (CEUR Workshop Proceedings), vol. 2651, pp. 10–28 (2020)

[WW12]   Wimmel, H., Wolf, K.: Applying CEGAR to the Petri Net state equation.
          Log. Meth. Comput. Sci. **8**(3) (2012). https://doi.org/10.2168/LMCS-8(3:
          27)2012

Reduction Using Induced Subnets to Systematically Prove Properties for Free-Choice Nets

Wil M. P. van der Aalst[1,2(✉)]

[1] Process and Data Science (Informatik 9), RWTH Aachen University,
Aachen, Germany
wvdaalst@pads.rwth-aachen.de
[2] Fraunhofer-Institut für Angewandte Informationstechnik (FIT),
Sankt Augustin, Germany

Abstract. We use sequences of t-induced T-nets and p-induced P-nets to convert free-choice nets into T-nets and P-nets while preserving properties such as well-formedness, liveness, lucency, pc-safety, and perpetuality. The approach is general and can be applied to different properties. This allows for more systematic proofs that "peel off" non-trivial parts while retaining the essence of the problem (e.g., lifting properties from T-net and P-net to free-choice nets).

Keywords: Petri nets · Free-choice nets · Net reduction · Lucency

1 Introduction

Although free-choice nets have been studied extensively, still new and surprising properties are discovered that cannot be proven easily [2]. This paper proposes the use of *T-reductions* and *P-reductions* to prove properties by reducing free-choice nets to either T-nets (marked graphs) or P-nets (state machines). These reductions are based on the notion of *t-induced T-nets* (denoted by $\square_N(t)$) and the notion of *p-induced P-nets* (denoted by $\odot_N(p)$). We propose to use such *reductions* to prove properties that go beyond well-formedness. This paper systematically presents T-reductions and P-reductions, and shows example applications.

Figure 1 illustrates the notion of induced subnets. The original net N has two proper induced T-nets (a) and two proper induced P-nets (b). If the original Petri net N is free-choice and well-formed, then the net after applying the corresponding reduction is still free-choice and well-formed. Think of the original net as an "onion" that is peeled off layer for layer until a T-net or P-net remains. We are interested in *properties that propagate through the different layers*, just like well-formedness. For example, we will show that all perpetual well-formed free-choice nets are lucent, i.e., the existence of a regeneration transition implies that there cannot be two markings enabling the same set of transitions.

© Springer Nature Switzerland AG 2021
D. Buchs and J. Carmona (Eds.): PETRI NETS 2021, LNCS 12734, pp. 208–229, 2021.
https://doi.org/10.1007/978-3-030-76983-3_11

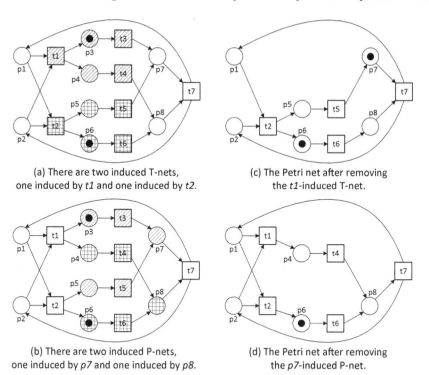

(a) There are two induced T-nets, one induced by *t1* and one induced by *t2*.

(c) The Petri net after removing the *t1*-induced T-net.

(b) There are two induced P-nets, one induced by *p7* and one induced by *p8*.

(d) The Petri net after removing the *p7*-induced P-net.

Fig. 1. A free-choice Petri net N has (a) two proper induced T-nets ($\Box_N(t1)$ and $\Box_N(t2)$) and (b) two proper induced P-nets ($\odot_N(p7)$ and $\odot_N(p8)$). The Petri nets after removing $\Box_N(t1)$ and $\odot_N(p8)$ are shown in (c) and (d).

The remainder of the paper is organized as follows. Section 2 discusses related work and Sect. 3 introduces some standard results and notations. Section 4 presents t-induced T-nets and p-induced P-nets and their characteristic properties. The general approach of using T- and P-reductions is presented in Sect. 5, followed by the application to some properties that go beyond known results like well-formedness (Sect. 6). Section 7 concludes the paper.

2 Related Work

For an introduction to free-choice nets and the main known results, we refer to [7,9]. The work presented in this paper is most related to the completeness proof of the reduction rules in [9]. Proper t-induced T-nets are similar to the CP-nets used in [9]. The use of reduction rules was first proposed and studied by Berthelot [6]. Desel provided reduction rules for free-choice nets without frozen tokens [8]. Indirectly related are the blocking theorem [12,15] and the notion of lucency in perpetual free-choice nets [2,4]. To get a deeper understanding of well-formed free-choice nets, we also refer to [11,14]. The problem addressed in this paper was inspired by questions originating from the process mining domain [1], e.g., see [10] for the application of traditional reduction rules in process discovery and see [4] for the relation between lucency and translucent event logs.

3   Preliminaries

This section introduces basic mathematical concepts and some well-known Petri net notions and results.

$\mathcal{B}(A)$ is the set of all *multisets* over some set A, e.g., $b = [x^3, y^2, z] \in \mathcal{B}(A)$ is a multiset with 6 elements ($|B| = 6$). We assume the standard multiset operators $\in$ (element), $\uplus$ (union), $\setminus$ (difference), $\leq$ (smaller or equal), and $<$ (smaller). $\sigma = \langle a_1, a_2, \ldots, a_n \rangle \in X^*$ denotes a *sequence* over X of length $|\sigma| = n$. $\sigma_i = a_i$ for $1 \leq i \leq |\sigma|$. $\langle \, \rangle$ is the empty sequence.

Definition 1 (Petri Net). *A Petri net is a tuple $N = (P, T, F)$ with P the set of places, T the set of transitions such that $P \cap T = \emptyset$, and $F \subseteq (P \times T) \cup (T \times P)$ the flow relation such that the graph $(P \cup T, F)$ is non-empty and weakly connected. A Petri net is non-trivial if $F \neq \emptyset$ (i.e., there is at least one place and one transition).*

Definition 2 (Marking). *Let $N = (P, T, F)$ be a non-trivial Petri net. A marking M is a multiset of places, i.e., $M \in \mathcal{B}(P)$. (N, M) is a marked net.*

The requirement that a marked net is non-trivial (i.e., $F \neq \emptyset$), together with the requirement that $(P \cup T, F)$ is weakly connected is there to avoid uninteresting border cases (nets without places or transitions cannot change state and unconnected parts can be analyzed separately). For a subset of places $X \subseteq P$: $M{\upharpoonright}_X = [p \in M \mid p \in X]$ is the marking *projected* on this subset. $M(X) = \sum_{p \in X} M(p) = |M{\upharpoonright}_X|$ is the total number of tokens in X.

A Petri net $N = (P, T, F)$ defines a directed graph with nodes $P \cup T$ and edges F. For any $x \in P \cup T$, $\bullet x = \{y \mid (y, x) \in F\}$ denotes the set of input nodes and $x \bullet = \{y \mid (x, y) \in F\}$ denotes the set of output nodes. The notation can be generalized to sets: $\bullet X = \{y \mid \exists_{x \in X} (y, x) \in F\}$ and $X \bullet = \{y \mid \exists_{x \in X} (x, y) \in F\}$ for any $X \subseteq P \cup T$.

Definition 3 (Elementary Paths and Circuits). *A path in a Petri net $N = (P, T, F)$ is a non-empty ($n \geq 1$) sequence of nodes $\rho = \langle x_1, x_2, \ldots, x_n \rangle$ such that $(x_i, x_{i+1}) \in F$ for $1 \leq i < n$. paths$(N) \subseteq (P \cup T)^*$ is the set of all paths in N. ρ is an elementary path if $x_i \neq x_j$ for $1 \leq i < j \leq n$ (i.e., no element occurs more than once). An elementary path is called a circuit if $(x_n, x_1) \in F$.*

A transition $t \in T$ is *enabled* in marking M of net N, denoted as $(N, M)[t\rangle$, if each of its input places $\bullet t$ contains at least one token. $en(N, M) = \{t \in T \mid (N, M)[t\rangle\}$ is the set of enabled transitions.

An enabled transition t may *fire*, i.e., one token is removed from each of the input places $\bullet t$ and one token is produced for each of the output places $t \bullet$. Formally: $M' = (M \setminus \bullet t) \uplus t \bullet$ is the marking resulting from firing enabled transition t in marking M of Petri net N. $(N, M)[t\rangle(N, M')$ denotes that t is enabled in M and firing t results in marking M'.

Let $\sigma = \langle t_1, t_2, \ldots, t_n \rangle \in T^*$ be a sequence of transitions $(n \geq 0)$. $(N, M)[\sigma\rangle(N, M')$ denotes that there is a set of markings $M_1, M_2, \ldots, M_{n+1}$ $(n \geq 0)$ such that $M_1 = M$, $M_{n+1} = M'$, and $(N, M_i)[t_i\rangle(N, M_{i+1})$ for $1 \leq i \leq n$. A marking M' is *reachable* from M if there exists a *firing sequence* σ such that $(N, M)[\sigma\rangle(N, M')$. $R(N, M) = \{M' \in \mathcal{B}(P) \mid \exists_{\sigma \in T^*} (N, M)[\sigma\rangle(N, M')\}$ is the set of all reachable markings.

Definition 4 (Live, Bounded, Safe, Dead, Deadlock-Free, Well-Formed). *A marked net (N, M) is* live *if for every reachable marking $M' \in R(N, M)$ and every transition $t \in T$ there exists a marking $M'' \in R(N, M')$ that enables t. A marked net (N, M) is* k-bounded *if for every reachable marking $M' \in R(N, M)$ and every $p \in P$: $M'(p) \leq k$. A marked net (N, M) is* bounded *if there exists a k such that (N, M) is k-bounded. A 1-bounded marked net is called* safe*. A place $p \in P$ is* dead *in (N, M) when it can never be marked (no reachable marking marks p). A transition $t \in T$ is* dead *in (N, M) when it can never be enabled (no reachable marking enables t). A marked net (N, M) is* deadlock-free *if each reachable marking enables at least one transition. A Petri net N is* structurally bounded *if (N, M) is bounded for any marking M. A Petri net N is* structurally live *if there exists a marking M such that (N, M) is live. A Petri net N is* well-formed *if there exists a marking M such that (N, M) is live and bounded.*

For particular subclasses of Petri nets, there are various relationships between structural properties and behavioral properties like liveness and boundedness [7]. In this paper, we focus on free-choice nets [9].

Definition 5 (P-net, T-net, and Free-choice Net). *Let $N = (P, T, F)$ be a Petri net. N is a* P-net *(also called a* state machine*) if $|{\bullet}t| = |t{\bullet}| = 1$ for any $t \in T$. N is a* T-net *(also called a* marked graph*) if $|{\bullet}p| = |p{\bullet}| = 1$ for any $p \in P$. N is a* free-choice net *if for any $t_1, t_2 \in T$: ${\bullet}t_1 = {\bullet}t_2$ or ${\bullet}t_1 \cap {\bullet}t_2 = \emptyset$. N is* strongly connected *if the graph $(P \cup T, F)$ is strongly connected, i.e., for any two nodes x and y there is a path leading from x to y.*

Definition 6 (Cluster). *Let $N = (P, T, F)$ be a Petri net and $x \in P \cup T$. The* cluster *of node x, denoted $[x]_c$ is the smallest set such that (1) $x \in [x]_c$, (2) if $p \in [x]_c \cap P$, then $p{\bullet} \subseteq [x]_c$, and (3) if $t \in [x]_c \cap T$, then ${\bullet}t \subseteq [x]_c$.*

Definition 7 (Subnet, Complement, P-Component, T-Component). *Let $N = (P, T, F)$ be a Petri net and $X \subseteq P \cup T$. $N{\restriction}_X = (P \cap X, T \cap X, F \cap (X \times X))$ is the subnet generated by X. $N \setminus\setminus X = (P \setminus X, T \setminus X, F \cap (((P \cup T) \setminus X) \times ((P \cup T) \setminus X)))$ is the complement generated by X. $N{\restriction}_X$ is a* P-component *of N if ${\bullet}p \cup p{\bullet} \subseteq X$ for $p \in X \cap P$ and $N{\restriction}_X$ is a strongly connected P-net. $N{\restriction}_X$ is a* T-component *of N if ${\bullet}t \cup t{\bullet} \subseteq X$ for $t \in X \cap T$ and $N{\restriction}_X$ is a strongly connected T-net. $PComp(N) = \{X \subseteq P \cup T \mid N{\restriction}_X \text{ is a P-component}\}$. $TComp(N) = \{X \subseteq P \cup T \mid N{\restriction}_X \text{ is a T-component}\}$.*

Definition 8 (P-cover, T-cover). *Let $N = (P, T, F)$ be a Petri net. N has a P-cover if $\bigcup PComp(N) = P \cup T$.[1] N has a T-cover if $\bigcup TComp(N) = P \cup T$.*

Theorem 1 (Coverability Theorems [9]). *Let $N = (P, T, F)$ be a well-formed free-choice net. $\bigcup PComp(N) = \bigcup TComp(N) = P \cup T$.*

Moreover, for any well-formed free-choice net N and marking M: (N, M) is live if and only if every P-component is marked in M (Theorem 5.8 in [9]).

The dual Petri net is the net where the role of places and transitions is swapped and the arcs are reversed.

Definition 9 (Dual Net). *Let $N = (P, T, F)$ be a Petri net. $N^{dual} = (T, P, F^{-1})$ with $F^{-1} = \{(x, y) \mid (y, x) \in F\}$ is the dual net of N.*

Note that $(N^{dual})^{dual} = N$. We also use the following well-known result [9,13].

Theorem 2 (Duality Theorem). *Let N be a Petri net and N^{dual} the dual net of N. N is a well-formed free-choice net if and only if N^{dual} is a well-formed free-choice net.*

4   Induced Subnets in Free-Choice Nets: Existence and Properties

We start by introducing the notion of t-induced T-nets, i.e., subnets fully defined by an initial transition t and all nodes that can be reached from t without visiting places with multiple input or multiple output transitions. Figure 1 highlights two induced T-nets: $\boxdot_N(t1) = \{t1, p3, p4, t3, t4\}$ and $\boxdot_N(t1) = \{t2, p5, p6, t5, t6\}$.

Definition 10 (t-Induced T-net). *Let $N = (P, T, F)$ be a Petri net and $t \in T$. $\boxdot_N(t) \subseteq P \cup T$ is the smallest set such that*

- $t \in \boxdot_N(t)$,
- $\{p' \in t'\bullet \mid |\bullet p'| = 1 \ \wedge \ |p'\bullet| = 1\} \subseteq \boxdot_N(t)$ *for any* $t' \in \boxdot_N(t) \cap T$, *and*
- $p'\bullet \subseteq \boxdot_N(t)$ *for any* $p' \in \boxdot_N(t) \cap P$.

$\boxdot_N(t)$ *are the nodes of the t-induced T-net of N that is denoted by $N_{\boxdot(t)} = N\!\restriction_{\boxdot_N(t)}$. $\overline{N_{\boxdot(t)}} = N \setminus\!\setminus \boxdot_N(t)$ is the complement of the t-induced T-net of N. $\boxdot_N(t)$ is proper if the complement $\overline{N_{\boxdot(t)}}$ is a non-trivial strongly-connected Petri net.*

Informally, a t-induced T-net can be viewed as the union of a set of elementary paths that all start in t and have non-branched places. A t-induced T-net is proper if after removal the net is strongly-connected. $\boxdot_N(t1)$ is a *proper $t1$-induced T-net* of the net N in Fig. 1(a), because removing all the nodes in $\boxdot_N(t1)$ leaves the strongly-connected Petri net $\overline{N_{\boxdot(t1)}}$ depicted in Fig. 1(c). Proper t-induced T-nets have the following properties.

---

[1] $\bigcup Q = \bigcup_{X \in Q} X$ for some set of sets Q.

Proposition 1 (Properties of Proper t-Induced T-net). *Let $N = (P, T, F)$ be a strongly-connected free-choice net and $\square_N(t)$ a proper t-induced T-net of N.*

(1) $N_{\square(t)}$ *is a T-net.*
(2) $\overline{N_{\square(t)}}$ *is free-choice.*
(3) *For all $p' \in \square_N(t) \cap P$: $\bullet p' \cup p' \bullet \subseteq \square_N(t)$.*
(4) *For all $t' \in \square_N(t) \cap (T \setminus \{t\})$: $\bullet t' \subseteq \square_N(t)$.*
(5) $\bullet t \subseteq P \setminus \square_N(t)$.
(6) *There is a $t' \in T \setminus \square_N(t)$ such that $\bullet t = \bullet t'$.*
(7) *For any path $\rho = \langle x_1, x_2, \ldots, x_n \rangle \in paths(N)$ such that $x_1 \notin \square_N(t)$ and $x_n \in \square_N(t)$: $t \in \{x_2, \ldots, x_n\}$.*
(8) *For any proper t'-induced T-net of N: $t' = t$ or $\square_N(t') \cap \square_N(t) = \emptyset$.*

Proof. (1) $N_{\square(t)}$ is a T-net, because, by construction, all added places have one input transition and one output transition, and only nodes connected to other nodes are added.

(2) Removing a node and all connected arcs cannot invalidate the free-choice property. The connections between the remaining places and transitions do not change.

(3) The t-induced T-net is transition bordered, i.e., for each place in $\square_N(t)$ the unique input transition and output transition are added.

(4) If $t' \in \square_N(t) \cap (T \setminus \{t\})$, then there is at least one input place $p' \in \bullet t' \cap \square_N(t)$ (by construction, transitions different from t are only added to $\square_N(t)$ after an input place was added). Assume t' has an input place outside $\square_N(t)$, i.e., $p'' \in \bullet t' \setminus \square_N(t)$. Since $\overline{N_{\square(t)}}$ is strongly-connected, there must be a $t'' \in p'' \bullet \setminus \square_N(t)$ (otherwise p' would be a sink place in $\overline{N_{\square(t)}}$). Since the net is free-choice, $\bullet t' = \bullet t''$ and $p' \in \bullet t''$. This contradicts with (3).

(5) Since N is strongly connected, there must be an arc from a node outside $\square_N(t)$ to a node inside $\square_N(t)$. Using (3) and (4), the node inside $\square_N(t)$ must be t. Hence, there is a place $p' \in \bullet t \setminus \square_N(t)$. Since $\overline{N_{\square(t)}}$ is strongly-connected, there must be a $t' \in p' \bullet \setminus \square_N(t)$ (otherwise p' would be a sink place in $\overline{N_{\square(t)}}$). Since the net is free-choice, $\bullet t = \bullet t'$. Assume that t has an input place inside $\square_N(t)$, then also t' has an input place inside $\square_N(t)$. This leads to a contradiction because the t-induced T-net is transition bordered. Hence, t cannot have an input place inside $\square_N(t)$, i.e., $\bullet t \subseteq T \setminus \square_N(t)$.

(6) The input places of t remain after removing $\square_N(t)$ and the complement is strongly connected. Hence, there must be a transition $t' \in T \setminus \square_N(t)$ such that $\bullet t = \bullet t'$.

(7) The only way to "enter" the t-induced T-net is through t. This directly follows from (3), (4), and (5).

(8) Assume $x \in \square_N(t) \cap \square_N(t')$. This implies that there must be an elementary non-branched path (i.e., places on the path have one input and one output transition) from t' to x. Using (3) and (4), we can follow this path backwards from x to t' and conclude that all the nodes belong to $\square_N(t)$, including t', i.e., $t' \in \square_N(t)$. Applying (4) once more assuming $t \neq t'$ shows that $\bullet t' \subseteq \square_N(t)$. However, using (6) we know that t' is involved in a choice, making the input places branching and thus leading to a contradiction. $\square$

A p-induced P-net is a subnet fully defined by a final place p and all nodes from which p can be reached without visiting transitions with multiple input or multiple output places. Figure 1 highlights two induced P-nets: $\odot_N(p7) = \{p7, t3, t5, p3, p5\}$ and $\odot_N(p8) = \{p8, t4, t6, p4, p6\}$.

Definition 11 (p-Induced P-net). *Let $N = (P, T, F)$ be a Petri net and $p \in P$. $\odot_N(p) \subseteq P \cup T$ is the smallest set such that*

- *$p \in \odot_N(p)$,*
- *$\{t' \in \bullet p' \mid |\bullet t'| = 1 \wedge |t'\bullet| = 1\} \subseteq \odot_N(p)$ for any $p' \in \odot_N(p) \cap P$, and*
- *$\bullet t' \subseteq \odot_N(p)$ for any $t' \in \odot_N(p) \cap T$.*

$\odot_N(p)$ are the nodes of the p-induced P-net of N which is denoted by $N_{\odot(p)} = N{\upharpoonright}_{\odot_N(p)}$. $\overline{N_{\odot(p)}} = N \setminus\!\setminus \odot_N(p)$ is the complement of the p-induced P-net of N. $\odot_N(p)$ is proper if the complement $\overline{N_{\odot(p)}}$ is a non-trivial strongly-connected Petri net.

Due to duality, symmetric properties can be found using similar reasoning.

Proposition 2 (Properties of Proper p-Induced P-net). *Let $N = (P, T, F)$ be a strongly-connected free-choice Petri net and $\odot_N(p)$ a proper p-induced P-net of N.*

(1) $N_{\odot(p)}$ is a P-net.
(2) $\overline{N_{\odot(p)}}$ is free-choice.
(3) For all $t' \in \odot_N(p) \cap T$: $\bullet t' \cup t'\bullet \subseteq \odot_N(p)$.
(4) For all $p' \in \odot_N(p) \cap (P \setminus \{p\})$: $p'\bullet \subseteq \odot_N(p)$.
(5) $p\bullet \subseteq T \setminus \odot_N(p)$.
(6) There is a $p' \in P \setminus \odot_N(p)$ such that $p\bullet = p'\bullet$.
(7) For any path $\rho = \langle x_1, x_2, \ldots, x_n \rangle \in paths(N)$ such that $x_1 \in \odot_N(p)$ and $x_n \notin \odot_N(p)$: $p \in \{x_1, x_2, \ldots, x_{n-1}\}$.
(8) For any proper p'-induced P-net of N: $p' = p$ or $\odot_N(p') \cap \odot_N(p) = \emptyset$.

Proof. Analogous to Proposition 1. □

Next, we show that a t-induced T-net corresponds to a t-induced P-net in the dual net (where t is a place). Recall that places and transitions are exchanged and the direction of all arcs is reversed in N^{dual}.

Lemma 1 (Duality Lemma for Induced Subnets). *Let $N = (P, T, F)$ be a Petri net.*

(1) For any $t \in T$: $\Box_N(t)$ is a (proper) t-induced T-net of N if and only if $\odot_{N^{dual}}(t)$ is a (proper) t-induced P-net of N^{dual}.
(2) For any $p \in P$: $\odot_N(p)$ is a (proper) p-induced P-net of N if and only if $\Box_{N^{dual}}(p)$ is a (proper) p-induced T-net of N^{dual}.

Proof. Let $N = (P, T, F)$ be a Petri net and $N' = (P', T', F') = N^{dual} = (T, P, F^{-1})$ the dual net. $\bullet$ is used for the pre and post sets in N and $\circ$ is used for the pre and post sets in N'. Note that $x \in \bullet y \Leftrightarrow (x, y) \in F \Leftrightarrow (y, x) \in F' \Leftrightarrow x \in y\circ$. Similarly, $x \in y\bullet \Leftrightarrow (y, x) \in F \Leftrightarrow (x, y) \in F' \Leftrightarrow x \in \circ y$. Using these insights and a pairwise comparison of the three rules in Definition 10 and Definition 11, the proof follows immediately. $\qquad\square$

Proposition 3 (Induced Subnets Relate to T/P-Components). *Let $N = (P, T, F)$ be a well-formed free-choice net. For any proper t-induced T-net $\boxdot_N(t)$, there exists a T-component $X \in TComp(N)$ such that $\boxdot_N(t) \subseteq X$. For any proper p-induced P-net $\odot_N(p)$, there exists a P-component $X \in PComp(N)$ such that $\odot_N(p) \subseteq X$.*

Proof. Let $\boxdot_N(t)$ be a proper t-induced T-net. t must be covered by some T-component X (Theorem 1). If t is included, then also $t\bullet$ is included in X. For the places in $t\bullet$ that have only one output transition, also these output transitions need to be included in X. For all transitions included, the input and output places must be included in X. Etc. Hence, $\boxdot_N(t) \subseteq X$. Let $\odot_N(p)$ be a proper p-induced P-net. We can apply the same reasoning since p is covered by some P-component X (Theorem 1). However, now the arcs are followed in the reverse direction to show that $\odot_N(p) \subseteq X$. $\qquad\square$

Next, we show that a well-formed free-choice net has at least two induced T-nets or is a T-net. The proof combines Proposition 7.11 in [9] with Lemma 1.2 in [15].

Lemma 2 (Existence of t-Induced T-nets). *Let $N = (P, T, F)$ be a well-formed free-choice net. N is either a T-net or there exist at least two different transitions $t_1, t_2 \in T$ such that $\boxdot_N(t_1)$ is proper and $\boxdot_N(t_2)$ is proper.*

Proof. N is covered by the set of T-components $TComp(N)$. Take a minimal $Q \subseteq TComp(N)$ such that $\bigcup Q = P \cup T$ (i.e., removing a T-component from Q leads to incomplete coverage of the net). Assume $|Q| \geq 2$ (otherwise N is a T-net). Create a spanning tree for the graph $G = (V, E)$ with $V = Q$ and $E = \{(X_1, X_2) \in Q \times Q \mid X_1 \cap X_2 \neq \emptyset\}$. Pick a T-component $X \in Q$ that is a leaf in the spanning tree (i.e., the remaining T-components in $Q' = Q \setminus \{X\}$ are still connected). There are at least two such leaf nodes, because $|Q| \geq 2$. Let $Y = X \setminus \bigcup Q'$ be the nodes only in X. $Y \neq \emptyset$ because Q was minimal. Let $Y' \subseteq Y$ be a maximal connected subset of Y. Obviously, Y' is a transition-bordered connected T-net. $N \setminus\!\!\setminus Y$ is strongly-connected because the remaining T-components in $Q' = Q \setminus \{X\}$ are still connected and inside a T-component all nodes are strongly-connected. The nodes in $Y \setminus Y'$ are not connected to Y' (due to maximality) and therefore connected to nodes in $\bigcup Q'$. Hence, $N \setminus\!\!\setminus Y'$ is strongly-connected. Moreover, there can only be one transition in Y' consuming tokens from $\bigcup Q'$. This implies that Y' corresponds to a proper induced T-net. (Note that we use the same reasoning as in Def. 7.7, Prop. 7.10, and Prop. 7.11 in [9].)

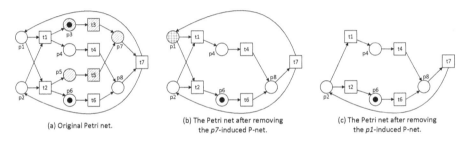

(a) Original Petri net.

(b) The Petri net after removing
the p7-induced P-net.

(c) The Petri net after removing
the p1-induced P-net.

Fig. 2. A well-formed free choice net is reduced in two steps into a P-net using $\gamma = \langle p7, p1 \rangle$.

We could have picked two different T-components $X_1, X_2 \in Q$ that are leaves in the spanning tree. Therefore, it is possible to find at least two different connected subsets that are non-overlapping. Hence, we can find two transitions $t_1, t_2 \in T$ such that both $\square_N(t_1)$ and $\square_N(t_2)$ are proper. $\qquad \square$

We can use a similar approach to show that a well-formed free-choice net has at least two induced P-nets or is a P-net. Consider the well-formed free-choice net in Fig. 2(a). This is not a P-net, so we can find at least two induced P-nets: $\odot_N(p7)$ and $\odot_N(p8)$. After removing the nodes in $\odot_N(p7)$, we get $N' = \overline{N_{\odot(p7)}} = N \setminus\!\setminus \odot_N(p7)$ shown in Fig. 2(b). In N' there are again at least two induced P-nets $\odot_{N'}(p1)$ and $\odot_{N'}(p2)$. After removing the nodes in $\odot_{N'}(p1)$, we obtain the P-net $\overline{N'_{\odot(p1)}}$ shown in Fig. 2(c).

Lemma 3 (Existence of p-Induced P-nets). *Let $N = (P, T, F)$ be a well-formed free-choice net. N is either a P-net or there exist at least two different places $p_1, p_2 \in P$ such that $\odot_N(p_1)$ is proper and $\odot_N(p_2)$ is proper.*

Proof. Let $N = (P, T, F)$ be a well-formed free-choice net. $N' = (P', T', F') = N^{dual} = (T, P, F^{-1})$ is the dual net. N' is also a well-formed free-choice net (Theorem 2). N is a P-net if and only if N' is a T-net. If N is not a P-net, then N' is not a T-net and there exists a $t \in T' = P$ such that $\square_{N'}(t)$ is a proper t-induced T-net of N' (apply Lemma 2). Using Lemma 1, this implies that $\odot_N(t)$ is a proper t-induced P-net of N for some place $t \in P$. A similar reasoning can be used to show that there are at least two proper induced P-nets $\odot_N(p_1)$ and $\odot_N(p_2)$. $\qquad \square$

Thus far, we ignored the marking of the free-choice net when removing an induced T-net or P-net. Removing an induced T-net and its tokens may destroy liveness. In Fig. 1, we had to "push out" the token in $p3$ to $p7$ to preserve liveness.

Proposition 4 (Pushed Out Markings Exist And Are Unique). *Let (N, M) be a strongly-connected marked free-choice net, $\square_N(t)$ a proper t-induced T-net of N, $\hat{T} = (\square_N(t) \cap T) \setminus \{t\}$, and $push(\square_N(t), M) = \{M' \in \mathcal{B}(P) \mid \exists_{\sigma \in (\hat{T})^*} (N, M)[\sigma\rangle(N, M') \wedge en(N, M') \cap \hat{T} = \emptyset\}$. $|push(\square_N(t), M)| = 1$.*

Proof. Follows directly from the properties listed in Proposition 1. For each transition $t' \in \hat{T} = (\square_N(t) \cap T) \setminus \{t\}$, there is an elementary path from t to t' where each place has one input transition and one output transition. Since only transitions in $\hat{T}$ are considered in $push(\square_N(t), M)$, the number of tokens on a path cannot increase, but decreases when t' fires. This applies to any $t' \in \hat{T}$, hence, after some time none of the transitions in $\hat{T}$ can fire anymore and we find a marking M' such that $en(N, M') \cap \hat{T} = \emptyset$. Since $N_{\square(t)}$ is a T-net, all interleavings lead to the same M'. $\square$

Since the "pushed out marking" is unique, we can update the marking after removing a t-induced T-net in a deterministic manner. When a p-induced P-net is removed, we can simply project the marking onto the remaining places.

Definition 12 (Updated Markings). *Let (N, M) be a marked Petri net, $N = (P, T, F)$, $\square_N(t)$ a proper t-induced T-net of N, and $\odot_N(p)$ a proper p-induced P-net of N.*

- *$mrk_\square(N, t, M) \in \{M' \upharpoonright_{P \setminus \square_N(t)} | M' \in push(\square_N(t), M)\}$ is the unique marking obtained by "pushing out" tokens as much as possible (see Proposition 4).*
- *$mrk_\odot(N, p, M) = M \upharpoonright_{P \setminus \odot_N(p)}$ is the unique marking obtained by removing the tokens in $\odot_N(p)$.*

Lemma 4 (Well-Formedness of $\overline{N_{\square(t)}}$). *Let $N = (P, T, F)$ be a well-formed free-choice net having a transition $t \in T$ such that $\square_N(t)$ is proper. $\overline{N_{\square(t)}} = (\overline{P}, \overline{T}, \overline{F})$ is the corresponding complement.*

(1) For any $\overline{M}, \overline{M}' \in \mathcal{B}(\overline{P})$, $\hat{M} \in \mathcal{B}(P)$, and $\sigma \in \overline{T}^*$: if $(\overline{N_{\square(t)}}, \overline{M})[\sigma\rangle(\overline{N_{\square(t)}}, \overline{M}')$, then $(N, \overline{M} \uplus \hat{M})[\sigma\rangle(N, \overline{M}' \uplus \hat{M})$.

(2) For any $M \in \mathcal{B}(P)$: if (N, M) is live and bounded, then $(\overline{N_{\square(t)}}, mrk_\square(N, t, M))$ is live and bounded.

(3) $\overline{N_{\square(t)}}$ is well-formed and free-choice.

Proof. Let $N = (P, T, F)$ be a well-formed free-choice net, $\square_N(t)$ a proper t-induced T-net, and $\overline{N_{\square(t)}} = (\overline{P}, \overline{T}, \overline{F})$. $\overline{N_{\square(t)}}$ is free-choice (apply Proposition 1(2)).

(1) If $(\overline{N_{\square(t)}}, \overline{M})[\sigma\rangle(\overline{N_{\square(t)}}, \overline{M}')$, then $(N, \overline{M})[\sigma\rangle(N, \overline{M}')$ (because $\overline{T} \subseteq T$ and $\bullet t$ and $t\bullet$ are the same for $t \in \overline{T}$ in both nets). Adding tokens cannot disable an enabled firing sequence. Hence, $(N, \overline{M} \uplus \hat{M})[\sigma\rangle(N, \overline{M}' \uplus \hat{M})$.

(2) Assume (N, M) is live and bounded. $M' \in push(\square_N(t), M)$ is the unique "pushed out marking" (see Proposition 4). Obviously, (N, M') is also live and bounded. Split M' into $\overline{M} = mrk_\square(N, t, M) = M' \upharpoonright_{P \setminus \square_N(t)}$ and $\hat{M} = M' \setminus mrk_\square(N, t, M)$, i.e., $M' = \overline{M} \uplus \hat{M}$. We need to show that $(\overline{N_{\square(t)}}, \overline{M})$ is live and bounded.

Using (1) we know that any firing sequence enabled in $(\overline{N_{\square(t)}}, \overline{M})$ is also enabled in $(N, \overline{M} \uplus \hat{M})$. Hence, $(\overline{N_{\square(t)}}, \overline{M})$ is bounded, because $(N, \overline{M} \uplus \hat{M})$ is bounded.

$(\overline{N_{\square(t)}}, \overline{M})$ is a bounded, strongly-connected, and free-choice. Using Theorem 4.31 in [9], we know that $(\overline{N_{\square(t)}}, \overline{M})$ is live if and only if $(\overline{N_{\square(t)}}, \overline{M})$ is deadlock-free. Assume $(\overline{N_{\square(t)}}, \overline{M})$ has a reachable deadlock $\overline{M}_D$. The corresponding reachable marking from (N, M') is $M_D = \overline{M}_D \uplus \hat{M}$ (recall $M' = \overline{M} \uplus \hat{M}$). The transitions in $\overline{T} \cup \{t\}$ are also disabled in (N, M_D) because the input places are unaffected (note that there is a $t' \in \overline{T}$ such that $\bullet t = \bullet t'$ that is disabled and so is t). The other transitions in $\hat{T} = (\square_N(t) \cap T) \setminus \{t\}$ are also dead because we started from a marking where tokens were "pushed out" until no transition in $\hat{T}$ was enabled anymore. Hence, also M_D is a dead reachable marking contradicting that (N, M') is live. Hence, $(\overline{N_{\square(t)}}, \overline{M})$ cannot have a reachable deadlock, implying that $(\overline{N_{\square(t)}}, \overline{M})$ is live.

(3) Because N is well-formed there is a marking M such that (N, M) is live and bounded. $\overline{N_{\square(t)}}$ is well-formed because $(\overline{N_{\square(t)}}, mrk_{\square}(N, t, M))$ is live and bounded (follows directly from (2)). $\qquad\square$

We can also show that removing a p-induced P-net does not jeopardize liveness and boundedness. Note that $mrk_{\odot}(N, p, M)$ is obtained by simply removing the tokens in $\odot_N(p)$ (Definition 12).

Lemma 5 (Well-Formedness of $\overline{N_{\odot(p)}}$). *Let $N = (P, T, F)$ be a well-formed free-choice net having a place $p \in P$ such that $\odot_N(p)$ is proper. $\overline{N_{\odot(p)}} = (\overline{P}, \overline{T}, \overline{F})$ is the corresponding complement.*

(1) For any $M, M' \in \mathcal{B}(P)$ and $\sigma \in T^*$: if $(N, M)[\sigma\rangle(N, M')$, then $(\overline{N_{\odot(p)}}, mrk_{\odot}(N, p, M))[\sigma\upharpoonright_{\overline{T}}\rangle(\overline{N_{\odot(p)}}, mrk_{\odot}(N, p, M'))$.
(2) $\overline{N_{\odot(p)}}$ is well-formed and free-choice.
(3) For any $M \in \mathcal{B}(P)$: if (N, M) is live and bounded, then $(\overline{N_{\odot(p)}}, mrk_{\odot}(N, p, M))$ is live and bounded.

Proof. Let $N = (P, T, F)$ be a well-formed free-choice net, $\odot_N(p)$ a proper p-induced P-net, and $\overline{N_{\odot(p)}} = (\overline{P}, \overline{T}, \overline{F})$. Recall that $mrk_{\odot}(N, p, M) = M \upharpoonright_{P \setminus \odot_N(p)} = M \upharpoonright_{\overline{P}}$ and $mrk_{\odot}(N, p, M') = M' \upharpoonright_{\overline{P}}$. In the proof, we use these more compact notations.

(1) If $(N, M)[t\rangle(N, M')$ and $t \in \overline{T}$, then $(\overline{N_{\odot(p)}}, M\upharpoonright_{\overline{P}})[t\rangle(\overline{N_{\odot(p)}}, M'\upharpoonright_{\overline{P}})$ because removing places cannot disable a transition. If $(N, M)[t\rangle(N, M')$ and $t \notin \overline{T}$, then we can ignore t, because t is not impacting places in $\overline{P}$ and $M\upharpoonright_{\overline{P}} = M'\upharpoonright_{\overline{P}}$. Iteration over all transitions in σ shows that indeed $(\overline{N_{\odot(p)}}, M\upharpoonright_{\overline{P}})[\sigma\upharpoonright_{\overline{T}}\rangle(\overline{N_{\odot(p)}}, M'\upharpoonright_{\overline{P}})$.
(2) Since $N = (P, T, F)$ is a well-formed free-choice net, $N^{dual} = (T, P, F^{-1})$ is a well-formed free-choice net (apply Theorem 2). Since $\odot_N(p)$ is a proper p-induced P-net of N, $\square_{N^{dual}}(p)$ is a proper p-induced T-net of N^{dual} (apply Lemma 1). Since $\square_{N^{dual}}(p)$ is proper and N^{dual} is well-formed, we can apply Lemma 4 to show that $\overline{N^{dual}_{\square(p)}}$ is well-formed. Moreover, $\odot_N(p) = \square_{N^{dual}}(p)$

(see proof of Lemma 4). Hence, $\overline{N_{\square(p)}^{dual}} = N^{dual} \setminus\!\setminus \square_{N^{dual}}(p) = N^{dual} \setminus\!\setminus$
$\odot_N(p) = (N \setminus\!\setminus \odot_N(p))^{dual} = (\overline{N_{\odot(p)}})^{dual}$ is well-formed. Since $(\overline{N_{\odot(p)}})^{dual}$ is
well-formed, also $\overline{N_{\odot(p)}}$ is well-formed (apply Theorem 2 again). Obviously,
$\overline{N_{\odot(p)}}$ is free-choice (use Proposition 2(2)).

(3) Both N and $\overline{N_{\odot(p)}}$ are well-formed and free-choice. Hence, both are structurally bounded and covered by P-components. Any P-component of $\overline{N_{\odot(p)}}$
is also a P-component N and initially marked in M because of liveness (use
Theorem 5.8 in [9]). Such a P-component is also marked in $M\!\restriction_{\overline{P}}$. Applying
Theorem 5.8 [9] in the other direction proves that $(\overline{N_{\odot(p)}}, M\!\restriction_{\overline{P}})$ is also live
because all P-components are marked.

$\square$

5   Approach: Using Induced Subnets for Reduction

Lemmata 4 and 5 show that iteratively removing proper induced T- and P-
nets preserves well-formedness, liveness, and boundedness. We first introduce
the approach based on *reductions using sequences of proper induced T- and P-
nets*. In Sect. 6, we apply this to properties like lucency and perpetuality.

Definition 13 (Reductions). *Let $N = (P, T, F)$ be a well-formed free-choice
net. A reduction of N is a sequence $\gamma = \langle x^1, x^2, \dots, x^n \rangle \in (P \cup T)^*$ such that
there exists a sequence of Petri nets denoted $nets_N(\gamma) = \langle N^0, N^1, \dots, N^n \rangle$ where
$N^0 = N$, and for any $i \in \{1, \dots, n\}$:*

- *$\square_{N^{i-1}}(x^i)$ is a proper x^i-induced T-net and $N^i = \overline{N^{i-1}_{\square(x^i)}}$ if $x^i \in T$.*
- *$\square_{N^{i-1}}(x^i)$ is a proper x^i-induced P-net and $N^i = \overline{N^{i-1}_{\odot(x^i)}}$ if $x^i \in P$.*

A reduction $\gamma = \langle x^1, x^2, \dots, x^n \rangle$ is nothing more than a sequence of proper
induced T- and P-nets. Figure 2 shows a two-step reduction $\gamma = \langle p7, p1 \rangle$ Note
that γ uniquely determines $nets_N(\gamma)$. Next, we consider different classes of reductions.

Definition 14 (Complete, T-, and P-Reductions). *Let $N = (P, T, F)$ be a
well-formed free-choice net having a reduction $\gamma = \langle x^1, x^2, \dots, x^n \rangle \in (P \cup T)^*$
with the corresponding sequence of Petri nets: $nets_N(\gamma) = \langle N^0, N^1, \dots, N^n \rangle$.[2]*

- *γ is x-preserving if $x \in P \cup T$ is a place/transition in the remaining net N^n.*
- *γ is a complete reduction if N^n is a T-net or a P-net.*
- *γ is a T-reduction if $\{x^1, x^2, \dots, x^n\} \subseteq T$ and N^n is a T-net.*
- *γ is a P-reduction if $\{x^1, x^2, \dots, x^n\} \subseteq P$ and N^n is a P-net.*

---

[2] The notions of T-reduction and P-reduction are unrelated to the "Desel rules" for
free-choice nets without frozen tokens [8]. We allow for "bigger steps" and can reduce
nets with frozen tokens (i.e., there may be an infinite firing sequence starting from
a strictly smaller marking).

The reduction $\gamma_1 = \langle p7, p1 \rangle$ illustrated by Fig. 2 is a complete P-reduction that is $t4$ preserving. $\gamma_2 = \langle p8, p2 \rangle$ is another complete P-reduction and $\gamma_3 = \langle t1 \rangle$ and $\gamma_4 = \langle t2 \rangle$ are complete T-reductions. Next, we show that such reductions always exist. Moreover, we can preserve any preselected node.

Lemma 6 (Existence of Reductions). *Let $N = (P, T, F)$ be a well-formed free-choice net. N has at least one T-reduction γ_T and at least one P-reduction γ_P. For any node $x \in P \cup T$ there is an x-preserving T-reduction and an x-preserving P-reduction.*

Proof. Let $N = (P, T, F)$ be a well-formed free-choice net. First, we construct a T-reduction $\gamma_T = \langle t^1, t^2, \ldots, t^n \rangle \in T^*$. If N is a T-net, then $\gamma_T = \langle \, \rangle$ (i.e., $n = 0$). If $N = N^0$ is not a T-net, then there exists a $t^1 \in T$ such that $\Box_{N^0}(t^1)$ is proper (Lemma 2). Next, we consider $N^1 = \overline{N^0}_{\Box(t^1)}$. If N^1 is a T-net, then $\gamma_T = \langle t^1 \rangle$ (i.e., $n = 1$). If N^1 is not a T-net, then there exists a $t^2 \in T$ such that $\Box_{N^1}(t^2)$. Etc. This is repeated until we encounter a T-net $N^n = \overline{N^{n-1}}_{\Box(t^n)}$. We can use the same approach to construct a P-reduction $\gamma_P = \langle p^1, p^2, \ldots, p^m \rangle \in P^*$. If N is a P-net, then $\gamma_P = \langle \, \rangle$. If not, we repeatedly apply Lemma 3 until we find a P-net.

Lemma 2 states that there exist at least two transitions t_1, t_2 such that $\Box_N(t_1)$ and $\Box_N(t_2)$ are proper. These are disjoint, i.e., $\Box_N(t_1) \cap \Box_N(t_2) = \emptyset$ (see Proposition 1(8)). Hence, in each step, we can pick an induced T-net not containing a particular node $x \in P \cup T$. The same applies to P-reductions (use Lemma 3 and Proposition 2(8)).   □

In Definition 12, we defined update functions for markings that preserve liveness and boundedness. These can be applied in sequence.

Definition 15 (Reduction of Marked Nets). *Let $N = (P, T, F)$ be a well-formed free-choice net having a reduction $\gamma = \langle x^1, x^2, \ldots, x^n \rangle \in (P \cup T)^*$ with the corresponding sequence of nets $nets_N(\gamma) = \langle N^0, N^1, \ldots, N^n \rangle$. In the context of $nets_N(\gamma)$, we denote $N^i = (P^i, T^i, F^i)$ for $i \in \{0, \ldots, n\}$. $mrks_{N,M}(\gamma) = \langle M^0, M^1, \ldots, M^n \rangle$ is such that $M = M^0$ and for any $i \in \{1, \ldots, n\}$:*

- *$M^i = mrk_{\Box}(N^{i-1}, x^i, M^{i-1})$ if $x^i \in T$.*
- *$M^i = mrk_{\odot}(N^{i-1}, x^i, M^{i-1})$ if $x^i \in P$.*

Reductions preserved liveness and boundedness, e.g., Fig. 2(c) is live and bounded because Fig. 2(a) is live and bounded.

Theorem 3 (Reduction Theorem). *Let (N, M) be a live and bounded free-choice net and $\gamma = \langle x^1, x^2, \ldots, x^n \rangle \in (P \cup T)^*$ a reduction of N. Let $nets_N(\gamma) = \langle N^0, N^1, \ldots, N^n \rangle$ and $mrks_{N,M}(\gamma) = \langle M^0, M^1, \ldots, M^n \rangle$ be the corresponding nets and markings. (N^i, M^i) is live and bounded and N^i is well-formed and free-choice for any $i \in \{0, \ldots, n\}$.*

Proof. Let (N, M) be a live and bounded free-choice net, $\gamma = \langle x^1, x^2, \ldots, x^n \rangle$ a reduction, $nets_N(\gamma) = \langle N^0, N^1, \ldots, N^n \rangle$, and $mrks_{N,M}(\gamma) =$

$\langle M^0, M^1, \ldots, M^n \rangle$. We use induction to prove that (N^i, M^i) is live and bounded and N^i is well-formed and free-choice for any $i \in \{0, \ldots, n\}$. If $i = 0$ this holds by definition. Assume $i \geq 1$, (N^{i-1}, M^{i-1}) is live and bounded, and N^{i-1} is well-formed and free-choice (induction hypothesis).

If $x^i \in T$, then $\Box_{N^{i-1}}(x^i)$ is proper, $N^i = \overline{N^{i-1}\Box_{(X^i)}}$, and $M^i = mrk_\Box(N^{i-1}, x^i, M^{i-1})$. Lemma 4 can be applied to show that (N^i, M^i) is live and bounded and N^i is well-formed and free-choice.

If $x^i \in P$, then $\odot_{N^{i-1}}(x^i)$ is proper, $N^i = \overline{N^{i-1}\odot_{(X^i)}}$, and $M^i = mrk_\odot(N^{i-1}, x^i, M^{i-1})$. Now, Lemma 5 can be applied to show that (N^i, M^i) is live and bounded and N^i is well-formed and free-choice. This completes the proof by induction. $\Box$

The reduction steps are commutative when both are applicable. Consider a reduction $\gamma = \langle x^1, x^2, \ldots, x^n \rangle$ of N, i and j such that $1 \leq i < j \leq n$, and $\gamma' = \langle x^1, x^2, \ldots, x^{i-1}, x^j, x^i, \ldots, x^{j-1}, x^{j+1}, \ldots, x^n \rangle$ (i.e., x^j is moved to the position before x^i). If $x^j \in T$ and $\Box_{N^{i-1}}(x^j)$ is proper or $x^j \in P$ and $\odot_{N^{i-1}}(x^j)$ is proper, then γ' is also a reduction of N.

6   Application of Reduction to Prove Perpetuality and Lucency

This section illustrates the usage of reductions. Well-formedness, liveness, and boundedness are preserved "downstream", i.e., these properties are preserved if the net is reduced. For example, N^j is well-formed if N^i is well-formed and $i < j$. We will show that less-common studied properties such as *pc-safeness* and *perpetuality* are also preserved "downstream". Other properties are preserved "upstream", i.e., these properties are preserved if the net is extended. We will use these "upstream" properties to convert results for T-nets or P-nets to free-choice nets (e.g., lucency). First, we introduce three properties that are preserved "downstream".

Definition 16 (Regeneration Transitions). *Let Petri net $N = (P, T, F)$ be a Petri net. Transition $t_r \in T$ is a* regeneration transition *of N if the marked Petri net $(N, [p \in \bullet t_r])$ is live and bounded.*

A *regeneration transition* t_r defines a *regeneration marking* $M_r = [p \in \bullet t_r]$. This can be viewed as a *structural property*: A net is *perpetual* if it has such a marking.

Definition 17 (Perpetual Nets [2]). *Petri net $N = (P, T, F)$ is a* perpetual net *if there exists at least one regeneration transition.*

In a *pc-safe marking* all P-components have precisely one token. Note that a safe marked net does not need to be pc-safe (see, for example, Fig. 6 in [2]).

Definition 18 (PC-Safely Marked Nets). *Let Petri net $N = (P, T, F)$ be a Petri net. $M \in \mathcal{B}(P)$ is a* pc-safe marking *of N if for any $X \in PComp(N)$: $M(X \cap P) = 1$, i.e., each P-component contains precisely one token. (N, M) is a* pc-safely marked net *if M is a pc-safe marking of N.*

In a marked perpetual well-formed free-choice net, regeneration markings can be reached *if and only if* the initial marking is pc-safe.

Lemma 7 (Perpetual Nets Are PC-Safely Marked). *Let $N = (P, T, F)$ be a perpetual well-formed free-choice net with regeneration transition $t_r \in T$. For any marking $M \in \mathcal{B}(P)$: M is pc-safe if and only if $[p \in \bullet t_r] \in R(N, M)$.*

Proof. $M_r = [p \in \bullet t_r]$. (N, M_r) is live and bounded because t_r is a regeneration transition. Take an arbitrary P-component $X \in PComp(N)$. $M_r(X \cap P) \neq 0$, because, otherwise, the transitions in $X \cap T$ would be dead contradicting liveness. $M_r(X \cap P) \not> 1$, because this implies that one of the input places of t_r has at least two tokens. Hence, M_r is pc-safe and all P-components contain precisely one input place of t_r. If $M_r \in R(N, M)$, then M needs to be pc-safe (the number of tokens in a P-component cannot change). Remains to show that M_r can be reached from M if M is pc-safe. (N, M) is live if M is pc-safe (use Theorem 5.8 in [9]). Hence, t_r can be enabled, proving that M_r is indeed reachable.   □

Next, we show that the properties just defined are preserved "downstream" *for any reduction* (i.e., also for mixtures of place- and transition-induced subsets).

Theorem 4 (Invariant Downstream Properties). *Let $N = (P, T, F)$ be a well-formed free-choice net having a reduction $\gamma = \langle x^1, x^2, \ldots, x^n \rangle$ with the corresponding sequence of nets $nets_N(\gamma) = \langle N^0, N^1, \ldots, N^n \rangle$.*

(1) If $t_r \in T$ is a regeneration transition of N (i.e., $(N, [p \in \bullet t_r])$ is live and bounded) and γ is t_r-preserving, then t_r is a regeneration transition of all nets in $nets_N(\gamma)$ (i.e., $(N^i, [p \in \bullet t_r])$ is live and bounded for any $i \in \{0, \ldots, n\}$).[3]

(2) If (N, M) is pc-safe, then all markings in $mrks_{N,M}(\gamma)$ are pc-safe.

(3) If N is perpetual, then all nets in $nets_N(\gamma)$ are perpetual.

Proof. Let $N = (P, T, F)$ be a well-formed free-choice net having a reduction $\gamma = \langle x^1, x^2, \ldots, x^n \rangle$ and $nets_N(\gamma) = \langle N^0, N^1, \ldots, N^n \rangle$.

(1) Assume $(N, [p \in \bullet t_r])$ is live and bounded and γ is t_r-preserving. We prove that $(N^i, [p \in \bullet t_r])$ is live and bounded for any $i \in \{0, \ldots, n\}$ using induction. If $i = 0$, this holds by definition ($N^0 = N$). Assume $i \geq 1$ and $(N^{i-1}, [p \in \bullet t_r])$ is live and bounded. If $x^i \in T$, then $\square_{N^{i-1}}(x^i)$ is proper, $N^i = \overline{N^{i-1}}_{\square(X^i)}$, and $M^i = mrk_\square(N^{i-1}, x^i, [p \in \bullet t_r]) = [p \in \bullet t_r]$, because t_r and $\bullet t_r$ are outside $\square_{N^{i-1}}(x^i)$ (γ is t_r-preserving).[3] We can apply Theorem 3 to show that $(N^i, [p \in \bullet t_r])$ is live and bounded. If $x^i \in P$, then $\odot_{N^{i-1}}(x^i)$ is proper, $N^i = \overline{N^{i-1}}_{\odot(X^i)}$, and $M^i = mrk_\odot(N^{i-1}, x^i, [p \in \bullet t_r]) = [p \in \bullet t_r]\lceil_{P^i}$. t_r is not removed because γ is t_r-preserving. Hence, also at least one input place of t_r remains. Therefore, $M^i = [p \in \bullet t_r]$ (note that $\bullet t_r$ may have been changed[3]) and $(N^i, [p \in \bullet t_r])$ is live and bounded (apply again Theorem 3). Hence, t_r is a regeneration transition of all nets in $nets_N(\gamma)$.

---

[3] Note that $\bullet t_r = \{p \mid (p, t_r) \in F^i\}$ depends on the net considered (here N^i).

(2) Assume (N, M) is pc-safe and $mrks_{N,M}(\gamma) = \langle M^0, M^1, \ldots, M^n \rangle$. Again we use induction and prove that (N^i, M^i) is pc-safe for any $i \in \{0, \ldots, n\}$. If $i = 0$, this holds by definition $((N^0, M^0) = (N, M)$ is pc-safe). Assume $i \geq 1$ and (N^{i-1}, M^{i-1}) is pc-safe (induction hypothesis). We need to show that (N^i, M^i) is pc-safe. Take an arbitrary P-component $X \in PComp(N^i)$, we need to show that $M^i(X \cap P) = 1$.

 – If $x^i \in P$, then $PComp(N^i) \subseteq PComp(N^{i-1})$ because for the remaining places the context did not change. Also the marking of the remaining places does not change, because $M^i = mrk_{\odot}(N^{i-1}, x^i, M^{i-1}) = M^i \upharpoonright_{P^i}$. Hence, $M^i(X \cap P) = 1$.

 – If $x^i \in T$, but $X \in PComp(N^{i-1})$, then nothing changed and $M^i(X \cap P) = 1$ (note that in a P-component all surrounding transitions are included, hence the marking of the places in X and their context, i.e., pre- and post-sets, did not change).

 – Assume $x^i \in T$ and $X \notin PComp(N^{i-1})$. Let $P_X = X \cap P$ be the places in the P-component X (these are outside the x^i-induced T-net) and $T_X = \square_{N^{i-1}}(x^i) \cap T$ the transitions in the x^i-induced T-net. $F_{in} = F^{i-1} \cap (P_X \times T_X)$ are the ingoing arcs and $F_{out} = F^{i-1} \cap (T_X \times P_X)$ are the outgoing arcs. Both sets need to have precisely one element, i.e., $F_{in} = \{(p_{in}, t_{in})\}$ and $F_{out} = \{(t_{out}, p_{out})\}$, and $t_{in} = x^i$. One of these two sets of arcs is non-empty because P_X must contain at least one place that was connected to a transition T_X and if one is non-empty the other one is also non-empty. Proposition 1(7) implies that $t_{in} = x^i$ and P_X cannot hold two input places of t_{in} because of Proposition 1(6). p_{in} is the unique input place in X. F_{out} cannot have multiple elements because N^{i-1} is well-formed and therefore structurally bounded. Consider now an elementary path $\rho = \langle t_{in}, p_1, \ldots, p_n, t_{out} \rangle \in (\square_{N^{i-1}}(x^i))^*$. Such a path must exist and the places are non-branching. $Y = X \cup \{x \in \rho\}$ is a P-component because Y is strongly connected, all places in Y are non-branching, and all input and output transitions are included. Hence, $Y \in PComp(N^{i-1})$ and $M^{i-1}(Y) = 1$ because (N^{i-1}, M^{i-1}) is pc-safe. Moreover, $M^{i-1}(Y) = M^i(X)$ (pushing out the tokens does not change the total number of tokens, and X must be marked in M^i). Hence, $M^i(X) = 1$.

(3) Assume that N is perpetual. To show that all nets in $nets_N(\gamma)$ are perpetual, the same approach can be used as in (1). The only difference is that there is not a fixed regeneration transition t_r that is preserved. Assume that $(N^{i-1}, [p \in \bullet t_r])$ is live and bounded. We need to show that there is a t'_r such that $(N^i, [p \in \bullet t'_r])$ is live and bounded. If $t_r \in N^i$ (i.e., the regeneration transition is outside the x^i-induced subset), then $t_r = t'_r$ and this transition remains a regeneration transition (as shown in (1)). If $t_r \notin N^i$, then we need to consider two cases:

 – If $x^i \in P$ and $t_r \notin N^i$, then we find a contradiction, because t_r, like any regeneration transition, should be in all P-components of N^{i-1}. This is impossible, because this implies $M^i = mrk_{\odot}(N^{i-1}, x^i, [p \in \bullet t_r]) = [\,]$.

 – If $x^i \in T$ and $t_r \notin N^i$, then pick $t'_r \in N^i$ such that $\bullet t'_r = \bullet x^i$. Proposition 1(6) shows that such a transition exists. t'_r is live in $(N^{i-1}, [p \in \bullet t_r])$.

Consider a reachable marking enabling t'_r and then "push out" as many tokens as possible using the same approach as in Proposition 4. Let M be the marking where t'_r and x^i are enabled and all other transitions in $\Box_{N^{i-1}}(x^i)$ are not. From M we must be able to enable the regeneration transition t_r by only firing transitions in $\Box_{N^{i-1}}(x^i)$ (other transitions can only influence the subnet through x^i). Therefore, all other places $P^i \setminus \bullet t'_r$ must be empty in M, showing that $(N^i, [p \in \bullet t'_r])$ is live and bounded.

Hence, using a similar approach as in (1) we showed that N^i is perpetual for any i. $\Box$

Next, we consider lucency, first defined in [2]. We are often interested in processes where the set of enabled actions uniquely defines the state, e.g., in the context of process mining or user-interface design [2,4]. In terms of Petri nets, this means that there cannot be two reachable marking enabling the same set of transitions.

Definition 19 (Lucency [2]). *Petri net $N = (P, T, F)$ is lucent if each pc-safe marking enables a unique set of transitions, i.e., for any two pc-safe markings M_1 and M_2: if $en(N, M_1) = en(N, M_2)$, then $M_1 = M_2$.*

After showing that well-formedness, liveness, boundedness, pc-safeness, and perpetuality are preserved "downstream", we show that lucency is preserved by traversing the reduction in "upstream" direction. This is non-trivial because even live and pc-safe free-choice nets may be non-lucent [2]. Therefore, we first present some results for perpetual nets, before using a T-reduction to prove that perpetuality implies lucency.

Lemma 8 (Identical Token Counts On Related Paths). *Let $N = (P, T, F)$ be a perpetual well-formed free-choice net. Let $M \in \mathcal{B}(P)$ be a pc-safe marking of N, $t_b, t_e \in T$ be two transitions, and $\rho_1 = \langle t_b, p_1^1, t_1^1, p_1^2, t_1^2, \ldots, p_1^m, t_e \rangle$ and $\rho_2 = \langle t_b, p_2^1, t_2^1, p_2^2, t_2^2, \ldots, p_2^n, t_e \rangle$ be two elementary paths leading from t_b and t_e covering places $P_1 = \{p_1^1, p_1^2, \ldots, p_1^m\}$ and $P_2 = \{p_2^1, p_2^2, \ldots, p_2^n\}$ such that for any $p \in P_1 \cup P_2$: $|\bullet p| = |p \bullet| = 1$. $M(P_1) = M(P_2)$, i.e., the number of tokens on both paths is identical.*

Proof. The number of tokens on both elementary paths ρ_1 and ρ_2 is only changed by t_b and t_e. All other transitions are either not connected to any place $p \in P_1 \cup P_2$ or move a token to the next place on the path. t_b adds a token to both paths and t_e removes a token from both paths. Hence, the difference $M'(P_1) - M'(P_2)$ remains constant for any $M' \in R(N, M)$.

Assume that $M(P_1) \neq M(P_2)$. This implies that $M'(P_1) \neq M'(P_2)$ for any $M' \in R(N, M)$. This includes $M_r(P_1) \neq M_r(P_2)$ for the regeneration marking $M_r = [p \in \bullet t_r]$ based on a regeneration transition t_r. Due to Lemma 7, M_r is pc-safe and can be reached from any pc-safe marking. Without loss of generality, we may assume $M_r(P_1) > M_r(P_2)$ (we can swap P_1 and P_2), i.e., there is a place $p_r \in \bullet t_r \cap P_1$ marked in the regeneration marking M_r. t_r cannot have two input places from P_1 because all places in P_1 have one output transition which is unique. Hence, $M_r(P_1) = 1$ implying that $M_r(P_2) = 0$. Hence, $M'(P_1) = M'(P_2) + 1$ for any $M' \in R(N, M)$. Because N is perpetual, all transitions are live, including t_b. After t_b fires, there is at least one token in P_2 until t_e fires. This implies that there are at least two tokens in P_1 until t_e fires. However, t_e cannot be reached without executing first t_r, but when executing t_r, p_r must be the only marked place in P_1 containing precisely one token leading to a contradiction. Hence, $M(P_1) = M(P_2)$.   □

We introduce *conflict-pairs* as "witnesses" of non-lucency. If a T-net is not lucent, then it must have a conflict-pair (Proposition 5).

Definition 20 (Conflict-Pair). *Let N be a Petri net. (M_1, M_2) is called a conflict-pair for N if (N, M_1) and (N, M_2) are pc-safely marked, $en(N, M_1) \cap en(N, M_2) = \emptyset$ (no transition is enabled in both markings), for all $t \in en(N, M_1)$: $M_2(\bullet t) \geq 1$, and for all $t \in en(N, M_2)$: $M_1(\bullet t) \geq 1$.*

Proposition 5 (Absence of Conflict-Pairs in T-nets Implies Lucency). *Let N be a perpetual well-formed T-net. If N is not lucent, then N has conflict-pairs.*

Proof. Assume N is not lucent, i.e., there are two pc-safe markings M_1 and M_2 such that $en(N, M_1) = en(N, M_2)$ and $M_1 \neq M_2$. Tokens in M_1 but not in M_2 are represented by ① and tokens in M_2 but not in M_1 are represented by ②. These ① and ② tokens can be viewed as "disagreement tokens", i.e., M_1 and M_2 disagree on the marking of the corresponding place. Tokens in both markings are denoted by $\bullet$ and are called "agreement tokens". We now synchronously modify the markings M_1 and M_2 by firing only transitions using "agreement tokens" ($\bullet$) and not consuming any of the "disagreement tokens" (① and ②). Because N is perpetual, there is regeneration transition $t_r \in T$. Since M_1 and M_2 are pc-safe, $M_r = [p \in \bullet t_r]$ can be reached by both. Consider a shortest firing sequence σ from M_1 to M_r: $(N, M_1)[\sigma\rangle(N, M_r)$. Try to execute the sequence without consuming any of the ① tokens. Transitions that need to consume "disagreement tokens" or that are disabled can be skipped. However, per cluster transitions are executed in the same order as in σ (note that if one transition in the cluster is enabled, all are). This is repeated until there are no transitions enabled using only "agreement tokens". This process can be formalized by considering the partially-ordered run corresponding to the firing sequence σ from M_1 to M_r. Remove all transition consuming "disagreement tokens" from the partially-ordered run and execute the run as far as possible. Let M_1' be the resulting marking and σ' the partial sequence such that $(N, M_1)[\sigma'\rangle(N, M_1')$. σ' can also be executed starting from M_2 since only agreement tokens are used. Let M_2' be such that $(N, M_2)[\sigma'\rangle(N, M_2')$. Also in M_2' all enabled transitions need to consume "disagreement tokens" (i.e., ② tokens).

(N, M_1') and (N, M_2') are pc-safely marked, $en(N, M_1') \cap en(N, M_2') = \emptyset$ (otherwise a transition using agreement tokens is enabled), for all $t \in en(N, M_1)$: $M_2(\bullet t) \geq 1$, and for all $t \in en(N, M_2)$: $M_1(\bullet t) \geq 1$ (because we did not produce new disagreement tokens, no transition is enabled based on disagreement tokens only).   □

The goal is to show that perpetual free-choice nets are lucent. To do this, we construct a T-reduction where perpetuality is preserved "downstream" and lucency is preserved "upstream". For the "upstream reasoning" we start from a T-net. Hence, we first show that any perpetual well-formed T-net is lucent (using conflict-pairs as witnesses of non-lucency and Lemma 8 to show that such witnesses cannot exist).

Theorem 5 (Perpetual T-nets Have No Conflict-Pairs). *Let* $N = (P, T, F)$ *be a perpetual well-formed T-net. N does not have any conflict-pairs.*

Proof. Let $N = (P, T, F)$ be a perpetual well-formed T-net with regeneration transition $t_r \in T$. $M_r = [p \in \bullet t_r]$ is a regeneration marking (i.e., $(N, [p \in \bullet t_r])$) is live and bounded). (N, M_r) is also pc-safe (Lemma 7). Assume N has a conflict-pair (M_1, M_2), i.e., (N, M_1) and (N, M_2) are pc-safely marked, $en(N, M_1) \cap en(N, M_2) = \emptyset$, for all $t \in en(N, M_1)$: $M_2(\bullet t) \geq 1$, and for all $t \in en(N, M_2)$: $M_1(\bullet t) \geq 1$. Note that for any $X \in PComp(N)$: $M_1(X) = M_2(X) = M_r(X) = 1$. Each circuit is a P-component of N (and vice versa) and contains precisely one token in any marking considered. This implies that each circuit includes t_r. $T_D = \{t \in T \setminus \{t_r\} \mid \exists_{p \in \bullet t} M_1(p) \neq M_2(p)\}$ are all transitions that disagree on at least one of the input places (excluding t_r). Note that $T_D \neq \emptyset$ (M_1 and M_2 disagree on at least one P-component, yielding two disagreeing transitions). Pick a disagreeing transition t_D such that there is no other disagreeing transition on a path from t_r to t_D. This is possible because each circuit includes t_r, i.e., there are no cycles not involving the regeneration transition. Without loss of generality we may assume that there is a place $p_D \in \bullet t_D$ such that $M_1(p_D) = 1$ and $M_2(p_D) = 0$. t_D must have at least one other input place p_A that is not just marked in M_1, i.e., $M_1(p_A) \leq M_2(p_A)$ (otherwise (M_1, M_2) is not a conflict-pair).

Now we can apply Lemma 8 using the elementary paths $\rho_1 = \langle t_b, p_1^1, t_1^1, p_1^2, t_1^2, \ldots, p_1^m, t_e \rangle$ and $\rho_2 = \langle t_b, p_2^1, t_2^1, p_2^2, t_2^2, \ldots, p_2^n, t_e \rangle$ with $t_b = t_r$, $t_e = t_D$, $p_1^m = p_D$, $p_2^n = p_A$, and $|\bullet p| = |p \bullet| = 1$ for any $p \in P_1 \cup P_2$. Hence, Lemma 8 implies that $M_1(P_1) = M_1(P_2)$ and $M_2(P_1) = M_2(P_2)$.

We picked t_D such that there is no other disagreeing transition on a path from t_r to t_D. Hence, M_1 and M_2 agree on $P_1 \setminus \{p_D\} = \{p_1^1, p_1^2, \ldots, p_1^{m-1}\}$ and $P_2 \setminus \{p_A\} = \{p_2^1, p_2^2, \ldots, p_2^{n-1}\}$, i.e., $M_1(p) = M_2(p)$ for all $p \in (P_1 \cup P_2) \setminus \{p_D, p_A\}$. $M_1(p_D) > M_2(p_D)$ and $M_1(p_A) \leq M_2(p_A)$. Therefore, $M_1(P_1) > M_2(P_1)$ and $M_1(P_2) \leq M_2(P_2)$. Combined with $M_1(P_1) = M_1(P_2)$ and $M_2(P_1) = M_2(P_2)$ this leads to a contradiction. Hence, (M_1, M_2) cannot be a conflict-pair of N. $\square$

Corollary 1 (Perpetual T-nets Are Lucent). *Let $N = (P, T, F)$ be a perpetual well-formed T-net. N is lucent.*

Proof. Follows directly from Proposition 5 and Theorem 5. $\square$

Starting from a perpetual well-formed free-choice net and a T-reduction, we show that lucency is preserved in the "upstream" direction. We first prove that the absence of conflict-pairs is preserved "upstream" and use Theorem 5 as the base case. To simplify the proof, we assume that a particular regeneration transition t_r is preserved, but this is not essential and this requirement could be dropped (see last part of Theorem 4).

Theorem 6 (T-Reduction Showing Absence of Conflict-Pairs). *Let N be a perpetual well-formed free-choice net having a regeneration transition $t_r \in T$ and a T-reduction $\gamma_T = \langle t^1, t^2, \ldots, t^n \rangle$ that is t_r preserving. None of the Petrinets in $nets_N(\gamma_T) = \langle N^0, N^1, \ldots, N^n \rangle$ has conflict-pairs.*

Proof. Assume that N is a perpetual well-formed free-choice net with regeneration transition $t_r \in T$ and the T-reduction $\gamma_T = \langle t^1, t^2, \ldots, t^n \rangle$ is t_r preserving (it is always possible to create such T-reduction). $nets_N(\gamma_T) = \langle N^0, N^1, \ldots, N^n \rangle$.

Using Theorem 4 we know that $N^i = (P^i, T^i, F^i)$ is perpetual for any $i \in \{0, \ldots, n\}$. We need to show that N^i has no conflict-pairs. We use induction in the reverse direction starting with $i = n$. Base case: N^n is a T-net and has no conflict-pairs (Theorem 5). Induction step: We need to show that if N^i has no conflict-pairs, N^{i-1} has no conflict-pairs. This is the same as showing that if N^{i-1} has conflict-pairs, N^i also has conflict-pairs. To simplify notation we introduce the shorthands: $N = N^{i-1} = (P, T, F)$, $N' = N^i = (P', T', F')$ and $t = t^i$, i.e., $\Box_N(t)$ is a proper t-induced T-net and $N' = \overline{N_{\Box(t)}}$.

Let (M_1, M_2) be a conflict-pair for N, i.e., (N, M_1) and (N, M_2) are pc-safely marked, $en(N, M_1) \cap en(N, M_2) = \emptyset$, for all $t' \in en(N, M_1)$: $M_2(\bullet t') \geq 1$, and for all $t' \in en(N, M_2)$: $M_1(\bullet t') \geq 1$. Based on (M_1, M_2) we construct (M_1', M_2') with $M_1' = mrk_\Box(N, t, M_1)$ and $M_2' = mrk_\Box(N, t, M_2)$. We need to show that (M_1', M_2') is a conflict-pair. (N', M_1') and (N', M_2') are pc-safely marked (use Theorem 4). The remaining requirements in Definition 20 are shown by case distinction.

If $M_1 \!\restriction \Box_N(t) = [\,]$ and $M_2 \!\restriction \Box_N(t) = [\,]$, then $M_1' = M_1$, $M_2' = M_2$, and (M_1', M_2') is indeed a conflict-pair for N' (it is easy to verify that the requirements in Definition 20 still hold).

If $M_1 \!\restriction \Box_N(t) \neq [\,]$ or $M_2 \!\restriction \Box_N(t) \neq [\,]$, then at least one transition in $\Box_N(t)$ has a token in its input place. Let $T_D = \{t' \in (T \cap \Box_N(t)) \setminus \{t\} \mid M_1(\bullet t') + M_2(\bullet t') \geq 1\}$ (i.e., all transitions have a marked input place in one of the two markings). Pick a transition $t_D \in T_D$ such that there is no other T_D transition on a path from t to t_D. This is possible because there are no cycles inside $\Box_N(t)$ and there is a path from t to any node in $\Box_N(t)$. If there would be a cycle, then the regeneration transition t_r needs to be in $\Box_N(t)$, which is not the case because t_r is preserved (actually, t_r is a regeneration transition of N'). See also Theorem 5, which uses similar reasoning.

One of the input places of t_D is marked in M_1 or M_2. Since (M_1, M_2) is a conflict-pair for N, t_D cannot be enabled in both. Hence, for at least one of the two markings M_1 or M_2, we can find two input places that "disagree" (check all cases using Definition 20). Without loss of generality, let us assume that $p_m, p_u \in \bullet t_D$, $p_m \in M_1$, and $p_u \notin M_1$, i.e., the input places p_m and p_u of t_D and marking are chosen such that p_m is marked and p_u is not. Moreover, all places on a path from t to these places are empty. Just like in Theorem 5 and Lemma 8, we create two elementary paths: $\rho_1 = \langle t_b, p_1^1, t_1^1, p_1^2, t_1^2, \ldots, p_1^m, t_e \rangle$

and $\rho_2 = \langle t_b, p_2^1, t_2^1, p_2^2, t_2^2, \ldots, p_2^n, t_e \rangle$, now with $t_b = t$, $t_e = t_D$, $p_1^m = p_m$, and $p_2^n = p_u$. All places on these two paths are empty in M_1 except $p_1^m = p_m$. This leads to a contradiction using Lemma 8, which states that the number of tokens on both paths should be identical. Hence, $M_1 \restriction \boxdot_N(t) = M_2 \restriction \boxdot_N(t) = [\ ]$, $M_1' = M_1$, $M_2' = M_2$, and (M_1', M_2') is indeed a conflict-pair for N'. □

Corollary 2 (Perpetual Free-Choice Nets Are Lucent). *All perpetual well-formed free-choice nets are lucent.*

Proof. Follows directly from Proposition 5 and Theorem 6. □

Corollary 2 corresponds to Theorem 3 in [2]. As pointed out earlier by the author in e.g. [3], the initial proof of Theorem 3 in [2] was incomplete and a repaired proof was provided [3]. When repairing the proof, the author discovered that the result also holds for non-well-formed perpetual free-choice nets. A detailed proof is given in [5]. This more general result uses a completely different approach and does not build upon existing results for well-formed free-choice nets.

Note that for any reduction $\gamma = \langle t^1, t^2, \ldots, t^n \rangle$ of a perpetual well-formed free-choice net all nets in $nets_N(\gamma) = \langle N^0, N^1, \ldots, N^n \rangle$ are lucent and free of conflict-pairs. Hence, it is also possible to provide alternative versions of Theorem 6 using a P-reduction and the fact that lucency trivially holds for perpetual P-nets.

The approach presented in this section can also be used to prove the so-called *blocking theorem* [12,15] which states that every cluster in a bounded and live free-choice system has a unique marking enabling the cluster. This can be seen as lucency for individual transitions without requiring perpetuality. To prove the blocking theorem, we first show that blocking markings exist by moving tokens towards the selected cluster (this is possible due to the free-choice properly). Moreover, the uniqueness of blocking markings is preserved "upstream" and holds for T-nets (similar to Theorem 5, but using the fact that in blocking markings all transitions outside the selected cluster have empty input places). A detailed proof is straightforward, but omitted for space reasons.

7   Conclusion

This paper proposed *reductions* based on sequences of proper t-induced T-nets and p-induced P-nets. Such a reduction can be used to transform any free-choice net into a T-net or P-net. Given an arbitrary reduction γ, properties are preserved "downstream" (e.g., well-formedness, liveness, pc-safety, and perpetuality) and "upstream" (e.g., lucency and the absence of conflict-pairs, assuming perpetuality). Using the framework, we could reconfirm classical and more recent results related to lucency and perpetuality in a systematic manner. The framework is general and can be used for other properties, e.g., it becomes straightforward to prove the well-known blocking theorem [12,15] using a T-reduction.

The theoretical work presented was driven by challenges in the field of process mining. Process discovery techniques greatly benefit from additional assumptions such as lucency and perpetuality [4]. Moreover, we want to extend our work on *interactive* and *incremental* process mining using t-induced T-nets and p-induced P-nets. An obvious limitation of the current framework is that well-formedness is preserved "downstream" but not "upstream". However, the approach can be adapted to work in the reverse direction (using P-covers and T-covers).

References

1. van der Aalst, W.M.P.: Process Mining: Data Science in Action. Springer, Berlin (2016). https://doi.org/10.1007/978-3-662-49851-4
2. van der Aalst, W.M.P.: Markings in perpetual free-choice nets are fully characterized by their enabled transitions. In: Khomenko, V., Roux, O.H. (eds.) PETRI NETS 2018. LNCS, vol. 10877, pp. 315–336. Springer, Cham (2018). https://doi.org/10.1007/978-3-319-91268-4_16
3. van der Aalst, W.M.P.: Markings in Perpetual Free-Choice Nets Are Fully Characterized by Their Enabled Transitions. CoRR, abs/1801.04315 (2018)
4. van der Aalst, W.M.P.: Lucent process models and translucent event logs. Fundamenta Informaticae **169**(1–2), 151–177 (2019)
5. van der Aalst, W.M.P.: Free-Choice Nets With Home Clusters Are Lucent (Aug 2020) (Under Review)
6. Berthelot, G.: Checking properties of nets using transformations. In: Rozenberg, G. (ed.) Advances in Petri Nets 1985. Lecture Notes in Computer Science, vol. 222, pp. 19–40. Springer, Berlin (1986). https://doi.org/10.1007/BFb0016204
7. Best, E., Wimmel, H.: Structure theory of petri nets. In: Jensen, K., van der Aalst, W.M.P., Balbo, G., Koutny, M., Wolf, K. (eds.) Transactions on Petri Nets and Other Models of Concurrency VII. LNCS, vol. 7480, pp. 162–224. Springer, Heidelberg (2013). https://doi.org/10.1007/978-3-642-38143-0_5
8. Desel, J.: Reduction and design of well-behaved concurrent systems. In: Baeten, J.C.M., Klop, J.W. (eds.) CONCUR 1990. LNCS, vol. 458, pp. 166–181. Springer, Heidelberg (1990). https://doi.org/10.1007/BFb0039059
9. Desel, J., Esparza, J.: Free Choice Petri Nets, vol. 40. Cambridge Tracts in Theoretical Computer Science. Cambridge University Press, Cambridge (1995)
10. Dixit, P.M., Verbeek, H.M.W., Buijs, J.C.A.M., van der Aalst, W.M.P.: Interactive data-driven process model construction. In: Trujillo, J.C., et al. (eds.) ER 2018. LNCS, vol. 11157, pp. 251–265. Springer, Cham (2018). https://doi.org/10.1007/978-3-030-00847-5_19
11. Esparza, J., Silva, M.: Circuits, handles, bridges and nets. In: Rozenberg, G. (ed.) ICATPN 1989. LNCS, vol. 483, pp. 210–242. Springer, Heidelberg (1991). https://doi.org/10.1007/3-540-53863-1_27
12. Gaujal, B., Haar, S., Mairesse, J.: Blocking a transition in a free choice net and what it tells about its throughput. J. Comput. Syst. Sci. **66**(3), 515–548 (2003)
13. Hack, M.H.T.: Analysis of Production Schemata by Petri Nets. Master's thesis, Massachusetts Institute of Technology, Cambridge, Massachusetts (1972)
14. Thiagarajan, P.S., Voss, K.: A fresh look at free choice nets. Inf. Control **61**(2), 85–113 (1984)
15. Wehler, J.: Simplified proof of the blocking theorem for free-choice petri nets. J. Comput. Syst. Sci. **76**(7), 532–537 (2010)

Model Checking of Synchronized Domain-Specific Multi-formalism Models Using High-Level Petri Nets

Michael Haustermann$^{(\boxtimes)}$, David Mosteller, and Daniel Moldt

Faculty of Mathematics, Informatics and Natural Sciences,
Department of Informatics, University of Hamburg, Hamburg, Germany
haustermann@informatik.uni-hamburg.de
https://www.inf.uni-hamburg.de/inst/ab/art/

Abstract. Complex systems require the use of different models that are linked with each other. Developers are naturally interested to show that their systems work. Domain practitioners, who work with domain-specific models, want to verify that their created models perform as desired. Correctness statements about the behavior of models are only possible if they have a clear semantics. Support is required for creating the semantics and also for checking properties of the model.

With the RMT approach, we make operational semantics usable for the domain-specific modeling languages (DSML) that are understandable to domain experts. High-level Petri nets as a target language of our transformational approach can be analyzed by the MoMoC CTL model checking tool. In this contribution MoMoC is extended and integrated with the RMT approach so that the results of verification based on the defined operational semantics can be applied to DSML.

The presented approach does not work equally well for all languages. However, it is well suited for languages with discrete states that can be uniquely named. Provided that they map well to Petri nets, questions about (reachable) states of multiple linked domain-specific models can be answered.

Keywords: Meta-modeling · Petri nets · Model checking ·
Verification · Multi-formalism · Graphical feedback · Reference Nets ·
CTL · Model synchronization · DSML

1 Introduction

Domain-specific modeling languages (DSML) are a popular way to lower the initial barrier and complexity for working with models. One problem with DSML is that while they are tailored to the domain practitioner, their functionality or correctness is not immediately transparent to them. This requires facilities for helping DSML users to interactively inspect their models and for proving that no problems exist. Interactive simulation is a suitable means for inspection

© Springer Nature Switzerland AG 2021
D. Buchs and J. Carmona (Eds.): PETRI NETS 2021, LNCS 12734, pp. 230–249, 2021.
https://doi.org/10.1007/978-3-030-76983-3_12

' and model checking is popular for verifying correctness because it allows to check different specifications without the requirement of individual algorithms for each property of a model. By providing specifications in the terminology of the corresponding domain, this is also possible for DSML users. Simulation and proof of properties require a clear understanding of the semantics of the language in the first place.

With a transformational semantics it becomes possible to specify a well-founded semantics for domain-specific languages. This enables the application of existing simulation and verification tools. However, the use of existing tools requires the adoption of the terminology and languages of the utilized tool. This requires that queries on the model as well as simulation and verification results must be transferred from one formalism to the other by the DSML user.

In previous contributions we addressed the definition of semantics for modeling languages as a transformation into Petri nets [8], which provide concepts for an integrated simulation in the graphical representation of the DSML [10] with the structured RENEW Meta-Modeling and Transformation (RMT) approach [11]. Our approach facilitates the combined execution of several formalisms [7]. With this paper, based on our previous results, we address the verification of the DSML languages. Our contribution is the integration of model checking into the RMT approach, which involves an extension of the framework and the adoption of the Modular Model Checker (MoMoC) [17]. By using a component-based approach, it becomes possible to establish a relation between DSML states and net markings that allows working with the models in their original representation, in simulations as well as in verification. This relation of states is used to formulate system specifications in the terminology that is well-known to the DSML user from his application domain.

With the goal of seamlessly integrating model checking of DSML into the RMT approach, we proceed as follows. Conforming to our agile approach we opt for transformational semantics with Reference Nets as target formalism (Sect. 2). Tool support is provided by the modeling and simulation environment RENEW [2], which is the technical basis of our approach and enables verification on Reference Nets with the MoMoC plugin (Sect. 3). The MoMoC tool itself is not part of this contribution, however we have extended MoMoC conceptually and technically to our needs. We subsequently present a running example to explain the RMT approach (Sect. 4) and follow this with a proposal for integrating a verification approach (Sect. 5). A remarkable feature of our solution is the support for graphical exploration of the reachability graph with a replay of witness paths and the possibility of inspecting the DSML's graphical representation in any state. Another feature is the support for multi-formalism verification, which we have integrated into the RMT approach. We demonstrate our results with the application to an elevator model (Sect. 6). Related results from the literature are presented (Sect. 7) before we summarize our results and give an outlook on future work (Sect. 8).

2   Reference Nets

The Reference Net formalism is a high-level Petri net formalism with support for modeling complex data structures, remote synchronization and Java integration. Some of the core features of Reference Nets will be presented in the following as they are relevant for this contribution. This section is adopted from a previous publication [9]. The Java features will not be discussed in detail as they are of minor interest in this context. A thorough introduction to Reference Nets with Java integration is in the RENEW manual [5]. RENEW provides full support for modeling and execution of Reference Nets and other modeling languages. An example of the application to software engineering can be found in the latest research paper on RENEW [2].

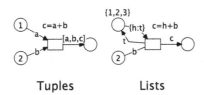

Fig. 1. Using collections in Reference Nets.

With high-level Petri nets, data in the form of colors is unified using unification expressions on the transitions with regard to the variable bindings on the edges. Reference Nets are not the only high-level Petri net formalism with support for collection types, however, Fig. 1 shows an example of how they are realized there. On the left side the variable c is calculated from the inputs a and b and all of the three variables are outputted in a tuple to the place on the right side. Tuples can be hierarchically nested to perform complex operations on them by the means of unification. Lists, as depicted in Fig. 1 on the right, are even more powerful as they permit iterative or recursive processing. In fact, in the depicted example the transition may fire three times, each time computing a new value for variable c from its head element (h) and variable b, which is only read and not removed by using a test arc.

Fig. 2. Synchronous channels of Reference Nets.

Synchronous channels, as exemplarily depicted in Fig. 2, control the firing of multiple transitions as one synchronized event. A synchronous channel consists of a pair of downlink (caller) and uplink (callee), so the reference must be known on one side of the channel. They may have arguments to transport information similar to the call of a function, but they have a slightly different notion as the unification of arguments enables a bidirectional exchange of data. In the depicted example the downlink (left side) calls the right side (uplink) in the local net instance (this). The first argument (m) on the downlink can be unified with the String ("match") on the uplink. The second argument (b) receives its actual parameter from the right side and the calculation of variable c is performed on the downlink transition on the left side before it flows through the channel to the uplink.

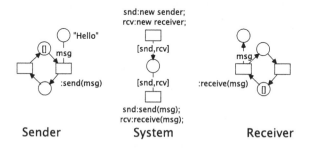

Fig. 3. Dynamic hierarchies: nets-within-nets [15].

Clearly, the expressiveness of synchronous channels is quite powerful. However, they unfold their real potential in combination with dynamic hierarchies and the *nets-within-nets* paradigm [15]. Using the new syntax shown in the upmost part of Fig. 3 Reference Nets can create instances of other net patterns. This enables dynamic hierarchies up to an arbitrary level of nesting. The depicted example is restricted to a level of two and models a message sending scenario. It shows a system of three components. In the center is a system net, which manages the instances of senders (left) and receivers (right). The latter communicate through the synchronous channels (send and receive) of the system net. Similar to the synchronous channels, the new constructor can also be parameterized with optional arguments. They are served by an uplink of the form :new(args) in the instantiated net.

While Reference Nets are well established for modeling and simulation of complex systems, possibilities for verification are still young. In the next section we show the MoMoC tool, which aims at making an advancement in terms of verification.

3   Modular Model Checker (MoMoC)

The Modular Model Checker (MoMoC) [17] is an explicit model checking tool for Reference Nets, which is integrated into the RENEW environment. Its focus is on a beginner-friendly user interface with graphical visualizations of reachability graphs and model checking results. The tool is under continuous development and the current version implements a simple CTL model checking algorithm.

A unique feature is the support for the nets-within-nets paradigm, which is inherently present in Reference Nets. This includes means for coping with problems that arise from the possibility of creating dynamic hierarchies. Since hierarchies are dynamic, net instances cannot be addressed directly. In general, it cannot be predicted whether certain net instances will be created at all and how many will be created. This means that in some state there may exist multiple net instances of one net pattern. These instances cannot be addressed directly because they are created dynamically and thus the identifiers are unknown a priori. A net instance that resides in a place of another net could be addressed by combining place names and the reference hierarchy but since one place may contain multiple instances of the same net pattern the identification of these instances is still ambiguous.

MoMoC handles this problem of net instance ambiguity [17] by extending CTL with a specific syntax to handle multiple net instances of the same pattern. The solution is the concept of net instance quantifiers, which are introduced to atomic propositions. With net instance quantifiers it becomes possible to determine whether a given proposition should hold in all instances of a given net pattern (with the operator !) or in one net instance at least (with the operator ?). The analysis is always started on a root net instance. The root net instance may have nested net instances, which are also treated during state space generation.

Another key feature of the MoMoC tool is its modular design that is intended for flexible extensibility. The tool is modular in a sense that parts of the model checking environment may be interchanged individually. The extensible interface allows to add additional *binding cores* to provide the firing semantics of a single transition, *storage managers* to implement the storage and access of explored net instances, as well as *procedures*, which provide the actual model checking algorithms. Additionally, multiple *result visualizers* may provide individual behaviors for visualizing the resulting artifacts of the algorithm.

The current result visualizer generates a reachability graph, which is graphically arranged with a simulated annealing algorithm. When applicable, the graph is colored according to the model checking result. Nodes are colored green if a formula holds and red otherwise. The checked CTL formula is presented in a tree-like representation, where the results of each individual subformula can be examined.

MoMoC has a specific syntax for atomic propositions because it needs to distinguish different net patterns and possibly multiple instances of each of these patterns. The upper half of Table 1 lists the atomic propositions provided by MoMoC. These include the net instance independent propositions *Deadlock* and *Fireable*. *Deadlock* is a global property, while *Fireable* refers to any instance of a specific transition. Furthermore, there are rules *NetInstance-Forall* and *NetInstanceExists*, which offer the means for dealing with the problem of net instance ambiguity by using net instance quantifiers. They contain

Table 1. MoMoC syntax of atomic propositions (Momoc Syntax: https://paose. informatik.uni-hamburg.de/paose/wiki/MoMoC).

Rule	Derivate	Example
Deadlock	DEADLOCK	DEADLOCK
Fireable	FIREABLE(Arg)	FIREABLE(T1)
NetInstanceForall	!(Arg, InstancePredicate)	!(NetA, m(p1) > 0)
NetInstanceExists	?(Arg, InstancePredicate)	?(NetB, m(p2) = 4)
InstancePredicate	CardinalityPredicate	
	ContentPredicate	
CardinalityPredicate	m(Arg) Op Number	m(p3) <= 1
ContentPredicate	c(Arg) contains String	c(p4) contains foo

an *InstancePredicate*, which refers to a specific net instance. An example would be the atomic proposition ?(NetA, *InstancePredicate*), which means that a net instance of the pattern with name *NetA* must exist in which the *InstancePredicate* holds. An *InstancePredicate* as depicted in the lower half of Table 1 can be a query on cardinalities of places (e.g. m(p1) >= 1).

With the extensions to MoMoC made for this contribution, queries on the contents of places become possible. For instance, c(p4) contains foo means that the place *p4* contains a token which has a string representation equal to foo. This is specifically necessary to individually address dynamically created net instances in the context of the presented approach to DSML verification. Furthermore, the modularity of MoMoC permits to add further atomic propositions. These atomic propositions can be combined in the usual way via Boolean operations and with well-known CTL operators each consisting of a path quantifier (A, E) and a path-specific quantifier (X, G, F, U).

4 RENEW Meta-Modeling and Transformation (RMT)

The RMT approach[1] (RENEW Meta-Modeling and Transformation) [8] is a model-driven approach for the agile development of DSML. It follows concepts from software language engineering (SLE, [4]) and enables a short development cycle to be appropriately applied in prototyping environments. The RMT approach defines a structured, iterative DSML development process that starts with the generation of an (empty) plugin. The initial models for defining the DSML are already included in the generated plugin and are extended step by step. Default values make configuration of various parameters optional and enable rapid deployment of initial prototypes and subsequent adjustment. The technical basis for the RMT approach is provided by the RMT framework, which builds upon the RENEW modeling and simulation environment. RENEW was originally

---

[1] The first paragraph of this section is adopted from our previous contribution [11].

designed as a Petri net editor and simulator and has evolved into an extensible integrated development environment (IDE) for various modeling techniques [2]. The RMT framework is particularly well-suited to develop languages with simulation feedback due to its lightweight approach to SLE and the tight integration with the extensible RENEW simulation engine, which supports the propagation of simulation events [10].

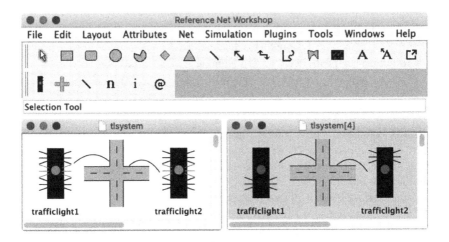

Fig. 4. Example model for a traffic light DSML in a RMT modeling tool.

In Fig. 4 a modeling tool is presented that was created following the RMT approach and using the RMT framework. It displays a DSML for modeling traffic light systems, which consist of traffic lights, crossroads and associations between them. The modeling tool was generated by the framework from a set of models and integrates into the RENEW modeling and simulation environment. Therefore the tool adds tool buttons to the menu bar and registers a new drawing type, from which drawings can be created, transformed and simulated. The left window shows a traffic light model that contains two traffic lights, a crossroad and associations respectively. The model was created by using the tools from the menu as depicted in the second row of icons in Fig. 4. We call this kind of model drawing a *pattern*, since multiple instances of the model may be created during runtime. In the pattern all traffic lights light up at the same time, because the visual states of the traffic lights are drawn as overlays in the simulation. We will go into details of this mechanism below. The right window shows the graphical representation of an instance of the traffic light model during simulation in a state where one of the traffic lights shows a green signal while the other traffic light signs red.

A tool like the one shown in Fig. 4 is generated as a RENEW plugin from a set of models. The basis of its specification is a meta-model, which describes the abstract syntax of a graphical modeling language. The meta-model of the traffic light DSML is not shown in this contribution in order to reduce redundancy with previous contributions. Examples of all the required models for the definition of the syntax can be found in earlier publications [8,11]. The concrete syntax is defined by creating graphical components within RENEW using the provided graphical primitives or alternatively by using style sheets. These graphical components contain additional annotations to specify the representation during execution.

Abstract and concrete syntax are connected by the tool configuration model, which additionally specifies basic properties of the resulting plugin, such as names for the plugin and its file type as well as the ordering of tool buttons. The semantics is provided by a transformation into Petri net formalisms, most prominently the Reference Net formalism. This has the advantage that it is possible to achieve fast results in a prototypical approach to the development of DSML. Reference Nets as a target formalism provide a well-founded mechanism for creating dynamic hierarchies and the synchronization of multiple models.

The transformation is done component-wise in the form of a $1 : n$ mapping from the DSML to the Petri net. A consequence of this approach is a local perspective on the behavior of the DSML constructs, reducing complexity for the language developer. This of course comes with the limitation that languages that have non-local semantics are not well supported. Accordingly, the RMT framework targets state-based languages with local semantics and is intended to provide the best possible support for them. Individual DSML constructs are mapped to net components (semantic components). A major role in the $1 : n$ approach comes to the join of two net components to specify their interaction. This results in the question of how to select connection points and how to connect or merge them. Various parameters for the configuration of the transformation make it possible to implement different variants of the interaction between DSML constructs. The basic behavior of the transformation is configured in a semantic mapping model. This includes how connection elements are specified and with which mechanism net components should be connected (merge of elements vs. connection with edges). Additional details, such as the identification of connection elements, are specified by annotations in the semantic components. They are described in the following using our running example.

Figure 5 shows the artifacts necessary for the visual representation and the semantics of the traffic light example. Figure 5a shows the graphical components of the traffic light DSML. It is supplemented by annotations that define how the traffic light should be represented at execution time. These annotations are provided in the form of key value attributes, which are presented in this paper as text annotations connected with a dotted line to the annotated graphical element. They are invisible in the artifacts that are actually used for generating the plugin. The UML note figures are only used for explanation in this paper. For the traffic light construct, each light is annotated with an `active-visible` attribute. This indicates a state dependent visualization. In this context, it means that the graphical element is only displayed in the *red* (or *yellow/green*) state. The state references markings of the underlying semantic component. In this

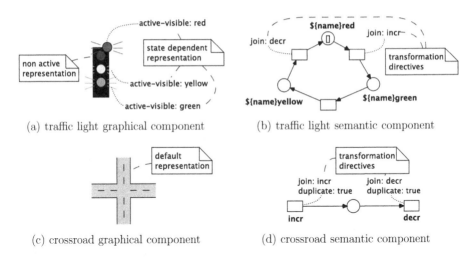

(a) traffic light graphical component   (b) traffic light semantic component

(c) crossroad graphical component   (d) crossroad semantic component

Fig. 5. Graphical and semantic component of the traffic light DSML.

simple case, the annotation `active-visible: red` refers to the marking where the place with the name *red* contains a token. We describe different variants for the specification of these state-dependent visualizations in an earlier publication [10]. The underlying graphical elements (unlit lamp) are displayed in all states.

The graphical component for the crossroad (cf. Fig. 5c) has no additional annotations. This results in using the default fallback highlighting behavior, where the DSML construct is considered active when at least one token is present in any of its places. The active state is then represented with a change of colors according to a color-dependent scheme to indicate a difference.

The semantic component in Fig. 5b is annotated differently. These annotations configure the transformation. In this case, the attribute `join: decr` resp. `join: incr` indicates that the respective elements are connected/merged with corresponding net elements with a matching attribute during transformation.

In the traffic light DSML, a traffic light can be connected to a crossroad by an association. The connection points are visible as well for the crossroad's semantic component in Fig. 5d, which contains the corresponding counterparts for the `join: decr` resp. `join: incr` annotations in the traffic light component. In addition, the two transitions have another transformation directive that specifies that these net elements are duplicated for each connection (`duplicate: true`). The transition with annotation `join: decr` from the traffic light's semantic component is merged with the transition with annotation `join: decr` from the crossroad's semantic component and for the transitions with annotations `join: incr` respectively. As the merging process is carried out for each of the associated constructs individually, with a replication of the merging element with annotation `duplicate: true`, both of the two traffic light constructs are connected to the crossroad.

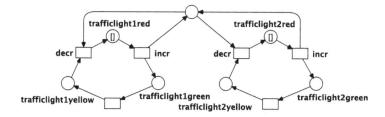

Fig. 6. Petri net generated from the traffic light model.

Figure 6 shows the Petri net that results from the transformation of the traffic light model from Fig. 4 by using the artifacts from Fig. 5. The resulting net is not displayed to the DSML user by default, it is only used as an execution object by the simulator. The models at runtime are displayed to the user in the original representation, as seen in Fig. 4 in the window on the right. These models serve as illustrations for the DSML verification presented in the next section.

5 DSML Verification

For our approach to DSML verification we intend to apply the core concepts of our approach for providing graphical simulation feedback, which are the component-based approach and the relating of DSML states and Petri net markings. A requirement for this approach is that the states of the evaluated modeling language are unique, disjoint and can be named individually.

In order to provide a meaningful overview of the system, states of DSML instances should be clearly distinguishable by the user by looking at the graphical representation. For state-based verification, which we address with our approach, it must therefore be possible to address individual states that can be identified from the graphical representation with a comprehensible and unique name. The interesting question is how to relate the DSML states to the Petri net markings so that the DSML user can formulate propositions over the model without the requirement of knowledge about the internal Petri net semantics.

One advantage of the component-based approach is that we benefit from the local perspective and reduce the task of relating the DSML state and the Petri net marking to relating a single DSML construct to its corresponding semantic component. This is implemented consistently with the integrated simulation by extending the semantic components. For semantic components however, the possible states (or the states that should be addressable) are to be defined explicitly. They are defined in the semantic components as atomic propositions (see Sect. 3) and assigned a name for reference. This is done during the development of DSML by the language developer, who must have knowledge of Petri nets in order to provide meaningful transformation semantics. The states defined in this way can then be referenced by the DSML user without knowledge of the underlying Petri nets. The labels of the states are parameterized at runtime so that the states of different DSML constructs can be queried individually.

For the traffic light construct introduced in the last section, the disjoint states are unique by design because there is only one token in the component that changes between colors. Additionally, overlapping states can be defined as a combination of multiple states. In regard of the traffic light, this should be the state *youcandrive*, which applies when the traffic light signs yellow or green.

The specification of states is based on the query language of the MoMoC tool and uses the same syntax. Specifications are written to the semantic components with the prefix MACRO. The definition of the state *youcandrive* has the following form, where ${name} is parameterized with the name of the DSML construct:

```
MACRO ${name}.youcandrive = (m(${name}yellow) = 1 OR
                             m(${name}green) = 1)
```

Macros like this can be used to formulate CTL queries to check a DSML model with MoMoC. The proposition language is based on the CTL variant presented in Sect. 3 where the *InstancePredicates* are references to the specified macros. This facilitates the query of states from the system by their provided state identifier.

For the traffic light model from Fig. 4 it is possible to check whether it can occur that the crossroad is entered in both directions simultaneously. The corresponding query looks like this:

```
AG ?(tlsystem, (NOT (trafficlight1.youcandrive AND
                     trafficlight2.youcandrive)))
```

Recall that the formula without the CTL operator AG is an atomic proposition (cf. Sect. 3). In this case the atomic proposition holds in a state if it holds in any of the net instances of the net pattern *tlsystem* due to the exists quantifier (?). Because we are inspecting only one instance of a traffic light system the predicate could just as well be quantified with the forall quantifier (!).

The extension of MoMoC that we developed for this contribution extracts the macros from the semantic components and parameterizes them during transformation. It generates the CTL specification that can be evaluated on the underlying Petri net, which looks as follows:

```
AG ?(tlsystem, (NOT ((m(trafficlight1yellow) = 1 OR
   m(trafficlight1green) = 1) AND (m(trafficlight2yellow) = 1 OR
   m(trafficlight2green) = 1))))
```

The model checking result consists of a tree-like view on the formula and its subformulas and a function to show a reachability graph that is colorized according to the result of a selected subformula.

Figure 7 shows two artifacts of the result from the verification of the traffic light model in Fig. 4. In Fig. 7a one can see the reachability graph generated by MoMoC, where the colors indicate whether the following subformula holds, green (true) and red (false):

```
?(tlsystem, (NOT (trafficlight1.youcandrive AND
                  trafficlight2.youcandrive)))
```

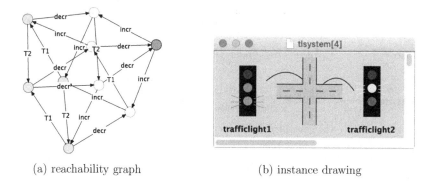

(a) reachability graph   (b) instance drawing

Fig. 7. Inspection of the model checking result.

It is plausible that the subformula does not hold in all states, because it is possible for one traffic light to sign green and the other one sign yellow at the same time (or both yellow/green). Consequently, the original formula quantified by AG does not hold in the initial state.

The nodes of the reachability graph store the model instances, which can be displayed when needed. Figure 7b shows a visualization of the lowermost state of the reachability graph, which is opened by double-clicking on the node. From the results, it can be seen that the semantics of the traffic light is not reasonably chosen for the crossroad scenario.

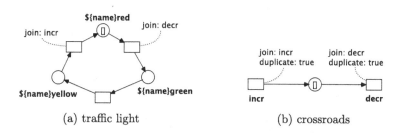

(a) traffic light   (b) crossroads

Fig. 8. Improved version of the semantic components.

For this reason, the semantic components in Fig. 8 have been adapted so that the traffic lights signal green or yellow in mutual exclusion. Mutual exclusion is achieved by inverting the edges that add or remove tokens in the crossroad components. This was realized by swapping the join annotations in the traffic light component. Another change is the new token added to the initial state of the crossroad. This may be interpreted as the place now implementing a capacity rather than activity on the crossroad.

Our modifications result in the generated Petri net in Fig. 9a, which shows that the desired behavior could be produced with the changes made to the semantic components. This is also reflected in the reachability graph generated

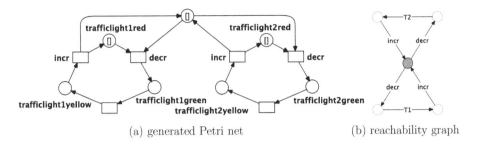

(a) generated Petri net   (b) reachability graph

Fig. 9. Generated Petri net that uses the improved semantics and the resulting reachability graph.

again in Fig. 9b, which has been stripped from the states that are undesired for the road crossing scenario.

The simple example presented in this section is intended to show how our approach can be used to support the development of DSML with execution semantics. The integrated simulation, in conjunction with the state-based analysis and inspectability of the system states, allows us to support the modelers in identifying errors and ambiguities in their models. Furthermore, problems do not only become identifiable in the models but also in the semantics of the language. Thus, the agile prototype-based approach to the development of DSML itself is supported. In the next section, we demonstrate how this approach can also be applied to systems in which multiple different model types are used in combination and are synchronized with each other.

6   Application to Multiple Formalism DSML Verification

The RMT framework is capable of executing multiple different DSML formalisms simultaneously and to enable communication between them using transformational Reference Net semantics. In this way, different views of a system can be combined to a coherent model. The prerequisite for this is a mapping of the formalisms involved to Reference Net implementations and the use of synchronous channels as described in Sect. 2. In our contribution to multi-formalism simulation based on Reference Nets, we developed a concept for synchronous and concurrent execution of multiple formalisms [7]. In this context, we have built a multi-formalism model of an elevator that uses finite automata for the elevator itself and the control panels on the different floors, and an activity diagram for the central control of the elevator. In this contribution we revisit the elevator model to perform DSML model checking on the model using the MoMoC integration that we developed for the RMT framework.

Figure 10 shows the elevator model, which slightly differs from the presentation in the previous contribution [7]. The system consists of three floors between which the elevator moves depending on the state (pressed/unpressed) of the control panels (Fig. 10c). The elevator itself (Fig. 10b) can be in the state

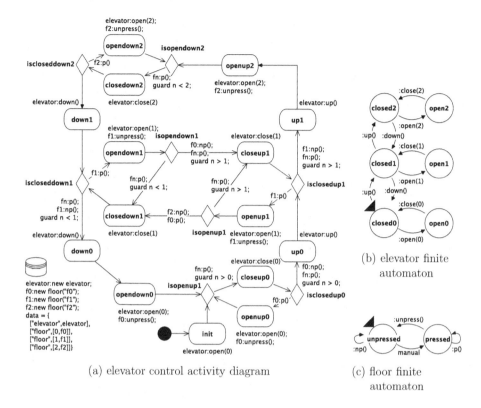

(a) elevator control activity diagram

(b) elevator finite automaton

(c) floor finite automaton

Fig. 10. A multi-formalism model of an elevator.

open/closed on one of the floors at a time. Any movement of the elevator is controlled by the central controller (Fig. 10a), which synchronizes the movement of the elevator with the state model. The controller is designed in a way that a change of direction is only possible if there are no pending requests in the current movement direction of the elevator. When the elevator has reached a floor, it opens (and closes) the door before it continues to move on.

The data component depicted in the bottom left of Fig. 10a manages the references to instances of the floor and elevator models and provides access to the model instances by their name. Tuples are employed in this context to manage the references to the model instances. One challenge in generating the state space of a Reference Net is to recognize the equivalence of net instance states. This is difficult because net instances may contain references to other net instances, which may produce cycles. Tuples and lists make the detection more difficult because they can also contain net instances. For efficient handling of references, we had to extend MoMoC to allow references in collections (lists and tuples) to be recognized and distinguished by the model checker.

Using the MoMoC integration for the RMT framework, we now show why the previously published model of the elevator is not an optimal design from an engineering point of view. A desired behavior of an elevator would be that in case someone is waiting on a floor and has pressed the button, the elevator should also arrive at that floor at some point and open the door. For such a request it is necessary to address a specific floor, which is not easily possible due to the problem of net instance ambiguity. To distinguish nets of the same type, they must be distinguishable by their content, because their ids are not known a priori. To this end, we have extended MoMoC with a feature that allows content queries in addition to cardinality queries. This makes it possible to store a unique identifier in a place of a net that can be referenced in CTL queries.

When creating instances of the floor model, unique names are assigned in each case. These are stored in a place of the semantic component and can be queried via a content predicate (`c(name) contains f0`). The query is simplified by a macro to `name.is f0`. Another macro allows to query if a finite automaton model is in a specific state, where the attribute `pressed.active` indicates that the respective automaton is in the state *pressed*. With the two macros above it is possible to specify the desired behavior using the CTL formula below.

```
AG ((NOT ?(floor, (name.is f2 AND pressed.active))) OR
    AF ?(elevator, open2.active))
```

With the provided formula, MoMoC performs an analysis on the elevator model using the DSML model checking procedure. The result from MoMoC indicates that the formula does not hold in the initial state of the system. The visualization tools can help to find and correct the problematic parts of the model.

The result of the analysis can be viewed and interactively inspected as shown in Fig. 11. In the lower area a small section of the reachability graph of the whole elevator system is shown. The graph is colored based on the negation of the sub-formula `AF ?(elevator, open2.active)`. A witness path for this subformula is highlighted, which is calculated for formulas quantified by a CTL operator during the execution of the model checking routine (procedure). Since tracing a path in a larger reachability graph is very difficult, we have implemented a tool to trace a witness path. This witness replay tool can be seen in the upper left part of the figure. It facilitates navigation through the state graph along the witness path and displays the graphical representation of the DSML model in the current state.

The determined path shows that it is possible to run a cycle in the elevator that never leads to an opening of the elevator door on the second floor. In the cycle, the constructs *isopenup1*, *closeup0*, *isclosedup0* and *openup0* can be run over and over again. This also applies if an elevator has been called from a floor above (*pressed*). The reason for this cycle is that the elevator can repeatedly open the door on the same floor if it was pressed there again. With the result, the model could now be adjusted, e.g. to introduce a limit for door openings on the same floor or a timer. Fairness constraints may also change the behavior of the elevator, but are not considered here. The relevant cycle can be visually

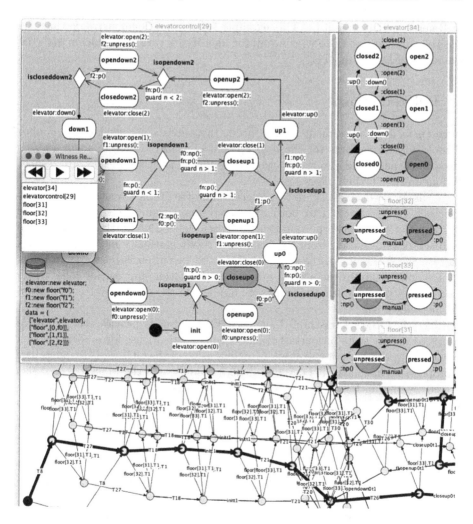

Fig. 11. Tools for inspecting the model checking result of the elevator example.

reproduced using the witness replay function. All model instances involved in the explored states can be examined.

In the upper part of the figure you can see the instance drawings in one of the states on the witness path. The highlighting in the elevator control (*elevator-control[29]*) indicates that the action *closeup0* is executed, while on the upper floor (*floor[32]*) the state *pressed* is valid. The other two floors are in the state *unpressed* but allow a transition through user interaction, indicated by the green arrow inscribed with `manual`. The elevator itself (*elevator[34]*) is currently in the state *open0*.

This example demonstrates the integration of multiple tools for interactive execution and analysis with the combination of multiple DSML formalisms based on Reference Net semantics. This example furthermore demonstrates that the selected combination of tools is useful for identifying various problems with the semantics of DSML models.

7   Related Work

Since there is an agreement on the need, there exists a considerable number of tools and frameworks for the development of DSML. Some of them address execution aspects and also, verification or validation is often an issue with respect to domain-specific languages.

A widely used approach is to use grammars for graph transformation. This approach can be applied to both integrated simulation and verification since the target of the transformation can be chosen depending on the use case. The feedback of information from the generated result is usually more difficult and likewise the dynamics of the feedback from the execution is missing. An interesting contribution in the area of graph transformation approaches comes from Varró since the transformation rules are applied on the model level as well as on the meta level [16].

Since the Object Constraint Language (OCL) already provides the syntactic means to specify state properties of model constructs, its use for model checking is evident. In the work of Bill et al. [1] an extension of the OCL by CTL operators is proposed. This makes it possible to check models with the MocOCL tool also offered by the authors, if they support OCL. A similar approach is taken by Mullins and Oarga [12] using a transformation of UML with OCL constraints to Abstract State Machines (ASM).

One contribution that puts an emphasis on the feedback of information from the model analysis comes from the work of Gerking et al. [3]. The approach is demonstrated on an example of MechatronicUML, a UML dialect for inter-car communication, using the UPPAAL model checking tool. The focus is on the propagation of counterexamples from the model checker into the DSML even if the relation is hard to integrate [3, p. 85]. We have decided to also address this issue with the present contribution and establish the relation between DSML and execution model by state annotations in the respective semantic components (cf. Sect. 5). This is only possible because we have a model-to-model transformation with an executable target model and restrict the possibilities of mapping constructs to a $1 : n$ mapping.

Another interesting result comes from Rusu and Lucanu [14]. They address executable DSMLs and transform their models to the $\mathbb{K}$ framework [13]. The transformation is rule-based and so is the operational semantics (using the $\mathbb{K}$ Model-Rewrite Language [14, p. 13]). The execution, however, is not interactive and does not automatically inherit advanced execution semantics in a way the presented approach benefits from dynamic hierarchies and true concurrency of the target formalism.

Meyers et al. pursue the goal of hiding formal methods from both the DSML user and the DSML developer with the argument that the DSML developer does not have to be an expert in formal methods either [6]. Both, the models and the system properties are to be represented in a domain-specific language, which is to be achieved by transformation between formalisms.

In our approach, a knowledge of formal languages is unavoidable for the language developer, because it is already necessary for the definition of the semantics. We use the same formalism for the definition of the semantics as for model checking. This favors our focus of a strong integration of interactive simulation and verification.

8 Conclusion

Our contribution addresses the problem that DSML users want to inspect their models for desired properties, which requires specific tools for the individual languages. Following up on the ongoing research on the RMT approach for developing DSML with execution semantics based on Reference nets, we now provide a concept for the integrated verification of DSML that is coherent with the RMT approach in a sense that it shares the same philosophy and suitability for agile and rapid prototype-based development. The core concept is the propagation of simulation and verification results from the underlying Petri net back to the graphical DSML models for interactive inspection and graphical exploration of the system.

Reference Nets as a target formalism provide the flexibility necessary to facilitate our approach to DSML verification for concurrent and dynamic systems (Sect. 2). The MoMoC tool, as a CTL model checker with support for the nets-within-nets paradigm and the net instance quantifiers, provides the basis for the verification of Reference Nets (Sect. 3). With RMT, we provide an agile, prototype-based approach to developing DSML with integrated simulation using annotated semantic components (Sect. 4). We have applied the approach taken to integrated simulation to the verification of DSML by establishing a relation between DSML states and Petri net markings also based on annotations of the semantic components (Sect. 5). This allows to specify states of DSML models that can be referenced in verification queries.

The MoMoC integration for the RMT framework provides the verification of model instances in the original representation. It supports the users and the developers of DSML with a graphical visualization of model checking results. The presented approach, with regard to Reference Nets as a target formalism, is particularly interesting for models in which multiple models of different types are involved that are related to each other (Sect. 6). The provided tools for inspection and visualization of verification results are particularly suitable for identifying undesired behavior in such systems, because they allow an interactive exploration of the reachability graph with a visualization of the DSML state. The generation of witness examples and the witness replay tool help to trace interesting paths found in complex state spaces.

One major topic for future work is the improvement of the performance and space consumption of the model checking algorithm. The currently implemented basic algorithm offers a lot of potential for optimization, especially with respect to the storage of net instances. However, these changes are not specific to our approach to DSML verification but to general MoMoC improvements that the original authors already address. An interesting direction of research regarding the dynamic hierarchies of nets within nets concerns the addressing of model instances. The net instance quantifiers could be made more flexible by parameterization, so that quantification is done over a subset of the instances of a net pattern. Using a graph query language, for example, restricting the search to net instances on specific places may be promising. Concerning the model driven approach to DSML verification furthermore increasing the flexibility of the query language should enhance its usability and applicability. The current implementation is still prototypical at this point. In addition, further visualization tools could provide additional value for users. These could include a structured arrangement of net instances or the step-by-step generation of the reachability graph to render only sections that are currently being explored.

References

1. Bill, R., Gabmeyer, S., Kaufmann, P., Seidl, M.: Model checking of CTL-extended OCL specifications. In: Combemale, B., Pearce, D.J., Barais, O., Vinju, J.J. (eds.) SLE 2014. LNCS, vol. 8706, pp. 221–240. Springer, Cham (2014). https://doi.org/10.1007/978-3-319-11245-9_13

2. Cabac, L., Haustermann, M., Mosteller, D.: Software development with Petri nets and agents: approach, frameworks and tool set. Sci. Comput. Program. **157**, 56–70 (2018). https://doi.org/10.1016/j.scico.2017.12.003

3. Gerking, C., Schäfer, W., Dziwok, S., Heinzemann, C.: Domain-specific model checking for cyber-physical systems. In: Famelis, M., Ratiu, D., Seidl, M., Selim, G.M.K. (eds.) Proceedings of the 12th Workshop on Model-Driven Engineering, Verification and Validation MoDeVVa@MoDELS'15, Ottawa, Canada. CEUR Workshop Proceedings, vol. 1514, pp. 18–27. CEUR-WS.org (2015). http://ceur-ws.org/Vol-1514/paper3.pdf

4. Kleppe, A.: Software Language Engineering: Creating Domain-Specific Languages Using Metamodels. Pearson Education, London (2008)

5. Kummer, O., Wienberg, F., Duvigneau, M., Cabac, L., Haustermann, M., Mosteller, D.: Renew - User Guide (Release 2.5.1). University of Hamburg, Faculty of Informatics, Theoretical Foundations Group, Hamburg (Nov 2020). http://www.renew.de/

6. Meyers, B., Vangheluwe, H., Denil, J., Salay, R.: A framework for temporal verification support in domain-specific modelling. IEEE Trans. Softw. Eng. **46**(4), 362–404 (2020). https://doi.org/10.1109/TSE.2018.2859946

7. Möller, P., Haustermann, M., Mosteller, D., Schmitz, D.: Model synchronization and concurrent simulation of multiple formalisms based on reference nets. In: Koutny, M., Kristensen, L.M., Penczek, W. (eds.) Transactions on Petri Nets and Other Models of Concurrency XIII. LNCS, vol. 11090, pp. 93–115. Springer, Heidelberg (2018). https://doi.org/10.1007/978-3-662-58381-4_5

8. Mosteller, D., Cabac, L., Haustermann, M.: Integrating petri net semantics in a model-driven approach: the renew meta-modeling and transformation framework. In: Koutny, M., Desel, J., Kleijn, J. (eds.) Transactions on Petri Nets and Other Models of Concurrency XI. LNCS, vol. 9930, pp. 92–113. Springer, Heidelberg (2016). https://doi.org/10.1007/978-3-662-53401-4_5

9. Mosteller, D., Haustermann, M., Dreschler-Fischer, L.S.: Graphical languages for functional reactive modeling based on petri nets. In: Köhler-Bußmeier, M., Kindler, E., Rölke, H. (eds.) Proceedings of the International Workshop on Petri Nets and Software Engineering, PNSE'20, Paris, France. CEUR Workshop Proceedings, vol. 2651, pp. 167–180. CEUR-WS.org (2020). http://ceur-ws.org/Vol-2651/paper11.pdf

10. Mosteller, D., Haustermann, M., Moldt, D., Schmitz, D.: Integrated simulation of domain-specific modeling languages with petri net-based transformational semantics. In: Koutny, M., Pomello, L., Kristensen, L.M. (eds.) Transactions on Petri Nets and Other Models of Concurrency XIV. LNCS, vol. 11790, pp. 101–125. Springer, Heidelberg (2019). https://doi.org/10.1007/978-3-662-60651-3_4

11. Mosteller, D., Haustermann, M., Moldt, D., Schmitz, D.: The RMT approach: a systematic approach to the development of DSML with integrated simulation based on petri nets. In: Koschmider, A., Michael, J., Thalheim, B. (eds.) 10th International Workshop on Enterprise Modeling and Information Systems Architectures, Kiel, Germany, May 14–15, 2020. CEUR Workshop Proceedings, vol. 2628, pp. 19–24. CEUR-WS.org (2020). http://ceur-ws.org/Vol-2628/paper3.pdf

12. Mullins, J., Oarga, R.: Model checking of extended OCL constraints on UML models in SOCLe. In: Bonsangue, M.M., Johnsen, E.B. (eds.) FMOODS 2007. LNCS, vol. 4468, pp. 59–75. Springer, Heidelberg (2007). https://doi.org/10.1007/978-3-540-72952-5_4

13. Rosu, G., Serbanuta, T.: An overview of the K semantic framework. J. Logic Algebraic Program. 79(6), 397–434 (2010). https://doi.org/10.1016/j.jlap.2010.03.012

14. Rusu, V., Lucanu, D.: A 𝕂-based formal framework for domain-specific modelling languages. In: Beckert, B., Damiani, F., Gurov, D. (eds.) FoVeOOS 2011. LNCS, vol. 7421, pp. 214–231. Springer, Heidelberg (2012). https://doi.org/10.1007/978-3-642-31762-0_14

15. Valk, R.: Petri nets as token objects. In: Desel, J., Silva, M. (eds.) ICATPN 1998. LNCS, vol. 1420, pp. 1–24. Springer, Heidelberg (1998). https://doi.org/10.1007/3-540-69108-1_1

16. Varró, D.: Automated formal verification of visual modeling languages by model checking. Softw. Syst. Model. 3(2), 85–113 (2004). https://doi.org/10.1007/s10270-003-0050-x

17. Willrodt, S., Moldt, D., Simon, M.: Modular model checking of reference nets: MoMoC. In: Köhler-Bußmeier, M., Kindler, E., Rölke, H. (eds.) Proceedings of the International Workshop on Petri Nets and Software Engineering, PNSE'20, Paris, France, June 24, 2020. CEUR Workshop Proceedings, vol. 2651, pp. 181–193. CEUR-WS.org (2020). http://ceur-ws.org/Vol-2651/paper12.pdf

Synthesis and Mining

Edge, Event and State Removal: The Complexity of Some Basic Techniques that Make Transition Systems Petri Net Implementable

Ronny Tredup[(✉)]

Universität Rostock, Institut Für Informatik, Theoretische Informatik,
Albert-Einstein-Straße 22, 18059 Rostock, Germany
ronny.tredup@uni-rostock.de

Abstract. In Petri net synthesis we ask whether a given transition system A can be implemented by a Petri net N. Depending on the level of accuracy, there are three ways how N can implement A: an *embedding*, the least accurate implementation, preserves only the diversity of states of A; a *language simulation* already preserves exactly the language of A; a *realization*, the most accurate implementation, realizes the behavior of A exactly. However, independent of the implementation sought, a corresponding net does not always exist. In this case, it was suggested to modify the input behavior −of course as little as possible. Since transition systems consist of states, events and edges, these components appear as the natural choice for modifications. In this paper we show that the task of converting an unimplementable transition system into an implementable one by removing as few states or events or edges as possible is NP-complete −regardless of what type of implementation we are aiming for.

1 Introduction

Petri nets are a widely accepted language for the modeling and validating of concurrent and distributed systems. In general, there are two ways to deal with the behavior of Petri nets: *Analysis* starts from a given Petri net and investigates if its behavior satisfies some properties. In *synthesis*, we deal with the opposite direction: starting from a regular behavior, given as a *transition system* (TS, for short), we try to find a Petri net that implements this behavior.

Synthesis of Petri nets has practical applications in numerous areas such as, for example, data and process mining [1,10], digital hardware design [8,9] and discovering of concurrency and distributability [4,5]. On the other hand, Petri net synthesis has also been the subject of theoretical studies that, for example, aim at characterizing the complexity of synthesis [2] or look for structural properties that classify a TS as implementable by subclasses of Petri nets such as, for example, *marked graphs*, and thus allow improved synthesis procedures [6].

© Springer Nature Switzerland AG 2021
D. Buchs and J. Carmona (Eds.): PETRI NETS 2021, LNCS 12734, pp. 253–273, 2021.
https://doi.org/10.1007/978-3-030-76983-3_13

TS have *states*, *events* and labeled *edges*, i.e., "source-event-target" triplets: the occurrence of the event at the source triggers a change of state to the target. They have an *initial state*, from which, triggered by an event-sequence, any other state is reachable. Petri nets have *places* containing *tokens* and an overall token distribution is considered as a *marking* (i.e., a global state) of the net; nets have *transitions*, connected with places, which possibly can *fire*, given the token distributions of their connected places allow the firing. A firing of a transition (locally) changes the token distribution (of its connected places) and thus (globally) the marking of the net. They have an *initial markig*, from which, triggered by the firing of a sequence of transitions, any other reachable marking is obtained. The global behavior of a net is captured by its *reachability graph*, which is a transition system, where reachable markings become states, transitions become events and edges correspond to "marking-transition-marking"-triplets. A Petri net N implements a TS A, if the events of A and the transitions of N coincide and, moreover, if (the states of) A and (the states of) the reachability graph of N can be related by a mapping, which satisfies certain requirements.

According to the mappings' properties, implementations with varying degrees of "accuracy" are possible: Such a mapping is first required to be a *simulation*, which means that every allowed sequence of events (starting at the initial state) can be simulated by a (fireable) sequence of transitions (starting at the initial marking). However, finding a net that allows a simulation is not a challenge: If N is the net without places that has a transition e for every event e of A, then N simulates A, since it can simply fire every sequence of events of A. Moreover, this net can obviously simulate *every* TS that has the same events as A. From this point of view, N simulates A with the greatest inaccuracy and every information about the (forbidden) original behavior is lost. On the other end of the spectrum, N simulates A most accurately if the simulation is an *isomorphism*: then N is an (exact) *realization* of the behavior defined by A. Unfortunately, not every TS can be realized by a Petri net. However, this is actually not always necessary depending on the application. Therefore, *embedding* and *language simulation* have been discussed as other implementations in the literature, which –in a certain sense– are less accurate, but still acceptable: an embedding preserves at least the diversity of states, that is, the simulation map is injective; a language simulation preserves exactly the allowed event sequences of A, that is, A and N are language-equivalent. Unfortunately, although these implementations are less restrictive, they also do not ensure the existence of a net sought. In order to achieve implementability, various techniques have been proposed in the literature that modify the components of the input behavior, i.e., its states, events and edges [3, 14]. One of the most discussed approaches among them is what is known as *label-splitting*: events are split into several (new) events and edges are relabeled so that edges that are initially labeled with the same event then are labeled with different events, which origin from the same event by splitting. This method is relatively well-understood from the practical point of view–in the sense of available algorithms [7, 8, 11]– and from the theoretical point of view as well. In particular, it was recently shown that achieving implementability by splitting as few events as possible is a problem hard to solve, namely NP-complete, regardless of the implementation sought [12, 13].

In a natural way, the question arises whether there are other suitable modifications to obtain implementable behavior. The answer is given by the nature of implementations itself: States of the input are related with reachable markings of the implementing net, and if the behavior is not implementable, then for some of its states no reachable marking may exist; hence, the *state-removal* of such states may lead to an implementable TS. Occurrences of events at (source-) states of the input correspond to the firing of transitions in markings that are associated with the sources; if the behavior is not implementable, then the firing of the transitions in the corresponding markings may not be possible; hence, removing such occurrences (i.e., the corresponding edges) may then yield an implementable behavior. If the latter is an option, then there are two distinct ways to put the focus on the removal: On the one hand, the modeler may allow the *edge-removal* of several edges that affects several events and, simultaneously, demand that some occurrences of every event remain –if this is possible; on the other hand, the modeler may come to the conclusion that some events are less interesting than others and thus prefer the complete *event-removal* of the former (i.e., all of their occurrences) and the complete preservation of the latter. Just like label-splitting, the removal of states, events and edges is a powerful transformation, since each of it is able to produce an implementable behavior: At the latest when A is degenerated to a single state, the resulting behavior is implementable. Surely, the latter is not desirable. Instead, we are interested in the corresponding optimization problems, that is, given a TS A, we are looking for a modification A' of A, such that the number κ of removed edges, events or states is as small as possible –depending on the technique applied. If we turn the number κ into a part of the input, then we obtain the corresponding decision version of the optimization problem. Obviously, if we can solve the optimization problem, then we can solve the decision version as well (with only polynomial overhead). Hence, the characterization of the computational complexity of the latter problem, provides a lower bound of the complexity of the former.

In this paper, we completely characterize the computational complexity of *state-removal*, *event-removal* and *edge-removal* for all thinkable implementations they can aim for, i.e., embedding, language-simulation and realization. In particular, we show that all of these decision problems are NP-complete and thus their optimization variant is also hard to solve.

This paper is structured as follows: Sect. 2 provides the basic notions and supports them with some examples. After that, Sect. 3, Sect. 4, and Sect. 5 provide the NP-completeness of *edge-removal*, *event-removal*, and *state-removal*, respectively. Finally, Sect. 6 briefly closes the paper.

2   Preliminaries

This section provides the basic notions that we use throughout the paper and supports them with examples. The overall starting point for the synthesis of Petri nets is a behavior that is given by a transition system:

Definition 1 (Transition System). *A (deterministic) transition system (TS, for short) $A = (S, E, \delta, \iota)$ consists of two disjoint sets of states S and events E and a partial transition function $\delta : S \times E \longrightarrow S$ and an initial state $\iota \in S$. An event e occurs at state s, denoted by $s\xrightarrow{e}$, if $\delta(s, e)$ is defined. By $\xrightarrow{e}\!\!\!\!\!/$ we denote that $\delta(s, e)$ is not defined. We abridge $\delta(s, e) = s'$ by $s\xrightarrow{e}s'$ and call the latter an edge with source s and target s'. By $s\xrightarrow{e}s' \in A$, we denote that the edge $s\xrightarrow{e}s'$ is present in A. A sequence $s_0\xrightarrow{e_1}s_1, s_1\xrightarrow{e_2}s_2, \dots, s_{n-1}\xrightarrow{e_n}s_n$ of edges is called a (directed labeled) path (from s_0 to s_n in A). A is called reachable, if there is a path from ι to s for every state $s \in S$. The language of A is the set of words $L(A) = \{e_1 \dots e_n \in E^* \mid \exists s \in S : \iota\xrightarrow{e_1}\dots\xrightarrow{e_n}s\} \cup \{\varepsilon\}$, where ε denotes the empty word.*

In the remainder of this paper, we always assume that TSs are *reachable*. In this paper, we relate TSs with the same set of events by so-called *simulations*:

Definition 2. *A simulation between a TS $A = (S, E, \delta, \iota)$ and a TS $B = (S', E, \delta', \iota')$ is a mapping $\varphi : S \to S'$ such that $\varphi(\iota) = \iota'$ and if $s\xrightarrow{e}s' \in A$, then $\varphi(s)\xrightarrow{e}\varphi(s') \in B$; φ is called an embedding, denoted by $A \hookrightarrow B$, if it is injective, that is, if $s \neq s'$, then $\varphi(s) \neq \varphi(s')$; φ is a language-simulation, denoted by $A \triangleright B$, if $s\xrightarrow{e}\!\!\!\!\!/$ implies $\varphi(s)\xrightarrow{e}\!\!\!\!\!/$; φ is an isomorphism, denoted by $A \cong B$, if it is bijective and $s\xrightarrow{e}s' \in A$ if and only if $\varphi(s)\xrightarrow{e}\varphi(s') \in B$.*

It is known from the literature that if $A \triangleright B$, then $L(A) = L(B)$ [3]; if $A \cong B$, then A and B are basically the same –possibly except for names of their states. A TS describes a behavior that is implementable or not. In the latter case, we apply the following *modifications* in order to obtain an implementable TS:

Definition 3 (Edge-Removal). *Let $A = (S, E, \delta, \iota)$ be a TS. A TS $B = (S', E', \delta', \iota)$ with state set $S' \subseteq S$ and event set $E' \subseteq E$ is an edge-removal of A if, for all $e \in E'$ and all $s, s' \in S'$, holds: if $s\xrightarrow{e}s' \in B$, then $s\xrightarrow{e}s' \in A$. By $\mathfrak{K} = \{s\xrightarrow{e}s' \in A \mid s\xrightarrow{e}s' \notin B\}$ we refer to the (set of) removed edges.*

Definition 4 (Event-Removal). *Let $A = (S, E, \delta, \iota)$ be a TS. A TS $B = (S', E', \delta', \iota)$ with state set $S' \subseteq S$ and event set $E' \subseteq E$ is an event-removal of A if for all $e \in E'$ the following is true: $s\xrightarrow{e}s' \in B$ if and only if $s\xrightarrow{e}s' \in A$ for all $s, s' \in S$. By $\mathfrak{E} = E \setminus E'$ we refer to the (set of) removed events.*

Definition 5 (State-Removal). *Let $A = (S, E, \delta, \iota)$ be a TS. A TS $B = (S', E', \delta', \iota)$ with states $S' \subseteq S$ and events $E' \subseteq E$ is a state-removal of A if the following two conditions are satisfied: (1) $s\xrightarrow{e}s' \in B$ if and only if $s\xrightarrow{e}s' \in A$ for all $e \in E'$ and all $s, s' \in S'$; (2) if $s\xrightarrow{e}s' \in A$ and $s\xrightarrow{e}s' \notin B$, then $s \notin S'$ or $s' \notin S'$. By $\mathfrak{S} = S \setminus S'$ we refer to the (set of) removed states.*

Notice that neither of these modifications is "functional", since, generally, there are several TS that can be considered as a suitable modification of A. Moreover, edge-removal is the most general modification introduced, since every

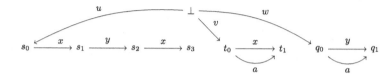

Fig. 1. The TS A.

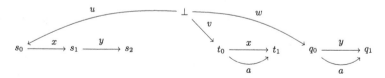

Fig. 2. The state-removal B of A that results by removing the state s_3.

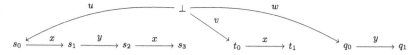

Fig. 3. The event-removal C of A that results by removing the event a.

event- or state-removal is also an edge-removal. However, not every edge-removal is an event-removal or a state-removal, not every event-removal is a state-removal, and not every state-removal is an event-removal. In particular, there are substantial differences between these modifications that focus on different aspects of the TS: If B is an edge-removal, then there could possibly be an event $e \in E'$ for which there is an edge $s \xrightarrow{e} s'$ in A that is not in B. In contrast, if B is an event-removal and $e \in E'$, then every e-labeled edge of A has to be present in B. Furthermore, if B is a state-removal, then an edge $s \xrightarrow{e} s'$ of A can only be missing in B if its source s or its target s' is removed. By contrast, the latter is not necessarily the case if B is an event- or an edge-removal.

Example 1. Consider the TS A of Fig. 1. The TS B of Fig. 2 is a state-removal of A resulting by removing the state s_3, i.e., $\mathfrak{S} = \{s_3\}$. B is also an edge-removal, where $\mathfrak{K} = \{s_2 \xrightarrow{x} s_3\}$. However, this TS is not an event-removal, since x belongs to B, but not all x-labeled edges of A are present. The TS C of Fig. 3 is an event-removal of A such that $\mathfrak{E} = \{a\}$. C is also an edge-removal and $\mathfrak{K} = \{t_0 \xrightarrow{a} t_1, q_0 \xrightarrow{a} q_1\}$, but is is not a state-removal.

Petri nets are the target model with which we want to implement TS:

Definition 6 (Petri Nets). *A Petri net* $N = (P, T, f, M_0)$ *consists of finite and disjoint sets of* places P *and* transitions T, *a (total)* flow $f : ((P \times T) \cup (T \times P)) \to \mathbb{N}$ *and an* initial marking $M_0 : P \to \mathbb{N}$. *A transition* $t \in T$ *can* fire *or* occur *in a marking* $M : P \to \mathbb{N}$, *denoted by* $M \xrightarrow{t}$, *if* $M(p) \geq f(p, t)$ *for all places* $p \in P$. *The firing of* t *in marking* M *leads* *to the marking $M'(p) = M(p) - f(p,t) + f(t,p)$ for all $p \in P$, denoted by $M \xrightarrow{t} M'$. This notation extends to sequences $w \in T^*$ and the* reachability set $RS(N) = \{M \mid \exists w \in T^* : M_0 \xrightarrow{w} M\}$ *contains all of N's reachable markings. The* reachability graph *of N is the TS $A_N = (RS(N), T, \delta, M_0)$, where, for every reachable marking M of N and transition $t \in T$ with $M \xrightarrow{t} M'$, the transition function δ of A_N is defined by $\delta(M,t) = M'$.*

Simulations between A and A_N define how a net N implements a TS A:

Definition 7 (Implementations). *If A is a TS and N a Petri net, then N is an* embedding *of A if $A \hookrightarrow A_N$; N is a* language-simulation *of A, if $A \triangleright A_N$, and N is a* realization *of A, if $A \cong A_N$. We say N* implements *A, if it is an embedding or a language-simulation or a realization of A.*

If a Petri net N implements a TS A, then the events of A are the transitions of N. We obtain the remaining components of N, that is, places, flow and initial marking, by regions of A:

Definition 8 (Region). *A* region *$R = (sup, con, pro)$ of a TS $A = (S, E, \delta, \iota)$ consists of the mappings* support *$sup : S \to \mathbb{N}$ and* con*sume and* pro*duce $con, pro : E \to \mathbb{N}$ such that if $s \xrightarrow{e} s'$ is an edge of A, then $con(e) \leq sup(s)$ and $sup(s') = sup(s) - con(e) + pro(e)$.*

Remark 1. It is essential that a region $R = (sup, con, pro)$ is *implicitly* completely defined by $sup(\iota)$, con and pro: Since A is reachable, there is a path $\iota \xrightarrow{e_1} \ldots \xrightarrow{e_n} s_n$ such that $s = s_n$ for every state $s \in S$. Consequently, we inductively obtain $sup(s_{i+1})$ by $sup(s_{i+1}) = sup(s_i) - con(e_{i+1}) + pro(e_{i+1})$ for all $i \in \{0, \ldots, n-1\}$ and $s_0 = \iota$. Hence, for the sake of simplicity, we often present regions only implicitly, since sup and thus R can be obtained from $sup(\iota)$, con and pro. For an even more compact presentation, for $c, p \in \mathbb{N}$, we group events with the same "behavior" together by $\mathcal{T}_{c,p}^R = \{e \in E \mid con(e) = c \text{ and } pro(e) = p\}$.

If there is an implementing net N for A, then each place correspond to a region $R = (sup, con, pro)$ of A: $con(e)$ and $pro(e)$ model $f(R, e)$ and $f(e, R)$ for all transitions e, respectively, and $sup(\iota)$ models the initial marking $M_0(R)$. In particular, every set of regions defines a *synthesized net*:

Definition 9 (Synthesized Net). *A set $\mathcal{R}$ of regions of TS $A = (S, E, \delta, \iota)$ defines the* synthesized net *$N_A^{\mathcal{R}} = (\mathcal{R}, E, f, M_0)$, where $f(R, e) = con(e)$ and $f(e, R) = pro(e)$ and $M_0(R) = sup(\iota)$ for all $R = (sup, con, pro) \in \mathcal{R}$ and $e \in E$.*

If the synthesized net is an embedding or a realization of A, then distinct states of A correspond to distinct markings of the net. The net $N_A^{\mathcal{R}}$ satisfies this requirement if the set $\mathcal{R}$ of regions prove the *state separation property*:

Definition 10 (State Separation Property). *A pair (s, s') of distinct states of TS $A = (S, E, \delta, \iota)$ defines a* states separation atom *(SSA). A region $R =*

(sup, con, pro) solves (s, s') if $sup(s) \neq sup(s')$. *We say a state s is* solvable *if, for every $s' \in S \setminus \{s\}$, there is a region that solves the SSA (s, s'). If every SSA or, equivalently, every state of A is solvable, then A has the* state separation property *(SSP)*.

If the net is a language-simulation or a realization, then the firing of a transition e must be inhibited at a marking M whenever the event e does not occur at the state s that correspond to M via φ. This is ensured if $\mathcal{R}$ witnesses the *event/state separation property*:

Definition 11 (Event/State Separation Property). *A pair (e, s) of event e and state s of TS $A = (S, E, \delta, \iota)$ such that $s \xrightarrow{e}\!\!\!\!/\;$ defines an event/state separation atom (ESSA). A region $R = (sup, con, pro)$ solves (e, s) if $sup(s) < con(e)$. We say an event e is* solvable *if, for every $s \in S$ with $s \xrightarrow{e}\!\!\!\!/\;$, there is a region that solves the ESSA (e, s). If every ESSA or, equivalently, every event of A is solvable, then A has the* event/state separation property *(ESSP)*.

A set $\mathcal{R}$ of regions of A is called a *witness* for the SSP or the ESSP (of A) if, for every SSA or ESSA, there is a region in $\mathcal{R}$ that solves it. The next lemma is based on [3, p. 162] and [3, p. 214 ff.] and discovers in which case the existence of a witness and the existence of an implementation are equivalent; this will allow us to formulate our decision problems rather on the notion of witnesses than on the notion of implementations:

Lemma 1 ([3]). *Let A be a TS and N a Petri net.*

1. $A \hookrightarrow A_N$ *if and only if there is a witness $\mathcal{R}$ for the SSP of A and $N = N_A^{\mathcal{R}}$;*
2. $A \triangleright A_N$ *if and only if there is a witness $\mathcal{R}$ for the ESSP of A and $N = N_A^{\mathcal{R}}$;*
3. $A \cong A_n$ *if and only if there is a witness $\mathcal{R}$ for both the SSP and the ESSP of A and $N = N_A^{\mathcal{R}}$.*
4. *Whether A has the SSP or the ESSP can be decided and, in case of a positive decision, a witness can be computed in polynomial time.*

Example 2. Let $A = (Z, E, \delta, \perp)$ be the TS of Fig. 1. The following implicitly defined region $R = (sup, con, pro)$ solves all SSA of A: $sup(\perp) = 8$ and $\mathcal{T}_{5,0}^R = \{v\}$, $\mathcal{T}_{7,0}^R = \{w\}$ and $\mathcal{T}_{1,0}^R = E \setminus \{v, w\}$. According to Remark 1, one obtains R explicitly: $sup(s_i) = 7 - i$ for all $i \in \{0, \dots, 3\}$ and $sup(t_0) = 3$, $sup(t_1) = 2$, $sup(q_0) = 1$ and $sup(q_1) = 0$. The set $\mathcal{R} = \{R\}$ witnesses the SSP of A and the net $N = N_A^{\mathcal{R}}$ is an embedding of A. Figure 4 shows N (top) and its reachability graph A_N (bottom). The injective simulation map φ is defined by $\varphi(\perp) = 8$, $\varphi(s_i) = 7 - i$ for all $i \in \{0, \dots, 3\}$ and $\varphi(t_0) = 3$, $\varphi(t_1) = 2$, $\varphi(q_0) = 1$ and $\varphi(q_1) = 0$.

Example 3. The TS $A = (Z, E, \delta, \perp)$ of Fig. 1 does not have the ESSP, since the ESSA $\alpha = (x, s_1)$ is not solvable. This can be seen as follows: Assume $R = (sup, con, pro)$ is a region that solves α, that is, $con(x) > sup(s_1)$. Since x occurs at s_0, we have $con(x) \leq sup(s_0)$. By $sup(s_1) = sup(s_0) - con(x) + pro(x)$ and $con(x) > sup(s_1)$, this implies $con(x) > pro(x)$. By $t_0 \xrightarrow{x} t_1$, this also implies

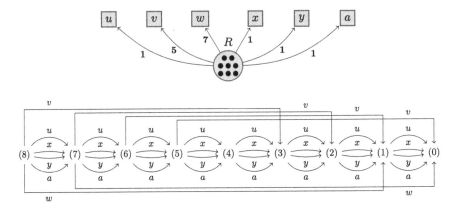

Fig. 4. The net $N = N_A^{\mathcal{R}}$ and its reachability graph A_N according to Example 2.

$sup(t_0) > sup(t_1)$ and thus $con(a) > pro(a)$ by $t_0 \overset{a}{\longrightarrow} t_1$. On the other hand, x occurs at s_2, which implies $con(x) \leq sup(s_2)$. By $con(x) > sup(s_1)$ and $s_1 \overset{y}{\longrightarrow} s_2$, this is only possible if $con(y) < pro(y)$, which implies $sup(q_0) < sup(q_1)$ and thus $con(a) < pro(a)$. This is a contradiction. Hence, α is not solvable.

Example 4. The TS of Fig. 1 does not have the ESSP, since the ESSA $\alpha = (x, s_1)$ is not solvable. However, for the TS B of Fig. 2, which is a state-removal for A, there is a region $R = (sup, con, pro)$ that solves the ESSA (x, s_1), which is implicitly defined as follows: $sup(\bot) = 2$ and $\mathcal{T}_{2,1}^R = \{x\}$ and $\mathcal{T}_{1,0}^R = \{a, y\}$ and $\mathcal{T}_{0,0}^R = E \backslash \{a, y, x\}$. One finds out that the remaining ESSA of B are also solvable. Hence, B has the ESSP and the SSP. If the modeler comes to the conclusion that the events x and y, their corresponding edges and all of their sources and targets are essential for the modeled behavior, then a realizable behavior can also be obtained as the event-removal C of A as defined in Fig. 3. A Region $R' = (sup', con', pro')$ solving (x, s_1) in C, is then implicitly defined as follows: $sup(\bot) = 1$ and $\mathcal{T}_{1,0}^{R'} = \{x\}$ and $\mathcal{T}_{0,1}^{R'} = \{y\}$ and $\mathcal{T}_{0,0}^{R'} = E \backslash \{x, y\}$.

3   The Complexity of Edge-Removal

According to Lemma 1, the question whether a particular implementation for a given TS exists is equivalent to the question whether the TS has the separation properties that correspond to the implementation. In this section, we are interested in modifying a TS into an implementable one by the removal of a bounded number of edges. In particular, we are interested in the computational complexity of the following decision problems:

EDGE-REMOVAL FOR EMBEDDING
Input: A TS $A = (S, E, \delta, \iota)$, a natural number κ.
Question: Does there exist an edge-removal B for A that has the SSP and
            satisfies $|\mathfrak{R}| \leq \kappa$?

EDGE-REMOVAL FOR LANGUAGE-SIMULATION
Input: A TS $A = (S, E, \delta, \iota)$, a natural number κ.
Question: Does there exist an edge-removal B for A that has the ESSP and
satisfies $|\mathfrak{K}| \leq \kappa$?

EDGE-REMOVAL FOR REALIZATION
Input: A TS $A = (S, E, \delta, \iota)$, a natural number κ.
Question: Does there exist an edge-removal B for A that has the ESSP and
the SSP and satisfies $|\mathfrak{K}| \leq \kappa$?

In [12], it is argued that EDGE-REMOVAL FOR EMBEDDING is NP-complete.
However, the complexity of the other two problems has not yet been characterized. The statement of following theorem closes this gap:

Theorem 1. EDGE-REMOVAL FOR LANGUAGE-SIMULATION *as well as* EDGE-
REMOVAL FOR REALIZATION *are NP-complete.*

It is easy to see that the addressed problems are in NP: If there is a sought
edge-removal for A, then a Turing machine can guess $\mathfrak{K}$ by a non-deterministic
computation in time polynomial in the size of the input. After that, the machine
can deterministically compute B and, since the size of B is bounded by the size
of A, it can compute a witness for the relevant property of B in time polynomial
in size of the input by Lemma 1. Hence, in order to complete the proof of
Theorem 1, it remains to prove the NP-hardness of the problems. This proof
bases on a reduction of the vertex cover problem that is well-known to be NP-
complete:

VERTEX COVER (VC)
Input: An undirected, unlabelled, finite Graph $G = (\mathfrak{U}, M)$ with a set
of vertices $\mathfrak{U} = \{X_0, \ldots, X_{n-1}\}$ and a set of arcs (2-subsets of $\mathfrak{U}$)
$M = \{M_0, \ldots, M_{m-1}\}$; a natural number λ.
Question: Does there exist a vertex cover of size at most λ for G, that is, a
set $Z \subseteq \mathfrak{U}$ such that $Z \cap \mathfrak{a} \neq \emptyset$ for all $\mathfrak{a} \in M$ and $|Z| \leq \lambda$?

Example 5. The instance $(G, 4)$ with $G = (\mathfrak{U}, M)$ such that $\mathfrak{U} = \{X_0, \ldots, X_5\}$
and $M = \{M_0, \ldots, M_8\}$, where $M_0 = \{X_0, X_1\}$, $M_1 = \{X_0, X_3\}$, $M_2 = \{X_0, X_5\}$, $M_3 = \{X_1, X_2\}$, $M_4 = \{X_1, X_5\}$, $M_5 = \{X_2, X_3\}$, $M_6 = \{X_2, X_4\}$, $M_7 = \{X_3, X_4\}$, $M_8 = \{X_4, X_5\}$, has the vertex cover $Z = \{X_0, X_2, X_3, X_5\}$
and thus allows a positive decision.

In the remainder of this paper, unless explicitly states otherwise, let (G, λ)
be an arbitrary but fixed input of VC, where $G = (\mathfrak{U}, M)$ has n vertices $\mathfrak{U} = \{X_0, \ldots, X_{n-1}\}$ and m arcs $M = \{M_0, \ldots, M_{m-1}\}$ such that $M_i = \{X_{i_0}, X_{i_1}\}$
and (without loss of generality) $i_0 < i_1$ for all $i \in \{0, \ldots, m-1\}$.

In order to prove Theorem 1, we reduce (G, λ) to a pair (A, κ) of TS A and
natural number κ as follows: If there is an edge-removal B of A that removes

at most κ edges and has the ESSP, then there is a vertex cover with at most λ elements for G. Conversely, if there is a vertex cover with at most λ elements for G, then there is an edge-removal B of A that removes at most κ edges and has both the ESSP and the SSP. Notice that such a reduction proves the hardness of both EDGE-REMOVAL FOR LANGUAGE-SIMULATION and EDGE-REMOVAL FOR REALIZATION. The announced TS A consists of several components. Just as it is common in the world of reductions, we refer to these components also as *gadgets*.

For a start, we define $\kappa = \lambda$. For every $i \in \{0, \ldots, m-1\}$, the TS A has the following gadget T_i that uses the vertices X_{i_0}, X_{i_1} of the arc M_i as events; and, for every $j \in \{0, \ldots, n-1\}$, the TS has the following gadget F_j that applies the vertex X_j as event and, moreover, has an a_ℓ-labeled edge that has the same direction as the X_j-labeled edge for all $\ell \in \{0, \ldots, \lambda\}$:

$$T_i = t_{i,0} \xrightarrow{X_{i_0}} t_{i,1} \xrightarrow{X_{i_1}} t_{i,2} \xrightarrow{X_{i_0}} t_{i,3} \qquad\qquad F_j = f_{j,0} \underset{\substack{a_0 \\ a_1 \\ \vdots \\ a_{\lambda-1} \\ a_\lambda}}{\overset{X_j}{\rightrightarrows}} f_{j,1}$$

Notice that, by $i_0 < i_1$ for all $i \in \{0, \ldots, m-1\}$, the construction of the gadgets is unique. Via the initial state $\perp$, the introduced gadgets are joined by several edges: for all $i \in \{0, \ldots, m-1\}$ and all $j \in \{0, \ldots, n-1\}$, the TS A has the edges $\perp \xrightarrow{w_i} t_{i,0}$ and $\perp \xrightarrow{y_j} f_{j,0}$, respectively. The resulting TS is $A = (S, E, \delta, \perp)$. Figure 5 provides the gadgets of the TS A that is based on the input of Example 5.

In the following, we prove the functionality of A. (Recall that $\mathfrak{K}$ refers to the set of edges removed from A, where the edge-removal is clear from the context).

Lemma 2. *If there is an edge-removal B of A that satisfies $|\mathfrak{K}| \leq \kappa$ and has the ESSP, then there is a vertex cover of size at most λ for G.*

Proof. Let $B = (S', E', \delta', \iota)$ be an edge-removal of A that satisfies $|\mathfrak{K}| \leq \kappa$ and has the ESSP. In the following, we argue that this implies that the set $Z = \{X \in \mathfrak{U} \mid s \xrightarrow{X} s' \in \mathfrak{K}\}$ defines a vertex cover with at most λ elements for G.

Let $i \in \{0, \ldots, m-1\}$ be arbitrary but fixed. There are two possibilities: either the gadget T_i is completely present in B or one of its edges is missing, that is, $\{t_{i,0} \xrightarrow{X_{i_0}} t_{i,1}, t_{i,1} \xrightarrow{X_{i_1}} t_{i,2}, t_{i,2} \xrightarrow{X_{i_0}} t_{i,3}\} \cap \mathfrak{K} \neq \emptyset$. In the latter case, we have already the situation that $M_i \cap Z \neq \emptyset$. We argue that this is also true in the former case and prove that then one of the edges $f_{i_0,0} \xrightarrow{X_{i_0}} f_{i_0,1}$ or $f_{i_1,0} \xrightarrow{X_{i_1}} f_{i_1,1}$ belongs to $\mathfrak{K}$ such that its corresponding event is a member of Z: Since T_i is completely present in B, the ESSA $\alpha = (X_{i_0}, t_{i,1})$

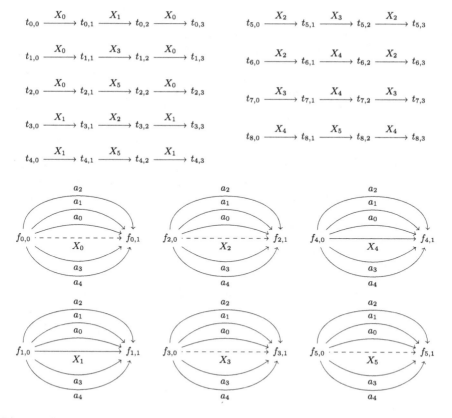

Fig. 5. The gadgets $T_0, \ldots, T_8$ and $F_0, \ldots, F_5$ of the TS A of Sect. 3 based on Example 5. Dashed lines correspond to edges that are removed according to the edge-removal B defined for the proof of Lemma 3, where the vertex cover is $\{X_0, X_2, X_3, X_5\}$.

is also present. Hence, there is a region $R = (sup, con, pro)$ that solves α, since B has the ESSP. We argue that assuming that both edges $f_{i_0,0} \xrightarrow{X_{i_0}} f_{i_0,1}$ and $f_{i_1,0} \xrightarrow{X_{i_1}} f_{i_1,1}$ are present in B yields a contradiction as follows: We will argue that $con(X_{i_0}) > pro(X_{i_0})$ and $con(X_{i_1}) < pro(X_{i_1})$ are simultaneously true. Since $sup(f_{i_0,1}) = sup(f_{i_0,0}) - con(X_{i_0}) + pro(X_{i_0})$ and $sup(f_{i_1,1}) = sup(f_{i_1,0}) - con(X_{i_1}) + pro(X_{i_1})$, this implies then $sup(f_{i_0,0}) > sup(f_{i_0,1})$ and $sup(f_{i_1,0}) < sup(f_{i_1,1})$. Since the set $\{a_0, \ldots, a_\lambda\}$ contains $\lambda + 1$ elements and $|\mathfrak{K}| \leq \kappa = \lambda$, there is an index $j \in \{0, \ldots, \lambda\}$ such that all a_j-labeled edges and thus particularly the edges $f_{i_0,0} \xrightarrow{a_j} f_{i_0,1}$ and $f_{i_1,0} \xrightarrow{a_j} f_{i_1,1}$ are present in B. Hence, by $sup(f_{i_0,0}) > sup(f_{i_0,1})$, we obtain $con(a_j) > pro(a_j)$ and, by $sup(f_{i_1,0}) < sup(f_{i_1,1})$, we get $con(a_j) < pro(a_j)$, which is a contradiction. Consequently, $\{f_{i_0,0} \xrightarrow{X_{i_0}} f_{i_0,1}, f_{i_1,0} \xrightarrow{X_{i_1}} f_{i_1,1}\} \cap \mathfrak{K} \neq \emptyset$ and thus $M_i \cap Z \neq \emptyset$. Since i was arbitrary, we have $M_i \cap Z \neq \emptyset$ for all $i \in \{0, \ldots, m-1\}$. Moreover, even if all edges of $\mathfrak{K}$ are pairwise labeled distinctly, there can be at most

$|\mathfrak{K}| \leq \kappa = \lambda$ distinct events be affected. This particularly implies $|Z| \leq \lambda$ and thus proves that Z defines a sought vertex cover for G. It remains to argue that $con(X_{i_0}) > pro(X_{i_0})$ and $con(X_{i_1}) < pro(X_{i_1})$: Since R solves α, we have $sup(t_{i,1}) < con(X_{i_0})$ and, by $t_{i,0}\xrightarrow{X_{i_0}}$, we have $con(X_{i_0}) \leq sup(t_{i,0})$. Together this implies $sup(t_{i,0}) > sup(t_{i,1})$ and thus $con(X_{i_0}) > pro(X_{i_0})$. On the other hand, X_{i_0} also occurs at $t_{i,2}$, which implies $con(X_{i_0}) \leq sup(t_{i,2})$ and thus $sup(t_{i,1}) < sup(t_{i,2})$. Hence, by $sup(t_{i,2}) = sup(t_{i,1}) - con(X_{i_1}) + pro(X_{i_1})$, the latter implies $con(X_{i_1}) < pro(X_{i_1})$. □

Conversely, if (G, λ) allows a positive decision, then so does (A, κ): Let $Z = \{X_{j_0}, \ldots, X_{j_{\lambda-1}}\} \subseteq \mathfrak{U}$ be a vertex cover for G. A suitable edge-removal $B = (S, E, \delta', \bot)$ is defined by removing, for every $\ell \in \{0, \ldots, \lambda - 1\}$, the edge $f_{j_\ell,0}\xrightarrow{X_{j_\ell}}f_{j_\ell,1}$ and nothing else: $\mathfrak{K} = \{f_{j_\ell,0}\xrightarrow{X_{j_\ell}}f_{j_\ell,1} \mid \ell \in \{0, \ldots, \lambda - 1\}\}$. Notice that B has the same states S and events E as A and satisfies $|\mathfrak{K}| \leq \kappa = \lambda$.

Fact 1. *The edge-removal B of A has the SSP.*

Proof. Let $i \in \{0, \ldots, m - 1\}$ be arbitrary but fixed. The following region $R_1 = (sup_1, con_1, pro_1)$ solves (s, s') for all $s \in \{t_{i,0}, \ldots, t_{i,3}\}$ and $s' \in S \setminus \{t_{i,0}, \ldots, t_{i,3}\}$: $sup_1(\bot) = 0$ and $\mathcal{T}_{0,1}^{R_1} = \{w_i\}$ and $\mathcal{T}_{0,0}^{R_1} = E \setminus \{w_i\}$. Similarly, if $j \in \{0, \ldots, n - 1\}$, then one defines a region that separates the states of F_j from the others. Hence, to complete the proof, it remains to consider the SSA (s, s'), where s and s' belong to the same gadget. The following region $R_2 = (sup_2, con_2, pro_2)$ solves all the corresponding atoms at once: $sup_2(\bot) = 0$ and $\mathcal{T}_{0,1}^{R_2} = E$. □

Fact 2. *The edge-removal B of A has the ESSP.*

Proof. Let $W = \{w_0, \ldots, w_{m-1}\}$ and $Y = \{y_0, \ldots, y_{n-1}\}$.

First of all, the following region $R_1 = (sup_1, con_1, pro_1)$ solves (a, s) for all $a \in W \cup Y$ and $s \in S \setminus \bot$: $sup(\bot) = 1$ and, for all $e \in E$, if $e \in W \cup Y$, then $(con(e), pro(e)) = (1, 0)$, otherwise $(con(e), pro(e)) = (0, 0)$.

Secondly, the following region $R_2 = (sup_2, con_2, pro_2)$ solves $(a, \bot)$ for all $a \in E \setminus (W \cup Y)$: $sup_2(\bot) = 0$ and $\mathcal{T}_{0,3}^{R_2} = W \cup Y$ and $\mathcal{T}_{1,0}^{R_2} = E \setminus (W \cup Y)$.

Hence, it remains to consider the ESSA (a, q) of B such that a and q belong to one of the gadgets. If a and q belong to *different* gadgets, then it is easy to see, that (a, q) is solvable by a region $R_3 = (sup_3, con_3, pro_3)$ such that $sup_3(\bot) = 0$ and, for all $e \in E$, if $e = a$, then $(con_3(e), pro_3(e)) = (1, 1)$; if $e \in W \cup Y$ such that there is a path from $\bot$ to a gadget that contains the event a, then $(con_3(e), pro_3(e)) = (0, 1)$, otherwise $(con_3(e), pro_3(e)) = (0, 0)$.

Consequently, in order to complete the proof of the lemma, it remains to consider the ESSA (a, q) of B such that a and q belong to the *same* gadget. Let $i \in \{0, \ldots, m - 1\}$ be arbitrary but fixed.

We first show that all ESSA provided by T_i are solvable. We start with the ESSA $(X_{i_0}, t_{i,1})$ and $(X_{i_0}, t_{i,3})$ and distinguish between $X_{i_0} \in Z$ and $X_{i_0} \notin Z$:

1. $X_{i_0} \in Z$. The following region $R_4 = (sup_4, con_4, pro_4)$ solves $(X_{i_0}, t_{i,1})$ and $(X_{i_0}, t_{i,3})$: $sup_4(\bot) = 0$; $T_{1,0}^{R_4} = \{X_{i_0}\}$ and $T_{0,1}^{R_4} = \{X_{i_1}\} \cup \{w_i\} \cup (\mathfrak{U} \setminus Z) \cup \{a_0, \ldots, a_\lambda\}$ and $T_{0,3}^{R_4} = E \setminus (T_{1,0}^{R_4} \cup T_{0,1}^{R_4})$.

2. $X_{i_0} \notin Z$ (implying $X_{i_1} \in Z$). The following region $R_5 = (sup_5, con_5, pro_5)$ solves $(X_{i_0}, t_{i,1})$ and $(X_{i_0}, t_{i,3})$: $sup_5(\bot) = 0$ and $T_{1,0}^{R_5} = (\mathfrak{U} \setminus Z) \cup \{a_0, \ldots, a_\lambda\}$ and $T_{0,1}^{R_5} = \{X_{i_1}\} \cup \{w_i\}$ and $T_{0,3}^{R_5} = E \setminus (T_{1,0}^{R_5} \cup T_{0,1}^{R_5})$.

Notice that $2con(X_{j_0}) + con(X_{j_1}) \leq 3$ for all $j \in \{0, \ldots, m-1\}$ such that R_4 and R_5 are well-defined. So far, we are already finished with X_{i_0}. We proceed with the ESSA $(X_{i_1}, t_{i,0})$ and consider the cases $X_{i_1} \in Z$ and $X_{i_1} \notin Z$ separately:

1. $X_{i_1} \in Z$. The following region $R_6 = (sup_6, con_6, pro_6)$ solves $(X_{i_1}, t_{i,0})$: $sup_6(\bot) = 1$ and $T_{1,0}^{R_6} = \{X_{i_1}, w_i\}$ and $T_{0,1}^{R_6} = \{X_{i_0}\} \cup (\mathfrak{U} \setminus Z) \cup \{a_0, \ldots, a_\lambda\}$ and $T_{0,3}^{R_6} = E \setminus (T_{1,0}^{R_6} \cup T_{0,1}^{R_6})$.

2. $X_{i_1} \notin Z$ (implying $X_{i_0} \in Z$). The following region $R_7 = (sup_7, con_7, pro_7)$ solves $(X_{i_1}, t_{i,0})$: $sup_7(\bot) = 1$ and $T_{1,0}^{R_7} = \{w_i\} \cup (\mathfrak{U} \setminus Z) \cup \{a_0, \ldots, a_\lambda\}$ and $T_{0,1}^{R_7} = \{X_{i_0}\}$ and $T_{0,3}^{R_7} = E \setminus (T_{1,0}^{R_7} \cup T_{0,1}^{R_7})$.

Restricted to T_i, it only remains to solve $(X_{i_1}, t_{i,2})$ and $(X_{i_1}, t_{i,2})$:

1. $X_{i_1} \in Z$. The following region $R_8 = (sup_8, con_8, pro_8)$ solves $(X_{i_1}, t_{i,2})$ and $(X_{i_1}, t_{i,2})$: $sup_8(\bot) = 1$ and $T_{1,0}^{R_8} = \{X_{i_1}\}$ and $T_{0,3}^{R_8} = (Y \cup W) \setminus \{w_i\}$ and $T_{0,0}^{R_8} = E \setminus (T_{1,0}^{R_8} \cup T_{0,3}^{R_8})$.

2. $X_{i_1} \notin Z$ (implying $X_{i_0} \in Z$). The following region $R_9 = (sup_9, con_9, pro_9)$ solves $(X_{i_1}, t_{i,2})$ and $(X_{i_1}, t_{i,2})$: $sup_9(\bot) = 1$ and $T_{1,0}^{R_9} = (\mathfrak{U} \setminus Z) \cup \{a_0, \ldots, a_\lambda\}$ and $T_{0,3}^{R_9} = (Y \cup W) \setminus \{w_i\}$ and $T_{0,0}^{R_9} = E \setminus (T_{1,0}^{R_9} \cup T_{0,3}^{R_9})$.

Altogether, this we have shown that the ESSA of T_i are solvable. Consequently, since i was arbitrary, it follows that all ESSA of B that originate from the gadgets $T_0, \ldots, T_{m-1}$ are solvable. Hence, it only remains to discuss the ESSA that come from the gadgets $F_0, \ldots, F_{n-1}$.

Let $i \in \{0, \ldots, n-1\}$ be arbitrary but fixed. The following region $R_{10} = (sup_{10}, con_{10}, pro_{10})$ solves $(e, f_{i,1})$ for all events e that occur in F_i: $sup(\bot) = 2$ and $T_{1,0}^{R_{10}} = \{y_i\} \cup (\mathfrak{U} \setminus Z) \cup \{a_0, \ldots, a_\lambda\}$ and $T_{0,3}^{R_{10}} = E \setminus T_{1,0}^{R_{10}}$. Since i was arbitrary, the fact follows.   □

By Fact 1 and Fact 2, we get the following Lemma 3, which, by Lemma 2 and the fact that the reduction is polynomial, completes the proof of Theorem 1:

Lemma 3. *If there is a vertex cover with at most λ elements for G, then there is an edge-removal B for A that satisfies $|\mathfrak{R}| \leq \kappa$ and has the ESSP and the SSP.*

4   The Complexity of Event-Removal

In this section, we are looking for implementable event-removals that remove only a bounded number of events. By the connection between implementations and separation properties stated by Lemma 1, the following problems arise:

EVENT-REMOVAL FOR EMBEDDING
Input: A TS $A = (S, E, \delta, \iota)$, a natural number κ.
Question: Does there exist an event-removal B for A that has the SSP and
satisfies $|\mathfrak{E}| \leq \kappa$?

EVENT-REMOVAL FOR LANGUAGE-SIMULATION
Input: A TS $A = (S, E, \delta, \iota)$, a natural number κ.
Question: Does there exist an event-removal B for A that has the ESSP
and satisfies $|\mathfrak{E}| \leq \kappa$?

EVENT-REMOVAL FOR REALIZATION
Input: A TS $A = (S, E, \delta, \iota)$, a natural number κ.
Question: Does there exist an event-removal B for A that has the ESSP
and the SSP and satisfies $|\mathfrak{E}| \leq \kappa$?

The following Sect. 4.1 characterizes the complexity of EVENT-REMOVAL FOR LANGUAGE SIMULATION and EVENT-REMOVAL FOR REALIZATION and, after that, Sect. 4.2 classifies the complexity of EVENT-REMOVAL FOR EMBEDDING.

4.1   Event-Removal Aiming at Language Simulation and Realization

The following theorem states the main result of this section:

Theorem 2. *Both* EVENT-REMOVAL FOR LANGUAGE SIMULATION *and* EVENT-REMOVAL FOR REALIZATION *are NP-complete.*

Obviously, the problems addressed by Theorem 2 belong to NP: If there is a fitting event-removal B for A, then a Turing machine T can guess $\mathfrak{E}$ non-deterministically in time polynomial in the size of A, since $|\mathfrak{E}| \leq |E|$. After that, T can deterministically compute B and a witness for the property in question in polynomial time by Lemma 1, since the size of B is bounded by the size of A.

Hence, to complete the proof of Theorem 2, it remains to prove the hardness part. In order to do that, we provide a reduction of VC that reduces the input (G, λ) to an instance (A, κ) with TS A and natural number κ: In particular, we first define $\kappa = \lambda$. Moreover, the TS A reuses, for all $i \in \{0, \ldots, m-1\}$ and for all $j \in \{0, \ldots, n-1\}$, the gadgets T_i and F_j, which have been introduced in Sect. 3. The TS A has the initial state $\bot$. In order to join the gadgets and, moreover, to ensure that an event-removal sought is initialized, we add for every state s of the gadgets an edge from $\bot$ to s, which is labeled by an event that labels no other edge of A: for all $i \in \{0, \ldots, m-1\}$ and all $j \in \{0, \ldots, 3\}$, we add the edge $\bot \xrightarrow{w_i^j} t_{i,j}$; for all $i \in \{0, \ldots, n-1\}$ and all $j \in \{0, 1\}$, we add the edge $\bot \xrightarrow{y_i^j} f_{i,j}$. The result is the TS $A = (S, E, \delta, \bot)$. In the following, we prove the functionality of A. (Recall that $\mathfrak{E}$ refers to the events removed from A.)

Lemma 4. *(1) If there is an event-removal B for A having the ESSP and satisfying $|\mathfrak{E}| \leq \kappa$, then there is a vertex cover with at most λ elements for G. (2) If G has a vertex cover of size at most λ, then there is an event-removal B for A that has the ESSP and the SSP and satisfies $|\mathfrak{E}| \leq \kappa$.*

Proof. (1): Let $B = (S', E', \delta', \perp)$ be a corresponding event-removal of A. Since $|\mathfrak{E}| \leq \kappa = \lambda$, there is an index $j \in \{0, \ldots, \lambda\}$ such that a_j and thus all a_j-labeled edges are present in B. Let $i \in \{0, \ldots, m-1\}$ be arbitrary but fixed. If $\{X_{i_0}, X_{i_1}\} \cap \mathfrak{E} = \emptyset$, then T_i is completely present in B and the edges $f_{i_0,0} \xrightarrow{X_{i_0}} f_{i_0,1}$ and $f_{i_1,0} \xrightarrow{X_{i_1}} f_{i_1,1}$ are present as well. Hence, similar to the arguments of the proof of Lemma 2, we obtain that the ESSA $(X_{i_0}, t_{i,1})$ is not solvable, which contradicts the ESSA of B. Consequently, $\{X_{i_0}, X_{i_1}\} \cap \mathfrak{E} \neq \emptyset$, which is true for all $i \in \{0, \ldots, m-1\}$, since i was arbitrary. Hence, by $|\mathfrak{E}| \leq \kappa = \lambda$, the set $Z = \mathfrak{U} \cap \mathfrak{E}$ defines a fitting vertex cover for G.

(2): Let Z be a fitting vertex cover, i.e., $|Z| \leq \lambda$. We get $B = (S, E', \delta', \perp)$ by removing the events of Z: $E' = E \setminus Z$ and if $e \in Z$, then the edge $s \xrightarrow{e} s'$ is not present in B. Notice that, by the y- and w-events, B preserves the states of A.

Let's prove the ESSP and the SSP: Let $W = \{w_0^0, \ldots, w_0^3, w_1^0 \ldots, w_{m-1}^3\}$ and $Y = \{y_0^0, y_0^1, y_1^0, \ldots, y_{n-1}^1\}$. The following region $R_1 = (sup_1, con_1, pro_1)$ shows that $(\perp, s)$ and (a, s) are solvable for all $s \in S \setminus \{\perp\}$ and all $a \in W \cup Y$: $sup_1(\perp) = 1$ and $\mathcal{T}_{1,0}^{R_1} = W \cup Y$ and $\mathcal{T}_{0,0}^{R_1} = E' \setminus \mathcal{T}_{1,0}^{R_1}$.

Moreover, the following $R_2 = (sup_2, con_2, pro_2)$ solves $(a, \perp)$ for all $a \in E' \setminus (W \cup Y)$: $sup_2(\perp) = 0$ and $\mathcal{T}_{0,1}^{R_2} = W \cup Y$ and $\mathcal{T}_{1,1}^{R_2} = E' \setminus (W \cup Y)$.

Consequently, in order to complete the proof, it remains to solve the ESSA (e, s) such that $e \in E' \setminus (W \cup Y)$ and the SSA (s, s') such that $s, s' \in S \setminus \{\perp\}$: We observe that there is a –more or less obvious– way to adopt the regions of Fact 1 and Fact 2, since the current TS A can be considered as an extension of the TS in Sect. 3, where, for every $i \in \{0, \ldots, m-1\}$ and every $j \in \{0, \ldots, n-1\}$, the events w_i^0 and y_j^0 of the current TS correspond to the events w_i and y_j of the one defined in Sect. 3, respectively. More exactly, if $R = (sup, con, pro)$ is a region defined for Fact 1 or Fact 2, then we can modify it to a region $R' = (sup', con', pro')$ of the current event-removal B as follows: $sup'(s) = sup(s)$ for all $s \in S$; for all $e \in E'$, if $e \in E' \setminus (W \cup Y)$, then $(con'(e), pro'(e)) = (con(e), pro(e))$; if $e = w_i^0$, then $(con'(e), pro'(e)) = (con(w_i), pro(w_i))$ for all $i \in \{0, \ldots, m-1\}$; if $e = y_i^0$, then $(con'(e), pro'(e)) = (con(y_i), pro(y_i))$ for all $i \in \{0, \ldots, n-1\}$; if $e \in \{w_i^1, \ldots, w_i^3 \mid i \in \{0, \ldots, m-1\}\} \cup \{y_i^1 \mid i \in \{0, \ldots, n-1\}\}$, then there is a unique state $s \in S \setminus \{\perp\}$ such that the edge $\perp \xrightarrow{e} s$ is present in B, and we define $(con(e), pro(e)) = (sup(\perp) - sup(s), 0)$ if $sup(\perp) > sup(s)$, and otherwise $(con(e), pro(e)) = (0, sup(s) - sup(\perp))$.

In particular, we obtain the solvability of the remaining SSA and ESSA of B by the regions defined for the proof of Fact 1 and Fact 2, respectively. $\square$

4.2   Event-Removal Aiming at Embedding

In this section, we show that finding a minimal event-removal is also hard if we are aiming at an embedding:

Theorem 3. EVENT-REMOVAL FOR EMBEDDING *is NP-complete.*

Similar to the arguments for Theorem 2, one argues that EVENT-REMOVAL FOR EMBEDDING is in NP. In order to prove the hardness part, we present a reduction of VC, that transforms the input (G, λ) into an instance (A, κ) as follows: For a start, $\kappa = \lambda$. For all $i \in \{0, \dots, m-1\}$, the TS A has the following gadget T_i at which the elements of $M_i = \{X_{i_0}, X_{i_1}\}$ occur as events:

$$T_i = \quad t_{i,0} \underset{X_{i_1}}{\overset{X_{i_0}}{\rightleftarrows}} t_{i,1}$$

Moreover, for all $j \in \{0, \dots, n-1\}$, the TS A reuses the gadget F_j introduced in Sect. 3. The initial state of A is $\perp$ and the following edges connect the gadgets and ensure that event-removals are initialized: for all $i \in \{0, \dots, m-1\}$ and all $j \in \{0,1\}$, we add the edge $\perp \overset{w_i^j}{\longrightarrow} t_{i,j}$, and for all $i \in \{0, \dots, n-1\}$ and all $j \in \{0,1\}$, we add the edge $\perp \overset{y_i^j}{\longrightarrow} f_{i,j}$. The result is the TS $A = (S, E, \delta, \perp)$.

Lemma 5. *There is an event-removal B for A that has the SSP and satisfies $|\mathfrak{E}| \leq \kappa$ if and only if there is a vertex cover with at most λ elements for G.*

Proof. $\Rightarrow$: Let $B = (S', E', \delta', \perp)$ be an event-removal for A that has the SSP and satisfies $|\mathfrak{E}| \leq \kappa$. Since $|\mathfrak{E}| \leq \kappa = \lambda$, there is an index $j \in \{0, \dots, \lambda\}$ such that the event a_j (and its edges) is present in B. Let $i \in \{0, \dots, m-1\}$ be arbitrary but fixed. If $\{X_{i_0}, X_{i_1}\} \cap \mathfrak{E} = \emptyset$, then $(t_{i,0}, t_{i,1})$ is an SSA of B. Since B has the SSP, there is a region $R = (sup, con, pro)$ that solves it, which implies either $sup(t_{i,0}) > sup(t_{i,1})$ or $sup(t_{i,0}) < sup(t_{i,1})$. The first case implies $con(X_{i_0}) > pro(X_{i_0})$ and $con(X_{i_1}) < pro(X_{i_1})$, and the second implies $con(X_{i_0}) < pro(X_{i_0})$ and $con(X_{i_1}) > pro(X_{i_1})$. Since the edges $f_{i_0,0} \overset{X_{i_0}}{\longrightarrow} f_{i_0,1}$ and $f_{i_1,0} \overset{X_{i_1}}{\longrightarrow} f_{i_1,1}$ and the a_j-labelled edges are present in B, both cases imply the contradiction $con(a_j) > pro(a_j)$ and $con(a_j) < pro(a_j)$. Hence, we get $\{X_{i_0}, X_{i_1}\} \cap \mathfrak{E} \neq \emptyset$. Since i was arbitrary and $|\mathfrak{E}| \leq \lambda$, the set $Z = \mathfrak{U} \cap \mathfrak{E}$ defines a vertex cover with at most λ elements for G.

$\Leftarrow$: Let Z be a corresponding vertex cover. We obtain $B = (S, E', \delta', \perp)$ in the obvious way by $E' = E \setminus Z$. Notice that B and A have actually the same state set. The following region $R = (sup, con, pro)$ solves all SSA of B: Firstly, we define $sup(\perp) = 0$ and $(con(e), pro(e)) = (1,0)$ for all $e \in \mathfrak{U} \setminus Z$ and all $e \in \{a_0, \dots, a_\lambda\}$; furthermore, for all $i \in \{0, \dots, m-1\}$, if $X_{i_0} \in E'$, then we define $sup(t_{i,0}) = 2i+2$ and $sup(t_{i,1}) = 2i+1$ and $(con(w_i^0), pro(w_i^0)) = (0, 2i+2)$ as well as $(con(w_i^1), pro(w_i^1)) = (0, 2i+1)$, otherwise (if $X_{i_0} \notin E'$) we define $sup(t_{i,0}) = 2i+1$ and $sup(t_{i,1}) = 2i+2$ and $(con(w_i^0), pro(w_i^0)) = (0, 2i+1)$ as well as $(con(w_i^1), pro(w_i^1)) = (0, 2i+2)$; moreover, for all $j \in \{0, \ldots, n-1\}$, we define $sup(f_{j,0}) = 2m + 2j + 2$ and $sup(f_{j,1}) = 2m + 2j + 1$ and $(con(y_j^0), pro(y_j^0)) = (0, 2m + 2j + 2)$ as well as $(con(y_j^1), pro(y_j^1)) = (0, 2m + 2j + 1)$. □

5 The Complexity of State-Removal

In this section, we are interested in finding implementable state-removals of A that remove only a restricted number of states. Again justified by Lemma 1, this task corresponds to the following decision problems:

STATE-REMOVAL FOR EMBEDDING
Input: A TS $A = (S, E, \delta, \iota)$, a natural number κ.
Question: Does there exist a state-removal B for A that has the SSP and satisfies $|\mathfrak{S}| \leq \kappa$?

STATE-REMOVAL FOR LANGUAGE-SIMULATION
Input: A TS $A = (S, E, \delta, \iota)$, a natural number κ.
Question: Does there exist a state-removal B for A that has the ESSP and satisfies $|\mathfrak{S}| \leq \kappa$?

STATE-REMOVAL FOR REALIZATION
Input: A TS $A = (S, E, \delta, \iota)$, a natural number κ.
Question: Does there exist a state-removal B for A that has the ESSP and the SSP and satisfies $|\mathfrak{S}| \leq \kappa$?

In the following Sect. 5.1, we show that state-removal aiming at embedding or realization is NP-complete. After that, we show that this is also true if we aim at language simulation in Sect. 5.2.

5.1 State-Removal Aiming at Embedding or Realization

Theorem 4. STATE-REMOVAL FOR EMBEDDING *and* STATE-REMOVAL FOR REALIZATION *are NP-complete.*

First of all, the problems are in NP: If there is a suitable state-removal, then the set $\mathfrak{S}$ can be guessed non-deterministically in polynomial time; after that, B and a witness for the corresponding separation property can be computed deterministically in polynomial time by Lemma 1. The proof of the hardness-part bases again on a reduction of VC that transforms the input (G, λ) into an instance (A, κ): We define $\kappa = \lambda$, and obtain the TS $A = (S, E, \delta, \bot)$ as follows:

- The states are defined by $S = \mathfrak{U} \cup \{\bot\}$, i.e., every vertex of G is a state in A;
- the events are defined by $E = \{a_0, \ldots, a_{n-1}\} \cup \{M_0, \ldots, M_{m-1}\}$, where, for every $i \in \{0, \ldots, m-1\}$, the event M_i is associated with the arc M_i of G;

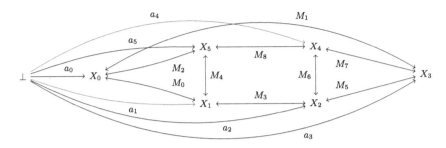

Fig. 6. The TS A of Sect. 5.1 based on Example 5. Red states and edges correspond to the state-removal B of Lemma 6 removing the states of $Z = \{X_0, X_2, X_3, X_5\}$.

- for all $i \in \{0, \ldots, n-1\}$, the TS A has the edge $\perp \xrightarrow{a_i} X_i$, and, for all $i \in \{0, \ldots, m-1\}$, it has the M_i-labelled edges $X_{i_0} \xrightarrow{M_i} X_{i_1}$ and $X_{i_1} \xrightarrow{M_i} X_{i_0}$.

Figure 6 shows the TS A that results from the input of Example 5. (For the following, recall that $\mathfrak{S}$ refers to the set of removed states.)

Lemma 6. *(1) If there is a state-removal B for A having the SSP and satisfying $|\mathfrak{S}| \leq \kappa$, then there is a vertex cover with at most λ elements for G.*
(2) If G has a vertex cover of size at most λ, then there is a state-removal B for A that has the ESSP and the SSP and satisfies $|\mathfrak{S}| \leq \kappa$.

Proof. (1): Let $B = (S', E', \delta', \perp)$ be a fitting state-removal. If there are equally labeled edges $s \xrightarrow{e} s'$ and $s' \xrightarrow{e} s$ in B, then B lacks the SSP: Let $R = (sup, con, pro)$ be an arbitrary region of B. By $s \xrightarrow{e} s'$, it holds $sup(s') = sup(s) - con(e) + pro(e)$ and, by $s' \xrightarrow{e} s$, it holds $sup(s) = sup(s') - con(e) + pro(e)$. If we subtract the latter equation from the former, we obtain $sup(s') - sup(s) = sup(s) - sup(s')$, which implies $sup(s') = sup(s)$. Since R was arbitrary, this shows that (s, s') is not solvable. Let $i \in \{0, \ldots, m-1\}$ be arbitrary but fixed. Since B has the SSP, by the former arguments, one of the edges $X_{i_0} \xrightarrow{M_i} X_{i_1}$ or $X_{i_1} \xrightarrow{M_i} X_{i_0}$ is missing in B. This implies $X_{i_0} \in \mathfrak{S}$ or $X_{i_1} \in \mathfrak{S}$, since B is a state-removal of A. Hence, by the arbitrariness of i, we have that $M_i \cap \mathfrak{S} \neq \emptyset$ for all $i \in \{0, \ldots, m-1\}$ and, by assumption, we also have $|\mathfrak{S}| \leq \kappa = \lambda$. This proves this direction.

(2): Let $Z \subseteq \mathfrak{U}$ be suitable a vertex cover for G. We obtain $B = (S', E', \delta', \perp)$ by defining $S' = S \setminus Z$ and by removing an edge $s \xrightarrow{e} s'$ of A whenever $\{s, s'\} \cap Z \neq \emptyset$ and the events e of A if there is no e-labelled edge in B. The result is a well-defined state-removal: for every $i \in \{0, \ldots, n-1\}$, if the state X_i is present in B, then it is reachable from $\perp$ by an a_i-labelled edge. Since Z is a vertex cover, for all $i \in \{0, \ldots, m-1\}$, one of the states X_{i_0}, X_{i_1} is removed. Consequently, neither $X_{i_0} \xrightarrow{M_i} X_{i_1}$ nor $X_{i_1} \xrightarrow{M_i} X_{i_0}$ are present in B. Hence, $\{M_0, \ldots, M_{m-1}\} \cap E' = \emptyset$. By construction, the other events occur at most once in A and thus in B. It is easy to see, that this implies the ESSP and the SSP for B. $\square$

5.2 State-Removal Aiming at Language Simulation

Removing as few states of a TS as possible in order to make it implementable is also hard if we are aiming at a language simulation:

Theorem 5. STATE-REMOVAL FOR LANGUAGE SIMULATION *is NP-complete.*

Similar to Theorem 4, one argues that STATE-REMOVAL FOR LANGUAGE SIMULATION is in NP. We prove the hardness part by extending the reduction of Sect. 5.1: We define $\kappa = \lambda$. The current TS A has all states, events and edges of the TS defined in Sect. 5.1. We want to achieve that if B is a state-removal of A having the ESSP, then one of X_{i_0}, X_{i_1} belongs to $\mathfrak{S}$ for all $i \in \{0, \ldots, m-1\}$. The following extension is sufficient: For all $i \in \{0, \ldots, m-1\}$ and for all $j \in \{0, \ldots, \lambda\}$, we add the path $\bot \xrightarrow{y_j^i} t_{i,j,0} \xrightarrow{M_i} t_{i,j,1}$. The result is $A = (S, E, \delta, \bot)$.

Lemma 7. *There is a state-removal B for A that has the ESSP and satisfies $|\mathfrak{S}| \leq \kappa$ if and only if there is a vertex cover with at most λ elements for G.*

Proof. $\Rightarrow$: Let B be a fitting state-removal for A. Let $i \in \{0, \ldots, m-1\}$ be arbitrary but fixed. Among others, the TS A has the $\lambda + 1$ M_i-labeled edges $t_{i,0,0} \xrightarrow{M_i} t_{i,0,1}, \ldots, t_{i,\lambda,0} \xrightarrow{M_i} t_{i,\lambda,1}$. By $|\mathfrak{S}| \leq \kappa = \lambda$, this implies that there is $j \in \{0, \ldots, \lambda\}$ such that the edge $t_{i,j,0} \xrightarrow{M_i} t_{i,j,1}$ is present in B. Since B has the ESSP, there is a region $R = (sup, con, pro)$ that solves $(M_i, t_{i,j,1})$. This implies $con(M_i) \leq sup(t_{i,j,0})$ and $con(M_i) > sup(t_{i,j,1})$ and, by $sup(t_{i,j,1}) = sup(t_{i,j,0}) - con(M_i) + pro(M_i)$, also $con(M_i) > pro(M_i)$. If both $X_{i_0} \xrightarrow{M_i} X_{i_1}$ and $X_{i_1} \xrightarrow{M_i} X_{i_0}$ are present in B, then, as just argued in Lemma 6 (1), we would have $sup(X_{i_0}) = sup(X_{i_1})$, implying the contradiction $con(M_i) = pro(M_i)$. Hence, $\{X_{i_0}, X_{i_1}\} \cap \mathfrak{S} \neq \emptyset$. Since i was arbitrary and $|\mathfrak{S}| \leq \lambda$, the claim follows.

$\Leftarrow$: Let Z be a suitable vertex cover for G. We get $B = (S', E', \delta', \bot)$ by defining $S' = S \setminus Z$ and removing the incident edges and events for which no edge remains. Obviously, B is a well-defined state-removal that does not contain $X_{i_0} \xrightarrow{M_i} X_{i_1}$ and $X_{i_1} \xrightarrow{M_i} X_{i_0}$. Let $a \in E'$ be arbitrary but fixed and let $\{q_0, \ldots, q_k\}$ be exactly the sources of a in B. One finds out that the following region $R = (sup, con, pro)$ is well-defined and solves (a, p) for all $p \in S' \setminus \{q_0, \ldots, q_k\}$: for all $s \in S'$, if $s \in \{q_0, \ldots, q_k\}$, then $sup(s) = 1$, otherwise $sup(s) = 0$; for all $e \in E'$, if $\xrightarrow{e} q$ for some $q \in \{q_0, \ldots, q_k\}$, then $(con(a), pro(e)) = (0, 1)$; if $q \xrightarrow{e}$ for some $q \in \{q_0, \ldots, q_k\}$, then $(con(a), pro(e)) = (1, 0)$; otherwise $(con(a), pro(e)) = (0, 0)$. Since a was arbitrary, the claim follows. □

6 Conclusion

In this paper, we show that converting an unimplementable TS into an implementable one by removing as few of its states, events or edges as possible, is intractable. This particularly solves a problem that was left open in [12]. Notice that the reductions for edge- and event-removal work also if these modifications are defined in a way that require all original states to be preserved. However, in general, they could then not always produce an implementable TS, since there are unimplementable trees. Future work may investigate the complexity of the problems from the point of view of parameterized complexity where κ is the parameter. It may also address other techniques of modifications that were suggested in the literature such as, for example, state- or event-refinement.

Acknowledgements. I would like to thank the anonymous reviewers for their detailed comments and valuable suggestions.

References

1. van der Aalst, W.M.P.: Process Mining - Discovery, Conformance and Enhancement of Business Processes. Springer (2011). https://doi.org/10.1007/978-3-642-19345-3
2. Badouel, E., Bernardinello, L., Darondeau, P.: Polynomial algorithms for the synthesis of bounded nets. In: Mosses, P.D., Nielsen, M., Schwartzbach, M.I. (eds.) CAAP 1995. LNCS, vol. 915, pp. 364–378. Springer, Heidelberg (1995). https://doi.org/10.1007/3-540-59293-8_207
3. Badouel, E., Bernardinello, L., Darondeau, P.: Petri Net Synthesis. TTCSAES. Springer, Heidelberg (2015). https://doi.org/10.1007/978-3-662-47967-4
4. Badouel, E., Caillaud, B., Darondeau, P.: Distributing finite automata through Petri net synthesis. Formal Asp. Comput. **13**(6), 447–470 (2002). https://doi.org/10.1007/s001650200022
5. Best, E., Darondeau, P.: petri net distributability. In: Clarke, E., Virbitskaite, I., Voronkov, A. (eds.) PSI 2011. LNCS, vol. 7162, pp. 1–18. Springer, Heidelberg (2012). https://doi.org/10.1007/978-3-642-29709-0_1
6. Best, E., Devillers, R.R.: Synthesis and reengineering of persistent systems. Acta Inf. **52**(1), 35–60 (2015). https://doi.org/10.1007/s00236-014-0209-7
7. Carmona, J.: The label splitting problem. Trans. Petri Nets Other Model. Concurr. **6**, 1–23 (2012). https://doi.org/10.1007/978-3-642-35179-2_1
8. Cortadella, J., Kishinevsky, M., Kondratyev, A., Lavagno, L., Yakovlev, A.: A region-based theory for state assignment in speed-independent circuits. IEEE Trans. Comput. Aided Des. Integr. Circuits Syst. **16**(8), 793–812 (1997). https://doi.org/10.1109/43.644602
9. Cortadella, J., Kishinevsky, M., Kondratyev, A., Lavagno, L., Yakovlev, A.: Logic Synthesis for Asynchronous Controllers and Interfaces. Springer, Berlin (2013)
10. de San Pedro, J., Cortadella, J.: Mining structured petri nets for the visualization of process behavior. In: Ossowski, S. (ed.) Proceedings of the 31st Annual ACM Symposium on Applied Computing, Pisa, Italy, 4–8 April, 2016. pp. 839–846. ACM (2016). https://doi.org/10.1145/2851613.2851645
11. Schlachter, U., Wimmel, H.: Relabelling LTS for petri net synthesis via solving separation problems. Trans. Petri Nets Other Model. Concurr. **14**, 222–254 (2019). https://doi.org/10.1007/978-3-662-60651-3_9
12. Schlachter, U., Wimmel, H.: Optimal label splitting for embedding an LTS into an arbitrary Petri net reachability graph is NP-complete. CoRR abs/2002.04841 (2020). https://arxiv.org/abs/2002.04841

13. Tredup, R.: Finding an optimal label-splitting to make a transition system petri net implementable: a complete complexity characterization. In: ICTCS. CEUR Workshop Proceedings, vol. 2756, pp. 131–144. CEUR-WS.org (2020)
14. Verbeek, H.M.W., Pretorius, A.J., van der Aalst, W.M.P., van Wijk, J.J.: Visualizing State Spaces with Petri Nets. Eindhoven University of Technology, Eindhoven, The Netherlands (2007). https://publications.rwth-aachen.de/record/715007

Synthesis of (Choice-Free) Reset Nets

Raymond Devillers[✉]

Département d'Informatique, Université Libre de Bruxelles, 1050 Brussels, Belgium
rdevil@ulb.ac.be

Abstract. Instead of synthesising a labelled transition system into a weighted Petri net, we shall here consider the larger class of nets with reset arcs, allowing to instanciate a larger class of transition systems. We shall also target an extension of choice-free nets with reset arcs, since choice-free nets appeared to be especially interesting in terms of properties, synthesis and implementation. In addition to a general algorithm, we shall analyse how to speed it up by reducing the number and complexity of the linear systems of constraints to be solved and how to set up a pre-synthesis phase. We shall also envisage how to implement the result of such a synthesis as a concurrent program.

Keywords: Labelled transition systems · Reset nets ·
Choice-freeness · Synthesis

1 Introduction

In order to validate a system, instead of analysing a model of the latter to check if it satisfies a set of desired properties, the synthesis approach tries to build a model "correct by construction" directly from those properties, and then to implement it. In particular, if the behaviour of a system is specified by a finite labelled transition system (LTS for short), more or less efficient algorithms have been developed to build a bounded weighted Petri net with a reachability graph isomorphic to (or close to) the given LTS [2,19]. It is also possible to target some subclasses of Petri nets [6], in particular choice-free nets and some of their specialisations [4,7,8,13] which present interesting features.

On the contrary, in order to extend a bit the power of the technique, we shall here consider a superclass of the classical Petri nets, by allowing reset arcs [1]. When one extends Petri nets, it is often the case that properties which are decidable for the latter (albeit sometimes with a huge complexity) become undecidable. And indeed, for reset nets, boundedness and reachability (in particular) are undecidable [14]. This increases the interest to avoid analysis techniques in favour of synthesis ones.

The paper is organised as follows. After recalling classical definitions, notations and properties in Sect. 2, we present presynthesis phases in Sect. 3, and then general algorithms to synthesise (choice-free) reset nets in Sect. 4. In the next sections, we analyse how to speed up the synthesis and to implement the resulting models. As usual, we conclude in the last section.

© Springer Nature Switzerland AG 2021
D. Buchs and J. Carmona (Eds.): PETRI NETS 2021, LNCS 12734, pp. 274–291, 2021.
https://doi.org/10.1007/978-3-030-76983-3_14

2 Classical Definitions, Notations and Properties

Definition 1. LTS, SEQUENCES AND REACHABILITY
A *labelled transition system with initial state, LTS* for short, is a quadruple $TS = (S, \rightarrow, T, \iota)$ where S is the set of *states*, T is the set of *labels*, $\rightarrow \subseteq (S \times T \times S)$ is the *transition relation*, and $\iota \in S$ is the *initial state*.
A label t is *enabled* at $s \in S$, written $s[t\rangle$, if $\exists s' \in S \colon (s, t, s') \in \rightarrow$, in which case s' is said to be *reachable* from s by the firing of t, and we write $s[t\rangle s'$.
Generalising to any (firing) sequences $\sigma \in T^*$, $s[\varepsilon\rangle$ and $s[\varepsilon\rangle s$ are always true, with ε being an empty sequence; and $s[\sigma t\rangle s'$, i.e., σt is *enabled* from state s and leads to s' if there is some s'' with $s[\sigma\rangle s''$ and $s''[t\rangle s'$.
A state s' is *reachable* from state s if $\exists \sigma \in T^* \colon s[\sigma\rangle s'$. The set of states reachable from s is noted $[s\rangle$.
□ 1

Definition 2. SOME PROPERTIES OF LTS
$TS = (S, \rightarrow, T, \iota)$ is *fully reachable* if $S = [\iota\rangle$, i.e., each state is reachable from the initial one.
TS is *forward deterministic* if $\forall s \in S, \forall t \in T \colon s[t\rangle s' \wedge s[t\rangle s'' \Rightarrow s' = s''$.
It is *backward deterministic* if $\forall s \in S, \forall t \in T \colon s'[t\rangle s \wedge s''[t\rangle s \Rightarrow s' = s''$. It is *deterministic* if it is both forward and backward deterministic, i.e., the successors or predecessors of a state are determined by the labels of the arcs.
TS is *quasi-persistent* if $\forall s, s_1, s_2 \in S \ \forall a \neq b \in T \colon s[a\rangle s_1 \wedge s[b\rangle s_2 \Rightarrow s_1[b\rangle \wedge s_2[a\rangle$, i.e., if there is a choice, it persists until both labels are performed
TS is *persistent* if $\forall s, s_1, s_2 \in S \ \forall a \neq b \in T \colon s[a\rangle s_1 \wedge s[b\rangle s_2 \Rightarrow s_1[b\rangle s'' \wedge s_2[a\rangle s''$ for some $s'' \in S$, i.e., it is quasi-persistent and the resulting states are the same.
TS is *reversible* if $\forall s \in [\iota\rangle \colon \iota \in [s\rangle$, i.e., every reachable state allows to go back to the initial state.
□ 2

Definition 3. PETRI NETS AND REACHABILITY GRAPHS.
A (finite, place-transition) *weighted Petri net*, or *weighted net*, is a tuple $N = (P, T, W)$ where P is a finite set of *places*, T is a finite set of *transitions*, with $P \cap T = \emptyset$, and W is a *weight* function $W \colon ((P \times T) \cup (T \times P)) \rightarrow \mathbb{N}$ giving the weight of each arc.
A *Petri net system*, or *system*, is a tuple $\mathcal{S} = (N, M_0)$ where N is a net and M_0 is the *initial marking*, a marking being a member of $P \rightarrow \mathbb{N}$ (hence a member of $\mathbb{N}^P$) indicating the number of *tokens* in each place.
A transition $t \in T$ is *enabled by* a marking M, denoted by $M[t\rangle$, if for all places $p \in P$, $M(p) \geq W(p, t)$. If t is enabled at M, then t can *occur* (or *fire*) in M, leading to the marking M' defined by $M'(p) = M(p) - W(p, t) + W(t, p)$; this is denoted by $M[t\rangle M'$. A marking M' is *reachable* from M if there is a sequence of firings leading from M to M'. The set of markings reachable from M is denoted by $[M\rangle$. The *reachability graph of* $\mathcal{S}$ is the labelled transition system $RG(\mathcal{S})$ with the set of vertices $[M_0\rangle$, the set of labels T, initial state M_0 and transitions $\{(M, t, M') \mid M, M' \in [M_0\rangle \wedge M[t\rangle M'\}$.
□ 3

A reset net (RPN for short) or system is an easy extension of the classical Petri nets and system, defined as follows:

Definition 4. RESET NETS AND SYSTEMS.

A (finite, place-transition, weighted) *reset net* is a tuple $N = (P, T, W, R)$ where (P, T, W) is a Petri net and $R \subseteq P \times T$ is a set of (undirected) reset arcs. A reset system is a reset net provided with an initial marking M_0: (P, T, W, R, M_0).

A transition $t \in T$ is *enabled by* a marking M, denoted by $M[t\rangle$, if for all places $p \in P$, $M(p) \geq W(p, t)$, i.e., it is enabled in the underlying Petri net. If t is enabled at M, then t can *occur* (or *fire*) in M, leading to the marking M' defined by $M'(p) = M(p) - W(p, t) + W(t, p)$ if $(p, t) \notin R$ and $M'(p) = W(t, p)$ otherwise, denoted by $M[t\rangle M'$. The latter case may be interpreted as follows: first, t absorbs $W(p, t)$ tokens from p, then erases the rest of the tokens in p, and finally produces $W(t, p)$ new tokens in p.

Like for Petri nets, a marking M' is *reachable* from M if there is a sequence of firings leading from M to M' and the set of markings reachable from M is denoted by $[M\rangle$. The *reachability graph of a reset system* $\mathcal{S} = (P, T, W, R, M_0)$ *is the labelled transition system* $RG(\mathcal{S})$ with the set of vertices $[M_0\rangle$, the set of labels T, initial state M_0 and transitions $\{(M, t, M') \mid M, M' \in [M_0\rangle \wedge M[t\rangle M'\}$.

A (reset) net is *pure* if $\forall p \in P, t \in T : W(p, t) \cdot W(t, p) = 0$, i.e., no transition both checks the presence of tokens in a place and produces tokens in that place. For any place $p \in P$, we shall denote $p^{\bullet} = \{t \in T | W(p, t) > 0\}$ (the set of transitions collecting tokens from p, also called successors or outputs of p) and $R(p) = \{t \in T | (p, t) \in R\}$ (the set of transitions resetting p). □ 4

Definition 5. BOUNDEDNESS

A (reset or Petri net) system $\mathcal{S}$ is *bounded* if $\exists k \in \mathbb{N} \ \forall p \in P \ \forall M \in [M_0\rangle :$ $M(p) \leq k$. It is k-bounded if $\forall p \in P \ \forall M \in [M_0\rangle : M(p) \leq k$. □ 5

A classical (and easy) result is that

Corollary 1. BOUNDED SYSTEM

A (reset or Petri net) system $\mathcal{S}$ is bounded iff its reachability graph $RG(\mathcal{S})$ is finite. □ 1

Among the very numerous subclasses of Petri net systems that have been considered in the literature, choice-free ones[1] (meaning there is no true choice to be performed when two or more transitions are enabled [20]; they have also been called output-nonbranching [5]) appeared very interesting in terms of properties, synthesis and implementation [8]. We shall thus introduce a similar subclass for reset nets.

---

[1] not to be confused with free-choice nets [9].

Definition 6. CHOICE-FREE SUBCLASSES

A Petri net is said *choice-free* if $\forall p \in P : |p^\bullet| \leq 1$.
A reset net will be said choice-free if $\forall p \in P : (|p^\bullet \cup R(p)| \leq 1)$. That is, each place has at most one successor transition and at most one resetting transition, and if they are both present they must be the same.

□ 6

In graphical representations, reset arcs will be drawn as (undirected) dotted lines, and as usual arcs with null weight are omitted. Figure 1 presents a reset net and the corresponding reachability graph for some initial marking. It is not choice free, while each place has a single output transition and a single reset arc, but not always the same: for instance p_2 has output a and is reset by c.

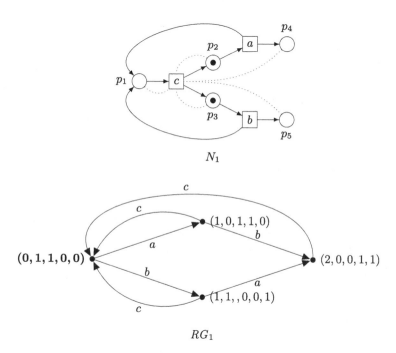

Fig. 1. A reset net system and its (finite) reachability graph for the initial marking specified in bold.

On the contrary, Figs. 2 and 3 present bounded choice-free reset net systems, and their reachability graphs (the initial markings are still respresented in bold).

Markings may be considered as a kind of vectors with indices in P, but we shall also consider vectors of transitions.

Definition 7. T-VECTORS

A *T-vector* is an element of $\mathbb{N}^T$.
The *support* of a vector is the set of the indices of its non-null components.

$$N_2 \qquad\qquad RG_2$$

Fig. 2. A simple choice-free reset net and its reachability graph

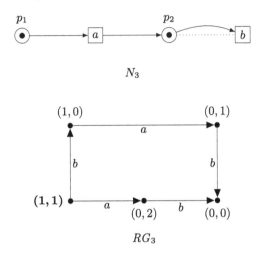

Fig. 3. Another bounded choice-free reset net and its reachability graph

A vector is called *prime* if the greatest common divisor of its components is one (i.e., its components do not have a common non-unit factor).

The *Parikh vector* $\Psi(\sigma)$ of a finite sequence $\sigma \in T^*$ of transitions is a T-vector counting the number of occurrences of each transition in σ, and the *support* of σ is the support of its Parikh vector, i.e., $supp(\sigma) = supp(\Psi(\sigma)) = \{t \in T \mid \Psi(\sigma)(t) > 0\}$.   □ 7

Definition 8. SYNTHESIS

Two LTS $TS_1 = (S_1, \rightarrow_1, T, \iota_1)$ and $TS_2 = (S_2, \rightarrow_2, T, \iota_2)$ are isomorphic if there is a bijection $\zeta \colon S_1 \rightarrow S_2$ with $\zeta(\iota_1) = \iota_2$ and $(s, t, s') \in \rightarrow_1 \Leftrightarrow (\zeta(s), t, \zeta(s')) \in \rightarrow_2$, for all $s, s' \in S_1$.

If an LTS TS is isomorphic to the reachability graph $RG(\mathcal{S})$ of some system $\mathcal{S}$, we say that $\mathcal{S}$ *solves* TS (or X-solves it, if X is the class of $\mathcal{S}$). A LTS is X-solvable if a system of class X solves it.

A synthesis is a procedure aimed at finding a solution from TS (when possible); it thus consists to keep the structure of the reachability graph of a system (dropping the exact values of the markings) in order to obtain a given LTS.   □ 8

3   Presynthesis

The presynthesis phase consists in checking that the given LTS satisfies some structural properties common to all the reachability graphs of the target class. These properties need of course to be easy to check. If a check fails, we may immediately reject the synthesis and produce a reason, easy to understand in general.

For a reset net synthesis, we may use the following:

Proposition 1. GENERAL PROPERTIES OF RESET NETS

The reachability graph of a bounded reset net system is finite, totally reachable and forward deterministic.

Proof: Finiteness results from Corollary 1. Forward determinism results from the firing rule, and total reachability from the definition of a reachability graph.
□ 1

Hence, if a LTS is not finite, or not totally reachable, or not forward deterministic (easy to check if the LTS is given explicitly), there is no reset net solution (and we know why). Note however that, contrary to what happens for usual Petri nets, it may happen that the reachability graph of a reset net is not backward deterministic. This may be observed on Fig. 2 for instance (N_2 is even a choice-free reset net): there are two arcs labelled a arriving at node (0).

For a choice-free reset net synthesis, we may use the following:

Proposition 2. GENERAL PROPERTIES OF CHOICE-FREE RESET NETS

The reachability graph of a bounded choice-free reset net system is finite, totally reachable, forward deterministic and quasi-persistent.

Proof: The first three properties result from Proposition 1.
Quasi-persistence results from the observation that, in a reset net system, for any place $p \in P$, the marking of p may only be decreased when firing t if $t \in p^\bullet \cup R(p)$. Hence, if the system is choice-free, the marking of p may only be decreased by a single transition. As a consequence, if t is enabled by some marking, it remains so at least until t is fired.
□ 2

However, contrary to what happened for usual Petri nets, it may happen that the reachability graph of a choice-free reset net system is not persistent: this is illustrated by Fig. 3, where N_3 is a choice-free reset net system, a and b are initially enabled, but $M_0[ab\rangle$ and $M_0[ba\rangle$ lead to different markings.

Many general properties of choice-free net system have been discovered and proposed for a pre-synthesis phase (see [8]). However, very few of them remain valid for choice-free reset net systems.

For instance, it is known that bounded choice-free nets always have home states in their reachability graphs, i.e., states that remain reachable whatever the evolution of the system (this is due to Keller's theorem [18]). This is not true for choice-free reset nets, as illustrated by Fig. 4.

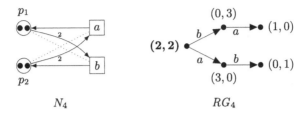

$$N_4 \qquad\qquad RG_4$$

Fig. 4. A choice-free reset system and its reachability graph without home state

Next, if the reachability graph of a bounded choice-free net is acyclic, all paths between two reachable markings have the same Parikh vector: Fig. 3 shows that this is no longer true for choice-free reset nets since there are two paths ab and bab from $(1,1)$ to $(0,0)$.

Moreover, it is known that, in the reachability graph of a bounded choice-free net, cycles are propagated Parikh-equivalently: if $s[\alpha\rangle s$ and $s[a\rangle s'$ for some $a \in T$ and $\alpha \in T^*$, then $s'[\beta\rangle s'$ with $\Psi(\alpha) = \Psi(\beta)$ for some $\beta \in T^*$. This is no longer true when we add reset arcs, as illustrated by Fig. 5. We may observe in this example that, while the initial cycle (a simple loop a) is not transported on the next state, it is however transported on the last state. We may then wonder if any cycle is eventually transported if we go further enough. This is not true however, as illustrated by Fig. 6: the initial cycle $babc$ is not transported if we perform the path ab since this leads to a dead end.

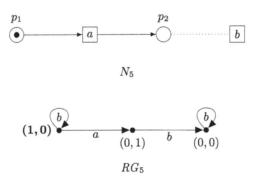

$$N_5$$

$$RG_5$$

Fig. 5. A bounded choice-free reset net and its reachability graph, where cycles are not always pushed forward

Finally, many interesting properties of bounded choice-free net systems are linked to the minimal Parikh vectors of non-empty cycles (a cycle with such a minimal Parikh vector is called *small*). However, when there are reset arcs, Parikh vectors are no longer characteristic of cycles. For instance, in Fig. 2, RG_2 has two paths a (hence with the same Parikh vector), but only one of them defines a cycle (in Fig. 5, there are three paths b, but only two of them are cycles). Hence

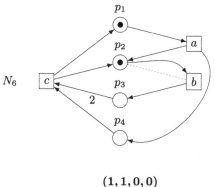

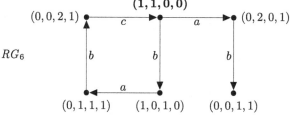

Fig. 6. A bounded choice-free reset net and its reachability graph, where a cycle is not at all pushed forward

we may suspect that the properties of small cycles will be different for reset net systems.

For instance, in a bounded choice-free net system, any cycle has a Parikh vector which is a sum of Parikh vectors of small cycles, but this is no longer true if we add reset arcs, as illustrated in Fig. 7: the cycles which do not visit twice some state are abc (and acb), $babc$ and $bacabc$, of which only the first one is small, but the Parikh vector $(1, 2, 1)$ of $babc$ is not a multiple of the minimal Parikh vector $(1, 1, 1)$ (note also that this example shows a bounded system where it is possible to reach a strictly larger marking: $(1, 2, 0, 0)$ may be reached from the initial marking $(1, 1, 0, 0)$; this explain why the classical Karp-Miller procedure [17] does not work for reset nets and reset arcs make boundedness undecidable).

The status of other classical properties of the reachability graphs of bounded choice-free nets is uncertain. For instance, it is not known if the Parikh vectors of small cycles remain prime, nor if they are either equal or disjoint.

4   General Algorithms

A classical technique to perform a Petri net synthesis is to start from a net with transitions only, then to add progressively new places in order to constrain the evolutions to get closer and closer to what is specified by the given LTS. Each added place must allow all the evolutions permitted by the specification, and exclude some forbidden situations that were allowed by the previously added

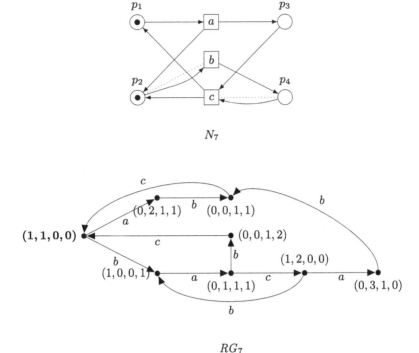

N_7

RG_7

Fig. 7. A bounded choice-free reset net and its reachability graph, where cycles do not have a base of small cycles

places. The process leaves some freeness in the way these new places are added, so that the result is usually not unique, and it may happen that some additions are not possible, meaning there is no solution to the considered synthesis problem.

Let $TS = (S, \rightarrow, T, \iota)$ be a given labelled transition system. The theory of regions [2] characterises the solvability of an LTS through the solvability of a set of *separation problems*. In case the LTS is finite, we have to solve $\frac{1}{2} \cdot |S| \cdot (|S| - 1)$ states separation problems and up to $|S| \cdot |T|$ event/state separation problems, as follows:

Definition 9. SEPARATION PROBLEMS

- A *(Petri net) region* of $(S, \rightarrow, T, \iota)$ is a triple $(\mathbb{M}, \mathbb{B}, \mathbb{F}) \in (S \rightarrow \mathbb{N}, T \rightarrow \mathbb{N}, T \rightarrow \mathbb{N})$ such that for all $s[t\rangle s' \in \rightarrow$, $\mathbb{M}(s) \geq \mathbb{B}(t)$ and $\mathbb{M}(s') = \mathbb{M}(s) - \mathbb{B}(t) + \mathbb{F}(t)$. A region models a place p, in the sense that $\mathbb{B}(t)$ models $W(p, t)$, $\mathbb{F}(t)$ models $W(t, p)$, and $\mathbb{M}(s)$ models the token count of p at the marking corresponding to s (and in particular, $\mathbb{M}(\iota)$ models the initial marking of p).
- A *states separation problem* (SSP for short) consists of a set of states $\{s, s'\}$ with $s \neq s'$, and it can be solved by a region (or place) distinguishing them, i.e., has a different number of tokens in the markings corresponding to the two states: $\mathbb{M}(s) \neq \mathbb{M}(s')$. There are $|S| \cdot (|S| - 1)/2$ such problems.

- An *event/state separation problem* (ESSP for short) consists of a pair $(s, t) \in S \times T$ with $\neg s[t\rangle$. For every such problem, one needs a region (or place) such that $\mathbb{M}(s) < \mathbb{B}(t)$. There are $|S| \cdot |T| - | \to |$ such problems.   □ 9

If the LTS is infinite, also the number of separation problems (of each kind) becomes infinite, but we need to find a finite set of regions solving all of them. Other techniques must then be searched for, instead of considering each separation problem separately, but here we shall restrict our attention to finite LTSs, i.e., to bounded solutions. Then, [10] showed that a Petri net synthesis problem is solvable iff each separation problem has a solution, and a possible solution to the synthesis problem is obtained by gathering all the places corresponding to those separation problem solutions.

For reset net synthesis problems, the situation is similar, but we need to consider reset regions:

Definition 10. RESET REGION

A *reset region* (RPN-region for short) of $(S, \to, T, \iota)$ is a tuple $(\mathbb{M}, \mathbb{R}, \mathbb{B}, \mathbb{F}) \in (S \to \mathbb{N}, 2^T, T \to \mathbb{N}, T \to \mathbb{N})$ such that for all $s[t\rangle s' \in \to$, $\mathbb{M}(s) \geq \mathbb{B}(t)$ and $\mathbb{M}(s') = \mathbb{M}(s) - \mathbb{B}(t) + \mathbb{F}(t)$ if $t \notin \mathbb{R}$, $\mathbb{F}(t)$ otherwise. A region models a place p (see Fig. 8), in the sense that $\mathbb{B}(t)$ models $W(p, t)$, $\mathbb{F}(t)$ models $W(t, p)$, $\mathbb{M}(s)$ models the token count of p at the marking corresponding to s, and $\mathbb{R}$ specifies which transitions belong to $R(p)$.   □ 10

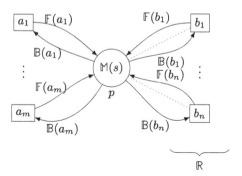

Fig. 8. A general reset region (pictured as a place p). $T = \{a_1, \ldots, a_m, b_1, \ldots, b_n\}$ and $\mathbb{R} = \{b_1, \ldots, b_n\}$. $\mathbb{M}(\iota)$ is the initial marking of p, and more generally $\mathbb{M}(s)$ is the marking of p corresponding to state s.

The proofs of [10] may be immediately adapted so that a finite, totally reachable, forward deterministic LTS admits a RPN-solutions iff each separation problem has a RPN-region solution, and a possible solution to the synthesis problem is obtained by gathering all the places corresponding to those separation problem solutions.

For each separation problem, we thus have to solve a system of $(1 + 2 \cdot | \to |)$ linear constraints: 1 to express the separation problem to consider (i.e., either

$M(s) \neq M(s')$ or $M(s) < B(t)$), $| \rightarrow |$ constraints expressing that a transition is possible, and the same number to express the resulting marking. There are $(|S|+2\cdot|T|)$ variables in $\mathbb{N}$: $|S|$ variables $M(s)$, $|T|$ variables $B(t)$ and $|T|$ variables $F(t)$. But for each case we may have to consider up to $2^{|T|}$ configurations for $\mathbb{R}$, as illustrated by Fig. 8.

For the synthesis of choice-free reset nets, instead of considering $2^{|T|}$ configurations, we only have to consider $2 \cdot |T|$ of them, as illustrated[2] on Fig. 9, each one having $1 + |T| + |S|$ variables. Note however that, for each event/state separation problem (s,t), we only have to consider 2 configurations since b must be t in this case. On the contrary, for a states separation problem (s,s'), b is not prescribed and we may need to consider all the possible configurations (in particular when the problem has no solution).

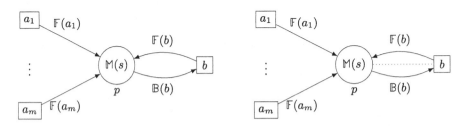

Fig. 9. The two general choice-free reset regions (pictured as a place p). $T = \{a_1,\ldots,a_m,b\}$ and $\mathbb{R}$ is either empty or $\{b\}$. $M(\iota)$ is the initial marking of p, and more generally $M(s)$ is the marking of p corresponding to state s.

For each separation problem and each configuration for $\mathbb{R}(p)$, the system of linear constraint to be solved is polynomial in the size of the LTS to be solved, and it is homogeneous (no independent term), so that instead of searching a solution in the integer domain we may work in the (non-negative) rational one and afterwards apply a multiplicative factor to get integer solutions. We may thus use a polynomial procedure, like the Karmarkar's one [16]. This may no longer be true if we add other constraints; for instance, if we search for a k-bounded solution, we may add $|S|$ constraints $M(s) \leq k$ (for each state $s \in S$): the size of the system remains polynomial, but it is no longer homogeneous, and the problem is NP-complete (see also [3,21]).

5   Acceleration

Since the complexity of a reset net synthesis may be quite high (still worse than for usual Petri net synthesis), it may be beneficial to use a divide and conquer strategy [11,12] and decompose the given LTS as a product or an articulation

---

[2] Note however that, since $B(b)$ may be null, several configurations of the left kind may intersect.

of simpler components: those decompositions were developed in the context of Petri net synthesis, but they apply as well to reset nets.

Concerning the choice-free case, several accelerations have been exploited in [7] for instance, based on the analysis of their reachability graphs. Unfortunately, most of them are no longer valid when we add reset arcs.

For instance, it has been shown that, for choice-free synthesis, states separation problems are irrelevant: if a LTS satisfies the pre-synthesis checks and all the event-state separation problems, then the states separation problems are automatically satisfied too. This is no longer the case for choice-free reset synthesis, as illustrated by Fig. 2: there is no event-state separation problems since there is a single label (a), enabled at each state; however, we need to separate the two states of the corresponding LTS.

Let us first consider the solution of the event-state separation problems for some event $b \in T$. It is not necessary to consider all of them in general, and it is sometimes possible to slightly simplify the systems of linear constraints to be solved.

Indeed, if we consider the two forms of places/regions detailed in Fig. 9, we may see that, if there is a cycle in the given LTS without the label b, all the labels a_i occurring in it must have a weight $\mathbb{F}(a_i) = 0$ (otherwise, the marking of p strictly increases around the cycle). As a consequence, if $\mathbb{C}(b)$ denotes the set of labels occurring in cycles without b, $\forall a \in \mathbb{C}(b) : \mathbb{F}(a) = 0$. Moreover, if $s[a\rangle s'$ for $a \in \mathbb{C}(b)$, we have $\mathbb{M}(s) = \mathbb{M}(s')$. If $\mathbb{C}(b) \neq \emptyset$, both properties reduce the number of variables to find when searching a region of the kinds exhibited in Fig. 9 (and it may happen that none is found, in which case the synthesis has no solution).

Now, if $s[a\rangle s'$ and $s[b\rangle$ with $a \neq b$, from the quasi-persistence (satisfied if the given LTS passed succesfully the presynthesis phase), we also have $s'[b\rangle$, so that we do not have to separate b from s' if we do not have to do it from s. If $a \in \mathbb{C}(b)$ and $s[a\rangle s'$, since $\mathbb{M}(s) = \mathbb{M}(s')$ for any region of the two kinds illustrated in Fig. 9, if $\neg s[b\rangle$ any place separating s from b will also separate s' from b; we may deduce that if $s'[b\rangle$, the synthesis by a choice-free reset net is then impossible (this could then be incorporated in the presynthesis phase). If $a \in \mathbb{C}(b)$, $\neg s[b\rangle$ and $\neg s'[b\rangle$, it is equivalent to separate b from s and to s'. Let us thus define the equivalence relation $s_1 \sim_b s_2$ generated by $\exists a \in \mathbb{C}(b) : s[a\rangle s'$. If $s_1 \sim_b s_2$ and $\neg s_1[b\rangle$, we must also have $\neg s_2[b\rangle$, but we only have to separate a single state from b in each equivalence class where b must be excluded. Moreover, it is not always necessary to separate all such equivalence classes from b.

Indeed, if $s[a\rangle s'$ with $a \in T \setminus (\{b\} \cup \mathbb{C}(b))$ and $\neg s'[b\rangle$, $\mathbb{M}(s) \leq \mathbb{M}(s')$ for any region of the two kinds illustrated in Fig. 9. As a consequence, any place separating b from s' will also separate it from s. Let us thus consider the graph whose nodes are the equivalence classes of $\sim_b$ which do not enable b, with an arc from c_1 to c_2 if there is $s_1[a\rangle s_2$ with $s_1 \in c_1$, $s_2 \in c_2$ and $a \in T \setminus (\{b\} \cup \mathbb{C}(b))$. From the discussion above, we only have to consider the event-state separation problems separating a member of a rightmost class (in this graph) from b.

More easily, when considering an event-state separation problem for b, it is always advisable to first look if some place previously devised for another separating problem for b does not already solve the present problem.

Let us now consider a states separation problem for $s \neq s'$. Again, we may first look if some of the regions constructed before does not already solve it.

From the discussion above, it occurs that if $s \sim_b s'$ for some $b \in T$, we should not search for a separating region with output or reset b, since then $\mathbb{M}(s) = \mathbb{M}(s')$: this reduces the burden of finding an adequate region, if any. And in particular, if $s \sim_b s'$ for any $b \in T$, it is not possible to separate s from s' and the synthesis fails (again, this could be incorporated in the presynthesis phase).

When a given LTS is reversible, it is known [5] that it has a classical choice-free solution iff it has a pure one, which reduces the number of unknowns in each linear system to be solved. We shall now see that this extends immediately when we add reset arcs.

Proposition 3. PREFIXED LTS

If each label occurring in some LTS occurs on a cycle around the initial state, then this LTS has a choice-free reset solution iff it has a pure one. Moreover, in the general schemes illustrated in Fig. 9, we may always assume $\mathbb{F}(b) = 0$.

Proof: First, we may observe that, if $t \in T$ but t does not occur as a label in the LTS, this may be obtained by a (pure) place without input, without (initial) token, and with a unique output transition t. Hence, in the following, without loss of generality we shall assume that each transition labels some arc in the given LTS.

Now, let $b \in T$ and let us consider a reset region $(\mathbb{M}, \mathbb{R}, \mathbb{B}, \mathbb{F})$ of the kind illustrated in Fig. 9, so that either $\mathbb{R} = \emptyset$ (non-resetting region) or $\mathbb{R} = \{b\}$ (region resetting b).

For any arc $s[b\rangle s'$ in the LTS, we may observe that $\mathbb{M}(s') \geq \mathbb{F}(b)$ in the non-resetting case, and $\mathbb{M}(s') = \mathbb{F}(b)$ in the resetting case. Moreover, if $s'[\sigma\rangle s''$ with $\sigma \in (T \setminus \{b\})^*$, $\mathbb{M}(s'') \geq \mathbb{M}(s')$. Since we assumed that, for some $s \in S$, $\exists \sigma \in T^* : s[b\sigma\rangle \iota$, we deduce $\mathbb{M}(\iota) \geq \mathbb{F}(b)$. And since $\forall s \in S \, \exists \sigma \in T^* : \iota[\sigma\rangle s$, we also have $\forall s \in S : \mathbb{M}(s) \geq \mathbb{F}(b)$.

Now, let $k = \min(\mathbb{F}(b), \mathbb{B}(b))$. Let us consider the object obtained from $(\mathbb{M}, \mathbb{R}, \mathbb{B}, \mathbb{F})$ by subtracting k from each $\mathbb{M}(s)$ as well as from $\mathbb{B}(b)$ and from $\mathbb{F}(b)$. It is easy to see from the previous property that this object is still a choice-free reset region, and that if the original region solves a (states or event/state) separation problem, the same is true for the new one. Since in the new region we may not have both $\mathbb{B}(b) > 0$ and $\mathbb{F}(b) > 0$, this leads to a pure solution if we apply this procedure to each region of a choice-free reset solution of the given LTS.

Finally, let us consider a pure non-resetting region for b (left of Fig. 9). If $\mathbb{F}(b) > 0$, we have $\mathbb{B}(b) = 0$, but then we cannot have a cycle with b (the marking would increase indefinitely while following the cycle), which contradicts the hypotheses. For a pure resetting region $(\mathbb{M}, \mathbb{R}, \mathbb{B}, \mathbb{F})$ for b (right of Fig. 9), if

$\mathbb{F}(b) > 0$, we have $\mathbb{B}(b) = 0$ and (from the argument above) $\forall s \in S : \mathbb{M}(s) \geq \mathbb{F}(b)$. Let us then consider the object obtained by subtracting $\mathbb{F}(b)$ from each $\mathbb{M}(s)$ and replacing $\mathbb{F}(b)$ by 0. It is easy to see that this object is still a choice-free reset region for b, and that if the original region solved a states separation problem (it cannot solve an event/state separation problem since $\mathbb{B}(b) = 0$), the same is true for the new one. We thus may assume in any case that $\mathbb{F}(b) = 0$.   □ 3

For instance, any LTS isomorphic to RG_6 in Fig. 6 has a pure choice-free reset solution, for instance N_6. The same is true for RG_7 in Fig. 7, and more generally we have the following corollary:

Corollary 2. PURE SOLUTIONS

If an LTS is reversible, it has a choice-free reset solution iff it has a pure one.

□ 2

6 Net Implementation

Synthesis may be considered as a simple step in an implementation process. From a behavioural specification (for instance in the form of a LTS), it allows to find (if possible) a model of a certain class, which may be considered as a structural specification presenting the adequate behaviour. It then remains to implement this model in a practical device, either hardware or software, or mixed.

In our case, since Petri nets and their extensions are especially devised to describe a distributed application, we may try to obtain a program, where each place corresponds to data structures (giving in particular the number of tokens in the place, but each token may provide other informations that will be used by the absorbing agent) and parallel processes for each transition. Since we consider models with 'black' tokens, the control flow will not rely on the information carried by the tokens, but this information may be exploited by the transition-process when it absorbs the needed tokens and fires. After or during the processing of these informations, the agent will produce some tokens in some places, possibly carrying some information that will be available in the future (but not for the control flow).

In general, a classical problem may occur when the agents check the availability of their needed tokens in a distributed way: it may happen that an agent observes that some input place (or all of them) has the needed tokens, but before the firing takes place and absorbs them, another agent does the same and absorbs the tokens before, disabling the first agent. In order to avoid this, a solution is to lock all accesses to memory when an agent tries to get its tokens, but this is not very distributed. Another solution is that each agent progressively locks all its input places, in some order compatible with the orders used by the other agents (putting all these local orders together must yield a – possibly partial – global order, to avoid deadlocks), but fixing this order is not exactly distributed either. During this locking, the transition may observe if the needed tokens are available, and if it is not the case it will be necessary to unlock all the locked places and retry later (this may induce some starvation phenomenon).

For choice-free nets, the situation is much more sympathetic, since there is no conflict in accessing the input places of each agent. Hence, if a transition observes that there are enough tokens in some place, this may not be changed by other transitions: the latter may only increase the set of tokens in the considered place. Note however that we could have problems if a transition is duplicated in several processes in order to implement some form of auto-concurrency [15]: we shall thus assume that auto-concurrency is not allowed in our systems. It may be necessary to lock accesses to each place however, in order to avoid intermixing absorptions and productions of tokens in the place by different parallel processes (classical problem when performing additions and/or subtractions in parallel, with local copies of the variables), but this may be done in a distributed way. Moreover, if a transition observes that there are not enough tokens in some input place, it is necessary to unlock the place and retry later, but it is never necessary to restart from the beginning. It is even possible to absorb needed tokens when their presence is observed, even if not all of them are there: it is never necessary to give back the absorbed tokens if the transition is blocked at some point, and starvations are equivalent to blockings. The productions and absorptions may be done concurrently in any order, provided the places are protected against simultaneous accesses, since additions and subtractions commute $(+i - j + k = -j + k + i)$. This does not create problems and does not perturb the evolutions of the underlying Petri net (but spurious intermediate markings may be created).

For general reset nets, since this generalises Petri nets, we encounter the same problems, and the same non-distributed solutions.

For choice-free reset nets, we have the same separation of the input places as for usual choice-free Petri nets, hence we have the same fact that checking in a distributed way that the needed tokens are available is not destroyed by the other agents. However, we have a problem with the production phase of each agent, due to the fact that commuting a reset and a production is not innocuous.

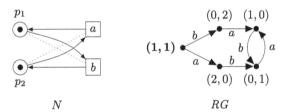

Fig. 10. A simple choice-free reset system and its reachability graph

This is illustrated by the example on Fig. 10. Initially, both a and b are enabled. Let us assume they observe it simultaneously, absorb the needed tokens and proceed as follows: a produces a token in p_1, b resets p_1, b produces a token in p_2 and a resets p_2; we get the marking $(0,0)$ which blocks the system, hence does not allow to still perform infinitely often ab or ba as specified by the reachability graph on the right.

Hence, for the reset and production phase of a transition, we must again either lock all the places, or progressively all its reset-output and output places in some well-defined order. However, in the latter case, when a place is locked by another transition, one only has to wait for its unlocking to proceed: it is never necessary to undo some modification, nor restart the phase, nor wait for an extra delay. When all the needed places have been modified we may unlock them and restart the input-phase.

The structure of each implemented transition could then be sketched as illustrated on Fig. 11. There are variants of this schema however; for instance, the absorption of the input tokens in some place may be performed progressively, without waiting they are all present simultaneously.

> **repeat**
> > **for** each input place p (i.e., such that $W(p,t) > 0$) in any order **do**
> > > end-collect = false
> > > **repeat**
> > > > lock place p
> > > > **if** there are not enough tokens $(M(p) < W(p,t))$
> > > > **then** unlock place p; wait for some time
> > > > **else** collect the needed tokens $(M(p) = M(p) - W(p,t))$;
> > > > > unlock place p; end-collect = true
> > > **until** end-collect
> > process the action of the transition
> > > (possibly using the hidden information of the black tokens)
> > **for** each output or reset place p (i.e., such that $t \in R(p)$ or $W(t,p) > 0$)
> > > in an adequate order **do**
> > > lock place p
> > > **if** p is reset by t (i.e., $t \in R(p)$)
> > > **then** erase the remaining tokens of p, if any (then $M(p) = 0$)
> > > **if** p is an output place for t
> > > **then** produce $W(t,p)$ tokens (possibly with adequate hidden
> > > > information, depending on the ones of the used tokens)
> > **for** each output or reset place p, in any order **do** unlock place p
> **until** we want to stop

Fig. 11. Sketch of a parallel subprogram implementing transition t

7   Concluding Remarks and Future Work

We succeeded in finding how to synthesise, when possible, a finite LTS into a Petri net when we allow reset arcs, either in the general case or in the choice-free case.

We explored how to realise a pre-synthesis phase, but some work has still to be accomplished. In particular, the status of two important properties has to be determined: the primality of small cycles and the disjointness of small cycles with non-identical Parikh vectors.

Region theory has been extended to cope with the addition of reset arcs, and the complexity of the separation problems has been delineated.

Some practical accelerations have been exhibited, but it is likely that some more could be discovered.

Finally, the way to implement a structural specification with reset arcs as a concurrent program has been analysed.

Of course, it should be possible to consider other superclasses of Petri nets (like the ones with inhibitor arcs or transfer arcs or a mixture of those various extensions), as well as other subclasses of those superclasses (similar to marked graphs or free-choice nets for instance) but the region approach assumes that the constraints are only linked to individual places, so the extensions may sometimes be delicate.

Acknowledgements. We want to thank the anonymous referees for their careful reading, as well as Eike Best for his encouragements, and some interesting examples.

References

1. Araki, T., Kasami, T.: Some decision problems related to the reachability problem for Petri nets. Theor. Comput. Sci. **3**(1), 85–104 (1976)
2. Badouel, E., Bernardinello, L., Darondeau, P.: Petri Net Synthesis. TTCSAES. Springer, Heidelberg (2015). https://doi.org/10.1007/978-3-662-47967-4
3. Badouel, E., Bernardinello, L., Darondeau, P.: The synthesis problem for elementary net systems is NP-complete. Theor. Comput. Sci. **186**(1–2), 107–134 (1997). https://doi.org/10.1016/S0304-3975(96)00219-8
4. Best, E., Devillers, R.: Characterisation of the state spaces of live and bounded marked graph Petri nets. In: 8th International Conference on Language and Automata Theory and Applications (LATA 2014), pp. 161–172 (2014). https://doi.org/10.1007/978-3-319-04921-2_13
5. Best, E., Devillers, R.: Synthesis of persistent systems. In: 35th International Conference on Application and Theory of Petri Nets and Concurrency (ICATPN 2014), pp. 111–129 (2014). https://doi.org/10.1007/978-3-319-07734-5_7
6. Best, E., Devillers, R., Erofeev, E., Wimmel, H.: Target-oriented Petri net synthesis. Fundamenta Informaticae **175**, 97–122 (2020). https://doi.org/10.3233/FI-2020-1949
7. Best, E., Devillers, R., Schlachter, U.: Bounded choice-free Petri net synthesis: algorithmic issues. Acta Inf. **55**(7), 575–611 (2018)
8. Best, E., Devillers, R.R., Erofeev, E.: A new property of choice-free Petri net systems. In: Application and Theory of Petri Nets and Concurrency - 41st International Conference, PETRI NETS 2020, Paris, France, 24–25 June 2020, Proceedings, pp. 89–108 (2020). https://doi.org/10.1007/978-3-030-51831-8_5
9. Desel, J., Esparza, J.: Free Choice Petri Nets, Cambridge Tracts in Theoretical Computer Science, vol. 40. Cambridge University Press, New York (1995)

10. Desel, J., Reisig, W.: The synthesis problem of Petri nets. Acta Inf. **33**(4), 297–315 (1996)
11. Devillers, R.: Products of transition systems and additions of petri nets. In: Desel, J., Yakovlev, A. (eds.) Proceedings 16th International Conference on Application of Concurrency to System Design (ACSD 2016), pp. 65–73 (2016). https://doi.org/10.1109/ACSD.2016.10
12. Devillers, R.: Articulation of transition systems and its application to Petri net synthesis. In: Application and Theory of Petri Nets and Concurrency - 40th International Conference, PETRI NETS 2019, Aachen, Germany, 23–28 June 2019, Proceedings, pp. 113–126 (2019)
13. Devillers, R., Hujsa, T.: Analysis and synthesis of weighted marked graph Petri nets. In: Application and Theory of Petri Nets and Concurrency - 39th International Conference, PETRI NETS 2018, Bratislava, Slovakia, 24–29 June 2018, Proceedings, pp. 19–39 (2018)
14. Dufourd, C., Finkel, A., Schnoebelen, P.: Reset nets between decidability and undecidability. In: Automata, Languages and Programming, 25th International Colloquium, ICALP'98, Aalborg, Denmark, 13–17 July 1998, Proceedings, pp. 103–115 (1998). https://doi.org/10.1007/BFb0055044
15. Grabowski, J.: On partial languages. Fundam. Informaticae **4**(2), 427–498 (1981)
16. Karmarkar, N.: A new polynomial-time algorithm for linear programming. Combinatorica **4**(4), 373–396 (1984)
17. Karp, R., Miller, R.: Parallel program schemata. J. Comput. Syst. Sci. **3**(2), 147–195 (1969)
18. Keller, R.M.: A fundamental theorem of asynchronous parallel computation. In: Feng, T. (ed.) Parallel Processing. LNCS, vol. 24, pp. 102–112. Springer, Heidelberg (1975). https://doi.org/10.1007/3-540-07135-0_113
19. Schlachter, U.: Over-approximative Petri net synthesis for restricted subclasses of nets. In: Language and Automata Theory and Applications - 12th International Conference, LATA 2018, Ramat Gan, Israel, 9–11 April 2018, Proceedings, pp. 296–307 (2018). https://doi.org/10.1007/978-3-319-77313-1_23
20. Teruel, E., Colom, J.M., Silva, M.: Choice-free petri nets: a model for deterministic concurrent systems with bulk services and arrivals. IEEE Trans. Syst. Man Cybern. Part A **27**(1), 73–83 (1997). https://doi.org/10.1109/3468.553226
21. Tredup, R.: Hardness results for the synthesis of b-bounded Petri nets. In: Application and Theory of Petri Nets and Concurrency - 40th International Conference, PETRI NETS 2019, Aachen, Germany, 23–28 June 2019, Proceedings, pp. 127–147 (2019). https://doi.org/10.1007/978-3-030-21571-2_9

Synthesis of Petri Nets with Restricted Place-Environments: Classical and Parameterized

Ronny Tredup[✉]

Universität Rostock, Institut Für Informatik, Theoretische Informatik,
Albert-Einstein-Stra ße 22, 18059 Rostock, Germany
ronny.tredup@uni-rostock.de

Abstract. Petri net synthesis consists in deciding for a given transition system A whether there exists a Petri net N whose reachability graph is isomorphic to A. In case of a positive decision, N should be constructed. Several works examined the synthesis of Petri net subclasses that restrict, for every place p of the net, the cardinality of its preset or of its postset or both *in advance* by *small* natural numbers ϱ and κ, respectively, such as, for example, (weighted) marked graphs and (weighted) T-systems and choice-free nets. In this paper, we study the synthesis aiming at Petri nets, which have such restricted place environments, from the viewpoint of classical and parameterized complexity: We first show that, for any fixed natural numbers ϱ and κ, deciding whether for a given transition system A there is a Petri net N such that (1) its reachability graph is isomorphic to A and (2) for every place p of N the preset of p has at most ϱ and the postset of p has at most κ elements is doable in polynomial time. Secondly, we introduce a modified version of the problem, namely ENVIRONMENT RESTRICTED SYNTHESIS (ERS, for short), where ϱ and κ are part of the input and show that ERS is NP-complete. Our methods also imply that ERS parameterized by $\varrho + \kappa$ is $W[2]$-hard.

1 Introduction

Petri net synthesis consists in deciding for a given transition system A whether there is a Petri net N such that the reachability graph of N is isomorphic to A. In the event of a positive decision, N should be constructed. Synthesis of Petri nets has applications in various fields: It is used, for example, to extract concurrency and distributability data from sequential specifications like transition systems or languages [6]. It is applied, for example, in the field of process discovery to reconstruct a model from its execution traces [1] and in supervisory control for discrete event systems [21], and it is used, for example, for the synthesis of speed-independent circuits [14].

The synthesis problem has been originally solved for the class of *Elementary net systems* [20], relying on *regions* of transition systems, and has been found to be NP-complete for this class in [4]. Later on, this solution was extended to

© Springer Nature Switzerland AG 2021
D. Buchs and J. Carmona (Eds.): PETRI NETS 2021, LNCS 12734, pp. 292–311, 2021.
https://doi.org/10.1007/978-3-030-76983-3_15

(pure) Petri nets for which, however, the synthesis problem is solvable in polynomial time [3]. Since then, many studies have been carried out on the synthesis
of structurally restricted subclasses of Petri nets, which aim at improved (pre-)
synthesis methods with regard to the specified subclass. The most investigated
subclasses of Petri nets include those that restrict the cardinality of the presets or the postsets of the places by *a priori* fixed *small* natural numbers ϱ
and κ, respectively. Among them are especially the so-called (weighted) *marked
graphs* [12] (every place has exactly one pre- and exactly one post-transition),
the (weighted) *T-systems* [7] (every place has at most one pre- and at most
one post-transition) and, as a generalization of both, the (weighted) *choice-free*
nets [25,26] (every place has at most one post-transition). These restrictions are
initially motivated by the fact that, from the theoretical point of view, the resulting net classes allow a rich and elegant theory with respect to their structure as
well as highly efficient analysis algorithms [7,16,22,26]. From the perspective of
practical applications, they are particularly useful in, for example, some applications like hardware design [13,14] or as a proper model for systems with bulk
services and arrivals [26]. On the other hand, as already mentioned, these classes
have also been the subject of research aiming at Petri net synthesis for many
years [8–11,18,19]. It turned out that these net classes provide some very useful features like, for example, persistency of their reachability graphs [26] that
–in the sense of complexity issues– allow improved synthesis procedures that –
instead on regions– rather rely on some basic structural properties of the input
transition system. Also the computational complexity of synthesis depending
on the desired subclass has been subject of this research: In [10], for example,
it has been shown that synthesis aiming at choice-free nets is polynomial, and
in [17], for example, it has been proved that synthesis aiming at weighted marked
graphs (or weighted T-systems) is polynomial when the input transition system
is circular.

In this paper, we extend the research on the computational complexity of
synthesis aiming at Petri nets with restricted place environments: We show that,
for any fixed natural numbers ϱ and κ, deciding whether for a given transition
system A there is a Petri net N such that (1) its reachability graph is isomorphic
to A and (2) for every place p of N the preset of p has at most ϱ and the
postset of p has at most κ elements is decidable in polynomial-time. In a natural
way, the question arises whether synthesis remains polynomial if the bounds
ϱ and κ are not fixed in advance, but are part of the input. In this paper,
we answer this question negatively and show that the corresponding decision
problem ENVIRONMENT RESTRICTED SYNTHESIS (ERS) is NP-complete. We
obtain this result by methods that give also information about the parameterized
complexity of ERS parameterized by $\varrho + \kappa$: On the one hand, the proof for
the membership of ERS implies that its parameterized version belongs to the
complexity class XP. On the other hand, the NP-hardness of ERS results from
a polynomial-time reduction of the well-known problem HITTING SET, which is
also a valid parameterized reduction. Since HITTING SET is $W[2]$-complete, this
implies that ERS parameterized by $\varrho+\kappa$ is $W[2]$-hard. Hence, $\varrho+\kappa$ is unsuitable
for FPT-approaches.

Further Related Work. For net classes for which the (underlying) unrestricted synthesis problem is already NP-complete as, for example, it is the case for b-bounded Petri nets [27] or an overwhelming amount of Boolean nets [30] the problem ERS (or its corresponding formulation) is also NP-complete. This can easily be shown by a trivial reduction from the unrestricted to the restricted problem. In [28], it has been shown that ERS, formulated for b-bounded Petri nets, is NP-complete even if $\kappa = 1$. Moreover, in [29], it has been argued that the corresponding problem, although being in XP, is $W[1]$-hard for these nets, when $\varrho + \kappa$ is considered as a parameter. In [31,32], it has been shown that the parametrized complexity of (the Boolean formulation of) ERS is $W[1]$-hard or $W[2]$-hard for a lot of Boolean Petri nets. However, neither of these results imply the ones provided by the current paper.

This paper is organized as follows. Section 2 introduces necessary definitions and provides some examples. After that, Sect. 3 provides the announced complexity results. Finally, Sect. 4 briefly closes the paper.

2   Preliminaries

In this section, we introduce relevant basic notions around Petri net synthesis and provide some examples.

Definition 1. (Transition Systems). *A (deterministic) transition system (TS, for short) $A = (S, E, \delta, \iota)$ is a directed labeled graph with the set of nodes S (called* states*), the set of labels E (called* events*), the partial transition function $\delta : S \times E \longrightarrow S$ and the initial state $\iota \in S$. Event e occurs at state s, denoted by $s \xrightarrow{e}$, if $\delta(s, e)$ is defined. By $s \xrightarrow{e} \!\!\!\!/\,$, we denote that e does not occur at s. We abridge $\delta(s, e) = s'$ by $s \xrightarrow{e} s'$ and call the latter an* edge*. By $s \xrightarrow{e} s' \in A$, we denote that the edge $s \xrightarrow{e} s'$ is present in A. We say A is* loop-free *if $s \xrightarrow{e} s' \in A$ implies $s \neq s'$. A sequence $s_0 \xrightarrow{e_1} s_1, s_1 \xrightarrow{e_2} s_2, \ldots, s_{n-1} \xrightarrow{e_n} s_n$ of edges is called a (directed labeled)* path *(from s_0 to s_n in A). A is called* initialized *if, for every state $s \in S$, we have $s = \iota$ or there is a path from ι to s.*

If a TS A is not explicitly defined, then we refer to its components by $S(A)$ (states), $E(A)$ (events) δ_A (function), ι_A (initial state). In this paper, we investigate whether a TS corresponds to the reachability graph of a Petri net. Since the latter are always initialized, we assume that all TS are initialized without explicitly mentioning this each time. Moreover, we consider TS A and B to be essentially the same when isomorphic:

Definition 2. (Isomorphic TS). *Two TS $A = (S, E, \delta, \iota)$ and $B = (S', E, \delta', \iota')$ with the same set of events are* isomorphic*, denoted by $A \cong B$, if there is a bijection $\varphi : S \to S'$ such that $\varphi(\iota) = \iota'$ and $s \xrightarrow{e} s' \in A$ if and only if $\varphi(s) \xrightarrow{e} \varphi(s') \in B$.*

Starting from a certain behavior that is defined by a transition system, we look for a machine that implements this behavior, namely a Petri net:

Definition 3. (Petri Nets). *A* Petri net $N = (P, T, f, M_0)$ *consists of finite and disjoint sets of* places P *and* transitions T, *a (total)* flow $f : ((P \times T) \cup (T \times P)) \rightarrow \mathbb{N}$ *and an* initial marking $M_0 : P \rightarrow \mathbb{N}$. *The* preset *of a place p is defined by* $^\bullet p = \{t \in T \mid f(t, p) > 0\}$ *and comprises the transitions that produce on p; the* postset *of p is defined by* $p^\bullet = \{t \in T \mid (p, t) > 0\}$ *and contains the transitions that consume from p. Notice that $^\bullet p \cap p^\bullet$ is not necessarily empty. For $\varrho, \kappa \in \mathbb{N}$, we say p is (ϱ, κ)-*restricted *if $|^\bullet p| \le \varrho$ and $|p^\bullet| \le \kappa$. A* transition $t \in T$ *can* fire *or* occur *in a marking $M : P \rightarrow \mathbb{N}$, denoted by $M \xrightarrow{t}$, if $M(p) \ge f(p, t)$ for all places $p \in P$. The firing of t in marking M leads to the marking $M'(p) = M(p) - f(p, t) + f(t, p)$ for all $p \in P$, denoted by $M \xrightarrow{t} M'$. This notation extends to sequences $w \in T^*$ and the* reachability set $RS(N) = \{M \mid \exists w \in T^* : M_0 \xrightarrow{w} M\}$ *contains all of N's reachable markings. The* reachability graph *of N is the TS $A_N = (RS(N), T, \delta, M_0)$, where for every reachable marking M of N and transition $t \in T$ with $M \xrightarrow{t} M'$ the transition function δ of A_N is defined by $\delta(M, t) = M'$.*

According to Definition 3, for every Petri net, there is a TS, that reflects the global behavior of the net, namely its reachability graph. However, not every TS is the behavior of a Petri net and thus the following decision problem arises:

SYNTHESIS
Input: A TS $A = (S, E, \delta, \iota)$.
Question: Does there exist a Petri net N such that $A \cong A_N$?

If SYNTHESIS allows a positive decision, then we want to construct N purely from A. Since A and A_N should be isomorphic, the events E of A become the transitions of N. The places, the flow and the initial marking of N originate from so-called *regions* of the TS A.

Definition 4. (Region). *A* region $R = (sup, con, pro)$ *of a TS $A = (S, E, \delta, \iota)$ consists of the mappings* support $sup : S \rightarrow \mathbb{N}$ *and* consume *and* produce $con, pro : E \rightarrow \mathbb{N}$ *such that if $s \xrightarrow{e} s'$ is an edge of A, then $con(e) \le sup(s)$ and $sup(s') = sup(s) - con(e) + pro(e)$. The preset of R is defined by $^\bullet R = \{e \in E \mid pro(e) > 0\}$ and its postset by $R^\bullet = \{e \in E \mid con(e) > 0\}$. For $\varrho, \kappa \in \mathbb{N}$, we say R is (ϱ, κ)-*restricted *if $|^\bullet R| \le \varrho$ and $|R^\bullet| \le \kappa$.*

Remark 1. Notice that if $R = (sup, con, pro)$ is a region of a TS $A = (S, E, \delta, \iota)$, then R can already be obtained from $sup(\iota)$, con, and pro: Since A is initialized, for every state $s \in S$, there is a path $\iota \xrightarrow{e_1} \dots \xrightarrow{e_n} s_n$ such that $s = s_n$. Hence, we inductively obtain $sup(s_{i+1})$ by $sup(s_{i+1}) = sup(s_i) - con(e_{i+1}) + pro(e_{i+1})$ for all $i \in \{0, \dots, n-1\}$ and $s_0 = \iota$. For brevity, we often use this observation and present regions only *implicitly* by $sup(\iota)$, con and pro. For an even more compact presentation, for $c, p \in \mathbb{N}$, we group events with the same "behavior" together by $\mathcal{T}_{c,p}^R = \{e \in E \mid con(e) = c \text{ and } pro(e) = p\}$.

Regions of the TS become places in a sought net if it exists: for a place $R = (sup, con, pro)$ of such a net, $con(e)$ defines $f(R, e)$, the number of tokens that e consumes from R, and $pro(e)$ defines $f(e, R)$, the number of tokens that e produces on R, and $sup(s)$ models (the number of tokens) $M(R)$ (that are on R) in the marking $\varphi(s) = M$ of N that corresponds to state s of A via the isomorphism φ between A and A_N.

Definition 5. (Synthesized Net). *Every set $\mathcal{R}$ of regions of a TS A defines the synthesized net $N_A^{\mathcal{R}} = (\mathcal{R}, E, f, M_0)$ with $f(R, e) = con(e)$, $f(e, R) = pro(e)$ and $M_0(R) = sup(\iota)$ for all $R = (sup, con, pro) \in \mathcal{R}$ and all $e \in E$.*

In order to ensure that the input behavior A is captured by the synthesized net N, meaning that A and A_N are identified by an isomorphism φ, on the one hand, we have to ensure that distinct states $s \neq s'$ of A correspond to distinct markings $\varphi(s) \neq \varphi(s')$ of N. In particular, we need A to have the *state separation property*, which means that its *state separation atoms* are solvable:

Definition 6. (State Separation). *A pair (s, s') of distinct states of A defines a state separation atom (SSA). A region $R = (sup, con, pro)$ solves (s, s') if $sup(s) \neq sup(s')$. We say a state s is solvable if, for every $s' \in S \setminus \{s\}$, there is a region that solves the SSA (s, s'). If every SSA or, equivalently, every state of A is solvable then A has the state separation property (SSP).*

On the other hand, we have to prevent the firing of a transition in a marking M, if its corresponding event does not occur at the state s of A that corresponds to M via the isomorphism φ, that is, if $s \xrightarrow{e} \not$, then $\xrightarrow{e} \not$, where $\varphi(s) = M$. In particular, A must have the *event/state separation property*, meaning that all *event/state separation atoms* of A are solvable:

Definition 7. (Event/State Separation). *A pair (e, s) of event $e \in E$ and state $s \in S$ such that $s \xrightarrow{e} \not$ defines an event/state separation atom (ESSA). A region $R = (sup, con, pro)$ solves (e, s) if $sup(s) < con(e)$. We say an event e is solvable if, for all $s \in S$ such that $s \xrightarrow{e} \not$, there is a region that solves the ESSA (e, s). If every ESSA or, equivalently, every event of A is solvable then A has the event state separation property (ESSP).*

Definition 8. (Admissible set). *A set $\mathcal{R}$ of regions of A is called admissible if it witnesses the SSP and the ESSP of A, that is, every SSA and ESSA of A is solvable by a region R of $\mathcal{R}$.*

The next lemma, borrowed from [5, p. 162], establishes the connection between the existence of an admissible set $\mathcal{R}$ of A and the existence of a Petri net N whose rechability graph is isomorphic to A. Notice that Petri nets correspond to the type of nets τ_{PT} in [5, p. 130].

Lemma 1. ([5]). *If A is a TS and N a Petri net, then $A \cong A_N$ if and only if there is an admissible set $\mathcal{R}$ of A and $N = N_A^{\mathcal{R}}$.*

By Lemma 1, deciding the existence of a sought net N for A is equivalent to deciding the existence of an admissible set $\mathcal{R}$ of A. Moreover, since the regions $R = (sup, con, pro)$ of $\mathcal{R}$ are places in $N = N_A^{\mathcal{R}}$ and the corresponding flow is defined by con and pro, the places of N are (ϱ, κ)-restricted if and only if every region $R \in \mathcal{R}$ is (ϱ, κ)-restricted. Eventually, this leads us to the following decision problem, which is the main subject of this paper:

ENVIRONMENT RESTRICTED SYNTHESIS

Input: A TS $A = (S, E, \delta, \iota)$ and two natural numbers ϱ and κ.

Question: Does there exist an admissible set $\mathcal{R}$ of A such that every region
$R \in \mathcal{R}$ satisfies $|{}^\bullet R| \le \varrho$ and $|R^\bullet| \le \kappa$?

Example 1. The TS A_1 of Fig. 1 has neither the SSP nor the ESSP: If $R = (sup, con, pro)$ is a region of A_1, then the edge $s_0 \overset{a}{\longrightarrow} s_0$ requires $sup(s_0) = sup(s_0) - con(a) + pro(a)$, implying $con(a) = pro(a)$. The latter implies $sup(s_1) = sup(s_2)$ by $sup(s_2) = sup(s_1) - con(a) + pro(a)$. Moreover, by $s_1 \overset{a}{\longrightarrow}$, we have $sup(s_1) \ge con(a)$ and thus $sup(s_2) \ge con(a)$. Consequently, since R was arbitrary, the SSA (s_1, s_2) and the ESSA (a, s_2) are not solvable.

Example 2. The TS A_2 of Fig. 1 has the ESSP by triviality, since the only event a occurs at all states of A_2, but not the SSP: The SSA (s_0, s_1) is not solvable, since any region $R = (sup, con, pro)$ of A_2 satisfies $sup(s_0) = sup(s_1) - con(a) + pro(a)$ and $sup(s_1) = sup(s_0) - con(a) + pro(a)$, which implies $sup(s_0) = sup(s_1)$.

Example 3. The TS A_3 of Fig. 1 has the ESSP and the SSP: The region $R_1 = (sup_1, con_1, pro_1)$, which, according to Remark 1, is implicitly given

$$A_1 : a \overset{}{\circlearrowleft} s_0 \overset{b}{\longrightarrow} s_1 \overset{a}{\longrightarrow} s_2 \qquad A_2 : s_0 \underset{a}{\overset{a}{\rightleftarrows}} s_1 \qquad A_3 : b \uparrow \begin{array}{ccc} s_2 & \overset{a}{\longrightarrow} & s_3 \\ & & \uparrow b \\ s_0 & \overset{a}{\longrightarrow} & s_1 \end{array}$$

Fig. 1. The TS A_1 (Example 1), A_2 (Example 2) and A_3 (Example 3).

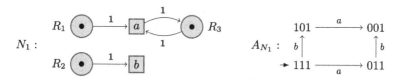

Fig. 2. Left: The Petri Net N_1 with initial marking $M_0(R_1)M_0(R_2)M_0(R_3) = 111$ and $(1, 1)$-restricted places. Right: The reachability graph A_{N_1}.

Fig. 3. Left: The Petri Net N_2 with initial marking $M_0(R_1)M_0(R_2) = 11$ and $(0, 1)$-restricted places. Right: The reachability graph A_{N_2}.

by $sup_1(s_0) = 1$ and $\mathcal{T}_{1,0}^{R_1} = \{a\}$ and $\mathcal{T}_{0,0}^{R_1} = \{b\}$, solves (a, s_1), (a, s_3), (s_0, s_1), (s_0, s_3), (s_2, s_1) and (s_2, s_3). We obtain R_1 explicitly by $sup_1(s_1) = sup_1(s_0) - con_1(a) + pro_1(a) = 0$ and $sup_1(s_2) = sup(s_0) - con_1(b) + pro_1(b) = 1$ and $sup_1(s_3) = sup_1(s_2) - con_1(a) + pro_1(a) = 0$.

Moreover, the region $R_2 = (sup_2, con_2, pro_2)$, which is defined by $sup_2(s_0) = sup_2(s_1) = 1$, $sup_2(s_2) = sup_2(s_3) = 0$ and $\mathcal{T}_{1,0}^{R_2} = \{b\}$ and $\mathcal{T}_{0,0}^{R_2} = \{a\}$, solves the remaining SSA (s_0, s_2) and (s_1, s_3) and ESSA (b, s_2) and (b, s_3) of A.

The TS A has also the region $R_3 = (sup_3, con_3, pro_3)$ defined by $sup_3(s_i) = 1$ for all $i \in \{0, 1, 2, 3\}$ and $\mathcal{T}_{1,1}^{R_3} = \{a\}$ and $\mathcal{T}_{0,0}^{R_3} = \{b\}$.

Since R_1 and R_2 solve all SSA and ESSA both of $\mathcal{R}_1 = \{R_1, R_2, R_3\}$ and $\mathcal{R}_2 = \{R_1, R_2\}$ are admissible sets of A. Figure 2 shows the synthesized net $N_1 = N_{A_3}^{\mathcal{R}_1}$ whose places are $(1, 1)$-restricted, but not $(0, 1)$-restricted, since $|{}^\bullet R_3| = |\{a\}| = 1$. The reachability graph A_{N_1} is sketched on the right hand side of Fig. 2 and it is isomorphic to A_3. The isomorphism φ is given by $\varphi(s_0) = 111$, $\varphi(s_1) = 011$, $\varphi(s_2) = 101$ and $\varphi(s_3) = 001$. However, the input $(A_3, 0, 1)$ for ENVIRONMENT RESTRICTED SYNTHESIS allows a positive decision, because the admissible set $\mathcal{R}_2$ satisfies $|{}^\bullet R| = 0$ and $|R^\bullet| = 1$ for all $R \in \mathcal{R}_2$. Figure 3 shows the synthesized net $N_2 = N_{A_3}^{\mathcal{R}_2}$ (left) and its reachability graph A_{N_2} (right).

3   The Computational Complexity of Environment Restricted Synthesis

The following theorem provides the main contribution of this paper:

Theorem 1. ENVIRONMENT RESTRICTED SYNTHESIS *is NP-complete.*

In order to prove Theorem 1, we have to show that ERS is in NP and that it is NP-hard. The following Sect. 3.1 is dedicated to the membership in NP. The corresponding proof basically extends the deterministic polynomial-time algorithm for the (unrestricted) SYNTHESIS to a non-deterministic algorithm for ERS. The applied methods also show that, for any fixed ϱ and κ, environment restricted synthesis can be done in polynomial-time. After that, Sect. 3.2 deals with the hardness part and provides a reduction of the problem HITTING SET.

3.1   Environment Restricted Synthesis is in NP

In this section, we show that ERS belongs to the complexity class NP. In order to obtain a fitting non-deterministic algorithm for ERS, we extend the deterministic approach for the (unrestricted) SYNTHESIS, which, for example, has been presented in [5]. The tractability of SYNTHESIS bases on the fact that the solvability of a single (state or event/state) separation atom α of a given TS $A = (S, E, \delta, \iota)$ is polynomial-time reducible to a system M of equations and inequalities with rational variables. Such a system can be solved in polynomial-time by Khachiyan's method and theorem [24, pp. 168–170]. Since A has at most $|E| \cdot |S| + |S|^2$ separation atoms, it follows that deciding the solvability of A by deciding the solvability of every atom via the solvability of its corresponding system M, yields a deterministic polynomial-time algorithm for SYNTHESIS. For a separation atom α, our approach extends the aforementioned system M non-deterministically to a system M'. The additional equations encode the restriction requirements: M' is solvable if and only if there is a properly restricted region that solves α. The size of M' is polynomially bounded by the size of A and can be solved with Khachiyan's method. Hence, the membership of ERS in NP then follows, since A has at most $|E| \cdot |S| + |S|^2$ atoms to solve.

In the following, unless explicitly stated otherwise, let (A, ϱ, κ) be an arbitrary but fixed input of ERS with TS $A = (S, \{e_1, \ldots, e_n\}, \delta, \iota)$, and let α be an arbitrary but fixed separation atom of A. In order to develop the announced approach, we first briefly recapitulate the deterministic approach to solve α by a region, which is not necessarily restricted. While we are only informally providing the intended functionality of all equations and inequalities presented, the formal proofs for the corresponding statements can be found in [5]. After that, we introduce the announced extension of this approach.

Recall that, by Remark 1, a region $R = (sup, con, pro)$ is completely defined by $sup(\iota)$ and con and pro. In particular, R can be identified with the element $\mathbf{y} \in \mathbb{Q}^{2n+1}$, defined by $\mathbf{y} = (sup(\iota), con(e_1), \ldots, con(e_n), pro(e_1), \ldots, pro(e_n))$. The reduction of the solvability of α to the solvability of a system M with rational variables can be sketched as follows: On the one hand, if R solves α, then $\mathbf{y}$ solves M. On the other hand, if $\mathbf{x}' \in \mathbb{Q}^{2n+1}$ is a solution of M, then there is a solution $\mathbf{x} = (x_0, x_1, \ldots, x_n, x_{n+1}, \ldots, x_{2n}) \in \mathbb{Z}^{2n+1}$ that implicitly defines a region $R = (sup, con, pro)$ by $sup(\iota) = x_0$ and $con(e_i) = x_i$ and $pro(e_i) = x_{i+n}$ for all $i \in \{1, \ldots, n\}$, which solves α. In particular, $\mathbf{x}$ can be obtained by multiplying $\mathbf{x}'$ by the common denominator of its entries. Hence, M is solvable if and only if α is solvable.

The aforementioned system M essentially consists of two parts: The first part consists of equations and inequalities that ensure that a solution can be interpreted as a region at all. Among others, this part encompasses equations that result from *fundamental cycles*, which are defined by *chords* of a *spanning tree* of A. The second part consists of an inequality that ensures that such a region actually solves α.

A spanning tree of a TS is a sub-TS whose underlying undirected and unlabeled graph is a tree in the common graph-theoretical sense that is rooted at ι:

Definition 9. (Spanning tree, chord). *A spanning tree $A' = (S, E', \delta', \iota)$ of the TS A is a loop-free TS such that, for all $s, s' \in S$ and all $e \in E'$ the following is satisfied: (1) if $s \xrightarrow{e} s' \in A'$, then $s \xrightarrow{e} s' \in A$; (2) either $s = \iota$ or there is exactly one directed labeled path P_s from ι to s in A'. An edge $s \xrightarrow{e} s'$ that is present in A but not in A' is called a* chord *(for A').*

In the following, let A' be an arbitrary but fixed spanning tree of A. By Definition 9, for every state $s \in S \setminus \{\iota\}$, there is exactly one path P_s from ι to s in A'. In order to count, for every event $e \in E$, the number of its occurrences along P_s, we use a *Parikh-vector*:

Definition 10. (Parikh-vector). *Let $s \in S$, and let $P_s = \iota \xrightarrow{e_{i_1}} \dots \xrightarrow{e_{i_m}} s$ be the unique path from ι to s in A' if $s \neq \iota$. The Parikh-vector ψ_s (of s in A') is the mapping $\psi_s : \{e_1, \dots, e_n\} \to \mathbb{N}$ that, for every $e \in \{e_1, \dots, e_n\}$, is defined by:*

$$\psi_s(e) = \begin{cases} |\{\ell \mid e_{i_\ell} = e \text{ and } \ell \in \{1, \dots, m\}\}|, & \text{if } s \neq \iota \\ 0, & \text{otherwise} \end{cases}$$

For convenience we identify $\psi_s = (\psi_s(e_1), \dots, \psi_s(e_n))$.

The chords for A' define so-called *fundamental cycles*:

Definition 11. *Let $t = s \xrightarrow{e} s'$ be a chord for A'. The fundamental cycle ψ_t (of t) is the mapping $\psi_t : \{e_1, \dots, e_n\} \to \mathbb{Z}$ that, for all $i \in \{1, \dots, n\}$, is defined by:*

$$\psi_t(e_i) = \begin{cases} \psi_s(e_i) - \psi_{s'}(e_i), & \text{if } e_i \neq e \\ \psi_s(e_i) + 1 - \psi_{s'}(e_i), & \text{otherwise} \end{cases}$$

For convenience, we identify $\psi_t = (\psi_t(e_1), \dots, \psi_t(e_n))$.

It is well-known that $\mathbb{Q}^n$ is a $\mathbb{Q}$-vector space and, for two elements $x = (x_1, \dots, x_n), y = (y_1, \dots, y_n) \in \mathbb{Q}^n$, we refer to the canonical (scalar) product by $x \cdot y = x_1 \cdot y_1 + \dots + x_n \cdot y_n$.

A region $R = (sup, con, pro)$ of A is completely defined by $sup(\iota)$ and con, pro: If $\iota \xrightarrow{a_1} s_1 \xrightarrow{a_2} \dots \xrightarrow{a_m} s$ is a path from ι to s in A, then $sup(s)$ is defined by

$$sup(s) = sup(\iota) - con(a_1) + pro(a_1) \dots - con(a_m) + pro(a_m) \qquad (1)$$

This path P_s is unqiue in the spanning tree A'. Hence, if we define $\mathbf{z} = (pro(e_1) - con(e_1), \dots, pro(e_n) - con(e_n))$, then Eq. 1 reduces to

$$sup(s) = sup(\iota) + \psi_s \cdot \mathbf{z} \qquad (2)$$

Armed with these notions, we are now able to introduce M. For the sake of simplicity, we first restrict ourselves to the case that α equals an (arbitrary but fixed) ESSA (e_j, q) of A, where $j \in \{1, \dots, n\}$. We will see later that the following arguments also apply to SSA.

For every fundamental cycle ψ_t of A', M has the following equation:

$$\psi_t \cdot (x_{n+1} - x_1, \dots, x_{2n} - x_n) = 0 \qquad (3)$$

Moreover, for all $s \in S$ and $i \in \{1, \dots, n\}$ with $s \xrightarrow{e_i}$ in A (where e_i is not necessarily defined at s in A'), it has the following inequalities:

$$0 \leq \underbrace{x_0}_{sup(\iota)} \qquad (4)$$

$$0 \leq \underbrace{x_0 + \psi_s \cdot (x_{n+1} - x_1, \dots, x_{2n} - x_n)}_{sup(s)} - \underbrace{x_i}_{con(e_i)} \qquad (5)$$

Finally, for the ESSA α, M has following inequality:

$$\underbrace{x_0 + \psi_q \cdot (x_{n+1} - x_1, \ldots, x_{2n} - x_n)}_{sup(q)} - \underbrace{x_j}_{con(e_j)} \le -1 \tag{6}$$

If R solves α, then $\mathbf{y} = (sup(\iota), con(e_1), \ldots, con(e_n), pro(e_1), \ldots, pro(e_n))$ solves M: For Eq. 3, this follows by Proposition 6.16 of [5]. For Eq. 4, Eq. 5 and Eq. 6, this follows by Eq. 2, the definition of regions, implying $sup(\iota) \ge 0$ and $con(e_i) \ge sup(s)$ for every state s of A at which e_i occurs, and the fact that R solves α, implying $sup(q) < con(e_j)$ and thus $sup(q) - con(e_j) \le -1$.

On the other hand, if $\mathbf{x} = (x_0, x_1, \ldots, x_n, x_{n+1}, \ldots, x_{2n})$ is an integer solution of this system, then $(sup(\iota), con(e_1), \ldots, con(e_n), pro(e_1), \ldots, pro(e_n)) = \mathbf{x}$ (implicitly) defines a region of A that solves α [5]: The Eqs. 3, ensure that defining $sup(s)$ for all $s \in S$ according to Eq. 2 yields a support sup that satisfies $sup(s') = sup(s) - con(e) + pro(e)$ for every edge $s \xrightarrow{e} s'$ of A. Equation 4 ensures a valid support value for ι. Equations 5 ensures $sup(s) \ge con(e_i)$ if the event e_i occurs at s in A and, by $sup(\iota) \ge 0$, the equations also ensure $sup(s) \ge 0$ for all $s \in S$. Hence, the result $R = (sup, con, pro)$ is a well-defined region of A. Moreover, Eq. 6 implies that R solves α.

Consequently, an integer vector $\mathbf{x}$ solves the system built by the Equations (1,2,3) if and only if $\mathbf{x} = (sup(\iota), con(e_1), \ldots, con(e_n), pro(e_1), \ldots, pro(e_n))$ for some region $R = (sup, con, pro)$ of A that solves α. As already mentioned above, the system is solvable if and only if it has an integer solution.

Now let's deduce the announced non-deterministic polynomial-time algorithm that decides whether α is solvable by a (ϱ, κ)-restricted region. If R is such a region, that is, $|{}^\bullet R| \le \varrho$ and $|R^\bullet| \le \kappa$, then there are at most ϱ indices $i_1, \ldots, i_\varrho \in \{1, \ldots, n\}$ and at most κ indices $j_1, \ldots, j_\kappa \in \{1, \ldots, n\}$, such that $con(e_{i_\ell}) > 0$ for all $\ell \in \{1, \ldots, \varrho\}$ and $pro(e_{j_k}) > 0$ for all $k \in \{1, \ldots, \kappa\}$, respectively. In particular, for all $i \in \{1, \ldots, n\} \setminus \{i_1, \ldots, i_\varrho\}$ and all $j \in \{1, \ldots, n\} \setminus \{j_1, \ldots, j_\kappa\}$, the vector $\mathbf{y}$ solves the following equations:

$$x_i = 0 \tag{7}$$
$$x_{j+n} = 0 \tag{8}$$

Conversely, let $\mathbf{x}'$ be a rational solution of the system M' that consists of the Eqs. 3 to Eq. 8. Since Eq. 7 and Eq. 8 are homogenous, the vector $\mathbf{x}$, obtained by multiplying $\mathbf{x}'$ with the common denominator of its entries, is a solution of M'. Moreover, by the discussion above, $\mathbf{x} = (sup(\iota), con(e_1), \ldots, con(e_n), pro(e_1), \ldots, pro(e_n))$ (implicitly) defines a region $R = (sup, con, pro)$ of A that solves α. In particular, $con(e_i) = 0$ for all $i \in \{1, \ldots, n\} \setminus \{i_1, \ldots, i_\varrho\}$ and $pro(e_j) = 0$ for all $j \in \{1, \ldots, n\} \setminus \{j_1, \ldots, j_\kappa\}$. Since $k \le \ell$ implies $n - (n-k) \le \ell$ for all $0 \le k, \ell \le n$, we have $|\{e \in E \mid con(e) > 0\}| \le \varrho$ and $|\{e \in E \mid pro(e) > 0\}| \le \kappa$. In particular, R is (ϱ, κ)-restricted.

Obviously, if R exists, then a Turing machine T can guess the indices $i_1, \ldots, i_\varrho$ and $j_1, \ldots, j_\kappa$ in a non-deterministic computation. After that, T can deterministically construct M' and compute an integer solution $\mathbf{x}$ of M'. The construction of M' (after the aforementioned indices are guessed) and the computation of a solution $\mathbf{x}$ of M' as well as the verification that $\mathbf{x}$ actually defines a sought region are doable in polynomial-time. Altogether, we have argued, that the solvability of an ESSA of A by a (ϱ, κ)-restricted region can be decided by a non-deterministic Turing-machine in polynomial-time.

Similar, one shows that this applies also for the case that α is an SSA (p, q). Instead of Eq. 6, the initial system M has then the following Eq. 9:

$$\psi_p \cdot (x_{n+1} - x_1, \ldots, x_{2n} - x_n) - \psi_q \cdot (x_{n+1} - x_1, \ldots, x_{2n} - x_n) \leq -1 \qquad (9)$$

It has been argued in [5] that the resulting system M has a solution if and only if α can be solved. In particular, Eq. 9 implies $sup(q) < sup(p)$. By [5, p. 214], such a region exists if and only if there is a region that implies implies $sup(p) < sup(q)$. Similarly to the former case, we get a non-deterministic procedure that decides if the SSA α is solvable by a restricted region. This implies that ERS belongs to NP, since there are at most $|S|^2 + |E| \cdot |E|$ atoms to solve.

Let $m = |A|$ denote the size of the input TS $A = (S, E, \delta, \iota)$, implying $|E| = n \leq m$. If ϱ and κ are fixed in advance, then there are at most $\binom{m}{\varrho}$ and $\binom{m}{\kappa}$ possibilities to choose $\{e_{i_1}, \ldots, e_{i_\varrho}\}$ and $\{e_{j_1}, \ldots, e_{j_\kappa}\}$, respectively. Hence, in order to decide whether α is solvable by a properly restricted region, we have to check the solvability of at most $\mathcal{O}(m^{\varrho+\kappa})$ systems of equations and inequalities according to the ones discussed above. For every system, its construction and the test of its solvability can be done deterministically in time polynomial in m. Moreover, every solution of a system implies a solving region and there are at most $|S| \cdot |E| + |S|^2$ separation atoms (and this number obviously depends polynomially on m). Finally, by Lemma 1, any admissible set $\mathcal{R}$ implies already a sought net $N = N_A^{\mathcal{R}}$. On the one hand, this implies that there is a constant c and an algorithm that (deterministically) solves ERS in time $\mathcal{O}(m^{\varrho+\kappa+c})$ and, on the other hand, the following corollary is implied:

Corollary 1. *For any fixed natural numbers ϱ and κ, there is a constant c and an algorithm that runs in time $\mathcal{O}(m^c)$ and decides whether for a given transition system A there is a Petri net N such that (1) the reachability graph of N is isomorphic to A and (2) every place p of N satisfies $|{}^\bullet p| \leq \varrho$ and $|p^\bullet| \leq \kappa$ and, in the event of a positive decision, constructs a sought net N.*

3.2   Environment Restricted Synthesis is NP-hard

In order to complete the proof of Theorem 1, it remains to argue that ERS is NP-hard. The proof of the NP-hardness bases on a polynomial-time reduction of the hitting set problem, which is known to be NP-complete from [23]:

HITTING SET (HS)
Input: A triple $(\mathfrak{U}, M, \lambda)$ that consist of a finite set $\mathfrak{U}$ and a set $M = \{M_0, \ldots, M_{m-1}\}$ of subsets of $\mathfrak{U}$ and a natural number λ.
Question: Does there exist a hitting set $\mathfrak{S}$ for $(\mathfrak{U}, M)$, that is, $\mathfrak{S} \subseteq \mathfrak{U}$ and $\mathfrak{S} \cap M_i \neq \emptyset$ for all $i \in \{0, \ldots, m-1\}$, that satisfies $|\mathfrak{S}| \leq \lambda$?

Example 4. The instance $(\mathfrak{U}, M, 3)$ such that $\mathfrak{U} = \{X_0, X_1, X_2, X_3\}$ and $M = \{M_0, \ldots, M_5\}$, where $M_0 = \{X_0, X_1\}$, $M_1 = \{X_0, X_2\}$, $M_2 = \{X_0, X_3\}$, $M_3 = \{X_1, X_2\}$, $M_4 = \{X_1, X_3\}$ and $M_5 = \{X_2, X_3\}$, allows a positive decision: $\mathfrak{S} = \{X_0, X_1, X_2\}$ is a fitting hitting set.

In the remainder of this section, until stated explicitly otherwise, let $(\mathfrak{U}, M, \lambda)$ be an arbitrary but fixed input of HS such that $\mathfrak{U} = \{X_0, \ldots, X_{n-1}\}$ and $M = \{M_0, \ldots, M_{m-1}\}$, where $M_i = \{X_{i_0}, \ldots, X_{i_{m_i-1}}\}$ (and thus $|M_i| = m_i$) for all $i \in \{0, \ldots, m-1\}$. For technical reasons, we assume without loss of generality that $i_0 < \cdots < i_{m_i-1}$ for the elements $X_{i_0}, \ldots, X_{i_{m_i-1}}$ of the set M_i for all $i \in \{0, \ldots, m-1\}$. Moreover, still for technical reasons, we assume that $\lambda \geq 5$. Notice that this is not a restriction of generality, since the hitting set problem is polynomial for every fixed λ [15].

Remark 2. Obviously, the input of Example 4 does not satisfy $\lambda \geq 5$. However, in order to be able to provide a complete example of the reduction despite the space restrictions, this input is deliberately chosen to be small.

The Reduction. In order to prove the hardness part of Theorem 1, we start from input $(\mathfrak{U}, M, \lambda)$ and construct an input (A, ϱ, κ) such that the elements of $\mathfrak{U}$ occur as events in the TS A. Moreover, by construction, the TS A has an ESSA α such that the following implication is true: If $R = (sup, con, pro)$ is a region such that $|{}^\bullet R| \leq \varrho$ and $|R^\bullet| \leq \kappa$ that solves α, then the set $\mathfrak{S} = \{X \in \mathfrak{U} \mid pro(X) > 0\}$ is a sought HS with at most λ elements for $(\mathfrak{U}, M)$. Consequently, if (A, ϱ, κ) allows a positive decision, then there is an admissible set of regions $\mathcal{R}$ whose pre- and postsets are accordingly restricted. In particular, there is a region $R \in \mathcal{R}$ that solves α and thus proves that $(\mathfrak{U}, M, \lambda)$ also allows a positive decision. Conversely, we argue that if $(\mathfrak{U}, M, \lambda)$ has a fitting hitting set, then there is an admissible set $\mathcal{R}$ of A such that $|{}^\bullet R| \leq \varrho$ and $|R^\bullet| \leq \kappa$ for all $R \in \mathcal{R}$. Altogether, this approach proves that $(\mathfrak{U}, M, \lambda)$ is a yes-instance if and only if (A, ϱ, κ) is a yes-instance.

First of all, we define $\varrho = 2\lambda$ and $\kappa = \lambda+1$. Notice that this implies $\varrho \geq 10$ and $\kappa \geq 6$, since we assume $\lambda \geq 5$. However, that does not restrict generality, since ERS is polynomial for any fixed integers. Figure 4 provides a complete example for the following construction, which is based on the input of Example 4. The TS A has, for every $i \in \{0, \ldots, m-1\}$, the following gadget T_i that represents the set $M_i = \{X_{i_0}, \ldots, X_{i_{m_i-1}}\}$ by using its elements as events:

$$T_i = \quad t_{i,0} \xrightarrow{\ k\ } t_{i,1} \xrightarrow{\ X_{i_0}\ } \cdots \xrightarrow{\ X_{i_{m_i-1}}\ } t_{i,m_i+1} \xrightarrow{\ k\ } t_{i,m_i+2}$$

In particular, T_0 provides $\alpha = (k, t_{0,1})$. Additionally, the states $t_{i,1}$ and $t_{i+1,1}$ are connected by an u_{i+1}-labeled edge $t_{i,1} \xrightarrow{u_{i+1}} t_{i+1,1}$ for all $i \in \{0, \ldots, m-2\}$. Moreover, for every $i \in \{0, \ldots, \lambda-1\}$, the TS A has the next gadget F_i and, for every $j \in \{1, \ldots, m-1\}$, it has the next gadget G_j that uses the event u_j again:

$$F_i = f_{i,0} \xleftarrow{\ \ z_i\ \ } f_{i,1}$$
$$\xrightarrow[\ \ k_i\ \ ]{\ \ k\ \ }$$

$$G_j = g_{j,0} \xleftarrow{\ \ u_j\ \ } g_{j,1}$$
$$\xrightarrow[\ \ v_j\ \ ]{}$$

The functional part of A is given by the introduced gadgets. The initial state of A is $\perp_0$. In order to connect the gadgets, we use the following edges that introduce fresh events and states: For every $i \in \{0, \ldots, m-1\}$, the TS A has the edges $\perp_i \xrightarrow{y_i} t_{i,0}$ and, if $i < m-1$, the edge $\perp_i \xrightarrow{w_{i+1}} \perp_{i+1}$; the TS A has the edge $\perp_0 \xrightarrow{a_0} \top_0$ and, for every $i \in \{0, \ldots, \lambda-1\}$, it has the edge $\top_i \xrightarrow{b_i} f_{i,0}$ and, if $i < \lambda-1$, it has the edge $\top_i \xrightarrow{a_{i+1}} \top_{i+1}$; the TS A has the edge $\perp_0 \xrightarrow{c_1} \triangle_1$ and, for every $i \in \{1, \ldots, m-1\}$, it has the edge $\triangle_i \xrightarrow{d_i} g_{i,0}$ and if $i < m-1$, then it has the edge $\triangle_i \xrightarrow{c_{i+1}} \triangle_{i+1}$. By S and E we refer to the (set of) states and (set of) events of A, respectively.

Lemma 2. *If there is an admissible set of (ϱ, κ)-restricted regions for A, then there is a hitting set with at most λ elements for $(\mathfrak{U}, M)$.*

Proof. Let $\mathcal{R}$ be an admissible set of A that satisfies $|{}^\bullet R| \leq \varrho$ and $|R^\bullet| \leq \kappa$ for all $R \in \mathcal{R}$. Since $\mathcal{R}$ is admissible, there is an accordingly restricted region that solves the atom $\alpha = (k, t_{0,1})$. Let $R = (sup, con, pro)$ be such a region, that is $con(k) > sup(t_{0,1})$ and $|{}^\bullet R| \leq 2\lambda$ and $|R^\bullet| \leq \lambda+1$. In the following, we argue that $\mathfrak{S} = \{X \in \mathfrak{U} \mid pro(X) > 0\}$ defines a sought hitting set for $(\mathfrak{U}, M)$.

Since k occurs at $t_{0,0}$ and R solves α, the following is true: (1) $con(k) \leq sup(t_{0,0})$ and (2) $sup(t_{0,1}) = sup(t_{0,0}) - con(k) + pro(k)$ and (3) $con(k) > sup(t_{0,1})$. By combining (1) and (2), we obtain $sup(t_{0,1}) \geq pro(k)$. If we combine $sup(t_{0,1}) \geq pro(k)$ with (3), then we get $con(k) > pro(k)$. For all $i \in \{0, \ldots, \lambda-1\}$, by $con(k) > pro(k)$ and $f_{i,0} \xrightarrow{k} f_{i,1}$, we conclude $sup(f_{i,0}) > sup(f_{i,1})$, since $sup(f_{i,1}) = sup(f_{i,0}) - con(k) + pro(k)$. This implies $con(k_i) > pro(k_i)$ as well as $con(z_i) < pro(z_i)$ and thus $k_i \in R^\bullet$ and $z_i \in {}^\bullet R$ for all $i \in \{0, \ldots, \lambda-1\}$. Since $k \in R^\bullet$, this implies already $|R^\bullet| = \lambda+1$. In particular, no further event can be a member of $R^\bullet$. Let $i \in \{1, \ldots, m-1\}$ be arbitrary but fixed. If $pro(u_i) > con(u_i)$, then we obtain $sup(g_{i,0}) > sup(g_{i,1})$ and thus $con(v_i) > pro(v_i)$ by $g_{i,1} \xrightarrow{u_i} g_{i,0}$ and $g_{i,0} \xrightarrow{v_i} g_{i,1}$. This would imply $v_i \in R^\bullet$ and $|R^\bullet| \geq \lambda+2$, a contradiction. Hence, we have $pro(u_i) \leq con(u_i)$. By $t_{i-1,1} \xrightarrow{u_i} t_{i,1}$, this implies $sup(t_{i-1,1}) \geq sup(t_{i,1})$. Since i was arbitrary, we get $sup(t_{0,1}) \geq sup(t_{1,1}) \geq \cdots \geq sup(t_{m-1,1})$. By $con(k) > sup(t_{0,1})$, this implies $con(k) > sup(t_{i,1})$ for all $i \in \{0, \ldots, m-1\}$. On the other hand, since $t_{i,m_i+1} \xrightarrow{k}$, we have $con(k) \leq sup(t_{i,m_i+1})$ for all $i \in \{0, \ldots, m-1\}$. Consequently, there has to be at least one event $X \in \{X_{i_0}, \ldots, X_{i_{m_i-1}}\}$ such that $pro(X) > 0$. This implies $\mathfrak{S} \cap M_i \neq \emptyset$ for all $i \in \{0, \ldots, m-1\}$, where $\mathfrak{S} = \{X \in \mathfrak{U} \mid pro(X) > 0\}$. Moreover, since $z_i \in {}^\bullet R$ for all $i \in \{0, \ldots, \lambda-1\}$ and $\mathfrak{S} \subseteq {}^\bullet R$ and $|{}^\bullet R| \leq 2\lambda$, we have that $|\mathfrak{S}| \leq \lambda$. This proves the claim and thus the lemma. $\square$

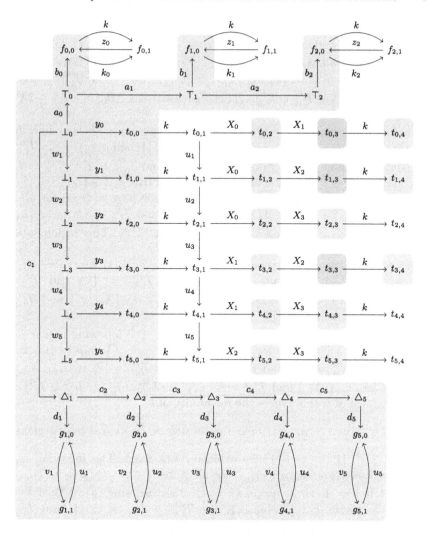

Fig. 4. The TS A with initial state $\perp_0$ that results from Example 4. Based on the 3-HS $\{X_0, X_1, X_2\}$ for $(\mathfrak{U}, M)$, the colored area sketches the region R_1 of Fact 1 that solves $(k, t_{0,1})$: the support of the states in the green colored area equals 1, states in the blue colored area have support 2, and the others have 0.

In order to show that all ESSA are solvable by (ϱ, κ)-restricted regions, provided there is a fitting hitting set for $(\mathfrak{U}, M)$, we treat the events of A individually. Recall that an event is solvable if all of its corresponding ESSA are solvable (Definition 7). Moreover, for a region $R = (sup, con, pro)$ of A, the set $\mathcal{T}_{c,p}^R$ summarizes the events $e \in E$ such that $(con(e), pro(e)) = (c, p)$ (Remark 1).

Fact 1. *If there is a hitting set with at most λ elements for $(\mathfrak{U}, M)$, then the event k is solvable by (ϱ, κ)-restricted regions.*

Proof. Let $\mathfrak{S}$ be a hitting set with at most λ elements for $(\mathfrak{U}, M)$.

The following region $R_0 = (sup_0, con_0, pro_0)$ solves (k, s) for all $s \in \bigcup_{i=1}^{m-1} S(G_i)$ and all $s \in \{\top_0, \ldots, \top_{\lambda-1}\}$ and all $s \in \{\triangle_1, \ldots, \triangle_{m-1}\}$: $sup_0(\bot_0) = 1$ and $\mathcal{T}_{1,1}^{R_0} = \{k\}$ and $\mathcal{T}_{0,1}^{R_0} = \{b_0, \ldots, b_{\lambda-1}\}$ and $\mathcal{T}_{1,0}^{R_0} = \{a_0, c_1\}$ and $\mathcal{T}_{0,0}^{R_0} = E \setminus (\mathcal{T}_{1,1}^{R_1} \cup \mathcal{T}_{0,1}^{R_1} \cup \mathcal{T}_{1,0}^{R_1})$. This region satisfies $|{}^\bullet R_0| = \lambda + 1 \leq 2\lambda$ and $|R_0^\bullet| = 3$.

The following region $R_1 = (sup_1, con_1, pro_1)$ solves $\alpha = (k, t_{0,1})$ and, moreover, (k, s) for all $s \in \{f_{i,1} \mid i \in \{0, \ldots, \lambda - 1\}\}$: $sup_1(\bot_0) = 1$; $\mathcal{T}_{1,0}^{R_1} = \{k, k_0, \ldots, k_{\lambda-1}\}$ and $\mathcal{T}_{0,1}^{R_1} = \mathfrak{S} \cup \{z_0, \ldots, z_{\lambda-1}\}$ and $\mathcal{T}_{0,0}^{R_1} = E \setminus (\mathcal{T}_{1,0}^{R_1} \cup \mathcal{T}_{0,1}^{R_1})$. This region satisfies $|{}^\bullet R_1| \leq 2\lambda$, since $|\mathfrak{S}| \leq \lambda$, and $|R_1^\bullet| = \lambda + 1$.

The following region $R_2 = (sup_2, con_2, pro_2)$ solves $\alpha = (k, s)$ for all $s \in \{t_{i,1}, t_{i,m_i+2} \mid i \in \{0, \ldots, m - 1\}\}$: $sup_2(\bot_0) = 2$ and $\mathcal{T}_{1,0}^{R_2} = \{k, k_0, \ldots, k_{\lambda-1}\}$ and $\mathcal{T}_{0,1}^{R_2} = \{z_0, \ldots, z_{\lambda-1}\}$ and $\mathcal{T}_{0,0}^{R_2} = E \setminus (\mathcal{T}_{1,0}^{R_2} \cup \mathcal{T}_{0,1}^{R_2})$. This region satisfies $|{}^\bullet R_1| = \lambda \leq 2\lambda$ and $|R_1^\bullet| = \lambda + 1$.

The next region $R_3 = (sup_3, con_3, pro_3)$ solves (k, s) for $s = \bot_0$ and all $s \in \{t_{0,2}, \ldots, t_{0,m_0}\}$: $sup_3(\bot_0) = 0$ and $\mathcal{T}_{1,1}^{R_3} = \{k\}$ and $\mathcal{T}_{0,1}^{R_3} = \{y_0, u_1, X_{0_{m_0-1}}, a_0, c_1\}\}$ and $\mathcal{T}_{0,2}^{R_3} = \{w_1\}$ and $\mathcal{T}_{1,0}^{R_3} = \{X_{0_0}, v_1\}$ and $\mathcal{T}_{0,0}^{R_3} = E \setminus (\mathcal{T}_{1,1}^{R_3} \cup \mathcal{T}_{0,1}^{R_3} \cup \mathcal{T}_{0,2}^{R_3} \cup \mathcal{T}_{1,0}^{R_3})$.

Let $i \in \{1, \ldots, m - 1\}$ be arbitrary but fixed. The following region $R_4 = (sup_4, con_4, pro_4)$ solves (k, s) for all $s \in \{\bot_i, t_{i,2}, \ldots, t_{i,m_i}\}$: $sup_4(\bot_0) = 2$ and $\mathcal{T}_{1,1}^{R_4} = \{k\}$ and $\mathcal{T}_{2,0}^{R_4} = \{w_i\}$ and $\mathcal{T}_{0,2}^{R_4} = \{w_{i+1}\}$ and $\mathcal{T}_{0,1}^{R_4} = \{y_i, v_i, u_{i+1}, X_{i_{m_i-1}}\}$ and $\mathcal{T}_{1,0}^{R_4} = \{u_i, v_{i+1}, X_{i_0}\}$ and $\mathcal{T}_{0,0}^{R_4} = E \setminus (\mathcal{T}_{1,1}^{R_4} \cup \mathcal{T}_{2,0}^{R_4} \cup \mathcal{T}_{0,2}^{R_4} \cup \mathcal{T}_{0,1}^{R_4} \cup \mathcal{T}_{1,0}^{R_4})$. By the arbitrariness of i, this proves the solvability of k. $\qquad\square$

Fact 2. *If $e \in \{u_1, \ldots, u_{m-1}\}$, then e is solvable by (ϱ, κ)-restricted regions.*

Proof. Let $i \in \{1, \ldots, m - 1\}$ be arbitrary but fixed. The following region $R_5 = (sup_5, con_5, pro_5)$ solves (u_i, s) for all states $s \in S \setminus (S(T_{i-1}) \cup \{g_{i-1,0}\})$ with $s \xrightarrow{u_i} \not$: If $i = 1$, then $sup_5(\bot_0) = 1$, otherwise $sup(\bot_0) = 0$; if $i = 1$, then $\mathcal{T}_{1,0}^{R_5} = \{u_i, w_i, v_{i-1}\} \cup \{a_0, c_1\}$, else $\mathcal{T}_{1,0}^{R_5} = \{u_i, w_i, v_{i-1}\}$; and $\mathcal{T}_{0,1}^{R_5} = \{w_{i-1}, u_i, d_{i-1}, v_i\}$ and $\mathcal{T}_{0,0}^{R_5} = E \setminus (\mathcal{T}_{1,0}^{R_5} \cup \mathcal{T}_{0,1}^{R_5})$.

Region $R_6 = (sup_6, con_6, pro_6)$ solves (u_i, s) for all $s \in \{\bot_{i-1}, t_{i-1,0}\}$ and if $i \geq 2$, then for $s \in \{g_{i-1,0}\}$: $sup_6(\bot_0) = 0$ and $\mathcal{T}_{1,1}^{R_6} = \{u_i\}$ and $\mathcal{T}_{0,1}^{R_6} = \{d_i, k, k_0, \ldots, k_{\lambda-1}\}$ and $\mathcal{T}_{1,0}^{R_6} = \{z_0, \ldots, z_{\lambda-1}\}$ and $\mathcal{T}_{0,0}^{R_6} = E \setminus (\mathcal{T}_{1,1}^{R_6} \cup \mathcal{T}_{0,1}^{R_6} \cup \mathcal{T}_{1,0}^{R_6})$. Notice that $|{}^\bullet R_6| = \lambda + 3 \leq 2\lambda$, since $\lambda \geq 6$, and $|R_6^\bullet| = \lambda + 1$.

Finally, the following region $R_7 = (sup_7, con_7, pro_7)$ solves (u_i, s) for all $s \in \{t_{i-1,2}, \ldots, t_{i-1,m_i+2}\}$: $sup_7(\bot_0) = 1$ and $\mathcal{T}_{1,1}^{R_7} = \{u_i\}$ and $\mathcal{T}_{1,0}^{R_7} = \{X_{i_0}\}$ and $\mathcal{T}_{0,0}^{R_7} = E \setminus (\mathcal{T}_{1,1}^{R_7} \cup \mathcal{T}_{1,0}^{R_7})$. This completes solving u_i and, by the arbitrariness of i, this proves the solvability for all $e \in \{u_1, \ldots, u_{m-1}\}$. $\qquad\square$

Fact 3. *If $e \in \{X_0, \ldots, X_{n-1}\}$, then e is solvable by (ϱ, κ)-restricted regions.*

Proof. Let $i \in \{0, \ldots, n - 1\}$ be arbitrary but fixed. Moreover, let $j, \ell \in \{0, \ldots, m - 1\}$ be arbitrary but fixed such that $X_i \notin M_j$ and $X_i \in M_\ell$.

The next region $R_8 = (sup_8, con_8, pro_8)$ solves (X_i, s) for $s = \perp_j$ and all $s \in \{t_{j,0}, \ldots, t_{j,m_j+2}\}$: If $j = 0$, then $sup_8(\perp_0) = 0$, otherwise $sup(\perp_0) = 1$; $\mathcal{T}_{1,1}^{R_8} = \{X_i\}$ and if $j > 0$, then $\mathcal{T}_{1,0}^{R_8} = \{w_j, u_j, v_{j+1}\} \cup \{a_0, c_1\}$, else $\mathcal{T}_{1,0}^{R_8} = \{w_j, u_j, v_{j+1}\}$; $\mathcal{T}_{0,1}^{R_8} = \{v_j, w_{j+1}, u_{j+1}, d_{j+1}\}$ and $\mathcal{T}_{0,0}^{R_8} = E \setminus (\mathcal{T}_{1,1}^{R_8} \cup \mathcal{T}_{0,1}^{R_8})$.

The next region $R_9 = (sup_9, con_9, pro_9)$ solves (X_i, s) for all $s \in \{\perp_\ell, t_{\ell,0}\}$: $sup_9(\perp_0) = 0$; $\mathcal{T}_{1,1}^{R_9} = \{X_i\}$ and $\mathcal{T}_{1,0}^{R_9} = \{z_0, \ldots, z_{\lambda-1}\}$ and $\mathcal{T}_{0,1}^{R_9} = \{k, k_0, \ldots, k_{\lambda-1}\}$ and $\mathcal{T}_{0,1}^{R_9} = E \setminus (\mathcal{T}_{1,1}^{R_9} \cup \mathcal{T}_{1,0}^{R_9} \cup \mathcal{T}_{0,1}^{R_9})$. Notice that $|R_9^\bullet| = \lambda + 1$ and $|^\bullet R_9| = \lambda + 1 \leq 2\lambda$.

Let $h \in \{0, \ldots, m_\ell - 1\}$ be the unique index such that $X_i = X_{\ell_h}$, that is, X_i is the "h-th element" of M_ℓ. The following region $R_{10} = (sup_{10}, con_{10}, pro_{10})$ solves (X_i, s) for all $s \in \{t_{\ell,h+1}, t_{\ell,m_\ell+2}\}$: $sup_{10}(\perp_0) = 1$ and $\mathcal{T}_{1,0}^{R_{10}} = \{X_i, a_0, c_1\}$ and $\mathcal{T}_{0,0}^{R_{10}} = E \setminus \mathcal{T}_{1,0}^{R_{10}}$.

It remains to discuss the case $X_i \neq X_{\ell_0}$, that is $h \geq 1$, which requires to solve (X_i, s) for all $s \in \{t_{\ell,1}, \ldots, t_{\ell,h}\}$. So let $s \in \{t_{\ell,1}, \ldots, t_{\ell,h}\}$ be arbitrary but fixed.

We distinguish between $\ell = 0$ and $\ell \geq 1$: If $\ell = 0$, then the following region $R_{11} = (sup_{11}, con_{11}, pro_{11})$ solves (X_i, s): $sup_{11}(\perp_0) = 0$ and $\mathcal{T}_{1,0}^{R_{11}} = \{X_i, v_1\}$ and $\mathcal{T}_{0,1}^{R_{11}} = \{w_1, u_1, d_1, X_{0_{h-1}}\}$ and $\mathcal{T}_{0,0}^{R_{11}} = E \setminus (\mathcal{T}_{1,0}^{R_{11}} \cup \mathcal{T}_{0,1}^{R_{11}})$.

If $\ell \geq 1$, then region $R_{12} = (sup_{12}, con_{12}, pro_{12})$ solves (X_i, s): $sup_{12}(\perp_0) = 1$ and $\mathcal{T}_{1,0}^{R_{12}} = \{X_i, w_\ell, u_\ell, v_{\ell+1}, a_0, c_1\}$ and $\mathcal{T}_{0,1}^{R_{12}} = \{X_{i_{h-1}}, w_{\ell+1}, u_{\ell+1}, v_\ell, d_{\ell+1}\}$ and $\mathcal{T}_{0,0}^{R_{12}} = E \setminus (\mathcal{T}_{1,0}^{R_{12}} \cup \mathcal{T}_{0,1}^{R_{12}})$. By the arbitrariness of h, this completes the solvability of (X_i, s) for all $s \in S(T_\ell)$.

The following region $R_{13} = (sup_{13}, con_{13}, pro_{13})$ solves (X_i, s) for all $s \in S \setminus (\bigcup_{j=0}^{m-1}(S(T_j) \cup \{\perp_j\}))$: $sup_{13}(\perp_0) = 1$ and $\mathcal{T}_{1,1}^{R_{13}} = \{X_i\}$ and $\mathcal{T}_{1,0}^{R_{13}} = \{a_0, c_1\}$ and $\mathcal{T}_{0,0}^{R_{13}} = E \setminus (\mathcal{T}_{1,1}^{R_{13}} \cup \mathcal{T}_{1,0}^{R_{13}})$. Since i, j, ℓ were arbitrary, we have the claim.$\square$

Fact 4. *If $e \in \{k_0, \ldots, k_{\lambda-1}\}$ or $e \in \{z_0, \ldots, z_{\lambda-1}\}$ or $e \in \{v_1, \ldots, v_{m-1}\}$ or $e \in \{w_1, \ldots, w_{m-1}\}$ or $e \in \{a_0, \ldots, a_{\lambda-1}\}$ or $e \in \{b_0, \ldots, b_{\lambda-1}\}$ or $e \in \{y_0, y_i, w_i, c_i, d_i \mid 1 \leq i \leq m-1\}$, then e is solvable by (ϱ, κ)-restricted regions.*

Proof. Let $i \in \{0, \ldots, \lambda - 1\}$ be arbitrary but fixed. The region R_1 of Fact 1 solves $(k_i, f_{i,1})$ and the region R_6 of Fact 2 solves $(z_i, f_{i,0})$. It is easy to see that (k_i, s) and (z_i, s) are suitably solvable for the remaining $s \in S \setminus \{f_{i,0}, f_{i,1}\}$. Since i was arbitrary, that proves the claim for all $e \in \{k_0, \ldots, k_{\lambda-1}, z_0, \ldots, z_{\lambda-1}\}$.

Let $i \in \{1, \ldots, m-1\}$ be arbitrary but fixed. The region R_6 or the region R_8 solves $(v_i, g_{i,1})$. It is easy to see that (v_i, s) is suitably solvable for all $s \in S \setminus \{g_{i,0}, g_{i,1}\}$. By their uniqueness, it is easy to see that the remaining events are also solvable by suitably restricted regions. The claim follows. $\square$

Altogether, the just presented facts prove that all ESSA of A are solvable by (ϱ, κ)-restricted regions, if there is a hitting set of size at most λ for $(\mathfrak{U}, M)$. Moreover, if (s, s') is an SSA of A, then either (s, s') is already solved by one of the presented regions or it is easy to see that a solving (ϱ, κ)-restricted region exists. Hence, we obtain the following lemma, which completes the proof of Theorem 1.

Lemma 3. *If there is a hitting set with at most λ elements for $(\mathfrak{U}, M)$, then A has an admissible set $\mathcal{R}$ of (ϱ, κ)-restricted regions.*

3.3   A Lower Bound for the Parameterized Complexity of ERS

By Theorem 1, the problem ERS is NP-complete. Hence, from the point of view of classical complexity theory, where we assume that P is different from NP, the problem is considered intractable, i.e., the worst-case time-complexity of any deterministic decision algorithm is above polynomial. However, measuring the complexity of the problem purely in the size of the input may let it appear harder than it actually is.

In *parameterized complexity* we deal with *parameterized problems*, where every input (x, k) has a distinguished part k, a natural number, that is called the parameter, and measure the complexity not only in terms of the input size n, but also in terms of the parameter k. For example, a natural parameter of ERS is $k = \varrho + \kappa$. By the results of Sect. 3.1, there is an algorithm that solves ERS in time $\mathcal{O}(n^{k+c})$, where c is a constant. In terms of parameterized complexity, this means that ERS parameterized by k belongs to the complexity class XP (for *slice-wise polynomial*). However, such algorithms are not considered as feasible, since n^{k+c} can be huge even for small k and moderate n. Hence, we are rather interested in algorithms where k does not appear in the exponent of n: We say a parameterized problem is *fixed parameter tractable* if has an algorithm with running time $\mathcal{O}(f(k)n^c)$, where f is a computable function that depends only on k, and c is a constant. Such algorithms are manageable even for large values of n, provided that $f(k)$ is relatively small and c is a small constant.

In order to obtain a successful parameterization, we need to have some reason to believe that the parameter is typically small, such that $f(k)$ can be expected to remain relatively small, too. In the absence of a benchmark that is specifically created for Petri net synthesis, we have analyzed the benchmark of the Model Checking Contest (MCC) [2], which contains both academic and industrial Petri nets. Their corresponding TS (reachability graphs) have usually (way) more than 10^7 states, which provides a lower bound for the length $n \geq |S| + |E|$ of an input TS $A = (S, E, \delta, \iota)$. We analyzed 878 Petri nets in total. For 395 of them (around 45%), we found that $\varrho + \kappa \leq 21$, and for still 308 of them (around 35%) we even found $\varrho + \kappa \leq 11$ such as, for example, `AutoFlight-PT-02a` and `CircadianClock-PT-100000` and `FMS-PT-50000`. In other words: in dependence on f, a synthesis algorithm with running time $f(k)n^c$ could possibly be useful for a third up to almost half of these nets. From this point of view, the parameterization of ERS by $\varrho + \kappa$ and the search for a fixed-parameter algorithm appear to be sensible. Unfortunately, we can provide strong evidence that such an algorithm does not exists. This is of practical relevance, since it prevents an algorithm designer to waste countless hours with the attempt to find a solution that most likely cannot be found.

From the viewpoint of classical complexity theory, a problem is considered as intractable if it is NP-hard. Analogously, in parameterized complexity, a problem is assumed not to be fixed-parameter-tractable if it is $W[i]$-hard for some $i \geq 1$.

We omit the formal definition of the complexity class $W[i]$ and rather refer to [15]. In order to show that a parameterized problem Q is $W[i]$-hard, we have to present a parameterized reduction from a known $W[i]$-hard problem P to Q. A *parameterized reduction* is an algorithm that transforms an instance (x, k) of P into an instance (x', k') of Q such that

1. (x, k) is in P if and only if (x', k') is in Q and
2. $k' \leq g(k)$ for some computable function independent of x, and
3. the running time is $f(k)|x|^c$ for some computable function f and constant c.

The problem HITTING SET parameterized by λ is known to be $W[2]$-hard (even $W[2]$-complete). Moreover, the reduction presented in Sect. 3.2 is a parameterized one, since $\varrho + \kappa = 3\lambda + 1$. This proves the following theorem, implying the fixed-parameter-intractability of ERS parameterized by $\varrho + \kappa$:

Theorem 2. ERS *parameterized by* $\varrho + \kappa$ *is* $W[2]$-*hard*.

4 Conclusion

In this paper, we investigated the computational complexity of synthesizing Petri nets for which the cardinality of the pre- and postset of their places is restricted by natural numbers ϱ and κ. We show that the problem is solvable in polynomial time for any fixed ϱ and κ. By way of contrast, if ϱ and κ are part of the input, then the resulting problem ERS is NP-complete. Moreover, we show that ERS parameterized by $\varrho + \kappa$ is $W[2]$-hard and thus most likely does not allow a fixed-parameter-algorithm. The presented reduction heavily relies on the fact that places (i.e. regions) need not to be pure, that is, t can be a pre- and a post-transition of a place at the same time. Future work could therefore focus on the problem restricted to pure Petri nets or on the search for other parameters putting the problem into FPT.

Acknowledgements. I would like to thank Karsten Wolf, who provided a summary of the data from the Model Checking Contest. Also, I'm very thankful to the anonymous reviewers for their detailed comments and valuable suggestions.

References

1. van der Aalst, W.M.P.: Process Mining - Discovery, Conformance and Enhancement of Business Processes. Springer, New York (2011). https://doi.org/10.1007/978-3-642-19345-3
2. Amparore, E., et al.: Presentation of the 9th edition of the model checking contest. In: Beyer, D., Huisman, M., Kordon, F., Steffen, B. (eds.) TACAS 2019. LNCS, vol. 11429, pp. 50–68. Springer, Cham (2019). https://doi.org/10.1007/978-3-030-17502-3_4
3. Badouel, E., Bernardinello, L., Darondeau, P.: Polynomial algorithms for the synthesis of bounded nets. In: Mosses, P.D., Nielsen, M., Schwartzbach, M.I. (eds.) CAAP 1995. LNCS, vol. 915, pp. 364–378. Springer, Heidelberg (1995). https://doi.org/10.1007/3-540-59293-8_207

4. Badouel, E., Bernardinello, L., Darondeau, P.: The synthesis problem for elementary net systems is NP-complete. Theor. Comput. Sci. **186**(1–2), 107–134 (1997). https://doi.org/10.1016/S0304-3975(96)00219-8

5. Badouel, E., Bernardinello, L., Darondeau, P.: Petri Net Synthesis. TTCSAES. Springer, Heidelberg (2015). https://doi.org/10.1007/978-3-662-47967-4

6. Badouel, E., Caillaud, B., Darondeau, P.: Distributing finite automata through Petri net synthesis. Formal Asp. Comput. **13**(6), 447–470 (2002). https://doi.org/10.1007/s001650200022

7. Best, E.: Structure theory of Petri nets: the free choice hiatus. In: Advances in Petri Nets. Lecture Notes in Computer Science, vol. 254, pp. 168–205. Springer, Berlin (1986). https://doi.org/10.1007/BFb0046840

8. Best, E., Darondeau, P.: A decomposition theorem for finite persistent transition systems. Acta Informatica **46**(3), 237–254 (2009)

9. Best, E., Devillers, R.R.: Synthesis of live and bounded persistent systems. Fundam. Informaticae **140**(1), 39–59 (2015)

10. Best, E., Devillers, R., Schlachter, U.: Bounded choice-free Petri net synthesis: algorithmic issues. Acta Informatica **55**(7), 575–611 (2017). https://doi.org/10.1007/s00236-017-0310-9

11. Best, E., Hujsa, T., Wimmel, H.: Sufficient conditions for the marked graph realisability of labelled transition systems. Theor. Comput. Sci. **750**, 101–116 (2018). https://doi.org/10.1016/j.tcs.2017.10.006

12. Commoner, F., Holt, A.W., Even, S., Pnueli, A.: Marked directed graphs. J. Comput. Syst. Sci. **5**(5), 511–523 (1971)

13. Cortadella, J., Kishinevsky, M., Lavagno, L., Yakovlev, A.: Deriving petri nets from finite transition systems. IEEE Trans. Comput. **47**(8), 859–882 (1998)

14. Cortadella, J., Kishinevsky, M., Kondratyev, A., Lavagno, L., Yakovlev, A.: A region-based theory for state assignment in speed-independent circuits. IEEE Trans. CAD Integr. Circ. Syst. **16**(8), 793–812 (1997). https://doi.org/10.1109/43.644602

15. Parameterized Algorithms. TTCSAES. Springer, Cham (2015). https://doi.org/10.1007/978-3-319-21275-3

16. Desel, J., Esparza, J.: Free Choice Petri Nets. Cambridge Tracts in Theoretical Computer Science, Cambridge University Press, New York (1995). https://doi.org/10.1017/CBO9780511526558

17. Devillers, R.R., Erofeev, E., Hujsa, T.: Efficient synthesis of weighted marked graphs with circular reachability graph, and beyond. CoRR abs/1910.14387 (2019). http://arxiv.org/abs/1910.14387

18. Devillers, R.R., Erofeev, E., Hujsa, T.: Synthesis of weighted marked graphs from constrained labelled transition systems: a geometric approach. Trans. Petri Nets Other Model. Concurr. **14**, 172–191 (2019). https://doi.org/10.1007/978-3-662-60651-3_7

19. Devillers, R.R., Hujsa, T.: Analysis and synthesis of weighted marked graph Petri nets: exact and approximate methods. Fundam. Inform. **169**(1-2), 1–30 (2019). https://doi.org/10.3233/FI-2019-1837

20. Ehrenfeucht, A., Rozenberg, G.: Partial (set) 2-structures. part I: basic notions and the representation problem. Acta Inf. **27**(4), 315–342 (1990). https://doi.org/10.1007/BF00264611

21. Holloway, L.E., Krogh, B.H., Giua, A.: A survey of Petri net methods for controlled discrete event systems. Discrete Event Dyn. Syst. **7**(2), 151–190 (1997). https://doi.org/10.1023/A:1008271916548

22. Hujsa, T., Delosme, J.-M., Munier-Kordon, A.: On the reversibility of well-behaved weighted choice-free systems. In: Ciardo, G., Kindler, E. (eds.) PETRI NETS 2014. LNCS, vol. 8489, pp. 334–353. Springer, Cham (2014). https://doi.org/10.1007/978-3-319-07734-5_18

23. Karp, R.M.: Reducibility among combinatorial problems. In: Miller, R.E., Thatcher, J.W. (eds.) Proceedings of a symposium on the Complexity of Computer Computations, held 20–22 March 1972, at the IBM Thomas J. Watson Research Center, Yorktown Heights, New York, USA. pp. 85–103, The IBM Research Symposia Series, Plenum Press, New York (1972). https://doi.org/10.1007/978-1-4684-2001-2_9

24. Rajan, A.: Theory of linear and integer programming, by alexander schrijver, Wiley, New York, 1986, 471 pp. price $71.95. Networks **20**(6), 801 (1990). https://doi.org/10.1002/net.3230200608

25. Teruel, E., Chrzastowski-Wachtel, P., Colom, J.M., Silva, M.: On weighted T-systems. In: Jensen, K. (ed.) ICATPN 1992. LNCS, vol. 616, pp. 348–367. Springer, Heidelberg (1992). https://doi.org/10.1007/3-540-55676-1_20

26. Teruel, E., Colom, J.M., Suárez, M.S.: Choice-free petri nets: a model for deterministic concurrent systems with bulk services and arrivals. IEEE Trans. Syst Man Cybern Part A **27**(1), 73–83 (1997). https://doi.org/10.1109/3468.553226

27. Tredup, R.: Hardness results for the synthesis of b-bounded petri nets. In: Donatelli, S., Haar, S. (eds.) PETRI NETS 2019. LNCS, vol. 11522, pp. 127–147. Springer, Cham (2019). https://doi.org/10.1007/978-3-030-21571-2_9

28. Tredup, R.: Synthesis of structurally restricted b-bounded petri nets: complexity results. In: Filiot, E., Jungers, R., Potapov, I. (eds.) RP 2019. LNCS, vol. 11674, pp. 202–217. Springer, Cham (2019). https://doi.org/10.1007/978-3-030-30806-3_16

29. Tredup, R.: Parameterized complexity of synthesizing b-bounded (m, n)-T-systems. In: Chatzigeorgiou, A., Dondi, R., Herodotou, H., Kapoutsis, C., Manolopoulos, Y., Papadopoulos, G.A., Sikora, F. (eds.) SOFSEM 2020. LNCS, vol. 12011, pp. 223–235. Springer, Cham (2020). https://doi.org/10.1007/978-3-030-38919-2_19

30. Tredup, R.: The complexity of synthesizing sf nop-equipped boolean petri nets from g-bounded inputs. Trans. Petri Nets Other Model. Concurr. **15**, 101–125 (2021)

31. Chen, J., Feng, Q., Xu, J. (eds.): TAMC 2020. LNCS, vol. 12337. Springer, Cham (2020). https://doi.org/10.1007/978-3-030-59267-7

32. Tredup, R., Erofeev, E.: On the parameterized complexity of synthesizing boolean petri nets with restricted dependency. In: Lange, J., Mavridou, A., Safina, L., Scalas, A. (eds.) Proceedings 13th Interaction and Concurrency Experience, ICE 2020, Online, 19 June 2020. EPTCS, vol. 324, pp. 78–95 (2020). https://doi.org/10.4204/EPTCS.324.7

Discovering Stochastic Process Models by Reduction and Abstraction

Adam Burke$^{(\boxtimes)}$ ⓘ, Sander J.J. Leemans ⓘ, and Moe Thandar Wynn ⓘ

Queensland University of Technology, Brisbane, Australia
{at.burke,s.leemans,m.wynn}@qut.edu.au

Abstract. In process mining, extensive data about an organizational process is summarized by a formal mathematical model with well-grounded semantics. In recent years a number of successful algorithms have been developed that output Petri nets, and other related formalisms, from input event logs, as a way of describing process control flows. Such formalisms are inherently constrained when reasoning about the probabilities of the underlying organizational process, as they do not explicitly model probability. Accordingly, this paper introduces a framework for automatically discovering stochastic process models, in the form of Generalized Stochastic Petri Nets. We instantiate this Toothpaste Miner framework and introduce polynomial-time batch and incremental algorithms based on reduction rules. These algorithms do not depend on a preceding control-flow model. We show the algorithms terminate and maintain a deterministic model once found. An implementation and evaluation also demonstrate feasibility.

Keywords: Stochastic Petri Nets · Process mining · Stochastic process discovery · Stochastic process mining

1 Introduction

Modelling is a way for us to understand and navigate the world; some thinkers argue it is the core activity of science [41]. Today's world, with its cheap computers and voluminous data, makes new forms and subjects of modelling possible. The last two decades have seen great progress in one form, process mining [1] – the analysis of organizational processes using computational techniques and large event logs. Process-mined models are then used to understand and improve organizations.

Stochastic process models, such as Stochastic Petri Nets [4], are well-established in fields from biology [26] to operations research [42] to describe evolving processes with complex causalities, and relative probabilities. *Stochastic process mining* discovers and analyzes stochastic process models. It is a relatively new area of research which aims to exploit the sophistication of stochastic models to advance our understanding of organizations and their frequent, or infrequent,

© Springer Nature Switzerland AG 2021
D. Buchs and J. Carmona (Eds.): PETRI NETS 2021, LNCS 12734, pp. 312–336, 2021.
https://doi.org/10.1007/978-3-030-76983-3_16

behaviour. Currently few discovery and conformance techniques exist. Importantly, these existing discovery techniques do not work well for a number of important real-world cases. They also often depend on control-flow models, limiting the use of stochastic information in the construction of the control-flow itself.

In this paper we introduce a new stochastic process discovery framework, *Toothpaste Miner*, which works with Generalized Stochastic Petri Nets (GSPNs) [4,25] via a data structure targeted at process mining, the Probabilistic Process Tree. Toothpaste Miner does "direct stochastic discovery", i.e., it does not rely on an initial control-flow discovery step, but calculates control-flow and stochastic aspects of the model using a common abstraction. It proceeds by the repeated application of reduction rules. We show polynomial-time computational complexity, termination and deterministic properties of these algorithms. An implementation, in Haskell, and its evaluation, against real-life event logs, shows the technique's practical relevance, and also that it trades off quality against more complex models and longer execution times.

In Sects. 2 and 3, below, we discuss related work and foundational concepts. The Toothpaste Miner discovery algorithms and transformation rules are introduced in Sect. 4, together with the Probabilistic Process Trees formalism. Incremental discovery and noise-management optimisations are discussed in Sect. 5. The implementation and evaluation are laid out in Sect. 6, before we conclude in Sect. 7.

2   Related Work

Important existing work in this area includes that on stochastic process mining, discovering Petri nets, and discovering probabilistic automata.

The stochastic process mining algorithms introduced in Sects. 4.2 and 5 are partially region-based. A number of process mining algorithms for region-based control-flow discovery exist [10]. The *Maximal Pattern Mining* algorithm [23] is a region-based algorithm which combines regular expression-like patterns in systematic ways, and helped inspire the loop and concurrency identification rules in Sect. 4.3. Other sources for rules are Petri net and Stochastic Petri net reductions [35,40] and the Inductive Miner [20], which uses process trees. The Probabilistic Process Trees introduced here extend process trees. (The term "stochastic process tree" is already used to refer to decision trees, e.g. in [13]).

Within process mining, existing stochastic process discovery techniques can be categorized as control-flow dependent, direct, or declarative. For control-flow dependent discovery, one key technique discovers Generally Distributed Stochastic Petri Nets (GDT_SPNs) after alignment-based repair [30,31]. Other techniques output Generalized Stochastic Petri Nets [9], trading some quality for faster execution times, or combine control-flow models using Bayesian inference [16]. Direct discovery techniques exist in the literature, as high level descriptions or algorithms [2,15,17,33] and for structures other than Petri nets. One recent discovery technique shows reduced error percentages by using a Bayesian network with non-classical probability [28]. It is however constrained to exclude loops and concurrency. Another recent technique [29] re-purposes the Direct Follows Graph Miner [21] to obtain a stochastic Direct Follows Model. Discovery of declarative stochastic process models has also seen good progress in recent years [5,24], though the difficulties of comparing control-flow and declarative models put them beyond the immediate scope of this article.

The problem of discovering probabilistic models from event data has three broad classes of existing solutions validated by empirical trials [37]: Bayesian inference, state merging, and parameter estimation.

Bayesian inference is a method where probability estimates are updated in a specific form of cumulative average in response to the introduction of new evidence. Bayesian inference on its own does not yield a structured and visualizable model, just probability estimates for particular events, so cannot be directly applied to process mining [8]. Probabilities obtained with Gibbs sampling [14] have recently been successfully combined with an input control flow model for stochastic process discovery [16].

State merging is exemplified by the ALERGIA algorithm [12], which discovers Stochastic Finite-State Automata through state merges in cubic time; alternatives such as MDI [36] achieve quadratic time complexity. ALERGIA can still be competitive in real-world trials [37]. Both ALERGIA and MDI construct an internal prefix automaton with weights. This general algorithmic structure is also used by our discovery algorithm in Sect. 4.2 and merge operators and rules in Sect. 4.3. We adopt a Petri net-based data structure used in process discovery algorithms, process trees [20], [1, p. 81], to manage state merges, instead of the prefix trees used in ALERGIA.

The RegPFA framework [8] uses parameter estimation to do process prediction. RegPFA uses an internal model for prediction based on Baum-Welch [3]. It outputs a noise-filtered Petri net model for user consumption, which emphasizes understandability against precision, and elides stochastic information.

To the best of our knowledge, the proposed techniques represent novel solutions for discovering stochastic process models. The framework uses a well-established process formalism (GSPNs) and supports loops and concurrency. Rather than annotating stochastic information after finding a control-flow model, it makes direct use of trace information from the event log to construct the stochastic aspect of the model in concert with the control-flow.

3   Preliminaries

Generalized Stochastic Petri Nets (GSPNs) [4,25] and Stochastic Deterministic Finite Automata [38,39] are well-established formalisms, and good overviews exist [4,25,38,39]. Definitions in this section are based on the process mining and Petri net literature [1, p. 80], [19]. We use $\mathbb{N}$ for natural numbers, $\mathbb{R}^+$ for positive real numbers, and $\mathbb{B}$ for booleans; $\bullet$ separates quantifiers and predicates.

Event logs. A process consists of various *activities*. Let the set A be an alphabet of activities for a process. A trace is a sequence of activities. Σ^* is the set of all possible traces over A. A *language* $\subseteq \Sigma^*$ is a set of traces. An event log L is a finite multiset of traces collating observations of the underlying process. Let $|L|$ represent the number of traces and $\|L\|$ the number of activities in the log. A log with ten traces of $\langle a, b \rangle$ and six traces of $\langle b, c \rangle$ is written $[\langle a, b \rangle^{10}, \langle b, c \rangle^{6}]$ following multiset notation in [32].

Definition 1 (Petri nets). *A Petri net [1, 19] is a tuple $PN = (P, T, F, M_0)$, where P is a finite set of places, T is a finite set of transitions, and $F: (P \times T) \to (T \times P)$ is a flow relation. A marking is a multiset of places $\subseteq P$ that indicate a state of the Petri net, with M_0 being the initial marking. A transition is enabled if every incoming place contains a token. A transition fires by changing the marking of the net to consume incoming tokens and producing tokens for its outgoing places. A net where no further transitions may fire has reached a terminal state and corresponds to a final marking.*

Definition 2 (Generalized Stochastic Petri Net (GSPN)). *A GSPN [4, 25] is a tuple $(P, T, F, M_0, W, T_i, T_t)$ such that (P, T, F, M_0) is a Petri net. Weight function $W: T \to \mathbb{R}^+$ assigns each transition a weight. T_i is a set of immediate transitions and T_t a set of timed transitions such that $T_i \cup T_t = T$ and $T_i \cap T_t = \varnothing$. If multiple transitions $T_e \subseteq T_i$ are enabled in a particular marking, the probability of a transition $t \in T_e$ firing is given by $\frac{W(t)}{\Sigma_{t' \in T_e} W(t')}$. Immediate transitions take priority over timed transitions. A timed transition, if enabled, fires according to an exponentially distributed wait time. Given a set of enabled timed transitions $T_e \subseteq T_t$, a transition t fires first with probability $\frac{W(t)}{\Sigma_{t' \in T_e} W(t')}$.*

Definition 3 (Generalized Stochastic Labelled Petri Net (GSLPN)). *A GSLPN [22] is a tuple $(P, T, F, M_0, W, T_i, T_t, \Sigma, \lambda)$ where $(P, T, F, M_0, W, T_i, T_t)$ is a GSPN. λ is a labelling function for transitions $\lambda: T \to \Sigma \cup \{\tau\}$ where $\tau \notin \Sigma$. When a transition $t \in T$ fires, if $\lambda(t) = a$ where $a \neq \tau$, the activity a has executed. $\lambda(t) = \tau$ is a silent transition where there is no evidence of activity execution. The set of GSLPNs with only immediate transitions is denoted by $\mathcal{G}$.*

Definition 4 (Stochastic Language). *A stochastic language $L_\Sigma : \Sigma^* \to [0, 1]$ is a function which denotes a probability for each trace such that $\Sigma_{t \in \Sigma^*} L_\Sigma(t) = 1$.*

Definition 5 (Stochastic Deterministic Finite Automaton). *A Stochastic Deterministic Finite Automaton (SDFA) [11] is a tuple (S, A, σ, p, s_0) where S is a set of states, A is an alphabet, $\sigma : S \times A \to S$ is a transition function, $p: S \times A \to [0, 1]$ maps state probability, and the initial state is $s_0 \in S$.*

SDFAs are a special case of Probabilistic Finite-State Automata (PFAs) [38] and SDFAs are also referred to as Deterministic Probabilistic Finite Automata (DPFA) [38]. All event logs can be represented by SDFAs [19].

4   Process Discovery by Model Reduction

In this section we first introduce *Probabilistic Process Trees* (PPTs), which add relative weights and some alternative operators to the process tree formalism, and can be translated straightforwardly to GSLPNs. We then describe a novel algorithm which uses Probabilistic Process Tree transformations for process model discovery in terms of general rule properties. Lastly, concrete transformation rules for manipulating these trees are introduced.

Table 1. Translation of PPTs to GSLPNs.

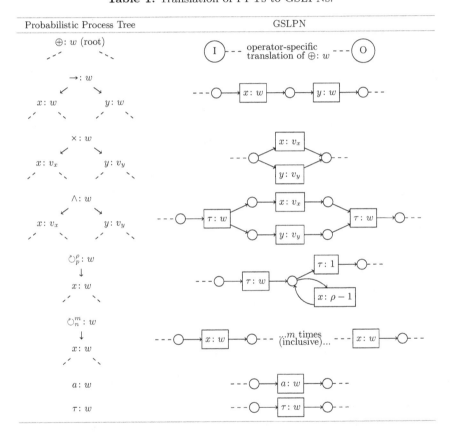

4.1   Probabilistic Process Trees

A *Probabilistic Process Tree* is view of a GSLPN, and a cousin of the process tree [1, p. 78] which allows probabilistic choice. As in Sect. 3, A refers to a finite set of activities with silent activity $\tau \notin A$. We first define trees recursively, followed by operators.

Definition 6 (Probabilistic Process Trees). *A Probabilistic Process Tree (PPT) is a tree of weighted nodes, where each node is denoted by $s\colon w$. The universe of PPTs over activity set A is $\mathcal{U}_A$:*

1. *A single activity. For $a \in A$, $a\colon w \in \mathcal{U}_A$.*
2. *A silent activity. For $\tau \notin A$, $\tau\colon w \in \mathcal{U}_A$.*
3. *An operator $\oplus$ over one or more child trees. Given $n \geqslant 1$, $U_1, ..., U_n \in \mathcal{U}_A$, then $\oplus(U_1, ..., U_n)\colon w \in \mathcal{U}_A$.*

Sub-trees. U_i is a *sub-tree* of its parent $U = \oplus(U_1, ...U_i, ..., U_n)\colon w$, and of all trees of which U is a sub-tree. Trees are *strictly equal*, $U_x = U_y$, when the structures are isomorphic and each node and sub-tree is equal, including weights. Activity powerset function *exp* gives the set of all activities in a given PPT.

In this paper, we consider operators $\oplus = \{\rightarrow, \wedge, \times, \circlearrowright_p, \circlearrowright_n\}$. These redefine and extend non-probabilistic process tree operators to include weight semantics.

Sequential operator. The $\rightarrow$ sequential operator executes its children in sequential order. To align weight with execution frequency, the weight of sequence components is the parent's weight. In the remainder of this paper, we may omit the child weights, writing $\rightarrow (x, y)\colon w$ rather than $\rightarrow (x\colon w, y\colon w)\colon w$.

Choice. The probabilistic choice operator $\times(U_1, ..., U_n)$ chooses one sub-tree $U_i = s_i\colon w_i$ for execution from its children, with probability $\frac{w_i}{\Sigma_{s_j\,:\,w_j \in \{U_1, ..., U_n\}} w_j}$.

Concurrency. The concurrency operator $\wedge$ indicates parallel composition. Each child must execute once with no constraint on order of execution.

Fixed Loops. The fixed loop operator $\circlearrowright_n^m (U)$ repeats its child tree m times. As for sequences, the weight of the loop child is the weight of the parent loop.

Probabilistic Loops. The probabilistic loop operator $\circlearrowright_p^\rho (U)$ executes the child tree with the probability of exiting at each iteration determined by the function $Pr\{\chi = 0\} = \frac{1}{\rho}$ where $\rho \in \mathcal{R}^+ \wedge \rho \geqslant 1$. The weight of the child node is the weight of the parent loop.

$$\forall x, y, w_1, w_2, w_3 \bullet \rightarrow (x\colon w_1, y\colon w_2)\colon w_3 \in \mathcal{U}_A \implies w_1 = w_2 = w_3$$

$$\forall s_1, .., s_n, w_1, .., w_n \bullet \times(s_1\colon w_1, ..., s_n\colon w_n)\colon w \implies w = \Sigma_{(s_j, w_j)} w_j$$

$$\forall s_1, .., s_n, w_1, .., w_n \bullet \wedge(s_1\colon w_1, ..., s_n\colon w_n)\colon w \implies w = \Sigma_{(s_j, w_j)} w_j$$

$$\forall m, x, w_1, w_2 \bullet m \in \mathbb{N} \wedge \circlearrowright_n^m (x\colon w_1)\colon w_2 \implies w_1 = w_2$$

$$\forall m, x, w_1, w_2 \bullet \rho \in \mathbb{R}^+ \wedge \circlearrowright_p^\rho (x\colon w_1)\colon w_2 \implies w_1 = w_2$$

Size. The size of a PPT is the number of nodes, denoted by $|U_A|$.

Example. One PPT, $pe = \times(\rightarrow (a\colon 2, b\colon 2)\colon 2, \rightarrow (b\colon 1, a\colon 1)\colon 1, \circlearrowright_p^3 (c\colon 1)\colon 1)\colon 4$, can be seen in Fig. 1, and has the stochastic language $\Sigma_E = [\langle a, b\rangle^{\frac{1}{2}}, \langle b, a\rangle^{\frac{1}{4}}, \langle c, c, c\rangle^{\frac{1}{4}}]$.

Translation to Generalized Stochastic Petri Nets. Having defined the syntax, and informally explained the meaning of PPT constructs, we now formally define the semantics. PPTs have equivalent constructs in a Generalized Stochastic Petri Net (GSPN), summarized in Table 1.

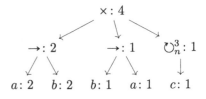
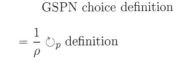

Fig. 1. Example PPT *pe*.

Note that in the probabilistic loop $\circlearrowleft_p^\rho$, each iteration is a Bernoulli trial with the number of iterations being a geometric variable. In the GSPN translation, this results in a transition weight of $\rho - 1$:

$$Pr(\circlearrowleft_p \text{ exit}) = \frac{w_{exit}}{\Sigma w_i} \qquad \text{GSPN choice definition}$$

$$= \frac{1}{1 + (\rho - 1)} \qquad = \frac{1}{\rho} \ \circlearrowleft_p \text{ definition}$$

The translation of PPTs to GSPNs, as described in Table 1, shows PPTs are a subset of Probabilistic Finite Automata [38]. Figure 2 gives the translation of example PPT *pe* into a GSPN, with a more sophisticated example shown in Fig. 4f.

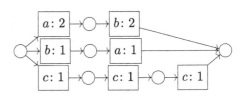

Fig. 2. Translation of *pe* to a GSPN.

4.2 A Discovery Algorithm Framework

The *Toothpaste framework* describes reduction algorithms that "squeeze" an initial trace model into a more summarized and useful form using transformation rules. The framework is illustrated in Fig. 3. After introducing component elements, an example instantiation is made in Definition 7.

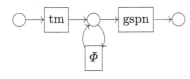

Fig. 3. Toothpaste framework.

Toothpaste miner algorithms first transform an event log into an internal PPT. They then repeatedly transform the PPT by applying transformation rules. These rules reduce, summarize, or restructure the tree towards a desirable form. When desired criteria for an output process model are met, the PPT is translated into a GSLPN as the final output. Criteria may include quality criteria such as fitness or precision thresholds, simplicity, or the preservation of certain critical trace paths in the final model. As the miner proceeds largely by reduction from an initial state which perfectly matches the event log, this allows for fine-grained control over what elements of the initial log are preserved in the final model.

Event Logs and Discovery. Given an event log $L = [t_1^{i_1}, ..., t_n^{i_n}]$, a trace model PPT is given by $tm : L \to \mathcal{U}_\mathcal{A}$, where each trace is converted by $st : \langle A \rangle \to \mathcal{U}_\mathcal{A}$.

$$st(\langle a_1, ..., a_m \rangle) = \to (a_1 : i_j, ..., a_m : i_j) : i_j$$
$$st(\langle a \rangle) = a : i_j$$
$$tm(L) = \times (st(t_1), ..., st(t_n)) : |L|$$

For example, $tm(\langle a, b \rangle, \langle c \rangle^3) = \times (\to (a, b) : 1, c : 3) : 4$.

Transformation. In the Toothpaste framework, discovery proceeds by the application of transformation rules to a PPT, yielding progressively improved models. For rule sequences, $+$ is used for concatenation and **ran** for the set of elements in a sequence. We instantiate the framework using *reduction rules*, those which reduce the total number of nodes in the tree, with specific rules in Sect. 4.3.

Function Φ applies a sequence of transformation rules to a PPT.

$$\Phi : \mathcal{U}_\mathcal{A} \times \langle \mathcal{U}_\mathcal{A} \to \mathcal{U}_\mathcal{A} \rangle \to \mathcal{U}_\mathcal{A}$$
$$\Phi(pt, \langle \rangle) = pt$$
$$\Phi(pt, \langle r \rangle) = r(pt)$$
$$\Phi(pt, \langle r \rangle + rs) = \Phi(r(pt), rs))$$

Φ_M finds a local reduction minima by applying Φ exhaustively.

$$\Phi_M : \mathcal{U}_\mathcal{A} \times \langle \mathcal{U}_\mathcal{A} \to \mathcal{U}_\mathcal{A} \rangle \to \mathcal{U}_\mathcal{A}$$
$$\Phi_M(pt, rs) = \begin{cases} \Phi_M(pt', rs) & \text{if } pt \neq pt' \\ pt & \text{otherwise} \end{cases}$$
$$\text{where } pt' = \Phi(pt, rs)$$

Both Φ and Φ_M are guaranteed to terminate when used with reduction rules, as the size of the input tree is monotonically decreasing. Informally, so long as rules are chosen to preserve fidelity to the log, a minimal reducible model is desirable. The degree to which fidelity is desirable can be controlled by which rules are provided to the discovery algorithm. If a given ruleset is not confluent [7, p10], finding a minimal model is not guaranteed and can depend on the sequence in which the rules are applied.

Definition 7 (Toothpaste miner)

Given reduction rule sequence rs and GSLPN translation function gspn,
$$dtm : L \times \langle \mathcal{U}_\mathcal{A} \to \mathcal{U}_\mathcal{A} \rangle \to \mathcal{G}$$
$$dtm(L, rs) = gspn \circ \Phi_M(tm(L), rs) \text{ is the direct toothpaste miner.}$$

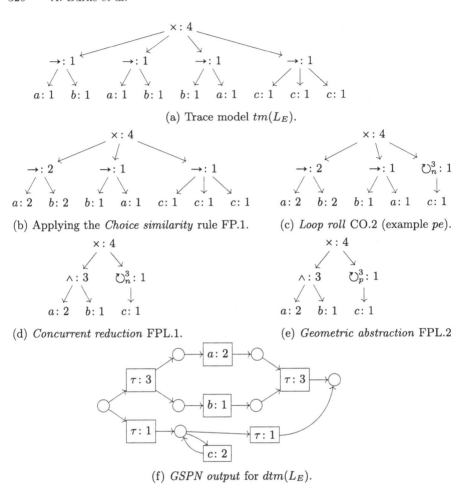

(a) Trace model $tm(L_E)$.

(b) Applying the *Choice similarity* rule FP.1.

(c) *Loop roll* CO.2 (example pe).

(d) *Concurrent reduction* FPL.1.

(e) *Geometric abstraction* FPL.2

(f) *GSPN output* for $dtm(L_E)$.

Fig. 4. Discovery example using dtm.

Example. An example of model discovery is in Fig. 4, applying (and previewing) rules from Sect. 4.3. The trace model in Fig. 4a has identical $\langle a, b \rangle$ traces consolidated in model Fig. 4b. The repeated c activities are summarized with a fixed loop in Fig. 4c. Concurrency of events a and b is identified in Fig. 4d, and a probabilistic loop is introduced in Fig. 4e. Finally the PPT is translated to a GSPN in Fig. 4f.

Complexity. The computational complexity of the Φ algorithms depend on the size of the PPT data structure. Function st produces a binary tree with $2|t_i| - 1$ nodes for each trace t_i. The full trace model produced by tm adds a choice node, for a total size (and memory complexity) of $\Sigma(2|t_i| - 1) + 1 = 2\|L\| - |L| + 1$ or $O(\|L\|)$.

The complexity of Φ depends on evaluating each node of the tree with reduction rules. If each sub-tree can be summarized with one traversal, the worst case is comparing each node to each other node, giving $O(\Phi(U_A, R)) = |U_A|^2 |R|$ comparisons. Writing U_L for the full PPT trace model produced by tm, this is limited by $(2||L||)^2 |R|$ or $O(\Phi(U_L, R)) = O(||L||^2 |R|)$.

Applying Φ exhaustively with Φ_M requires executing this process a number of times. So long as the rule list R is solely reduction rules, then the size of the tree is monotonically decreasing. The worst case for time complexity is then also the best case for model size reduction and is bounded by the size of the trace model, $2||L||$. This yields $O(\Phi_M(L, R)) = O(||L||^3 |R|)$. Translation to a GSPN with the $gspn$ function is linear in the size of the tree (see Table 1). The overall worst-case time complexity is then dominated by the cubic term and is $O(||L||^3 |R|)$.

4.3   Transformation Rules

Probabilistic Process Trees may be manipulated using transformation rules. In this section we organize and classify rules two ways: by information-preservation (using quality criteria), and by impact on determinism. After introducing some useful functions for merging and scaling PPTs, we then explain specific rules.

Classification by Quality Criteria. Transformation rules are classified using process model quality criteria of fitness, precision, and simplicity [1, p. 189], and by the criteria of stochastic information loss. Standard process model quality criteria are control-flow criteria: they do not take the stochastic perspective into account.

Consider log L with language L_L and stochastic language Σ_L, and model M with language L_M and stochastic language Σ_M. Fitness is given by $ft(L_M, L_L) = \frac{|L_M \cap L_L|}{|L_L|}$. In defining precision, we have to account for infinitely many traces, due to the $\circlearrowright_p$ construct [34]. Low-probability traces which are longer than the number of events in the log are filtered out in truncated language L_{TM}:

$$L_{TM} = \{t \in L_M \mid |t| < ||L|| \vee \Sigma_M(t) > \epsilon\} \text{ where } 0 < \epsilon \ll 1$$

Precision is then given by $pn(L_M, L_L) = \frac{|L_{TM} \cap L_L|}{|L_{TM}|}$.

As a categorization tool for process model transformation rules, fitness and precision are helpful in showing the loss or retention of information, even if they are insensitive to stochastic information. The classification of reduction rules is of particular interest, and necessary to maintain the monotonically simplifying property of the discovery algorithm in Sect. 4.2. We categorize reduction rules in four cuts: Preserving Compression, Fitness- and Precision-Preserving, Fitness-Preserving, and Simplifying Lossy, as seen in Fig. 5. Another useful category of rules, Preserving but Non-Simplifying, change model structure without reducing the size of the model.

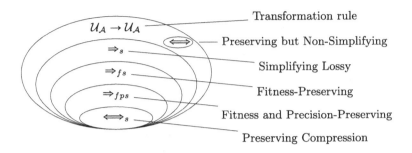

Fig. 5. Rule categories by information preservation.

No Loss Of Fitness Or Precision Without Loss of Stochastic Information.
There are no categories of "Fitness- and Stochastic Information-Preserving But
Precision-Reducing" or "Precision- and Stochastic Information-Preserving but
Fitness Reducing" transformation rules. Let stochastic fidelity between stochas-
tic languages Σ_1, Σ_2 be $\forall t \in T \bullet \Sigma_1(t) = \Sigma_2(t)$. Models that no longer have
stochastic fidelity have experienced stochastic information loss.

Theorem 1. *It is not possible to maintain fitness and reduce precision without
also losing stochastic information from a model. Given log L, models M and M',
and corresponding stochastic languages $\Sigma_M, \Sigma_{M'}$:*

$$ft(M, L) = ft(M', L) \wedge pn(M, L) > pn(M', L) \implies \exists t \bullet \Sigma_M(t) \neq \Sigma_{M'}(t)$$

Proof. Let Σ_L be the stochastic language for a trace model, which then has full
stochastic information from the log. Let M' be a second model covering language
$L_{M'}$ such that $ft(M, L) = ft(M', L) \wedge pn(M, L) > pn(M', L)$. By the fitness
definition, $\frac{|L_M \cap L_L|}{|L_L|} = \frac{|L_{M'} \cap L_L|}{|L_L|}$. As precision decreases, there must therefore be
at least one new trace $t \in M' \wedge t \notin M$, which is equivalent to $\Sigma_M(t) = 0$ and
$\Sigma_{M'}(t) > 0$. As probabilities must sum to one, some other trace $s \in M$ must
have reduced in probability. $\exists s \in L_M \bullet \Sigma_{M'}(s) < \Sigma_M(s)$.
Case 1: Stochastic fidelity had been retained. $\Sigma_M(s) = \Sigma_L(s) \neq \Sigma_{M'}(s)$.
Case 2: Stochastic fidelity already lost. $\Sigma_M(s) \neq \Sigma_L(s)$. The trace t holds no
information for restoring stochastic information on s, as t is not an element of
the log L or covered by the original model M. □

If $\frac{|M'|}{|M|} < \frac{|L \cap M'|}{|L \cap M|}$, and a rule reduces fitness, then precision increases. This
defines a sub-category of Simplifying Lossy Rules, however no useful concrete
rules were found in this sub-category.

Classification by Determinism. PPTs are not constrained to describe deterministic languages. Non-determinism arises when the next symbol in a trace will
satisfy multiple paths through a process tree. This can be shown trivially with the
tree $\times(a: 1, a: 2)$. The trace model produced by function tm may also produce

a non-deterministic tree, for example $tm([\langle a \rangle, \langle a, b \rangle]) = \times(a\colon 1, \to (a, b)\colon 1)\colon 2$. Determinism is a desirable property in an output model: it makes problems such as parsing and calculating the most probable path easier [38], and some important stochastic conformance techniques are constrained to deterministic models [19]. In this section we describe functions for calculating the determinism of a model, and how rules may preserve determinism.

As the operators $\to, \circlearrowright_n$, and $\circlearrowright_p$ are all sequential in form, the only PPT operators which may introduce non-determinism are $\times$ and $\wedge$. The function β reports whether a tree is deterministic.

Let $\alpha\colon \mathcal{U_A} \to \mathcal{P}(A)$ identify starting symbols

$$\alpha(a\colon w) = \{a\} \text{ where } a \in A$$

$$\alpha(\oplus(U_1, ..., U_n)\colon w) = \begin{cases} \alpha(U_1) \text{ where } \oplus \in \{\to, \circlearrowright_n, \circlearrowright_p\} \\ \alpha(U_1) \cup ... \cup \alpha(U_n) \text{ where } \oplus = \times \\ \exp(U_1) \cup ... \cup \exp(U_n) \text{ where } \oplus = \wedge \end{cases}$$

Let $\alpha_{st}\colon \mathcal{U_A} \to \mathcal{P}(A)$ identify non-determinant sub-tree symbols

$$\alpha_{st}(a\colon w) = \varnothing \text{ where } a \in A$$

$$\alpha_{st}(\oplus(U_1, ..., U_n)\colon w) = \begin{cases} \alpha(U_1) \cap ... \cap \alpha(U_n) \text{ where } \oplus \in \{\times, \wedge\} \\ \alpha(\oplus(U_1, ..., U_n)\colon w) \text{ otherwise} \end{cases}$$

Let $\beta\colon \mathcal{U_A} \to \mathbb{B}$ identify whether a tree is deterministic

$$\beta(U) \text{ where } a \in A$$

$$\beta(\oplus(U_1, ..., U_n)\colon w) = \begin{cases} \forall i \bullet \beta(U_i) \text{ where } \oplus \in \{\to, \circlearrowright_n, \circlearrowright_p\} \\ \alpha_{st}(\oplus(U_1, ..., U_n)\colon w) = \varnothing \text{ where } \oplus \in \{\times, \wedge\} \end{cases}$$

$\beta(U)$ only has to visit each node at most once, so can complete in $O(|U|)$ time.

As PPTs are a subset of Probabilistic Finite Automata (PFAs), Deterministic Probabilistic Process Trees (DPPTs) are Stochastic Deterministic Finite Automata (SDFAs) [38]. SDFAs are not closed under union [38] and DPPTs combinations are not closed under $\times$; trivially, $\times(a\colon 1, a\colon 2)$ combines two deterministic sub-trees. However DPPTs have the useful property of being closed under certain subsets of transformation rules.

Definition 8 (β-trap). *A β-trap is a transformation rule which preserves determinism:* $\forall U \in \mathcal{U_A}, tr\colon \langle \mathcal{U_A} \to \mathcal{U_A} \rangle \bullet tr$ *is a β-trap* $\iff \beta(U) \implies \beta(tr(U))$

We use α and β to classify rules by their impact on determinism in Table 2.

Theorem 2. *DPPTs are closed under reduction by β-trap rules, so will not introduce non-determinism to a deterministic model.*

Proof. From the definition of β and β-trap, $\beta(U) \implies \beta(tr(U))$, so the composition of β-trap rules is itself a β-trap. □

Table 2. PPT transformation rule classification by impact on determinism, given $U' = tr(U)$ for some rule tr.

Classification	Definition	Description
α-reducing	$\alpha(U') \subseteq \alpha(U) \wedge \alpha_{st}(U') \subseteq \alpha_{st}(U)$	Separates and restricts relevant symbols
Non-optional	No $\times$ or $\wedge$	Special case of α-stable
α-stable	$\alpha(U') = \alpha(U)$	No change to relevant symbols
β-trap	$\beta(U) \implies \beta(U')$	Never introduces non-determinism. Superset of preceding types

DPPTs are closed under β-trap rule composition, but not tree composition.

Theorem 3. *Application of α-stable or α-reducing rules to a sub-tree preserves the determinism of the parent tree, but a β-trap rule may not.*

Proof. Consider possible transformation rule $nb(\rightarrow (a:1, a:1, b:1)) = \times(a:1, b:1)$. (Note this is not a rule used for our discovery algorithm.) The rule preserves determinism, as $\beta(nb(x)) \wedge \beta(x)$. However, $\alpha(\times(a:1,b:1)) = \{a,b\} \supset \alpha(\rightarrow (a:1, a:1, b:1)) = \{a\}$.

For α-stable and α-reducing rules tr, for process tree $\oplus(U_1, ..., U_n)$, as $\forall i \bullet \alpha(tr(U_i)) \subseteq \alpha(U_i)$, the intersection $\alpha(U_i) \cap \alpha(U_j)$ does not increase, so $\beta(U) \implies \beta(tr(U))$, the definition of a β-trap. $\qquad \square$

These results do not guarantee the discovery of a deterministic model, but they do maintain the determinism of a reduced model once one is discovered.

Concrete Rules. The remainder of this section concerns concrete rules. We use $a, b \in A$ for activities and $u_1, u_2 \in U$ to represent PPTs or sub-trees. Weights are represented by $v_i, w_i \in \mathbb{R}^+$. Unweighted operators or activities are represented by x, y such that $x: w, y: v \in \mathcal{U}_A$. In each subsection, we introduce the information-preservation category, then each rule in the category. The impact on determinism, using the system in Table 2, is also shown.

Helper Functions. These functions and relations help define transformation rules.

A scaling function, $\Gamma(U, \gamma)$ multiplies every weight in the tree by γ. Γ preserves α-stability. PPTs are *similar*, denoted $\cong_c$, when only weights need to be changed to make them strictly equal.

A stochastic merge function, Ψ, combines two similar trees by adding weights, $\circlearrowright_p$ repetitions being scaled by relative weight. Ψ preserves α-stability, and also the control-flow fitness and precision of the input process trees, but loses stochastic information, unless $x = y$ for $\Psi(x, y)$.

Parameter Consolidation (Preserving but non-simplifying). The following associativity rules allow all PPTs to be transformed into equivalent trees with at most two children per operator. In our Toothpaste miner algorithms, such rules must only be used in combination with reduction rules, to maintain monotonicity of reduction; they act as meta-rules which allow other rules to be stated more concisely. We denote these rules as $\Longleftrightarrow$. They are information-preserving and perform no compression from left-to-right. They are all α-stable.

PC.1 $\rightarrow (s_1, s_2, ..., s_n): w \Longleftrightarrow \rightarrow (s_1, \rightarrow (s_2, ..., s_n): w): w$

PC.2 $\times (s_1: w_1, s_2: w_2, ..., s_n: w_n): v$
$\Longleftrightarrow \times (s_1: w_1, \times (s_2: w_2, ..., s_n: w_n): v - w_1): v$

PC.3 $\wedge (s_1: w_1, s_2: w_2, ..., s_n: w_n): v$
$\Longleftrightarrow \wedge (s_1: w_1, \wedge (s_2: w_2, ..., s_n: w_n): v - w_1): v$

PC.4 The $\times$ and $\wedge$ operators are commutative. $\oplus(u, v) = \oplus(v, u)$
The remaining rules are accordingly stated using at most two operator parameters.

Preserving Compressions. The following rules are information-preserving reduction rules, achieving compression by using a smaller tree to describe the same stochastic language. They are denoted with $\Longleftrightarrow_s$ and are all non-optional deterministic.

CO.1 *Fixed loop identity.* $\circlearrowleft_n^1 (u) : w \Longleftrightarrow_s u : w$. This is used in reverse in FPL.7.

CO.2 *Fixed Loop roll.* $\rightarrow (x, x): w \Longleftrightarrow_s \circlearrowleft_n^2 (x): w$
$\rightarrow (x, \circlearrowleft_n^m (x)): w \Longleftrightarrow_s \circlearrowleft_n^{m+1} (x): w$
$\rightarrow (\circlearrowleft_n^m (x), x): w \Longleftrightarrow_s \circlearrowleft_n^{m+1} (x): w$

CO.3 *Silent sequence.* $\rightarrow (u, \tau): w \Longleftrightarrow_s \rightarrow (\tau, u): w \Longleftrightarrow_s u: w$

CO.4 *Silent concurrency.* $\wedge (u: w, \tau: v): w + v \Longleftrightarrow_s \Gamma(u, \frac{w+v}{w})$

CO.5 *Fixed loop nesting.* $\circlearrowleft_n^n (\circlearrowleft_n^m (u)): w \Longleftrightarrow_s \circlearrowleft_n^{nm} (u): w$

Fitness and Precision-Preserving With Stochastic Information Loss. For these rules, stochastic information is preserved only where sub-trees are strictly equal, as for Ψ. They are denoted with $\Rightarrow_{fps}$. The determinism properties vary by rule.

FP.1 *Choice similarity reduction.* Merge choices between structurally similar trees. $\times (u_1, u_2) \Rightarrow_{fps} \Psi(u_1, u_2)$ where $u_1 \cong_c u_2$. This rule is α-stable.

FP.2 *Choice folding.* Pull up a common prefix into the head of a new sequence.
$\times ((\rightarrow (u_{x1}, u_2): w_1), (\rightarrow (u_{x2}, u_3) : w_2) : w_1 + w_2$
$\Rightarrow_{fps} \rightarrow (\Psi(u_{x1}, u_{x2}), \times (u_2, u_3)) : w_1 + w_2$ where $u_{x1} \cong_c u_{x2}$. This rule is α-reducing. (Illustrated in Fig. 6).

FP.3 *Choice folding suffixes* $\times ((\rightarrow (u_1, u_{y1}) : w_1), (\rightarrow (u_2, u_{y2}) : w_2)) : w_1 + w_2$
$\Rightarrow_{fps} \rightarrow (\times (u_1, u_2), \Psi(u_{y1}, u_{y2})) : w_1 + w_2$ where $u_{y1} \cong_c u_{y2}$. This rule is α-stable.

FP.4 *Choice skip.* A common head is pulled into a sequence with a choice between the tail and a silent activity. $\times (x_1: w_1, \rightarrow (x_2, y): w_2) : v$
$\Rightarrow_{fps} \rightarrow (\Psi(x_1: w_1, x_2: w_2), \times (y: w_2, \tau: w_1)) : v$ where $x_1: w_1 \cong_c x_2: w_2$. This rule is α-reducing.

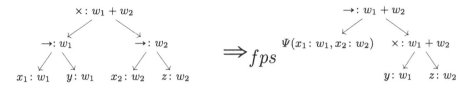

Fig. 6. *Choice folding* transformation rule for shared prefixes FP.2.

FP.5 *Choice skip suffix.* $\times(x_1 : w_1, \rightarrow (y, x_2) : w_2) : v$
$\Rightarrow_{fps} \rightarrow (\times(y : w_2, \tau : w_1), \Psi(x_1 : w_1, x_2 : w_2)) : v$ where $x_1 : w_1 \cong_c x_2 : w_2$.
This rule is α-stable.

FP.6 *Concurrent similarity reduction.* Similar concurrent subtrees reduce to
repetition. $(\wedge(x_1 : w, x_2 : v) : w + v)$
$\Rightarrow_{fps} \circlearrowleft_n^2 (\Psi(x_1 : w, x_2 : v)) : w + v$ where $x_1 \cong_c x_2$. This rule is α-stable.

FP.7 *Concurrent subsumption.* Sequences already recognized as concurrent are
pulled under that pattern. $\times(\rightarrow (x_1, y_1) : w_1, \wedge(x_2 : w_2, y_2 : w_3) : v)$
$\Rightarrow_{fps} \wedge(\Psi(x_1 : w_1, x_2 : w_2), \Gamma(\Psi(y_1 : w_1, y_2 : w_3), \frac{w_3}{w_1 + w_3})) : v$ where $x_1 \cong_c$
$x_2 \wedge y_1 \cong_c y_2$. This rule is α-stable.

Fitness-Preserving Lossy Reductions. These rules preserve control-flow fitness
of the input model with respect to a given log, but may reduce precision and
stochastic information. They are denoted with $\Rightarrow_{fs}$.

FPL.1 *Concurrent reduction from choice sequences.* Concurrency is inferred
when permutations of a given two-step sequence are seen. Generalizing
concurrency involves re-scaling the weights of the merged sub-trees.
$\times(\rightarrow (u_{x1}, u_{y1}) : w, \rightarrow (u_{y2}, u_{x2}) : v) : w + v$
$\Rightarrow_{fs} \wedge(\Gamma(\Psi(u_{x1} : w, u_{x2} : v), \frac{w}{w+v}), \Gamma(\Psi(u_{y1} : w, u_{y2} : v), \frac{v}{w+v}) : w + v$
where $u_{x1} \cong_c u_{x2} \wedge u_{y1} \cong_c u_{y2}$.
When sub-trees are compound (that is, not activities), and applied across
partial parameters, as in the binary statement here, there is concurrency generalization from a sample rather than complete evidence of
concurrency. E.g., traces $\langle a, b, c \rangle, \langle b, a, c \rangle, \langle c, a, b \rangle$ are sufficient to reduce
to $\wedge(a : 1, b : 1, c : 1)$. As all activities are already present in both children
before the application of the rule, concurrent reduction is α-stable.

FPL.2 *Geometric Abstraction.* This rule combines fixed loops into a single prob-
abilistic loop. Consider $\times(\circlearrowleft_n^{m_1} (u_1) : w_1, ..., \circlearrowleft_n^{m_n} (u_n) : w_n) : v$ where
$\forall i, j \leqslant n \bullet u_i \cong_c u_j$. By definition of $\times$, $v = \Sigma_1^n w_i$. These loops are

used as samples in a geometric probability distribution.

$$\text{Probability of exit } P = \frac{\Sigma_1^n w_i}{\Sigma_1^n m_i w_i}$$

$$\text{Mean repetitions } \bar{\rho} = \frac{1}{P} = \frac{\Sigma_1^n m_i w_i}{\Sigma_1^n w_i}$$

Helper function μ averages n loops using a scaled fold with Ψ

$$\bar{u} = \mu(\circlearrowright_n^{m_1}(u_1): w_1, ..., \circlearrowright_n^{m_n}(u_n): w_n) =$$
$$\circlearrowright_p^{\bar{\rho}}(\Gamma(\Psi(\Gamma(u_1, m_1), \Psi(..., \Gamma(u_n, m_n)))), P): v$$

Then $\times(\circlearrowright_n^{m_1}(u): w_1, ..., \circlearrowright_n^{m_n}(u): w_n): v$
$$\Rightarrow_{fs} \mu(\circlearrowright_n^{m_1}(u): w_1, ..., \circlearrowright_n^{m_n}(u): w_n).$$

Geometric abstraction is non-optional deterministic.

FPL.3 *Choice Loop roll.* A tree is always a loop of length one, so can be incorporated in a loop by averaging. $\times(x_1: w_1, \circlearrowright_p^\rho(x_2): w_2)$
$\Rightarrow_{fs}\circlearrowright_p^{\bar{\rho}}(\mu(\circlearrowright_p^1(x_1: w_1), \circlearrowright_p^\rho(x_2): w_2))): w_1 + w_2$ given $x_1 \cong_c x_2$ and where $\bar{\rho} = \frac{w_1 + \rho w_2}{w_1 + w_2}$. This rule is α-stable.

FPL.4 *Fixed Loop of Probability loops.* The sum of geometric distributions is a negative binomial distribution. This rule approximates with a geometric distribution of the same mean. $\circlearrowright_n^m(\circlearrowright_p^\rho(x)): w$
$\Rightarrow_{fps}\circlearrowright_p^{\rho(m-1)}(x): w$ where $m > 1$.
The $m = 1$ case is handled by Fixed loop identity CO.1. This rule is in the non-optional determinism category.

FPL.5 *Probability Loop of Fixed loops.* $\circlearrowright_p^\rho(\circlearrowright_n^m(x)): w \Rightarrow_{fps}\circlearrowright_p^{m\rho}(x): w$.
This rule is in the non-optional determinism category.

FPL.6 *Probabilistic Loop Nesting.* $\circlearrowright_p^{\rho_1}(\circlearrowright_p^{\rho_2}(x)): w \Rightarrow_{fps}\circlearrowright_p^{\rho_1\rho_2}(x): w$ by the product of expectations. This rule is in the non-optional determinism category.

FPL.7 *Loop Similarity Normalization.* Loops are not similar to their subtrees by $\cong_c$. However loops and their children can be usefully consolidated with some loss of information. Noting that $\circlearrowright_n^1(u): w \Longleftrightarrow u: w$, we define loop similarity $\cong_L: \mathcal{U}_A \leftrightarrow \mathcal{U}_A$:

$$u_1 \cong_L \circlearrowright_p^\rho(u_2) \Leftrightarrow u_1 \cong_c u_2$$

$$u_1 \cong_L \circlearrowright_n^m(u_2) \Leftrightarrow u_1 \cong_c u_2$$

Reduction rules using $\cong_c$ each have a Fitness-Preserving Lossy ($\Rightarrow_{fs}$) variation using $\cong_L$ and replacement tree $\bar{u}$.

For rule parameters $x_1: w_1 \cong_L \circlearrowright_p^\rho(x_2: w_2): w_2$

$$\circlearrowright_p^{\rho'}(\bar{u}) = \mu(\circlearrowright_p^1(x_1: w_1), \circlearrowright_p^\rho(x_2: w_2))$$

The consolidated tree $\bar{u}$ may replace u_1 and u_2 in a transformation rule tr where $u_1 \cong_L u_2$ and the resulting rule application still results in tree size reduction. Loop similarity is non-optional deterministic, but note the rule it is applied to may have a weaker determinism category, which will become the category of the final rule.

Simplifying Lossy Reductions. The last rule category abstracts or summarizes a PPT, from both a control flow and stochastic perspective, at the expense of control-flow fitness and precision. Such rules are useful for the management of noise and for allowing a user to tune the detail of the model for their specific use case. They are denoted with $\Rightarrow_s$, with one rule in this category.

SL.1 *Choice Pruning.* $\times(x\colon w_1, y\colon w_2)\colon v \Rightarrow_s x\colon v$ where $\frac{w_2}{v} < \epsilon$ for some supplied probability threshold ϵ. This rule is α-reducing.

5 Incremental Discovery and Optimisations

A simple Toothpaste Miner was introduced in Sect. 4.2, but it can be limited when applied to streams, very large event logs, or noisy data. Incremental process model discovery is of interest for streams of events and very large event logs, e.g. in [18]. Better models may be achieved through other optimizations while maintaining tractability. Management of noise is another key process mining challenge [1, p. 185], which alternative rulesets can address within the overall Toothpaste Miner framework.

5.1 Incremental Discovery

The Φ_Δ algorithm adds a new trace to the existing model and applies reduction rules: $\Phi_\Delta\colon \mathcal{U}_A \times \mathcal{U}_A \times \langle \mathcal{U}_A \to \mathcal{U}_A \rangle \to \mathcal{U}_A$.

$$\Phi_\Delta(x\colon w, tt\colon v, rs) = \Phi_M(\times(x\colon w, tt\colon v)\colon w + v, rs)$$

Definition 9 (Incremental Toothpaste Miner) *Given trace $t \in \langle A \rangle$, rule sequence rs, existing model $U_M \in \mathcal{U}_A$, and functions gspn and tm per Definition 7, incremental miner dinc: $\langle A \rangle \times \langle \mathcal{U}_A \to \mathcal{U}_A \rangle \times \mathcal{U}_A \to (\mathcal{G} \times \mathcal{U}_A)$ is:*

$$dinc(t, rs, U_M) = (gspn(U_{M+1}), U_{M+1}), \ given \ U_{M+1} = \Phi_\Delta(U_M, st(t), rs)$$
$$dinc(t, rs, \varnothing) = (gspn(st(t)), st(t))$$

An entire event log L may be presented as a stream to *dinc*, resulting in repeated invocations of Φ_Δ.

Definition 10 (Repeated Incremental Toothpaste Miner).

$$di\colon L \times \langle \mathcal{U}_A \to \mathcal{U}_A \rangle \to \mathcal{G}$$
$$(di(L, rs), pt) = dinc(t, rs, di(L - \{t\}, rs)), \ for \ some \ t \in L$$
$$(di(\llbracket t^1 \rrbracket, rs), pt) = dinc(t, rs, \varnothing)$$

5.2   Incremental Complexity

As for other Φ and dtm algorithms, the time complexity of di is dependent on the size of the tree. For the first trace t_1, the size of the initial model is $2|t_1| - 1$. In the worst case, the size of the process model increases over time, so that subsequent traces t_i add $2|t_i|$ nodes each. The time for each run of Φ_Δ is $2|t_i|^2|R|$. The overall worse case size for a log L is then

$$O_{mem}(di(L, rs)) = \Sigma_{i=1}^{|L|} 2|t_i||R| = 2|R|\Sigma_{i=1}^{|L|}|t_i|$$
$$= O(|R| \cdot ||L||), \text{ by the definition of} ||L||$$

An upper bound for time complexity can be found using Lemma 1.

Lemma 1. *For* $A \subseteq \mathbb{N}$, $\Sigma_{a\in A}a^2 \leqslant (\Sigma_{a\in A}a)^2$.

Proof. By induction over $|A|$; initial case $|A| = 1$, for which $a^2 = (a)^2$,

$$\text{Show } \Sigma_{a\in A}a^2 \leqslant (\Sigma_{a\in A}a)^2 \implies \Sigma_{a\in A\cup\{b\}}a^2 \leqslant (\Sigma_{a\in A\cup\{b\}}a)^2$$
$$(\Sigma_{a\in A\cup\{b\}}a)^2 = (\Sigma_{a\in A}a + b)^2 = (\Sigma_{a\in A}a)^2 + 2b(\Sigma_{a\in A}a) + b^2$$
$$\Sigma_{a\in A}a^2 \leqslant (\Sigma_{a\in A}a)^2 \implies \Sigma_{a\in A}a^2 + b^2 \leqslant (\Sigma_{a\in A}a)^2 + b^2 + 2b(\Sigma_{a\in A}a)$$

□

Then, $O_{time}(di(L, rs)) = \Sigma_{i=1}^{|L|} 2|t_i|^2|R| = 2|R|\Sigma_{i=1}^{|L|}|t_i|^2 \leqslant 2|R|(\Sigma_{i=1}^{|L|}|t_i|)^2 \leqslant 2|R|||L||^2$, by definition of $||L||$. So $O_{time}(di(L, rs)) \leqslant O(||L||^2|R|)$.

Notably, the time complexity of the incremental algorithm di is quadratic, rather than the cubic complexity of reducing the entire trace model at once with dtm. Informally, the model is of a smaller size for more of the execution of the algorithm, with the time savings compounding. An important design trade-off remains, as stochastic information loss occurs with most reduction rules, and some classes of rules cause more information loss than others.

5.3   K Retries

Finding the minimal model with $\Phi(pt, rs)$ would require checking all $|rt|!$ permutations of reduction rules, so becomes intractable even for relatively small collections of rules. Rather than using the first full reduction, as in Definition 7, exploring an alternative K permutations, for small constant K, may yield a smaller model, without impacting computational complexity.

$$R_K = \{r \in \langle \mathcal{U}_A \to \mathcal{U}_A \rangle \mid \mathbf{ran}\ r = \mathbf{ran}\ rs\} \wedge |R_K| = K$$
$$\Phi_{MK}(pt, rs) = pt'$$
$$\text{where } |pt'| = \mathbf{min}(\{c \in \mathbb{N} \mid \exists p \in \mathcal{U}_A, rs' \in R_K \wedge c = |\Phi(p, rs')|\})$$

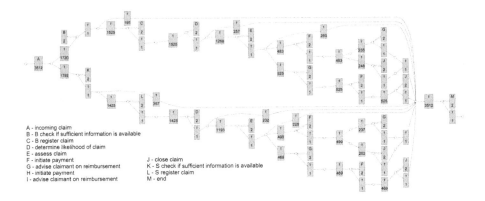

A - incoming claim
B - B check if sufficient information is available
C - B register claim
D - determine likelihood of claim
E - assess claim
F - initiate payment
G - advise claimant on reimbursement
H - initiate payment
I - advise claimant on reimbursement

J - close claim
K - S check if sufficient information is available
L - S register claim
M - end

Fig. 7. *Teleclaims* process model.

5.4   Noise and Lossy Rules

As discussed in Sect. 4.3, reduction rules may be categorized according to the information loss they cause and their impact on process model quality criteria. This can impact incremental or full-trace reduction algorithms, and information losses may also compound with repeated rule applications. This is most severe for the incremental algorithm *dinc*, as summary models are local on the log presented so far, and do not benefit from the context of the full log.

As an example, the *Choice Pruning* rule SL.1 removes low probability activities. However, probabilities of activities will fluctuate when few traces have been processed, and this may remove nodes which are actually well-represented across a full log. Accordingly, discovery algorithm variant *dc* takes a cleanup ruleset as a separate parameter, and performs a penultimate cleanup phase, applying the cleanup rules only once.

Definition 11 (Toothpaste Miner with Cleanup (TMC))

Given primary ruleset rs and cleanup ruleset cl,

$$dc\colon L \times \langle \mathcal{U}_A \to \mathcal{U}_A \rangle \times \langle \mathcal{U}_A \to \mathcal{U}_A \rangle \to \mathcal{G}$$

$$dc(L, rs, cl) = \Phi_K(\Phi(di(L, rs), cl), rs) \text{ is the TMC discovery algorithm}$$

The allocation of rules to main or cleanup rulesets is implementation-dependent.

6   Implementation and Evaluation

A prototype *Toothpaste Miner* was evaluated against existing stochastic process mining techniques using real-world logs. Results show good log conformance based on Earth-Movers distance [22], in feasible execution times, at the cost of more complex models.

6.1 Implementation

The prototype has been implemented in Haskell[1], extending other open source process mining tools [27]. Conversion from XES to a simpler, delimited text log format is done using Python and pm4py [6]. The implementation uses binary trees, to exploit the pattern-matching capabilities of Haskell. It maintains $\wedge$ and $\times$ nodes in lexical order for cheaper comparisons, and to limit traversal distance for similarity rules such as *Choice similarity* FP.1. Haskell allows for concise expression of transformation rules, as in Listing 1.1.

Listing 1.1. Choice similarity FP.1

```
choiceSim :: (Eq a) ⇒ PRule a
choiceSim (Node2 Choice x y n)
  | x =˜= y = merge x y
choiceSim x = x
```

The extensions of incremental discovery and K retries are not included in the prototype. The noise-reducing choice pruning rule SL.1 is not included, and loop similarity FPL.7 is only partially applied. All fixed loops $\circlearrowleft_n$ are converted to probabilistic loops $\circlearrowleft_p$. Rules in Sect. 4.3 are otherwise included.

6.2 Evaluation Design

In order to evaluate the potential practical use of our technique, we compare it to established stochastic discovery techniques in the literature. K-fold cross validation ($k = 5$) was used on three logs, and results compared using stochastic quality criteria, simplicity and computation time, summarized in Table 3.

Table 3. Evaluation design

Logs	Techniques	Measures	Environment
BPIC 2013 closed	Toothpaste (this paper)	tEMSC 0.8 [22]	Windows 10
Sepsis	GDT_SPN Discovery [30]	Entity count	2.3 GHz CPU
Teleclaims [1]	walign-inductive [9]	Computation time	50 Gb memory
	walign-split [9]		GHC 8.8.4
	wfreq-inductive [9]		JDK 1.8.0_222
	wfreq-split [9]		Python 3.8.3
	wpairscale-split [9]		
	wpairscale-inductive [9]		
	Trace model		

We evaluated all stochastic process discovery techniques where, to our knowledge, a public implementation was available. GDT_SPN discovery [30], weight estimator [9] and Inductive Miner [20] implementations are from ProM

---

[1] Source code is accessible at https://github.com/adamburkegh/toothpaste.

6.9 (development branch Jan 2021). Stochastic weight estimator combinations were chosen where previous results [9] had shown meaningful differences. Logs were selected to represent multiple domains. All logs are publicly available[2]. BPIC2013 and Sepsis are real-life logs, and Teleclaims is an established dataset [1, p. 243].

Stochastic quality measures used were Earth Movers' Distance (tEMSC) [22] with 0.8 probability mass. Other stochastic measures such as [19] were restricted to deterministic nets. Entity count is used to measure complexity.

6.3   Results and Discussion

A selection of evaluation results[3] are shown in Table 4. The output from the full Teleclaims log is shown in Fig. 7. For some k-fold logs and models, Earth Movers' Distance errored due to memory limits, or the calculation was timed out after 5 h. No value reflects no result from any k-fold log; otherwise, partial results from the remaining logs have been used.

The Toothpaste Miner prototype shows a trade-off of improved Earth Movers' Distance against longer running times and higher model complexity. This is analogous to the trade-off of fitness and precision against complexity and run-time often seen with region-based control-flow miners. For reference log Teleclaims, a human-readable process model was discovered in four seconds.

Some rules did not fire during the evaluations, which suggests the value of optimizations in Sect. 5.4, where certain rules are only applied during later phases of discovery. Real-life logs show sensitivity to the ordering of choice rules versus log formation rules, with marked model differences depending on rule sequence. This may reflect the partial implementation of loop similarity FPL.7 in the prototype. For Teleclaims (see Fig. 7), some redundancy is apparent due to prototype limitations in similarity identification for larger sub-trees.

On more challenging real-life logs, the prototype returned within two minutes, a shorter time than the existing GDT_SPN technique. For the BPIC2013 log, the additional complexity did not achieve a marked quality improvement. For the Sepsis log, it was able to retain a very similar $tEMSC$ to the trace model with a smaller entity footprint, though the final model is not human-readable. The reductions used may form part of other discovery or conformance strategies, say as a post-processing step.

---

[2] BPIC2013 and sepsis logs: https://data.4tu.nl/. Teleclaims: http://www.processmining.org/event_logs_and_models_used_in_book.

[3] Full results are available at https://github.com/adamburkegh/toothpaste.

Table 4. Evaluation results.

Miner	Log	Duration (ms)	Entities	tEMSC 0.8
GDT_SPN discovery	BPIC2013 closed	203	25	0.387
Toothpaste	BPIC2013 closed	3071	315	0.668
Trace	BPIC2013 closed	31	2417	0.9334
walign-inductive	BPIC2013 closed	194	6	0.6756
walign-split	BPIC2013 closed	56	14	0
wfreq-inductive	BPIC2013 closed	160	6	0.7346
wfreq-split	BPIC2013 closed	18	14	0.6188
wpairscale-inductive	BPIC2013 closed	18	6	0.7364
wpairscale-split	BPIC2013 closed	15	14	0.8753
GDT_SPN discovery	Sepsis	10147434	95	–
Toothpaste	Sepsis	84551	2992	0.7019
Trace	Sepsis	25	5877	0.7029
walign-inductive	Sepsis	497	40	–0
walign-split	Sepsis	478	51	0.3645
wfreq-inductive	Sepsis	178	40	–
wfreq-split	Sepsis	25	51	0.5108
wpairscale-inductive	Sepsis	34	40	0.6664
wpairscale-split	Sepsis	37	51	–
GDT_SPN discovery	Teleclaims	453	60	–
Toothpaste	Teleclaims	3564	121	–
Trace	Teleclaims	61	17754	0.9771
walign-inductive	Teleclaims	238	28	0.3529
walign-split	Teleclaims	137	56	0.0931
wfreq-inductive	Teleclaims	203	28	0.5105
wfreq-split	Teleclaims	59	56	0.6301
wpairscale-inductive	Teleclaims	62	28	0.5143
wpairscale-split	Teleclaims	68	56	0.6299

7   Conclusion

Stochastic Petri Nets are powerful modelling tools with wide applicability. Automatically discovered stochastic process models, in turn, can help understand and improve organizations. In this paper we presented the *Toothpaste Miner* framework for discovering and reasoning about stochastic process models in the context of process mining. We shared both batch and incremental discovery algorithms, showed they were computationally tractable, and would maintain determinism in models once such a model was discovered. A classification scheme relating transformation rules and process mining quality measures was articulated. Lastly we discussed an implementation of the discovery technique, with an empirical evaluation showing close to trace-model levels of similarity to real-life logs, with significantly less required model entities.

Future work in this area may investigate algorithms guaranteed to output deterministic stochastic models, simpler process models with better human-readability, and solutions where the choice of ruleset allows for constraining solutions by particular quality parameters. Other extensions could support additional statistical distributions and timed transitions.

References

1. van der Aalst, W.: Process Mining: Data Science in Action, 2nd edn. Springer-Verlag, Heidelberg (2016). https://doi.org/10.1007/978-3-662-49851-4_1
2. Anastasiou, N., Horng, T.C., Knottenbelt, W.: Deriving generalised stochastic Petri net performance models from high-precision location tracking data. In: Proceedings of the 5th International ICST Conference on Performance Evaluation Methodologies and Tools, pp. 91–100. ICST (Institute for Computer Sciences, Social-Informatics and Telecommunications Engineering) (2011)
3. Baum, L.E., Petrie, T., Soules, G., Weiss, N.: A maximization technique occurring in the statistical analysis of probabilistic functions of Markov chains. Ann. Math. Stat. **41**(1), 164–171 (1970), publisher: JSTOR
4. Bause, F., Kritzinger, P.: Stochastic Petri Nets: An Introduction to the Theory. Vieweg+Teubner Verlag (2002)
5. Bellodi, E., Riguzzi, F., Lamma, E.: Statistical relational learning for workflow mining. Intell. Data Anal. **20**(3), 515–541 (2016)
6. Berti, A., van Zelst, S.J., van der Aalst, W.M.P.: Process mining for python (PM4Py) : bridging the gap between process- and data science. In: ICPMD 2019, ICPM Demo Track 2019 : proceedings of the ICPM Demo Track 2019, co-located with 1st International Conference on Process Mining (ICPM 2019) : Aachen, Germany, 24–26 June 2019 / edited by Andrea Burattin (Technical University of Denmark, Kgs. Lyngby, Denmark), Artem Polyvyanyy (The University of Melbourne, Melbourne, Australia), Sebastiaan van Zelst (Fraunhofer Institute for Applied Information Technology (FIT), Sankt Augustin, Germany). CEUR workshop proceedings, vol. 2374, pp. 13–16. RWTH Aachen, Aachen, Germany (June 2019), backup Publisher: 1st International Conference on Process Mining, Aachen (Germany), 24 June 2019–24 June 2019
7. Bezem, M., Klop, J., Barendsen, E., de Vrijer, R., Terese: Term Rewriting Systems. Cambridge Tracts in Theoretical Computer Science. Cambridge University Press, Cambridge (2003)
8. Breuker, D., Matzner, M., Delfmann, P., Becker, J.: Comprehensible Predictive Models for Business Processes. MIS Q. **40**(4), 1009–1034 (2016)
9. Burke, A., Leemans, S.J.J., Wynn, M.T.: Stochastic process discovery by weight estimation. In: 2020 International Conference on Process Mining (ICPM) press (2020)
10. Carmona, J., Cortadella, J., Kishinevsky, M.: New region-based algorithms for deriving bounded petri nets. IEEE Trans. Comput. **59**(3), 371–384 (2010)
11. Carrasco, R.C.: Accurate computation of the relative entropy between stochastic regular grammars. RAIRO-Theor. Inform. Appl. **31**(5), 437–444 (1997)

12. Carrasco, R.C., Oncina, J.: Learning stochastic regular grammars by means of a state merging method. In: Carrasco, R.C., Oncina, J. (eds.) ICGI 1994. LNCS, vol. 862, pp. 139–152. Springer, Heidelberg (1994). https://doi.org/10.1007/3-540-58473-0_144
13. Fünfgeld, S., Holzäpfel, M., Frey, M., Gauterin, F.: Stochastic forecasting of vehicle dynamics using sequential Monte Carlo simulation. IEEE Trans. Intell. Veh. **2**(2), 111–122 (2017)
14. Geman, S., Geman, D.: Stochastic relaxation, gibbs distributions, and the bayesian restoration of images. IEEE Trans. Pattern Anal. Mach. Intell. **PAMI-6**(6), 721–741 (1984)
15. Hu, H., Xie, J., Hu, H.: A novel approach for mining stochastic process model from workflow logs. J. Comput. Inf. Syst. **7**(9), 3113–3126 (2011)
16. Janssenswillen, G., Depaire, B., Faes, C.: Enhancing discovered process models using Bayesian inference and MCMC. In: Proceedings of the 2020 BPI Workshop (2020)
17. Leclercq, E., Lefebvre, D., El Medhi, S.O.: Identification of timed stochastic petri net models with normal distributions of firing periods. IFAC Proc. **42**(4), 948–953 (2009)
18. Leemans, S.J.J., Fahland, D., van der Aalst, W.M.P.: Scalable process discovery with guarantees. In: Gaaloul, K., Schmidt, R., Nurcan, S., Guerreiro, S., Ma, Q. (eds.) CAISE 2015. LNBIP, vol. 214, pp. 85–101. Springer, Cham (2015). https://doi.org/10.1007/978-3-319-19237-6_6
19. Leemans, S.J.J., Polyvyanyy, A.: Stochastic-aware conformance checking: an entropy-based approach. In: Dustdar, S., Yu, E., Salinesi, C., Rieu, D., Pant, V. (eds.) CAiSE 2020. LNCS, vol. 12127, pp. 217–233. Springer, Cham (2020). https://doi.org/10.1007/978-3-030-49435-3_14
20. Leemans, S.J.J., Fahland, D., van der Aalst, W.M.P.: Discovering block-structured process models from event logs - a constructive approach. In: Colom, J.-M., Desel, J. (eds.) PETRI NETS 2013. LNCS, vol. 7927, pp. 311–329. Springer, Heidelberg (2013). https://doi.org/10.1007/978-3-642-38697-8_17
21. Leemans, S.J., Poppe, E., Wynn, M.T.: Directly follows-based process mining: exploration & a case study. In: 2019 International Conference on Process Mining (ICPM), pp. 25–32, June 2019
22. Leemans, S.J.J., Syring, A.F., van der Aalst, W.M.P.: Earth movers' stochastic conformance checking. In: Hildebrandt, T., van Dongen, B.F., Röglinger, M., Mendling, J. (eds.) BPM 2019. LNBIP, vol. 360, pp. 127–143. Springer, Cham (2019). https://doi.org/10.1007/978-3-030-26643-1_8
23. Liesaputra, V., Yongchareon, S., Chaisiri, S.: Efficient process model discovery using maximal pattern mining. In: Motahari-Nezhad, H.R., Recker, J., Weidlich, M. (eds.) BPM 2015. LNCS, vol. 9253, pp. 441–456. Springer, Cham (2015). https://doi.org/10.1007/978-3-319-23063-4_29
24. Maggi, F.M., Montali, M., Peñaloza, R.: Probabilistic conformance checking based on declarative process models. In: Herbaut, N., La Rosa, M. (eds.) CAiSE 2020. LNBIP, vol. 386, pp. 86–99. Springer, Cham (2020). https://doi.org/10.1007/978-3-030-58135-0_8
25. Marsan, M.A., Balbo, G., Bobbio, A., Chiola, G., Conte, G., Cumani, A.: The effect of execution policies on the semantics and analysis of stochastic Petri nets. IEEE Trans. Softw. Eng. **15**(7), 832–846 (1989)
26. Matsuno, H., Doi, A., Nagasaki, M., Miyano, S.: Hybrid petri net representation of gene regulatory network. In: Biocomputing 2000, pp. 341–352. World Scientific, December 1999

27. Mokhov, A., Carmona, J., Beaumont, J.: Mining conditional partial order graphs from event logs. In: Koutny, M., Desel, J., Kleijn, J. (eds.) Transactions on Petri Nets and Other Models of Concurrency XI. LNCS, vol. 9930, pp. 114–136. Springer, Heidelberg (2016). https://doi.org/10.1007/978-3-662-53401-4_6

28. Moreira, C., Haven, E., Sozzo, S., Wichert, A.: Process mining with real world financial loan applications: improving inference on incomplete event logs. PLOS ONE **13**(12), e0207806 (2018)

29. Polyvyanyy, A., Moffat, A., García-Bañuelos, L.: An Entropic Relevance Measure for Stochastic Conformance Checking in Process Mining. arXiv e-prints 2007. arXiv:2007.09310 (Jul 2020)

30. Rogge-Solti, A., van der Aalst, W.M.P., Weske, M.: Discovering stochastic petri nets with arbitrary delay distributions from event logs. In: Lohmann, N., Song, M., Wohed, P. (eds.) BPM 2013. LNBIP, vol. 171, pp. 15–27. Springer, Cham (2014). https://doi.org/10.1007/978-3-319-06257-0_2

31. Rogge-Solti, A., Weske, M.: Prediction of business process durations using non-Markovian stochastic Petri nets. Inf. Syst. **54**, 1–14 (2015)

32. Secretary, I.C.: Information technology - Z formal specification notation - Syntax, type system and semantics. Standard, International Organization for Standardization, Geneva, CH (March 2002), volume: (2002)

33. Silva, R., Zhang, J., Shanahan, J.G.: Probabilistic workflow mining. In: Proceedings of the Eleventh ACM SIGKDD International Conference on Knowledge Discovery in Data Mining, pp. 275–284. KDD 2005. Association for Computing Machinery, Chicago, Illinois, USA, August 2005

34. Tax, N., Lu, X., Sidorova, N., Fahland, D., van der Aalst, W.M.P.: The imprecisions of precision measures in process mining. Inf. Process. Lett. **135**, 1–8 (2018). https://doi.org/10.1016/j.ipl.2018.01.013

35. Thierry-Mieg, Y.: Structural reductions revisited. In: Janicki, R., Sidorova, N., Chatain, T. (eds.) PETRI NETS 2020. LNCS, vol. 12152, pp. 303–323. Springer, Cham (2020). https://doi.org/10.1007/978-3-030-51831-8_15

36. Thollard, F., Dupont, P., De La Higuera, C.: Probabilistic DFA inference using Kullback-Leibler divergence and minimality, pp. 975–982, June 2000

37. Verwer, S., Eyraud, R., De La Higuera, C.: PAutomaC: a probabilistic automata and hidden Markov models learning competition. Mach. Learn. **96**(1–2), 129–154 (2014)

38. Vidal, E., Thollard, F., de la Higuera, C., Casacuberta, F., Carrasco, R.: Probabilistic finite-state machines - part II. IEEE Trans. Pattern Anal. Mach. Intell. **27**(7), 1026–1039 (2005)

39. Vidal, E., Thollard, F., Higuera, C.d.l., Casacuberta, F., Carrasco, R.C.: Probabilistic finite-state machines - part I. IEEE Trans. Pattern Anal. Mach. Intell. **27**(7), 1013–1025 (2005)

40. Wang, X., Chen, G., Zhao, Q., Guo, Z.: Reduction of stochastic petri nets for reliability analysis. In: 2007 8th International Conference on Electronic Measurement and Instruments, pp. 1-222–1-226, August 2007, iSSN: null

41. Weisberg, M.: Simulation and Similarity : Using Models to Understand the World. Oxford University Press, Oxford (2013)

42. Zhou, M., Venkatesh, K.: Modeling, Simulation, and Control of Flexible Manufacturing Systems: A Petri Net Approach. World Scientific (1999)

Reachability and Partial Order

Efficient Algorithms for Three Reachability Problems in Safe Petri Nets

Pierre Bouvier[✉] and Hubert Garavel[✉]

Univ. Grenoble Alpes, INRIA, CNRS, Grenoble INP LIG, 38000 Grenoble, France
{pierre.bouvier,hubert.garavel}@inria.fr

Abstract. We investigate three particular instances of the marking coverability problem in ordinary, safe Petri nets: the Dead Places Problem, the Dead Transitions Problem, and the Concurrent Places Problem. To address these three problems, which are of practical interest, although not yet supported by mainstream Petri net tools, we propose a combination of static and dynamic algorithms. We implemented these algorithms and applied them to a large collection of 13,000+ Petri nets obtained from realistic systems—including all the safe benchmarks of the Model Checking Contest. Experimental results show that 95% of the problems can be solved in a few minutes using the proposed approaches.

1 Introduction

The present article focuses, in the framework of ordinary, safe Petri nets, on three related problems: the *Dead Place Problem*, which searches for all places that are never marked, the *Dead Transition Problem*, which searches for all transitions that can never fire, and the *Concurrent Places Problem*, which searches for all pairs of places that get a token in some reachable marking.

The two former problems characterize those parts of a net that are never active, and are thus relevant for simplifying complex Petri nets, especially those generated automatically from higher-level formalisms such as process calculi. The latter problem characterizes those parts of a net that can be simultaneously active, and plays a crucial role in the conversion of an ordinary, safe Petri net into an equivalent network of automata (see, e.g., [2]), an operation that opens the way to a compositional expression of Petri nets using process calculi.

The present article is organized as follows. Section 2 introduces the three problems, discussing their practical usefulness and theoretical complexity. Section 3 explains why mainstream model checkers are not optimal for these problems and presents the dedicated software tools that implement our algorithms, as well as the comprehensive set of models used to evaluate these algorithms. Section 4 describes various algorithms for the Dead Place Problem and the Dead Transition Problem, and reports about their performance when applied to the set of models. Section 5 does the same as Sect. 4 for the Concurrent Places Problem. Finally, Sect. 6 gives concluding remarks.

© Springer Nature Switzerland AG 2021
D. Buchs and J. Carmona (Eds.): PETRI NETS 2021, LNCS 12734, pp. 339–359, 2021.
https://doi.org/10.1007/978-3-030-76983-3_17

2   Problem Statement

2.1   Basic Definitions

We briefly recall the usual definitions of Petri nets and refer the reader to classical surveys for a more detailed presentation of Petri nets.

Definition 1. *A (marked) Petri Net is a 4-tuple* (P, T, F, M_0) *where:*

1. *P is a finite, non-empty set; the elements of P are called* places.
2. *T is a finite set such that $P \cap T = \varnothing$; the elements of T are called* transitions.
3. *F is a subset of $(P \times T) \cup (T \times P)$; the elements of F are called* arcs.
4. *M_0 is a non-empty subset of P; M_0 is called the* initial marking.

Notice that the above definition only covers *ordinary* nets (i.e., it assumes all arc weights are equal to one). Also, it only considers *safe* nets (i.e., each place contains at most one token), which enables the initial marking to be defined as a subset of P, rather than a function $P \to \mathbb{N}$ as in the usual definition of P/T nets. We now recall the classical firing rules for ordinary safe nets.

Definition 2. *Let (P, T, F, M_0) be a Petri Net.*

- *A marking M is defined as a set of places ($M \subseteq P$). Each place belonging to a marking M is said to be* marked *or, also, to* possess *a token.*
- *The* pre-set *of a transition t is the set of places ${}^\bullet t \overset{\text{def}}{=} \{p \in P \mid (p, t) \in F\}$.*
- *The* post-set *of a transition t is the set of places $t^\bullet \overset{\text{def}}{=} \{p \in P \mid (t, p) \in F\}$.*
- *A transition t is* enabled *in some marking M iff ${}^\bullet t \subseteq M$.*
- *A transition t can* fire *from some marking M_1 to another marking M_2 iff t is enabled in M_1 and $M_2 = (M_1 \setminus {}^\bullet t) \cup t^\bullet$, which we write as $M_1 \overset{t}{\longrightarrow} M_2$.*
- *A marking M is* reachable *from the initial marking M_0 iff $M = M_0$ or there exist $n \geq 1$ transitions $t_1, ..., t_n$ and $(n-1)$ markings $M_1, ..., M_{n-1}$ such that $M_0 \overset{t_1}{\longrightarrow} M_1 \overset{t_2}{\longrightarrow} M_2 ... M_{n-1} \overset{t_n}{\longrightarrow} M$, which we write as $M_0 \overset{*}{\longrightarrow} M$.*

We then recall the basic definition of a NUPN, referring the interested reader to [5] for a complete presentation of this model of computation.

Definition 3. *A (marked)* Nested-Unit Petri Net *(acronym: NUPN) is a 8-tuple $(P, T, F, M_0, U, u_0, \sqsubseteq, \mathsf{unit})$ where (P, T, F, M_0) is a Petri net, and where:*

5. *U is a finite, non-empty set such that $U \cap T = U \cap P = \varnothing$; the elements of U are called* units.
6. *u_0 is an element of U; u_0 is called the* root *unit.*
7. *$\sqsubseteq$ is a binary relation over U such that $(U, \sqsupseteq)$ is a tree with a single root u_0, where $(\forall u_1, u_2 \in U)\ u_1 \sqsupseteq u_2 \overset{\text{def}}{=} u_2 \sqsubseteq u_1$; intuitively[1], $u_1 \sqsubseteq u_2$ expresses that unit u_1 is transitively nested in or equal to unit u_2.*

---

[1] $\sqsubseteq$ is reflexive, antisymmetric, transitive, and u_0 is the greatest element of U for $\sqsubseteq$.

8. unit *is a function* $P \rightarrow U$ *such that* $(\forall u \in U \setminus \{u_0\})$ $(\exists p \in P)$ unit $(p) = u$; *intuitively,* unit $(p) = u$ *expresses that unit* u *directly contains place* p.

Because NUPNs merely extend Petri nets by grouping places into units, they do not modify the Petri-net firing rules for transitions: all the concepts of Definition 2 for Petri nets also apply to NUPNs, so that Petri-net properties are preserved when NUPN information is added. Finally, we recall a few useful NUPN concepts [5]:

Definition 4. *Let* $N = (P, T, F, M_0, U, u_0, \sqsubseteq, \text{unit})$ *be a NUPN.*

- *The predicate* disjoint $(u_1, u_2) \stackrel{\text{def}}{=} (u_1 \not\sqsubseteq u_2) \wedge (u_2 \not\sqsubseteq u_1)$ *characterizes pairs of units neither equal nor nested one in the other.*
- *A marking* $M \subseteq P$ *is said to be* unit safe *iff* $(\forall p_1, p_2 \in M)$ $(p_1 \neq p_2) \Rightarrow$ disjoint $(\text{unit}(p_1), \text{unit}(p_2))$, *i.e., all the places of a unit-safe marking are contained in disjoint units.*
- *The NUPN* N *is said to be* unit safe *iff its underlying Petri net* (P, T, F, M_0) *is safe and all its reachable markings are unit safe.*
- N *is said to be* trivial *iff its number of leaf units equals its number of places, meaning that* N *carries no more NUPN information than a Petri net.*

2.2   The Dead Places Problem

Definition 5. *Let* (P, T, F, M_0) *be a Petri Net. A place* $p \in P$ *is* dead *if there exists no reachable marking containing* p.

A more general definition is given in [4, Def. 4.16], where a place p is said to be dead at a marking M if it is not marked at any marking reachable from M. Our definition only considers the case where the net is safe and p is dead at the initial marking M_0.

In the present article, we will carefully avoid using the word *live*, because it is not the negation of *dead* according to the standard definition given in Petri-net literature, namely: a place p is live if, for any reachable marking M, there exists a marking M' that contains p and is reachable from M. Thus, any live place is not dead, but a non-dead place is not necessary live.

Definition 6. *Given a Petri net* (P, T, F, M_0), *the* Dead Places Problem *consists in finding all the dead places. Equivalently, this problem consists in computing a vector of* $|P|$ *bits such that, for each place* p, *the bit corresponding to* p *in the vector is equal to one iff* p *is dead.*

2.3   The Dead Transitions Problem

Definition 7. *Let* (P, T, F, M_0) *be a Petri Net. A transition* $t \in T$ *is* dead *if there exists no reachable marking in which* t *is enabled.*

The above definition is a particular case, for the initial marking M_0, of the definition given in [4, Def. 4.16], where a transition t is dead at a marking M if t is not enabled in any marking reachable from M.

Again, to avoid confusion, we will not use the word *live* throughout the present paper, as there exist multiple notions of liveness for transitions, e.g., *L1-live*, *L2-live*, *L3-live*, and *L4-live* [13, Sect. IV.C]; in our context, the negation of *dead* is not *live*, which means *L4-live* and is a too strong property, but *L1-live*, also known as *quasi-live*.

Definition 8. *Given a Petri net (P, T, F, M_0), the* Dead Transitions Problem *consists in finding all the dead transitions. Equivalently, this problem consists in computing a vector of $|T|$ bits such that, for each transition t, the bit corresponding to t in the vector is equal to one iff t is dead.*

Notice that this problem is different from the *deadlock freeness* problem: a net has a deadlock if there exists a reachable marking in which no transition is enabled, whereas a net has a dead transition if there exists a transition that is not enabled in any reachable marking.

Notice that this problem is also different from the *net quasi-liveness* problem: asking whether a net is quasi-live only calls for a Boolean answer, whereas the Dead Transitions Problem calls for a vector of Booleans. The latter problem is more general, as the net is quasi-live iff the solution of the Dead Transitions Problem is a vector of zeros. In practice, when studying a complex net, quasi-liveness is not sufficient, as one needs to know the exact set of dead transitions.

In practice, the Dead Places and Dead Transition Problems are relevant for several reasons. These problems are instances, with respect to Petri nets, of the more general *dead code* problem in software engineering. Dead code is generally considered as a nuisance for readability and maintenance, so that most software methodologies recommend to get rid of dead code. This is also the case for Petri nets in industrial automation, as the Grafcet specification prohibits Sequential Function Charts containing "unreachable" branches (i.e., Petri nets with dead places or dead transitions).

In model-checking verification, dead places and dead transitions are likely to increase the cost (in memory and time) of verification. Moreover, many global properties of a net can be changed to true or to false just by adding dead places and/or dead transitions. This may disturb structural analyses or net transformations, by invalidating certain "good" properties (e.g., free choice) and thus lead to incorrect transformations or prevent the application of efficient algorithms relying on such properties. It is therefore important to detect and eliminate dead places and dead transitions to only consider a truly "minimal" Petri net.

2.4   The Concurrent Places Problem

Definition 9. *Let (P, T, F, M_0) be a Petri Net. Two places p_1 and p_2 are concurrent, which we write as $p_1 \parallel p_2$, if there exists a reachable marking M containing both p_1 and p_2.*

Proposition 1. *The relation $\parallel$ is symmetric and quasi-reflexive. It is reflexive iff the net has no dead place.*

Proof. Symmetry and quasi-reflexivity follow from Definition 9. As for reflexivity, a place is concurrent with itself iff it is not dead. Notice that both the relation $\parallel$ and its negation $\nparallel$ are neither transitive nor intransitive (since they are not irreflexive).

The relation $\parallel$ can be found in the literature under various names: *coexistency defined by markings* [9, Sect. 9], *concurrency relation* [7,10–12,15], *concurrency graph* [16], etc. These definitions differ by details, such as the kind of Petri nets considered or the handling of reflexivity, i.e., whether and when a place is concurrent with itself or not. For instance, [9] defines that $p \parallel p$ is always false, while [10] defines that $p \parallel p$ is true iff there exists a reachable marking in which place p has at least two tokens.

Although the concurrency relation can easily be defined on non-ordinary, non-safe nets (see, e.g., [10]), this is of limited interest. For instance, consider a Petri net that is a state machine, with an initial marking M_0 containing a single place p_0, and all the places reachable from p_0: if there is initially a single token in p_0, each place is non-concurrent with each other, but as soon as there is initially more than one token in p_0, all the places become pairwise concurrent.

Definition 10. *Given a Petri net (P, T, F, M_0), the* Concurrent Places Problem *consists in finding all pairs of concurrent places. Equivalently, since the concurrency relation is symmetric, this problem consists in computing a half matrix of $|P|(|P| + 1)/2$ bits such that, for each two places p_1 and p_2, the corresponding bit in the half matrix is equal to one iff $p_1 \parallel p_2$.*

As mentioned above, solving the Concurrent Places Problem is practically useful, e.g., for decomposing a Petri net into a network of automata [2].

2.5   Complexity

Proposition 2. *The Dead Places Problem is a subproblem of the Dead Transition Problem.*

Proof. Given the set of dead transitions, one can easily compute the set of dead places. Indeed, each non-dead place belongs to the initial marking and/or the post-set of at least one non-dead transition. Notice that the converse does not hold, as the set of dead transitions cannot be directly inferred from the set of dead places.

Proposition 3. *The Dead Places Problem is a subproblem of the Concurrent Places Problem.*

Proof. The diagonal of the Concurrent Places half matrix is the negation of the Dead Places vector.

Proposition 4. *The Dead Places Problem, the Dead Transition Problem, and the Concurrent Places Problem are subproblems of the marking coverability problem, which is the problem of deciding whether a given marking is included in some reachable marking of a given Petri net.*

Proof. Given a set of places M, let $R(M)$ be the predicate: *does it exist a reachable marking containing all the places of M?* The Dead Places Problem can be expressed as: for each place p of the net, decide $R(\{p\})$. The Dead Transitions Problem can be expressed as: for each transition t, decide $R(^\bullet t)$. The Concurrent Places Problem can be expressed as: for each two places p_1 and p_2, decide $R(\{p_1, p_2\})$.

Proposition 5. *On safe nets, the Dead Places Problem, the Dead Transition Problem, and the Concurrent Places Problem are PSPACE-complete.*

Proof. One knows from [3, Th. 15] that the marking coverability problem is PSPACE-complete on safe nets. Although the Dead Places Problem is a subproblem of the marking coverability problem (see Proposition 4), its complexity is not lower: given a net N, let N' be the net consisting of N to which one adds a new place p and a new transition t such that $^\bullet t = M$ and $t^\bullet = \{p\}$; if N is safe, then N' is also safe; deciding $R(p)$ in N' requires to decide whether M is coverable in N. Given that the Dead Place Problem is a subproblem of the Dead Transition Problem (see Proposition 2) and of the Concurrent Places Problem (see Proposition 3), the two latter problems are also PSPACE-complete on safe nets.

2.6 Complete vs Incomplete Solutions

Because the three aforementioned problems are PSPACE-complete, there will always be Petri nets large enough to prevent the computation of exhaustive solutions on a given computer. Rather than an "all or nothing" approach (in which an algorithm is considered to fail if it cannot compute all the dead places, dead transitions, or concurrent places of a given net), one should consider more pragmatic approaches in which an algorithm may stop or be halted after computing only a part of the solution, still leaving some results unknown.

Concretely, this means that the vectors of dead places and dead transitions, and the half matrices of concurrent places should contain three-valued logical results (zero, one, or unknown) rather than mere Boolean results. A solution is said to be *incomplete* if it contains unknown values, or *complete* otherwise. An efficient algorithm should produce as few unknown results as possible in a given lapse of time.

For the Dead Places Problem and Dead Transitions Problems, an incomplete solution can be turned into a complete one by replacing all unknown values by zeros, meaning that places and transitions are considered to be dead only if this has been positively proven.

For the Concurrent Places Problem, the elimination of unknown values depends on the context. For instance, the decomposition of safe Petri nets into automata networks [2] can work with incomplete half matrices, in which it replaces unknown values by ones—meaning that two places are assumed by default to be concurrent unless proven otherwise. Under such a pessimistic assumption, the algorithms that produce zeros in the half matrix are clearly more useful than the algorithms that produce ones. However, they may exist other applications based upon optimistic assumptions about concurrency, for which the latter kind of algorithms would be preferable.

3   Implementation and Experimentation

3.1   Relation to Temporal Logic

Given that our three reachability problems are particular instances of the marking coverability problem, one way to solve these problems is to encode them in temporal logic (e.g., CTL or LTL formulas) and to submit them to an existing model checker for Petri nets, e.g., one of those competing every year at the Model Checking Contest [1]. However, such an approach is not practical:

- The number of temporal-logic formulas required for each problem is linear or quadratic in the size of the Petri net—respectively, $|P|$, $|T|$, and $|P|(|P|+1)/2$. This is generally too large to be done manually, so one has to develop ad hoc tools that build these formulas (generating a huge file or a large number of small files), invoke a model checker on each of these formulas, then collect and aggregate the results of these invocations.
- Invoking a model checker repeatedly to evaluate hundreds or thousands of formulas on the same Petri net is inefficient. At each invocation, the formula and the net are parsed, checked for correctness, and translated from concrete syntax to internal abstract representation: most of these steps are redundant given that all formulas are alike and correct by construction.

We thus believe that, although the three aforementioned problems can be reduced to the evaluation of temporal-logic formulas, it is better to express these problems at a higher level, namely by equipping Petri-net tools with built-in options (e.g., -dead-places, -dead-transitions, and -concurrent-places) dedicated to these problems. Not only such options would be easier for tool users, but they would also give tool developers more freedom to choose the most efficient approach(es).

For a tool based on some temporal logic, such options would allow a major boost in performance: (i) they could produce the formulas in the most appropriate temporal logic supported by the tool; (ii) they could generate the formulas directly in the internal abstract representation, thus saving the cost of writing intermediate files and later parsing these files; (iii) they would also save the cost of correctness checking, since the generated formulas would be known to be correct by construction; (iv) the Petri-net model would be read and analyzed only once; (v) because the tool knows in advance how many formulas are to be evaluated on this Petri net, it can try applying preliminary simplifications (e.g., structural reductions) and sophisticated optimizations to this model; (vi) the tool may profitably consider using *global model checking* (i.e., building the set of reachable markings first, then evaluating all the formulas on this state space) whereas, for a single formula, *local model checking* (i.e., on-the-fly evaluation that only explores a fragment of the state space relevant to the formula) is often the default choice.

If temporal logics are one possible way to address the three aforementioned problems, they are not the only way. The remainder of this article presents alternative approaches, whose algorithms are implemented in software tools.

3.2   Software Tools and File Formats

We implemented our proposed algorithms in two different tools:

- CÆSAR.BDD[2] is a verification tool for Petri nets and NUPNs. It is written in C (10,400 lines of code), uses the CUDD library for Binary Decision Diagrams, and is available as part of the CADP toolbox. CÆSAR.BDD provides many functionalities, among which solutions for the three aforementioned problems. For instance, it is routinely used to remove dead transitions from large interpreted Petri nets automatically generated from specifications written in higher-level languages such as LOTOS, LNT, AADL, etc.
- ConcNUPN is a prototype written in Python (730 lines of code) that implements algorithms for the three aforementioned problems and a few other functionalities. It is used to quickly prototype new ideas and to cross-check the results produced by CÆSAR.BDD.

Both tools take as input a file in the NUPN format[3] [5, Annex A], which provides a concise, human-readable representation for ordinary, safe Petri nets. Two translators[4] automatically convert the NUPN format to the standard PNML format [8] and vice versa.

Depending on the option given (`-dead-places`, `-dead-transitions`, or `-concurrent-places`), the tools produce as output a vector or a half matrix. Both are encoded in the same textual format [6, Sect. 8 and A] made up of one or more lines containing the characters "0", "1", and ".", the latter representing unknown values. If the input net is an non-trivial NUPN, some values "0" and "." in the half matrix may be replaced by other characters giving additional information about the NUPN structure. Because half matrices can get large, a run-length compression scheme[5] is used, which, on average, divides by 2.4 the size of vectors and by 214 the size of half matrices—a reduction factor of 4270 was even observed on very large half matrices.

3.3 Data Sets

To perform our experiments, we used a collection of 13,116 models in NUPN format. Most of these models are derived from "realistic" specifications (i.e., nets obtained from industrial problems described by humans in high-level languages, rather than randomly generated Petri nets). Our collection contains all the ordinary, safe models from the former PetriWeb collection and from the 2020 edition of Model Checking Context[6].

A statistical survey confirms the diversity of our collection. The upper part of Table 1 gives the percentage of models that satisfy various (structural and behavioural) net properties, including the percentage of nets that are non-trivial NUPNs known to be unit safe (and thus, safe) by construction. The lower part of the table gives information about the size of the models: number of places, transitions, and arcs, as well as *arc density*, which we define as the number of arcs divided by twice the product of the number of places and the number of transitions, i.e., the amount of memory needed to store the arc relation as a pair of place×transition matrices.

Table 1. Structural, behavioural, and numerical properties of our models

Property	Yes	No	Property	Yes	No
Pure	62.9%	37.1%	Connected	94.0%	6.0%
Free Choice	41.3%	58.7%	Strongly connected	14.3%	85.7%
Extended free choice	42.7%	57.3%	Conservative	16.5%	83.5%
Marked graph	3.5%	96.5%	Sub-conservative	29.7%	70.3%
State machine	12.1%	87.9%	Non trivial and unit safe	67.7%	32.3%

Feature	Min value	Max value	Average	Median	Std deviation
#places	1	131,216	282.4	15	2,690
#transitions	0	16,967,720	9,232.8	20	270,287
#arcs	0	146,528,584	72,848	55	2,141,591
Arc density	0.0%	100.0%	14.5%	9.4%	0.2

4 Algorithms for Dead Places and Dead Transitions

This section discusses various algorithms for the Dead Places and Dead Transition Problems, both of which are addressed together as they are largely similar.

4.1 Marking Graph Exploration

The easiest (at least, conceptually) way of computing the dead places and dead transitions of a safe Petri net is to follow the global model checking approach

---

[6] http://mcc.lip6.fr/models.php.

mentioned in Sect. 3.1 by first exploring all the reachable markings, and then examining whether, during this exploration, some places have never been marked or some transitions have never been fired.

If the state space can be explored entirely, one obtains the complete vectors for dead places and dead transitions. If the state space is too large for being exhaustively generated, these vectors only contain zeros and unknown values, but no ones.

Actually, one can often avoid generating the state space entirely, still obtaining complete vectors. This can be achieved using algorithmic *shortcuts*, which stop the exploration as soon as all the information needed has been determined, following the idea of *on-the-fly* verification. For dead transitions, there is one shortcut: the exploration can stop if each transition has been fired at least once, meaning that the net is quasi-live. For dead places, there are two shortcuts: the exploration can stop if each place has been marked, meaning that there are no dead places, or if each transition has been fired at least once, meaning that a place never marked so far cannot receive a token from further transition firings.

These ideas have been implemented as follows in the CÆSAR.BDD tool. Reachable markings are represented using Binary Decision Diagrams, as symbolic exploration proved, in the case of safe Petri nets, to be much more efficient than explicit-state exploration. If the input net is a non-trivial NUPN and is known to be unit-safe by construction, CÆSAR.BDD takes advantage of this information to significantly reduce the number of Boolean variables needed to encode reachable markings [5, Sect. 6]. The user can limit state-space construction by specifying a timeout or giving an upper bound on the depth of exploration. Observing which transitions have been fired is easy, as CÆSAR.BDD fires each transition separately (like in explicit-state exploration) rather than building a single BDD that encodes all the transition relation. Shortcuts are simply implemented using decreasing counters that store the number of remaining unknown values in the solution vectors.

4.2 Structural Rules

Marking-graph exploration is a brute-force approach, which may fail on large nets, yielding incomplete vectors. We now examine complementary algorithms of a lower complexity, which take as input a vector containing unknown values and produce as output the same vector in which some unknown values have been replaced by zeros or ones—meaning that all previously known values are preserved. In this section, we present such an algorithm based on eight simple "structural" rules (Proposition 6–13) that can decide whether certain places or transitions are dead or not.

Proposition 6. *Any place belonging to the initial marking M_0 is not dead.*

Proposition 7. *Any transition having no input place and no output place is not dead.*

The two next rules exploit the properties of safeness (we only consider Petri nets that are expected to be safe) and unit safeness (a large proportion of our models are known to be unit safe by construction—see Sect. 3.3) to characterize certain classes of dead transitions.

Proposition 8. *If the net is safe, any transition whose input places form a strict subset of the output places is dead.*

Notice that, if a transition has no input place and at least one output place, the net is not safe, as this transition can fire indefinitely often, accumulating tokens in its output places.

Proposition 9 (from [5] Proposition 8). *If the net is unit safe, any transition having at least two input (resp. two output) places located in two non-disjoint NUPN units is dead.*

The four last rules allow to propagate, in certain cases, the fact that a place (resp. a transition) is dead or not dead to its adjacent transitions (resp. places).

Proposition 10 (from [4] Proposition 4.17(3)). *If a place p is dead, all the transitions of $^\bullet p \cup p^\bullet$ are also dead.*

Proposition 11 (contrapositive of Proposition 10). *If a transition t is not dead, all the places of $^\bullet t \cup t^\bullet$ are also not dead.*

Proposition 12. *If a transition t is dead, any place p such that $^\bullet t = \{p\}$ is also dead.*

Proposition 13 (contrapositive of Proposition 12). *If a place p is not dead, any transition t such that $^\bullet t = \{p\}$ is also not dead.*

The algorithm below applies Proposition 6–13 iteratively until saturation. The vector of dead places is represented by P_1 and P_0, which are, respectively, the sets of places known to be dead and not dead, the places of $P \setminus (P_0 \cup P_1)$ being unknown. Similarly, the vector of dead transitions is represented by two sets T_1 and T_0. Before the algorithm starts, these four sets have been either initialized to $\varnothing$ or filled in by some other algorithm(s), such as the one of Sect. 4.1.

```
1    P_0 := P_0 ∪ M_0                                          — from Proposition 6
2    T_0 := T_0 ∪ {t ∈ T | •t = t• = ∅}                        — from Proposition 7
3    T_1 := T_1 ∪ {t ∈ T | (•t ⊆ t•) ∧ (•t ≠ t•)} ∪           — from Proposition 8 and 9
4        {t ∈ T | (∃(p_1, p_2) ∈ (•t × •t) ∪ (t• × t•)) ¬disjoint (unit (p_1), unit (p_2))}
5    P' := ∅ ; T' := ∅ ; assert P_0 ≠ ∅
6    while P' ≠ P_0 ∪ P_1 loop
7        assert P' ⊊ P_0 ∪ P_1
8        for p ∈ (P_0 ∪ P_1) \ P' loop
9            if p ∈ P_1 then
10               T_1 := T_1 ∪ •p ∪ p•                          — from Proposition 10
11           else if p ∈ P_0 then
```

12                    $T_0 := T_0 \cup \{t \in T \mid {}^\bullet t = \{p\}\}$ — *from Proposition 13*
13   **end loop**
14          $P' := P_0 \cup P_1$
15   **assert** $T' \subseteq T_0 \cup T_1$
16   **for** $t \in (T_0 \cup T_1) \setminus T'$ **loop**
17   **if** $t \in T_0$ **then**
18                  $P_0 := P_0 \cup {}^\bullet t \cup t^\bullet$ — *from Proposition 11*
19   **else if** $t \in T_1 \wedge |{}^\bullet t| = 1$ **then**
20                  $P_1 := P_1 \cup {}^\bullet t$ — *from Proposition 12*
21   **end loop**
22          $T' := T_0 \cup T_1$
23 **end loop**

4.3   Linear Over-Approximation

Definition 11. *Let M_1 and M_2 be two markings, and t a transition. We write $M_1 \overset{t}{\Longrightarrow} M_2$ iff t is enabled in M_1 (i.e., ${}^\bullet t \subseteq M$) and $M_2 = M_1 \cup t^\bullet$.*

The relation differs from the usual firing relation $M_1 \overset{t}{\longrightarrow} M_2$ (cf. Definition 2) in that the latter uses $(M_1 \setminus {}^\bullet t)$ instead of M_1. Thus, when a transition t fires according to Definition 11, each input place of t keeps its token, while each output place of t gets a token. Said otherwise, $\Longrightarrow$ behaves exactly as $\longrightarrow$ if one assumes that each marked place holds an infinite number of tokens (hence, $M_1 \setminus {}^\bullet t = M_1$).

Proposition 14. *If a marking M is reachable from the initial marking M_0, i.e., $M_0 \overset{*}{\longrightarrow} M$, there exists a marking M' such that $M_0 \overset{*}{\Longrightarrow} M'$ and $M \subseteq M'$.*

Proof. By induction on firing sequences $M_0 \overset{t_1}{\longrightarrow} M_1 \overset{t_2}{\longrightarrow} M_2 \ldots M_{n-1} \overset{t_n}{\longrightarrow} M$.

The algorithm below is based on the contrapositive of Proposition 14. It performs a marking-graph exploration, starting from M_0 and using $\Longrightarrow$ instead of $\longrightarrow$. During the exploration, each place, once marked, never loses its token, so that the state space can be simply represented using the set P' of visited places and the state P'' of explored places (with $P' \subseteq P''$). We speed up transition firings by attaching to each (non-dead) transition t a counter $c[t]$ containing the number of input places of t that have not been marked yet, so that t becomes fireable when $c[t]$ drops to zero. When the exploration completes, any place that has not been marked is dead (and thus added to P_1) and any transition that has not been fired is dead (and thus added to T_1). The converse is not true: unless each transition has a single input place, the algorithm may overlook some dead places and/or dead transitions, as it over-approximates the number of tokens and the set of fireable transitions.

1 $P' := \varnothing$
2 $P'' := P_0 \cup M_0$
3 **for** $t \in T \setminus T_1$ **loop** $c[t] := |{}^\bullet t|$

```
4   while P' ≠ P'' loop
5       assert (P' ⊊ P'') ∧ (P'' ∩ P_1 = ∅) ∧ (∀t ∈ T \ T_1) (c[t] = |•t \ P'|)
6       let p = oneof (P'' \ P')
7       P' := P' ∪ {p}
8       for t ∈ p• \ T_1 loop
9           c[t] := c[t] − 1
10          if c[t] = 0 then P'' := P'' ∪ t•
11      end loop
12  end loop
13  assert (P' = P'') ∧ (P' ∩ P_1 = ∅) ∧ (∀t ∈ T \ T_1) (c[t] = |•t \ P'|)
14  P_1 := P_1 ∪ (P \ P')
15  T_1 := T_1 ∪ {t ∈ T \ T_1 | c[t] > 0}
```

At line 14, the new set of dead places $(P \setminus P')$ contains the initial set of dead places P_1 if Proposition 10 has been applied before executing this algorithm.

4.4   Ordering of Algorithms

Let A_1, A_2, and A_3 denote the algorithms presented in Sects. 4.1, 4.2, and 4.3, respectively. Let A_2 be split into two successive parts: $A_2 = A_2'; A_2''$, where A_2' comprises lines 1 to 4 (Proposition 6–9) and A_2'' comprises lines 5 to 23 (Proposition 10–13). It is easy to see that, after executing any of these algorithms, applying it immediately again never decreases the number of unknown values. But two successive executions of some algorithm may be fruitful if another algorithm has been successfully applied in between.

Thus, the next question is: in which order, and how many times, should these algorithms be applied? Our experiments suggest that executing them in the order $(A_2; A_3; A_1; A_2'')$ is likely to give the best results, based upon the fact that A_1, which is the most expensive algorithm, can take advantage of the information pre-computed by $(A_2; A_3)$. Namely, A_1 can avoid trying to fire those transitions known to be dead, and it can enhance the effectiveness of the algorithmic shortcuts defined in Sect. 4.1 by generalizing their triggering conditions: instead of checking if all places have been marked or all transitions fired at least once, A_1 can merely check if the solution vector contains no more unknown values, which better takes into account the existence of dead places or dead transitions, if any.

Finally, between each two algorithms, one also checks the number of remaining unknown values in the vector being computed, and the execution sequence stops if this number drops to zero. For instance, A_1 will not be applied if the prior execution of $(A_2; A_3)$ has produced a complete solution.

4.5   Experimental Results

We applied these algorithms to compute the dead places and dead transitions of the 13,116 models presented in Sect. 3.3. Our experiments are parameterized by a duration t, which is the number of seconds allocated to algorithm A_1 to symbolically explore (a fragment of) the graph of reachable markings—using the

CUDD library for Binary Decision Diagrams. If t is zero, only the initial marking is explored and no transition is fired by A_1.

Our experiments reveal that at least 16.2% (resp. 15.9%) of the models contain dead places (resp. transitions), and that at least 20.4% (resp. 37.7%) of dead places (resp. transitions) are globally present among the models. Such important ratios confirm the practical relevance of detecting and eliminating dead places and transitions as a means to reduce the complexity of Petri nets.

Table 2. Experimental results for dead places and dead transitions

Problem	Value of t	0	5	10	15	30	45	60	120	180	240	300
Dead places	% Complete vectors	44.6	93.0	93.6	93.8	94.4	94.6	95.1	95.3	95.4	95.5	95.6
	% Unknowns values	48.9	33.5	32.0	31.3	28.9	28.3	27.9	27.1	26.5	25.9	25.8
	% Vector completion	69.3	97.0	97.3	97.5	97.7	97.9	97.9	98.1	98.1	98.2	98.2
Dead trans.	% Complete vectors	29.3	92.3	92.9	93.2	93.7	94.0	94.1	94.4	94.7	94.9	95.0
	% Unknowns values	68.7	65.0	63.5	62.0	61.0	59.3	57.8	54.6	45.2	39.9	29.8
	% Vector completion	50.9	95.8	96.2	96.4	96.7	96.8	96.9	97.1	97.2	97.3	97.3

Table 2 provides, for various values of t, three metrics about the computation of dead places (resp. transitions). The first metrics ("% complete vectors") gives the percentage of models for which the solution vector can be completely computed within t seconds. The second metrics ("% unknowns values") gives the percentage of unknown values that remain after t seconds in all the computed solution vectors. The third metrics ("% vector completion") gives the mean, over all models, of percentage of known values in the computed solution vectors.

The first metrics shows, for $t = 0$, that algorithms A_2 and A_3 alone are sufficient to completely handle 44.6% (resp. 29.3%) of the models; but algorithm A_1, as soon as turned on, gives an major boost, pushing the percentage of models completely solved to 93.0% (resp. 92.3%) of the models; from there, increasing the value of t slowly increases this percentage; applying algorithm A_2'' again after A_1 fully solves 0.1% (resp. 0.1%) more models. The second metrics gives quite similar results concerning the resolution of unknown values, although the influence of A_1 is not as strong as with the first metrics; the percentage of unknown values does not quickly converge down to zero, due to a small number of large models that remain incompletely solved for long, with thousands of unknown values. The third metrics corroborates the first one, but exhibits higher success percentages, as each model counts for the proportion of known values in its solution vector rather than for a binary value (one for a fully complete model, and zero otherwise).

Additional measurements indicate that: (i) the shortcuts of algorithm A_1 are effective, as they are triggered for more than 82.6% (resp. 79.6%) of the models;

(ii) 94.8% (resp. 94.0) of the models are fully solved in less than one second; the other models are (incompletely) processed in less than $1.56 \times t$ (resp. $1.48 \times t$) seconds; (iii) all nets having less than 74 places, 92 transitions, and 366 arcs can be fully processed with $t = 60$ (resp. $t = 180$).

5 Algorithms for Concurrent Places

This section presents various algorithms for the Concurrent Places Problem. In the sequel, we consider the elements of the half matrix as (unordered) pairs of places. The diagonal elements of the half matrix are also considered as pairs, even if both elements of these pairs are equal.

5.1 Marking Graph Exploration

Concurrent places can be determined by an exploration of reachable markings similar to the one presented in Sect. 4.1, based upon the fact that the places of each reachable marking are pairwise concurrent. If the state space can be generated exhaustively, one obtains a complete half matrix; otherwise, if the state space is too large for being explored entirely, the half matrix contains only ones and unknown values, but no zeros.

There are no obvious algorithmic shortcuts, apart from halting the exploration if the half matrix gets entirely full of known values, which only occurs if all unknown values turn out to be equal to one, since the exploration, as long as it has not been done exhaustively, only produces ones but not zeros. In practice, such a situation is unlikely, as Sect. 5.6 shows that, statistically, there are much less ones than zeros in half matrices.

5.2 Structural Rules

We now study the case where the state-space exploration of Sect. 5.1 has not been exhaustively performed, and propose complementary algorithms, with a lower complexity, that help reducing the number of unknown values in the half matrix. The two following rules exploit the information gained about non-dead places and transitions during marking-graph construction.

Proposition 15. *The places of the initial marking M_0 are pairwise concurrent.*

Proposition 16. *If a transition is not dead, its input places (resp. output places) are pairwise concurrent.*

We then apply algorithm A_2, i.e., the structural rules of Sect. 4.2, to identify (a subset of) the dead places and dead transitions. Based on this information, further unknown values can be eliminated from the half matrix.

Proposition 17. *(1) A non dead place is concurrent with itself. (2) A dead place is non concurrent with any other place, including itself.*

Proposition 18. *If a dead transition has two (distinct) input places, these places are non concurrent.*

The next rule exploits the assumption that the Petri nets considered are safe.

Proposition 19. *If a transition t (dead or not) has a single input place p, this place is non concurrent with any output place of t different from p.*

Proof. By contradiction: if there exists a reachable marking M containing p and some output place of t, the net is unsafe, as t can fire from M. Notice that, if t is dead, p is dead too, and the result follows directly from Proposition 17(2).

The previous rule can be easily generalized to sequences of transitions having each a single input place. It is implemented using a transitive-closure algorithm.

Proposition 20. *For any path $(p_1, t_1, p_2, t_2, ..., p_n, t_n, p_{n+1})$ such that each transition t_i has a single input place p_i and at least one output place p_{i+1}, the places p_1 and p_{n+1} are non concurrent if they are distinct.*

The last rule exploits the unit-safeness property for those nets known to be unit safe by construction.

Proposition 21 (from [5] Proposition 6). *If the net is a unit-safe NUPN, any two distinct places located in non-disjoint units are non concurrent. Formally:*

$$(\forall p_1 \in P)\, (\forall p_2 \in P)\, (p_1 \neq p_2) \wedge \neg\mathsf{disjoint}\, (\mathsf{unit}\, (p_1), \mathsf{unit}\, (p_2)) \Rightarrow p_1 \nparallel p_2$$

In particular, any two distinct places located in the same unit are non concurrent.

5.3   Quadratic Under-Approximation

From now on, if P' and P'' are two sets of places, we write $P' \otimes P''$ for the set of (unordered) pairs of places defined as $\{\{p', p''\} \mid (p' \in P') \wedge (p'' \in P'')\}$, assuming that the set notation $\{p', p''\}$ actually denotes a singleton if $p' = p''$. We represent the half matrix of concurrent places by R_0 and R_1, which are, respectively, the sets of pairs of places known to be non concurrent and concurrent, the pairs of $(P \otimes P) \setminus (R_0 \cup R_1)$ being unknown. For instance, the structural rules of Proposition 15, 16, and 17(1) can be summarized as follows:

$$R_1 := R_1 \cup (M_0 \otimes M_0) \cup \bigcup_{t \in T_0} \left((^\bullet t \otimes {}^\bullet t) \cup (t^\bullet \otimes t^\bullet) \right) \cup \bigcup_{p \in P_0} \{\{p\}\}$$

In this section, we propose an algorithm to detect more concurrent places. This algorithm starts from the set R_1 computed during prior phases and extends this set by examining all transitions having one or two input places, namely by combining Proposition 13, 16, and 18 together with the following result:

Proposition 22. *If two distinct places p_1 and p_2 are concurrent, p_2 is also concurrent with each output place of any transition t such that $^\bullet t = \{p_1\}$.*

The algorithm below stores, in a set R', pairs of places found to be concurrent. The algorithm is said to perform a *quadratic* approximation because each visited marking M is abstracted away and represented by its set of concurrent pairs $M \otimes M$—contrary to algorithm A_3 of Sect. 4.3, which performs a *linear* approximation by storing only the set of places that appear in at least one visited marking. The algorithm performs an *under*-approximation because it may miss exploring certain concurrent pairs that are actually reachable.

```
1   R' := ∅
2   while R' ≠ R₁ loop
3         assert R' ⊊ R₁
4         let {p₁,p₂} = oneof (R₁ \ R')              — possibly with p₁ = p₂
5         R' := R' ∪ {{p₁,p₂}}
6         for t ∈ T | •t = {p₁,p₂} loop
7               assert (1 ≤ |•t| ≤ 2) ∧ (t ∉ T₁) — from Proposition 13 and 18
8               R₁ := R₁ ∪ (t• ⊗ t•)                   — from Proposition 16(b)
9         end loop
10        for t ∈ T | (•t = {p₁}) xor (•t = {p₂}) loop
11              assert (|•t| = 1) ∧ (p₁ ≠ p₂) ∧ (t ∉ T₁)  — from Proposition 13
12              R₁ := R₁ ∪ (({p₁,p₂} \ •t) ⊗ t•)        — from Proposition 22
13        end loop
14  end loop
```

5.4   Quadratic Over-Approximation

Our last algorithm is based upon the works of Kovalyov and Esparza, who proposed various algorithms [10–12] of polynomial complexity that compute a least fix-point for three rules derived from the Petri-net token game, and produce an over-approximation of the concurrency relation (i.e., a superset of concurrent pairs) from which one can safely obtains a subset of R_0, the set of non-concurrent pairs. Our algorithm below evolves their algorithms in several ways: (i) it does not assume that all places and all transitions are not dead and, instead, exploits the pre-existing set T_1 of dead transitions; (ii) it requires the input nets to be safe and uses this assumption to produce more accurate results by discarding unsafe markings, whereas the algorithms of Kovalyov and Esparza handle non-safe nets, with the alternative definition of diagonal values discussed in Sect 2.4 above; (iii) to get better and faster results, our algorithm reuses the sets of pairs R_0 and R_1 precomputed, e.g., by the algorithms of Sect. 5.1 to 5.3, whereas the algorithms of Kovalyov and Esparza start with no prior knowledge about concurrent pairs, i.e., $R_0 = R_1 = \varnothing$. We now formalize the over-approximation (which we call *quadratic* due to its memory cost) that underlies all these algorithms.

Definition 12. *Let R and R' be two sets containing pairs of places, t a transition, and M a marking. We write $R \xrightarrow{t} R'$ if we have $•t \otimes •t \subseteq R$ and $R' = R \cup (t• \otimes t•) \cup \{\{p\} \otimes t• \mid (p \in P \setminus •t) \wedge (\{p\} \otimes •t \subseteq R)\}$. We write $R \xRightarrow{*m} M$ if there exists R' such that $R \xRightarrow{} R'$ and $M \otimes M \subseteq R'$.*

In contrast with the firing relation $\xrightarrow{t}$ defined between two markings (cf. Definition 2), this relation $\xRightarrow{}$ is defined between two sets of pairs. With $\xrightarrow{}$, the state space is the set of all reachable markings M, whereas, with $\xRightarrow{t}$, the (abstracted) state space is the union of all sets of pairs $M \otimes M$, for each reachable marking M.

Proposition 23. *In a safe Petri net, if a marking M is reachable from the initial marking M_0 (i.e., $M_0 \xrightarrow{t} M$), then $M_0 \otimes M_0 \xRightarrow{*m} M$.*

Proof. By induction on firing sequences from $M_0 \otimes M_0$.

Our algorithm starts from $M_0 \otimes M_0$, to which the known concurrent pairs of R_1 are added, and explores, using two variables R' and R'', the state space of all pairs that can be reached by firing the relation $\xRightarrow{*}$ for all non-dead transitions. The non-concurrent pairs of R_0 are systematically excluded from the state space. Upon termination, all pairs that have not been explored are non concurrent for sure, and can thus be added to R_0. To speed up calculations, we reuse the counter $c[t]$ of Sect. 4.3, which now stores how many pairs of $({}^\bullet t \times {}^\bullet t)$ have not been yet proven concurrent; a transition is considered to be fireable when its counter drops to zero. For the conciseness of the algorithm, we introduce an auxiliary function $\text{fire}\,(M, t, R) \stackrel{\text{def}}{=} (M \otimes {}^\bullet t \subseteq R) \wedge ((M \otimes {}^\bullet t) \cap R_0 = \varnothing)$.

```
1   R' := ∅ ; R'' := (M₀ ⊗ M₀) ∪ R₁
2   T₁ := T₁ ∪ {t ∈ T | ((•t ⊗ •t) ∩ R₀ ≠ ∅) ∨ ((t• ⊗ t•) ∩ R₀ ≠ ∅)}
3   for t ∈ T \ T₁ loop c[t] := |•t| × (|•t| + 1)/2
```

```
1    while R' ≠ R'' loop
2        assert (R' ⊊ R'') ∧ (R'' ∩ R₀ = ∅)
3        assert (∀t ∈ T \ T₁) c[t] = |(•t ⊗ •t) \ R'|
4        let {p₁, p₂} = oneof (R'' \ R')              — possibly with p₁ = p₂
5        R' := R' ∪ {{p₁, p₂}}
6        for t ∈ T \ T₁ | {p₁, p₂} ⊆ •t loop
7            c[t] := c[t] − 1
8            if c[t] = 0 then
9                for p ∈ (P \ •t) ∪ t• | fire ({p}, t, R'') loop
10                   assert (p ∉ •t \ t•) ∧ (M₀ ⊗ M₀ ⟹*m {p} ∪ •t)
11                   R'' := R'' ∪ ({p} ⊗ t•)
12               end loop
13           end if
14       end loop
15       for t ∈ T \ T₁ | (c[t] = 0) ∧ ((p₁ ∈ •t) xor (p₂ ∈ •t)) ∧
16           fire (({p₁, p₂} \ •t), t, R'') loop
17           assert (|{p₁, p₂} \ •t| = 1) ∧ (M₀ ⊗ M₀ ⟹*m {p₁, p₂} ∪ •t)
18           R'' := R'' ∪ (({p₁, p₂} \ •t) ⊗ t•)
19       end loop
20       assert (∀t ∈ T \ T₁) (∀p ∈ (P \ •t) ∪ t•) (c[t] = 0) ∧ fire ({p}, t, R')
21               ⇒ ({p} ⊗ t• ⊆ R'')
```

22 **end loop**
23 **assert** $(R' = R'') \wedge (R' \cap R_0 = \varnothing) \wedge (R_0 \subseteq (P \otimes P) \setminus R')$
24 $R_0 := (P \otimes P) \setminus R'$

5.5 Ordering of Algorithms

Let C_1, C_2, C_3, and C_4 denote the algorithms presented in Sects. 5.1, 5.2, 5.3, and 5.4, respectively. Each of these algorithms needs to be applied only once. Analysis of dependencies suggests that these algorithms are best applied in the following order $(C_1; C_2; C_3; C_4)$, knowing that C_2 also invokes algorithm A_2 of Sect. 4.2. This execution sequence stops as soon as the number of unknown values in the half matrix drops to zero.

5.6 Experimental Results

We applied these algorithms to compute the half matrices of concurrent places for the 13,116 models presented in Sect. 3.3. As for algorithm A_1 in Sect. 4.5, the execution of algorithm C_1 is parameterized by a maximum duration t allocated to the symbolic exploration of reachable markings. Because the computation of concurrent places for large models can be much longer than the computation of dead places and dead transitions, each execution run was bounded by a timeout of 4000 s, which, for various values of t, hits at most 0.82% of our models.

Our experiments reveal that, over the 26,577,437,180 pairs of places present in all half matrices of the models not interrupted by the timeout, 4.0% are concurrent, 67.0% are non-concurrent, the others being unknown.

Table 3 reuses the three metrics of Table 2 by adapting them from vectors to half matrices. The first metrics shows that 94.0% of the models can be completely solved for $t = 60$, which is slightly less than in Table 2, although the Concurrent Places Problem usually requires more CPU time. The second metrics decreases more slowly than in Table 2 and seems to stabilize at a much higher percentage of unknown values, which can be explained by the quadratic size of a few large, incomplete half matrices. However, for most models, the third metrics converges to a high completion rate similar to those of Table 2.

Additional measurements indicate that: (i) alone, the algorithms C_2, C_3 and C_4 can completely handle 51.0% of the models for $t = 0$ but, as soon as algorithm C_1 is turned on, it performs so well on the models whose reachable markings can be fully explored that algorithms C_2, C_3 and C_4 only contribute for 1% to the number of complete half matrices; (ii) however, on large models that can not be fully explored using Binary Decision Diagrams, the algorithms C_2, C_3 and C_4 play a greater role by eliminating 22.5% of unknown values, in addition to the 33.8% already eliminated by C_1; (iii) all nets having less than 66 places, 64 transitions, and 256 arcs can be completely processed with $t = 60$.

Table 3. Experimental results for concurrent places

Value of t	0	5	10	15	30	45	60	120	180	240	300	360	420
% Complete matrices	51.0	91.6	92.2	92.5	93.0	93.6	94.0	94.2	94.4	94.5	94.6	94.7	94.7
% Unknowns values	45.0	44.7	44.7	44.4	44.4	43.7	43.7	43.7	43.6	43.6	43.6	43.6	43.6
% Matrix completion	81.6	96.3	96.6	96.8	97.0	97.1	97.2	97.3	97.4	97.4	97.4	97.5	97.5

6   Conclusion

In the present article, we studied three reachability problems that, although practically useful, are not well supported by mainstream Petri-net tools. As we could not find a unified algorithm for addressing each problem, we proposed instead a combination of methods (static vs dynamic state-space exploration, exact vs approximate solution, polynomial or exponential cost), which we implemented in two software tools (written in C and in Python) that cross check each other. We observed that our approach statistically performs well on a large number of realistic models but, given that the three problems are PSPACE-complete, there will always exist models too large for being processed in reasonable time.

Future work should focus on refined algorithms capable of handling even larger models, either by computing solutions with less unknown values or by providing equivalent results in shorter time. Among the various approaches that might be profitably applied, we can mention invariants, semi-flows, partial orders, stubborn sets (e.g., [14, Sect. 11]), SAT solving, explicit-state model checking (to specifically spot the unknown values that remain in a solution vector or half matrix computed by other algorithms), and net reductions (in this respect, we can already mention a recent tool named Kong[7] that computes concurrent places by invoking our CÆSAR.BDD tool in combination with net reductions based on polyhedral abstractions). To foster such research, we suggest [6] that our three problems, possibly extended to unsafe nets and/or colored nets, become integral part of Model Checking Contest.

Acknowledgements. The experiments of Sect. 5.6 have been performed using the French Grid'5000 testbed.

References

1. Amparore, E., et al.: Presentation of the 9th edition of the Model Checking Contest. In: Beyer, D., Huisman, M., Kordon, F., Steffen, B. (eds.) TACAS 2019. LNCS, vol. 11429, pp. 50–68. Springer, Cham (2019). https://doi.org/10.1007/978-3-030-17502-3_4

---

[7] https://github.com/nicolasAmat/Kong.

2. Bouvier, P., Garavel, H., Ponce-de-León, H.: Automatic decomposition of Petri nets into automata networks – a synthetic account. In: Janicki, R., Sidorova, N., Chatain, T. (eds.) PETRI NETS 2020. LNCS, vol. 12152, pp. 3–23. Springer, Cham (2020). https://doi.org/10.1007/978-3-030-51831-8_1

3. Cheng, A., Esparza, J., Palsberg, J.: Complexity Results for 1-Safe Nets. Theoret. Comput. Sci. **147**(1–2), 117–136 (1995)

4. Desel, J., Esparza, J.: Free Choice Petri Nets, Cambridge Tracts in Theoretical Computer Science, vol. 40. Cambridge University Press, Cambridge (1995)

5. Garavel, H.: Nested-unit Petri nets. J. Logical Algebraic Methods Program. **104**, 60–85 (2019)

6. Garavel, H.: Proposal for Adding Useful Features to Petri-Net Model Checkers, December 2020. https://arxiv.org/abs/2101.05024

7. Garavel, H., Serwe, W.: State space reduction for process algebra specifications. Theoret. Comput. Sci. **351**(2), 131–145 (2006)

8. ISO/IEC: High-level Petri Nets - Part 2: Transfer Format. International Standard 15909–2:2011, International Organization for Standardization, Geneva (2011)

9. Janicki, R.: Nets, sequential components and concurrency relations. Theoret. Comput. Sci. **29**, 87–121 (1984)

10. Kovalyov, A.: Concurrency relations and the safety problem for Petri nets. In: Jensen, K. (ed.) ICATPN 1992. LNCS, vol. 616, pp. 299–309. Springer, Heidelberg (1992). https://doi.org/10.1007/3-540-55676-1_17

11. Kovalyov, A.: A polynomial algorithm to compute the concurrency relation of a regular STG. In: Yakovlev, A., Gomes, L., Lavagno, L. (eds.) Hardware Design and Petri Nets, chap. 6, pp. 107–126. Springer, Boston, MA, USA, January 2000. https://doi.org/10.1007/978-1-4757-3143-9_6

12. Kovalyov, A., Esparza, J.: A polynomial algorithm to compute the concurrency relation of free-choice signal transition graphs. In: Proceedings of the 3rd Workshop on Discrete Event Systems (WODES 1996), Edinburgh, Scotland, UK, pp. 1–6 (1996)

13. Murata, T.: Petri nets: analysis and applications. Proc. IEEE **77**(4), 541–580 (1989)

14. Schmidt, K.: Stubborn sets for standard properties. In: Donatelli, S., Kleijn, J. (eds.) ICATPN 1999. LNCS, vol. 1639, pp. 46–65. Springer, Heidelberg (1999). https://doi.org/10.1007/3-540-48745-X_4

15. Semenov, A., Yakovlev, A.: Combining partial orders and symbolic traversal for efficient verification of asynchronous circuits. In: Ohtsuki, T., Johnson, S. (eds.) Proceedings of the 12th International Conference on Computer Hardware Description Languages and their Applications (CHDL 1995), Makuhari, Chiba, Japan. IEEE (1995)

16. Wiśniewski, R., Karatkevich, A., Adamski, M., Kur, D.: Application of comparability graphs in decomposition of Petri nets. In: Proceedings of the 7th International Conference on Human System Interactions (HSI 2014), Costa da Caparica, Portugal. IEEE (2014)

A Lazy Query Scheme for Reachability Analysis in Petri Nets

Loïg Jezequel[1,3]([✉]), Didier Lime[2,3], and Bastien Sérée[2,3]

[1] Université de Nantes, Nantes, France
[2] École Centrale de Nantes, Nantes, France
[3] LS2N, UMR CNRS 6004, Nantes, France
{loig.jezequel,didier.lime,bastien.Seree}@ls2n.fr

Abstract. In recent works we proposed a lazy algorithm for reachability analysis in networks of automata. This algorithm is optimistic and tries to take into account as few automata as possible to perform its task. In this paper we extend the approach to the more general settings of reachability analysis in unbounded Petri nets and reachability analysis in bounded Petri nets with inhibitor arcs. We consider we are given a reachability algorithm and we organize queries to it on bigger and bigger nets in a lazy manner, trying thus to consider as few places and transitions as possible to make a decision. Our approach has been implemented in the ROMEO model checker and tested on benchmarks from the model checking contest.

Keywords: Reachability analysis · Unbounded Petri nets · Inhibitor arcs · Lazy algorithms

1 Introduction

In recent works [8,9] we proposed an algorithm for reachability analysis in networks of automata. This algorithm is called lazy as it tries to use as few automata as possible to complete its task. To that extent, it is a non-trivial instance of a general principle that has been implemented in many approaches (e.g. program slicing [21]). In practice, on many benchmarks this approach proved to be efficient: the LARA tool (which implements our approach) used only a small portion of the automata in the network to conclude about reachability. Runtime comparisons with LOLA [22] (in a non-timed setting [8]) and UPPAAL [2] (in a timed setting [9]) were also frequently in favor of LARA.

Networks of automata are in fact a subclass of Petri nets as they can be syntactically transformed into safe Petri nets. Extending our lazy reachability algorithm to larger classes of Petri nets is thus a natural next step in our work. Moreover, it is of particular interest for us as it will allow to implement lazy reachability in the ROMEO model checker [16], developed in our research team. In fact, ROMEO works on models even more expressive than Petri nets, where reachability is not always decidable. In this paper we focus on unbounded Petri

D. Buchs and J. Carmona (Eds.): PETRI NETS 2021, LNCS 12734, pp. 360–378, 2021.
https://doi.org/10.1007/978-3-030-76983-3_18

nets and bounded Petri nets with inhibitor arcs, two subclasses of these models for which reachability is decidable.

Reachability analysis in Petri nets (or equivalent models such as vector addition systems) has been widely studied. The problem is known to be decidable in general [12,13,15,17]. Efficient techniques exist for performing it in the particular case of bounded nets, that is nets with a finite state space. One can notice, for example, Petri net unfolding [6,18] or variations around it [3,4], partial order techniques [7], and decision diagram based approaches [5,19].

Here we follow a different approach and do not propose a *standalone* reachability algorithm but rather, given such an algorithm, we propose a scheme to use its results on subnets that are built incrementally from the reachability property by adding only places and transitions that are required to make a decision. This is why we call the approach lazy. Compared to [8], the main challenges we address here are (1) that the *components* of the system are less well-defined in a Petri net than in a network of finite automata, and (2) that the state-space is infinite in general. Note also that even if a net is bounded, its subnets might not be. We propose an algorithm for reachability in plain Petri nets, and also show how to deal with inhibitor arcs in the bounded case.

This paper is organized as follows. We start by giving some definitions and notations in Sect. 2. Then we present our algorithm for lazy reachability analysis in Petri nets in Sect. 3 and show its validity. After that, we show how this approach can be transposed to perform reachability analysis for the class of bounded Petri nets with inhibitor arcs in Sect. 4. Finally, in Sect. 5 we report on an implementation of our algorithm in the model checker ROMEO and give experimental results obtained from a run of our tool on all the benchmarks from the 2020 edition of the model checking contest [10,11].

2   Definitions and Notations

We define Petri nets and their semantics, as well as the central notion of reachability of markings in Petri nets. Then, we define the notion of subnets and partial markings, that we use later to perform reachability analysis on a Petri net without considering it in its entirety.

2.1   Petri Nets

Definition 1 (Petri net). *A Petri net is a tuple $N = (P, T, F, m_0)$ where P and T are disjoint finite sets of places and transitions respectively, $F : P \times T \cup T \times P \to \mathbb{N}$ is a flow function, and $m_0 : P \to \mathbb{N}$ is called the initial marking.*

In a net N, for any $x \in P \cup T$, we define ${}^\bullet x = \{y \ : \ F(y, x) \neq 0\}$ the *preset* of x and $x^\bullet = \{y \ : \ F(x, y) \neq 0\}$ the *postset* of x. We can extend this postset (resp preset) concept to subsets of P or T by doing the union of the postsets (resp presets) of each element of the considered subset.

In a net N, any function $m : P \to \mathbb{N}$ is called a *marking* of N. A transition $t \in T$ is *fireable* from a marking m if and only if $\forall p \in {}^\bullet t, m(p) \geq F(p, t)$. In this case, firing t from m leads to the new marking m' such that $\forall p \in P, m'(p) = m(p) - F(p,t) + F(t,p)$. We denote it by $m \xrightarrow{t} m'$. Given a sequence $\omega = t_1, \ldots, t_n$ of transitions, we define $m \xrightarrow{\omega} m'$ if there exist markings $m_1, \ldots m_{n-1}$ such that $m \xrightarrow{t_1} m_1, \forall 2 \leq i \leq n-1, m_{i-1} \xrightarrow{t_i} m_i$, and $m_{n-1} \xrightarrow{t_n} m'$.

Definition 2 (Reachability). *A marking m is said to be* reachable *in N if and only if there exists a sequence of transitions ω such that $m_0 \xrightarrow{\omega} m$.*

Definition 3 (Boundedness). *A Petri net is said to be k-bounded, for a given k, if for every reachable marking m and every place p, we have $m(p) \leq k$. A Petri net is said to be* bounded, *if there exists a k such that it is k-bounded.*

2.2 Subnets and Partial Markings

In the following, we will perform reachability analysis on parts of a Petri net: not all the places and transitions of the net will be considered. This is formalized through the notion of subnet.

Definition 4 (Subnet). *A* Subnet *N' of a Petri net $N = (P, T, F, m_0)$ is a tuple (P', T', F', m'_0) such that $P' \subseteq P$, $T' \subseteq T$, $F' = F_{|P',T'}$, and $m'_0 = m_{0|P'}$.*

Given a subnet N' of a net N, and for any $x \in P' \cup T'$, we define $^{\bullet N}x = \{y : F(y, x) \neq 0)\}$ and $x^{N\bullet} = \{y : F(x, y) \neq 0\}$ (that is, intuitively, the preset and postset taken in N rather than in N').

We introduce two notions of completeness with respect to a net N for a subnet N'. They will be central in our algorithms and their proofs. The notion of P-completeness expresses that N' contains all the places from N that are used as preconditions for enabling transitions in N'.

Definition 5 (P-completeness). *A subnet N' of a net N is said to be P-complete when $\forall t \in T', ^{\bullet N}t \subseteq P'$.*

In the other way around, the notion of T-completeness expresses that N' contains all the transitions from N that can add tokens on places in N'.

Definition 6 (T-completeness). *A subnet N' of a net N is said to be T-complete when $\forall p \in P', ^{\bullet N}p \subseteq T'$.*

Partial marking will be used to express reachability objectives that do not concern all the places in a net.

Definition 7 (Partial marking). *For a Petri net $N = (P, T, F, m_0)$, any function $m_p : P \to \mathbb{N} \cup \{\star\}$ is called a* partial marking *of N.*

Intuitively, a partial marking is a marking which is not fully specified: when $m_p(p) = \star$ for some $p \in P$ it means that this value is left unspecified. For a partial marking m_p of a net N, we define $supp(m_p) = \{p \in P : m_p(p) \neq \star\}$. A marking m such that $m(p) = m_p(p)$ for any $p \in supp(m_p)$ is said to *realize*

m_p. We can notice that a every marking m of a net N is a partial marking such that $supp(m) = P$. For a net N with a subnet N', and a (partial) marking m of N, we denote by m' the (partial) marking of N' such that $m' = m_{|P'}$ and call it the *submarking* of m in N'.

Definition 8 (Reachability). *A partial marking m_p is said to be reachable in N if and only if there exists a marking m that realizes m_p and is reachable in N.*

Finally, a third notion of completeness, m-completeness, is defined for partial markings. It expresses the fact that, for a given marking m of N', all the transitions from N that can affect this marking by reducing its value for some place are included in N'.

Definition 9 (m-completeness). *A subnet N' of a net N is said to be m-complete with m a marking of N' when $\forall p \in supp(m), p^{N\bullet} \subseteq T'$.*

3   Lazy Reachability Analysis in Petri Nets

In this section we propose an algorithm which, given a Petri net N and a marking m in N, decides whether or not m is reachable in N. However, this algorithm consists in a heuristic way to perform reachability queries on smaller nets, which proves more efficient in some cases than a query on the full net. It therefore requires an algorithm to perform reachability on Petri nets, used as a black-box. The technique works on unbounded nets, provided that the reachability black-box handles them.

We start by demonstrating the concept of our algorithm – in particular, in which way it is lazy – on two examples, then we formalize the algorithm, and finally we prove its validity.

3.1   Preliminary Example

Consider the Petri net of Fig. 1 – the places are represented by circles, the transitions by squares, the flow function by arrows, the initial marking by black dots. We look at two reachability questions on this net: (Q1) is it possible to reach a partial marking m_1 so that $m_1(p_2) = 1$, and $m_1(p_3) = 1$? (Q2) is it possible to reach a partial marking m_2 so that $m_2(p_4) = 3$?

Let us focus on (Q1) first. To this end, consider the subnets N_1 and N_2 of Fig. 2. These two subnets were built from N by considering exactly the places in the support of m_1. If the initial marking of each of these subnets had been a submarking of m_1, then the answer to (Q1) would immediately be a yes. This is not the case however, so we cannot conclude yet.

Consider N_1 first. Our objective with this subnet is to find whether or not it is possible to reach some marking m so that $m(p_2) = 1$. As this is not the case initially, one needs to find how to increase the marking of p_2. This can only be done by using transitions t so that $p_2 \in t^\bullet$. We thus add these transitions to our

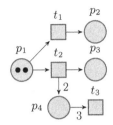

Fig. 1. A Petri net N.

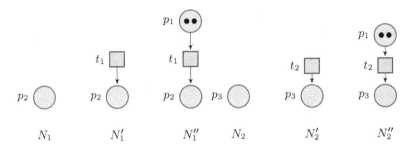

Fig. 2. Six subnets incrementally built from p_2 and p_3.

subnet, leading to N_1' (which is T-complete). Now, some m so that $m(p_2) = 1$ is reachable, however, we cannot yet conclude that this is the case in the full net N because we do not have all the presets of the transitions we use. So we add the places in these presets, leading to the new subnet N_1'' (which is P-complete). From this subnet, one can conclude that some m so that $m(p_2) = 1$ is reachable in N.

A similar process allows to build N_2'' and prove that some m so that $m(p_3) = 1$ is reachable in N. However, having obtained these two results does not guarantee that m_1 is reachable in N because N_1'' and N_2'' overlap (the place p_1 appears in both), which may lead to conflicts between the transition sequences found in these two subnets. We thus merge the subnets (on common places and transitions, here only p_1), which leads to the subnet of Fig. 3. In this subnet m_1 is reachable. Moreover – as we prove later – because this subnet is P-complete, m_1 is also reachable in N.

The place p_4 and the transition t_3 were never included in the subnets considered. This is why we call our algorithm lazy: it omits the places and transitions that are not useful for its analysis.

Let us now focus on (Q2). In this case our analysis will start from the subnet N_3 of Fig. 4. For similar reasons as before, we first add the transitions that can put tokens in p_4, leading to N_3'. In this subnet, no marking m so that $m(p_4) = 3$ is reachable (because there is always an even number of tokens in p_4). However, markings with $m(p_4) \geq 3$ are reachable. Hence, we add the transitions that can remove tokens from p_4, leading to N_3''. In this subnet, such an m is reachable

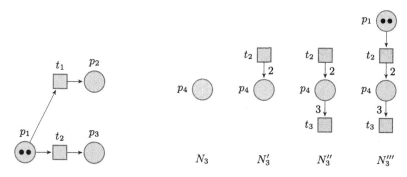

Fig. 3. Merging of N_1'' and N_2''.   **Fig. 4.** Four subnets built from p_4.

(for example by firing t_2 three times and then t_3 one time). As before, in order to conclude one needs to verify that (at least some of) the sequences allowing to reach such an m are fireable in the original net. For that, the places in the presets of t_2 and t_3 need to be added, leading to N_3'''. In this subnet it is not possible to fire t_2 three times. Moreover, this subnet is T-complete and so, as we prove later, if no m such that $m(p_4) = 3$ is reachable in N_3''', no such m is reachable at all in N.

3.2   An Algorithm for Lazy Reachability Analysis in Petri Nets

The formalization of the ideas presented in the above examples leads to Algorithm 1[1]. It is a lazy algorithm that, given a net N and a (partial) marking m, tells whether or not m is reachable in N. This algorithm works on subnets of N. These subnets are identified by their sets of places and transitions.

Algorithm 1 starts with subnets built from a partition of the set of places involved in m. This allows to handle each part of the objective separately as long as they do not interact. Initially, each element of the List $LNets$ is a pair (P, T) representing one of these subnets. The algorithm does two main tasks: concretisation (addition of places and transitions to subnets) and merging (union of interacting subnets). At each iteration of its main loop it does at least one of those two tasks.

Concretisation. Concretisation consists in expanding one subnet. If the partial objective is not reachable one adds new transitions to add new ways to reach it. If the partial objective is reachable, one needs to add new places to ensure that the transitions used in the subnet can also be used in the original net (i.e. to ensure that their full preset is taken into account).

More formally, the objective of the concretisation is to ensure that each subnet verifies the completeness notion of Definition 10. If this is the case, then

---

[1] It uses the classical list data structure. The length of a list L is given by $length(L)$. The kth element of L is $L[k]$.

Algorithm 1. Lazy algorithm checking if a marking m is reachable in a Petri net $N = (P, T, F, m_0)$

1: **choose** a partition $\{P_1, ..., P_p\}$ of $supp(m)$
2: $LNets \leftarrow [(P_1, \emptyset), ..., (P_p, \emptyset)]$
3: $Complete \leftarrow false$
4: $Consistent \leftarrow true$
5: **while not** $Complete$ **or not** $Consistent$ **do**
6:     $Complete \leftarrow \forall k, LNets[k]$ is $complete$
7:     **if not** $Complete$ **then**
8:         **optional unless** $Consistent$
9:             $mayHaveSol \leftarrow Concretise(LNets, m)$
10:             **if not** $mayHaveSol$ **then**
11:                 **return** false
12:             **end if**
13:         **end option**
14:     **end if**
15:     $Consistent \leftarrow LNets$ is consistent
16:     **if not** $Consistent$ **then**
17:         **optional unless** $Complete$
18:             $Merge(LNets)$
19:         **end option**
20:     **end if**
21: **end while**
22: **return** true

---

m is reachable in N (provided that there is not interaction with other subnets, which is ensured by the notion of consistency described below). If at least one subnet cannot be made complete, then it is granted that m is not reachable in N.

Definition 10. *Let $N = (P, T, F, m_0)$ be a Petri net and m a marking. A subnet N' of N is* complete *(with respect to N and m) if N' is P-complete and the submarking m' is reachable in N'.*

Remark that completeness can be effectively checked provided reachability can be checked. The rest of the conditions is syntactic.

Concretisation can be implemented as described in Algorithm 2, which alternately adds places and transitions to a subnet.

Merging. Merging consists in replacing two subnets in $LNets$ by a single subnet obtained by union of places and transitions sets. Merging is needed when two subnets share places, as in this case the solutions to the reachability problem found in these subnets can interfere. For the same interference reason, merging is also needed when one of the subnets contains a transition whose postset contains a place involved in the submarking of m (the reachability objective) in another subnet. The fact that two subnets may interfere is formalised through the notion of consistency in Definition 11.

Algorithm 2. Auxiliary function $Concretise(LNets, m)$ for Algorithm 1

1: **choose** k such that $LNets[k]$ is not complete
2: $(P_k, T_k) \leftarrow LNets[k]$
3: $m' \leftarrow m_{|P_k}$
4: **if not** $Reachable(LNets[k], m')$ **then**
5:     **choose** T_k' such that $T_k \subset T_k' \subseteq supp(m')^{N\bullet} \cup {}^{\bullet N}P_k$
6:     **if not possible then**
7:         **return** false
8:     **else**
9:         $LNets[k] \leftarrow (P_k, T_k')$
10:     **end if**
11: **else**
12:     **choose** P_k' such that $P_k \subset P_k' \subseteq {}^{\bullet N}T_k$
13:     $LNets[k] \leftarrow (P_k', T_k)$
14: **end if**
15: **return** true

Definition 11. *The list of subnets* $LNets = [(P_1, T_1), ..., (P_n, T_n)]$ *is consistent if*

1. $\forall k \neq \ell, (P_k \cap P_\ell) = \emptyset,$ *and*
2. $\forall k \neq \ell, T_k^{N\bullet} \cap supp(m_{|P_\ell}) = \emptyset.$

Remark that the definition of consistency is completely syntactic and can therefore be checked effectively.

3.3   Proof of the Algorithm

We now prove the correctness of Algorithm 1. Propositions 1 and 2 together prove the soundness of Algorithm 1, while Proposition 3 proves its completeness. The proofs of these propositions are based on lemma for which proofs are presented in this part.

Proposition 1. *If Algorithm 1 returns false, then m is not reachable in N.*

Proof. The only way for Algorithm 1 to return false is at line 11. It implies that the previous call to the Concretise function (Algorithm 2) returned false. This can only occur at line 7 of Algorithm 2.

In this case, it must not be possible to choose T' such that $T \subset T' \subseteq supp(m')^{N\bullet} \cup {}^{\bullet N}P$ (line 5). In other words, the subnet $LNets[k]$ considered is m'-complete (Definition 9) and T-complete (Definition 6). Moreover, the current marking m' must not be reachable in the subnet $LNets[k]$ (line 4) and is a submarking of m (line 3). Hence, by applying Lemma 1 below, m is not reachable in N.   □

Lemma 1. *Let N' be a subnet of a Petri net N. Assume that N' is T-complete. Let m' be a marking of N', that is not reachable. If N' is m'-complete, then no marking m of N such that m' is the submarking of m in N' is reachable in N.*

Proof. Let m be a reachable marking of N, with $m_0 \xrightarrow{\omega} m$. Denote by n the number of transitions in ω. We prove by induction on n that m' is reachable in N'. By the contrapositive, this proves the Lemma.

Induction hypothesis for all n, if m is reachable in N and there is a sequence ω of n transitions such that $m_0 \xrightarrow{\omega} m$, then the submarking m' of m in N' is reachable in N'.

Initialisation when $n = 0$, the only possible m is m_0 (it must be reachable with 0 transitions). Thus, simply take $m' = m'_0$, which is, by construction the submarking of m_0 in N' and is obviously reachable in N'.

Induction. Let us consider some $n > 0$ and assume that the induction hypothesis is verified for $n - 1$. First, remark that ω can always be split in two parts: ω' of length $n-1$ and t_n (a single transition) such that $m_0 \xrightarrow{\omega'} \tilde{m} \xrightarrow{t_n} m$ in N for some marking $\tilde{m}$. From the induction hypothesis, the submarking $\tilde{m}'$ of $\tilde{m}$ is reachable in N'. Four cases are then possible. (1) $t_n^{N\bullet} \cap P' = \emptyset$ and $^{\bullet N}t_n \cap supp(m') = \emptyset$, in which case $\tilde{m}' = m'$ and so m' is reachable in N'. (2) $t_n^{N\bullet} \cap P' \neq \emptyset$ and $^{\bullet N}t_n \cap supp(m') = \emptyset$, in which case, as N' is $T - complete$, t_n must be in T'. Moreover, as t_n is fireable in N from $\tilde{m}$, it must be fireable as well in N' from $\tilde{m}'$. The effects of t_n on the places of P' are the same in N and N', so $\tilde{m}' \xrightarrow{t_n} m'$ in N'. Hence, m' is reachable in N'. (3) $t_n^{N\bullet} \cap P' = \emptyset$ and $^{\bullet N}t_n \cap supp(m') \neq \emptyset$, in which case, as N' is m'-complete, t_n must be in T'. Then, using the same arguments as in case (2), m' is reachable in N'. (4) $t_n^{N\bullet} \cap P' \neq \emptyset$ and $^{\bullet N}t_n \cap supp(m') \neq \emptyset$, in which case, as N' is T-complete and m'-complete, t_n must be in T'. Again, using the same arguments as in case (2), m' is reachable in N'. In each case, m' is reachable, and so the induction hypothesis is also verified for n, which concludes the induction. $\qquad\qquad\square$

Proposition 2. *If Algorithm 1 returns true, then m is reachable in N.*

Proof. The only way for Algorithm 1 to return true is at line 22. This implies that it goes out of the while loop. It means that both *Complete* and *Consistent* are true (line 5). So, each element of the list $LNets$ is complete according to Definition 10 (line 6). Hence, for any k, $LNets[k]$ is P-complete and the submarking m_k of m in $LNets[k]$ is reachable in $LNets[k]$. By Lemma 2, the partial marking of N whose support is exactly the same as the support of m_k is reachable in N, using the same sequence of transitions ω_k as in $LNets[k]$. Moreover, the list $LNets$ is consistent according to Definition 11 (line 15). Hence, for any k, ℓ the sets of places of $LNets[k]$ and $LNets[\ell]$ are disjoint (part 1. of Definition 11), as $LNets[k]$ is P-complete, this implies that no transition from $LNets[k]$ can reduce the marking of a place from $LNets[\ell]$. Moreover, no transition from $LNets[k]$ can increase the marking of a place from the support of m_ℓ (part 2. of Definition 11). As a consequence, the concatenation of all the ω_k allows to reach the objective marking m in N. $\qquad\qquad\square$

Lemma 2. *Let N' be a P-complete subnet of a Petri net N. Let m' be a partial marking of N' and let m be a partial marking of N so that $supp(m) = supp(m')$ and for all $p \in supp(m)$, $m(p) = m'(p)$. If there exists a sequence of transitions ω such that $m'_0 \xrightarrow{\omega} m'$ in N', then $m_0 \xrightarrow{\omega} m$ in N.*

Proof. We proceed by induction.

Induction Hypothesis. For all n, if a partial marking m' is reachable in N' by a sequence ω of n transitions, then there exists a partial marking m so that $supp(m) = supp(m')$, for all $p \in supp(m)$, $m(p) = m'(p)$ and m is reachable in N by the sequence ω.

Initialisation. When $n = 0$, any partial marking m' reachable by an empty sequence of transitions must realize m'_0, hence the partial marking m such that $\forall p \in P', m(p) = m'(p)$ and $\forall p \in P\backslash P', m(p) = \star$ must also be realized by m_0 and is thus reachable. This marking m is obviously such that $supp(m) = supp(m')$, which concludes the initialisation.

Induction. Let consider some $n > 0$ and assume that the induction hypothesis is verified for $n-1$. First remark that ω can always be split in two parts: ω' of length $n - 1$ and t_n (a single transition) such that $m'_0 \xrightarrow{\omega'} \tilde{m}'_c \xrightarrow{t_n} m'_c$ in N' for some marking $\tilde{m}'_c$ (not a partial one) and some marking m'_c that realizes m'. From the induction hypothesis the only partial marking $\tilde{m}$ so that $supp(\tilde{m}) = supp(\tilde{m}'_c)$ is reachable in N by the sequence ω', so there exists $\tilde{m}_c$ a marking that realizes $\tilde{m}$ and is reached by ω'. Moreover, as N' is P-complete, if t_n is fireable from $\tilde{m}'_c$ in N', then t_n is fireable from $\tilde{m}_c$ in N (all the preconditions of t_n appear in N'). Firing t_n in N leads to a marking m_c so that $\forall p \in P', m_c(p) = m'_c(p)$, by definition of a subnet. Hence, taking for m the partial marking such that $\forall p \in supp(m'), m(p) = m_c(p) = m'_c(p) = m'(p)$ and $\forall p \in P\backslash supp(m'), m(p) = \star$ concludes the induction. $\square$

Proposition 3. *Algorithm 1 always terminates and returns true or false.*

In order to prove Proposition 3 we define a relation over the lists $LNets$ involved in an execution of Algorithm 1. We show that this relation is an order relation and use this fact to conclude about the termination of the algorithm.

Definition 12. *Let $LNets_1$ and $LNets_2$ be two lists involved in an execution of Algorithm 1. We write $LNets_1 <_\ell LNets_2$ if and only if:*

- *$length(LNets_1) < length(LNets_2)$ or*
- *$length(LNets_1) = length(LNets_2)$ and $\exists 1 \leq k \leq length(LNets_2)$ such that $\forall 1 < i < k, LNets_1[i] = LNets_2[i]$ and $LNets_1[k] <_n LNets_2[k]$,*

where, for two subnets N_1 and N_2 of a net N, we have $N_1 <_n N_2$ if and only if:

- $P_1 \supset P_2$ *or*
- $P_1 = P_2$ *and* $T_1 \supset T_2$.

If $LNets_1 <_\ell LNets_2$ or $LNets_1 = LNets_2$ we write $LNets_1 \leq_\ell LNets_2$.

Lemma 3. *The relation $\leq_\ell$ of Definition 12 is an order relation.*

Proof. We prove that $\leq_\ell$ is reflexive, antisymmetric, and transitive.

Reflexive. This is a direct consequence of the fact that equality is reflexive.

Antisymmetric. Assume that $LNets_1 \leq_\ell LNets_2$ and $LNets_2 \leq_\ell LNets_1$. Suppose that $LNets_1 <_\ell LNets_2$. If $length(LNets_1) < length(LNets_2)$, then $length(LNets_1) \neq length(LNets_2)$ and $length(LNets_2) < length(LNets_1)$ cannot be true, so neither $LNets_2 <_\ell LNets_1$ nor $LNets_2 = LNets_1$ can be true, so $LNets_2 \leq_\ell LNets_1$ cannot be true. If $length(LNets_1) = length(LNets_2)$ and $LNets_1[k] < LNets_2[k]$ for some k with $LNets_1[i] = LNets_2[i]$ for any $i < k$, then either (1) $P_1 \subset P_2$ or (2) $P_1 = P_2$ and $T_1 \subset T_2$. In case (1), then $P_2 \neq P_1$ and $P_2 \subset P_1$ cannot be true, moreover $length(LNets_2) < length(LNets_1)$ cannot be true. So $LNets_2 <_\ell LNets_1$ nor $LNets_2 = LNets_1$ can be true, so $LNets_2 \leq_\ell LNets_1$ cannot be true. In case (2), then $P_2 = P_1$ but $T_2 \neq T_1$ and $T_2 \subset T_1$ cannot be true, moreover $length(LNets_2) < length(LNets_1)$ cannot be true. So $LNets_2 <_\ell LNets_1$ nor $LNets_2 = LNets_1$ can be true, so $LNets_2 \leq_\ell LNets_1$ cannot be true. In all cases if $LNets_1 <_\ell LNets_2$ then $LNets_2 \leq_\ell LNets_1$ cannot be true. Thus, as $LNets_1 \leq_\ell LNets_2$, one necessarily gets $LNets_1 = LNets_2$. This proves that $\leq_\ell$ is antisymmetric.

Transitive. Assume that $LNets_1 \leq_\ell LNets_2$ and $LNets_2 \leq_\ell LNets_3$. We show that $LNets_1 \leq_\ell LNets_3$. If $LNets_1 = LNets_2$ or $LNets_2 = LNets_3$, this is clearly true. Thus, assume $LNets_1 <_\ell LNets_2$ and $LNets_2 <_\ell LNets_3$. If $length(LNets_1) < length(LNets_2)$, then, as $length(LNets_2) \leq length(LNets_3)$, one gets $length(LNets_1) < length(LNets_3)$, thus $LNets_1 \leq_\ell LNets_3$. In the case where $length(LNets_1) = length(LNets_2)$, two cases are possible: (1) $length(LNets_2) < length(LNets_3)$, then $length(LNets_1) < length(LNets_3)$ is clearly true, and so $LNets_1 \leq_\ell LNets_3$, (2) $length(LNets_2) = length(LNets_3)$. In this second case, there exists k such that $LNets_1[k] <_n LNets_2[k]$ with $\forall i < k, LNets_1[i] = LNets_2[i]$ and k' such that $LNets_2[k'] <_n LNets_3[k']$ with $\forall i < k', LNets_2[i] = LNets_3[i]$. We need to distinguish three cases: (a) k < k', (b) k = k', and (c) k > k'. In case (a), one gets $LNets_1[k] <_n LNets_3[k]$ and $\forall i < k, LNets_1[i] = LNets_3[i]$, thus $LNets_1 \leq_\ell LNets_3$. In case (b), one gets $\forall i < k, LNets_1[i] = LNets_3[i]$ and $LNets_1[k] <_n$

$LNets_2[k] <_n LNets_3[k]$. The transitivity of $<_n$ (immediately obtained by transitivity of $\subset$) is sufficient to conclude that $LNets_1[k] <_n LNets_3[k]$, and thus $LNets_1 \leq_\ell LNets_3$. In case (c), one gets $LNets_1[k'] <_n LNets_3[k']$ and $\forall i < k', LNets_1[i] = LNets_3[i]$, thus $LNets_1 \leq_\ell LNets_3$. □

Proof. (of Proposition 3). The only return statements in Algorithm 1 are at line 22 and line 11. At line 22 the algorithm returns true and at line 11 it returns false. This implies that the only possible return values are true and false. It remains to prove that the algorithm terminates.

We prove that the successive values of $LNets$ in Algorithm 1 are strictly decreasing with respect to the order relation $\leq_\ell$. As there exists a minimal element (the empty list) with respect to this relation, this suffices to prove the termination.

Assume that an iteration of the main loop (while loop at line 5) starts. This results in, at least, one call to *Concretise* or one call to *Merge*. Thus, if we show that $LNets$ strictly decreases with respect to $\leq_\ell$ after a call to either of these functions, the termination is given by the above argument.

A call to *Merge* strictly decreases the length of $LNets$. So, by the first point of Definition 12, $LNets$ strictly decreases with respect to $\leq_\ell$.

A call to *Concretise* does not modify the length of $LNets$ and modifies exactly one element e of $LNets$ (or returns false, in which case Algorithm 1 terminates). The modified element is chosen so that its set of transitions is strictly increased (line 5 of Algorithm 2) or its set of places is strictly increased (line 12 of Algorithm 2). In either case, the modified element e' is such that $e' <_n e$. This ensures that $LNets$ strictly decreases with respect to $\leq_\ell$ and concludes the proof. □

4   Lazy Reachability Analysis with Inhibitor Arcs

We now transpose the previous results from Petri nets to Petri nets with inhibitor arcs. In such nets, places – when marked – can prevent transitions from being fired. The results of the previous section were correct in all Petri nets, bounded or not. In this section, this will no longer be the case as reachability is known for not being decidable in unbounded Petri nets with inhibitor arcs [1,20].

4.1   From Petri Nets to Petri Nets with Inhibitor Arcs

We start by formally defining Petri nets with inhibitor arcs.

Definition 13 (Petri net with inhibitor arcs). *A Petri net with inhibitor arcs is a tuple* $N_I = (P, T, F, I, m_0)$ *where* (P, T, F, m_0) *is a Petri net and* $I : P \times T \to \mathbb{N} \cup \{\infty\}$ *is an* inhibition function.

Markings, presets and postsets are defined similarly in Petri nets with and without inhibitor arcs. In a Petri net with inhibitor arcs N_I, for any $t \in T$, we define $°t = \{p \in P : I(p, t) \neq \infty\}$ the *inhibition set* of t. We extend this notion to sets T of transitions by union of inhibition sets. A transition

$t \in T$ is *fireable* from a marking m if and only if $\forall p \in {}^\bullet t, m(p) \geq F(p,t)$ and $\forall p \in {}^\circ t, m(p) < I(p,t)$. The result of firing t is similar as for Petri nets without inhibitor arcs, only the fireability condition changes. Reachability is thus also similarly defined in these two kinds nets.

Definition 14 (Subnet with inhibitor arcs). *A subnet N_I' of a Petri net with inhibitor arcs $N_I = (P, T, F, I, m_0)$ is a tuple (P', T', F', I', m_0') such that (P', T', F', m_0') is a subnet of (P, T, F, m_0) and $I' = I_{|P',T'}$.*

Given a subnet N_I' of a Petri net with inhibitor arcs N_I, and for any $t \in T'$, we define ${}^{\circ_{N_I}}t = \{p \in P \ : \ I(p,t) \neq \infty)\}$ (that is, intuitively, the inhibition set taken in N_I rather than in N_I').

The notions of P-completeness, T-completeness, partial marking, reachability of partial markings, and m-completeness remain the same in presence of inhibitor arcs. However, we introduce two other notions of completeness with respect to a net N_I for a subnet N_I'. The notion of PI-completeness expresses that N_I' contains all the places from N_I that may inhibit transitions in N_I'.

Definition 15 (PI-completeness). *A subnet N_I' of a net N_I is said to be PI-complete when $\forall t \in T', {}^{\circ_{N_I}}t \subseteq P'$.*

The notion of TI-completeness expresses that N_I' contains all the transitions from N_I that may remove tokens from places that inhibit transitions in N_I'.

Definition 16 (TI-completeness). *A subnet N_I' of a net N_I is said to be TI-complete when $\forall p \in {}^\circ T', p^{N_I \bullet} \subseteq T'$.*

4.2 An Algorithm for Lazy Reachability Analysis with Inhibitor Arcs

The basic principles of our lazy reachability analysis algorithm for Petri nets with inhibitor arcs are the same as for the case where there are no inhibitor arcs. In fact, the main algorithm that we use is still Algorithm 1 – we simply rename N as N_I to make it clear that it has inhibitor arcs – and the merging does not change. However, we use a different concretisation function (Algorithm 3) as well as the following definitions for completeness and consistency.

Definition 17. *Let $N_I = (P, T, F, I, m_0)$ be a Petri net with inhibitor arcs and m a marking. A subnet N_I' of N_I is complete with respect to N_I and m if (N_I' is P-complete, PI-complete and) the submarking m' is reachable in N_I'.*

Since we still need to check reachability, we must now assume that N_I is bounded, though we do not need to know the bound. Even when N_I is bounded, its subnets may be unbounded. In the concretisation function, we will nonetheless build our subnets so that they are bounded by construction.

Definition 18. *The list of subnets $LNets = [(P_1, T_1), ..., (P_n, T_n)]$ is consistent if*

Algorithm 3. Auxiliary function $Concretise(LNets, m)$ for Algorithm 1

1: **choose** k such that $LNets[k]$ is not complete
2: $(P_k, T_k) \leftarrow LNets[k]$
3: $m' \leftarrow m_{|P_k}$
4: **choose** T'_k such that $T_k \subset T'_k \subseteq {}^{\bullet N_I} P_k \cup supp(m')^{N_I \bullet} \cup ({}^{\circ} T_k)^{N_I \bullet}$
5: **if not possible then**
6:     **return** false
7: **else**
8:     $P'_k \leftarrow {}^{\bullet N_I} T'_k \cup {}^{\circ N_I} T'_k$
9:     $LNets[k] \leftarrow (P'_k, T'_k)$
10: **end if**
11: **return** true

---

1. $\forall k \neq \ell, (P_k \cap P_\ell) = \emptyset$, and
2. $\forall k \neq \ell, T_k^{N \bullet} \cap supp(m_{|P_\ell}) = \emptyset$.
3. $\forall k \neq \ell, T_k^{N \bullet} \cap {}^{\circ} T_\ell = \emptyset$.

The main difference with the concretisation function that was used for Petri nets with no inhibitor arcs is that one does not distinguish between the case where transitions should be added and the case where places should be added. Indeed, if one allows for adding transitions without their preconditions, this may result in unbounded subnets. In presence of inhibitor arcs this prevents for checking reachability. One can remark, however, that if all the preconditions of all transitions of a subnet are also part of this subnet, then this subnet is necessarily bounded (if the original net was bounded). This is expressed by Proposition 4 below. Thus, in Algorithm 3, one adds transitions as before to the considered subnet but then one always adds all the preconditions of the newly added transitions.

Remark 1. This explains the parenthesis in Definition 17: all the subnets considered are always P-complete and PI-complete.

Finally, notice that transitions that may remove tokens from inhibition places are also considered when adding transitions, as they can enable new transitions firings.

Proposition 4. *Let $N_I = (P, T, F, I, m_0)$ be a bounded Petri net with inhibitor arcs. Let $N'_I = (P', T', F', I', m_{0|P'})$ be a subnet of N_I. If $P' \supseteq {}^{\bullet N_I} T'_k \cup {}^{\circ N_I} T'_k$, then N'_I is bounded.*

Proof. Since N_I is bounded, let k be the corresponding bound. Assume N'_I is not k-bounded. Then there exists a transition firing sequence $\omega = t_1, \ldots, t_n$, some marking m', and some place $p \in P'$ such that in N'_I, we have $m_0 \xrightarrow{m'}$ and $m'(p) > k$.

We prove by induction on the length n of ω that it is also fireable in N_I and that if m (resp. m') is the marking obtained in N_I (resp. in N'_I) after firing ω, then $m_{|P'} = m'$.

First suppose $n = 0$, then the property holds trivially. Now assume it holds for some sequence $t_1, \ldots, t_n$, with $n \geq 0$, and consider an additional transition $t_{n+1} \in T'$. Let m_n (resp. m'_n) be the marking obtained in N_I (resp. N'_I) after firing $t_1, \ldots, t_n$. By the induction hypothesis $m_{n|P} = m'_n$. By construction the preset and inhibitor preset of t_{n+1} in N_I are included in P' and therefore since t_{n+1} is fireable in N'_I from m'_n, it is fireable in N_I from m_n and the effect of those firings on the places in P' is the same.

Since N_I is k-bounded, ω cannot be fireable in N_I so we have a contradiction and N'_I is therefore bounded. $\square$

4.3   Proof of the Algorithm

We now prove the correctness of Algorithm 1 when used with the concretisation function of Algorithm 3 on a bounded Petri net with inhibitor arcs.

First, remark that completeness of the algorithm is achieved essentially for the same reasons as in the previous case.

Proposition 5. *Algorithm 1 always terminates and returns true or false when used with the concretisation function of Algorithm 3 on a bounded Petri net with inhibitor arcs.*

Proof. The proof of Proposition 3 also works here as the order relation of Definition 12 does not depends on the arcs of the nets but only on their places and transitions. The only difference in the proof is to remark that a call to *Concretise* always increases the set of transitions. $\square$

It remains to prove to soundness of the algorithm, which is achieved by proving Propositions 6 and 7.

Proposition 6. *If Algorithm 1 used with the concretisation function of Algorithm 3 on a bounded Petri net N_I with inhibitor arcs returns false, then m is not reachable in N_I.*

Proof. As before, the only way for Algorithm 1 to return false is at line 11. It implies that the previous call to the Concretise function (Algorithm 3) returned false. This can only occur at line 6 of Algorithm 3.

In this case, it must not be possible to choose T'_k such that $T_k \subset T'_k \subseteq {}^{\bullet N_I} P_k \cup supp(m')^{N_I \bullet} \cup ({}^{\circ}T_k)^{N_I \bullet}$ (line 4). In other words, the subnet $LNets[k]$ considered is m'-complete (Definition 9), T-complete (Definition 6), and TI-complete (Definition 16). Moreover, the current marking m' must not be reachable in the subnet $LNets[k]$ due to Remark 1 and is a submarking of m. Hence, by applying Lemma 4 below, m is not reachable in N_I. $\square$

Lemma 4. *Let N'_I be a subnet of a bounded Petri net with inhibitor arcs N_I. Assume that N'_I is T-complete and TI-complete. Let m' be a marking of N'_I, that is not reachable. If N'_I is m'-complete, then no marking m of N_I such that m' is the submarking of m in N'_I is reachable in N_I.*

Proof. Let m be a reachable marking of N_I, with $m_0 \xrightarrow{\omega} m$. Denote by n the number of transitions in ω. We can prove by induction on n that m' is reachable in N_I'. By the contrapositive, this proves the Lemma.

The induction is in fact similar to the one used for proving Lemma 1. The only difference is in the induction step where – in cases (2), (3), and (4) – one could imagine that t_n would be fireable in N_I but not in N_I', due to inhibitor arcs. However, the fact N_I' is TI-complete prevents this: any place p that could inhibit t_n must have its full postset in N_I' and so, if, at some point in ω, p is marked, and later it is unmarked by some transition t, then t must be a transition from N_I'. □

Proposition 7. *If Algorithm 1 used with the concretisation function of Algorithm 3 on a bounded Petri net N_I with inhibitor arcs returns true, then m is reachable in N_I.*

Proof. The only way for Algorithm 1 to return true is at line 22. This implies that it goes out of the while loop. It means that both *Complete* and *Consistent* are true (line 5). So, each element of the list *LNets* is complete according to Definition 17 (line 6). Hence, for any k, $LNets[k]$ is P-complete, PI-complete, and the submarking m_k of m in $LNets[k]$ is reachable in $LNets[k]$. By Lemma 5, the partial marking of N_I whose support is exactly the same as the support of m_k is reachable in N_I, using the same sequence of transitions ω_k as in $LNets[k]$. Moreover, the list *LNets* is consistent according to Definition 18 (line 15). Hence, for any k, ℓ the sets of places of $LNets[k]$ and $LNets[\ell]$ are disjoint (part 1. of Definition 18), as $LNets[k]$ is P-complete, this implies that no transition from $LNets[k]$ can reduce the marking of a place from $LNets[\ell]$. Moreover, no transition from $LNets[k]$ can increase the marking of a place from the support of m_ℓ (part 2. of Definition 18) or the marking of a place that inhibits a transition of $LNets[\ell]$ (part 3. of Definition 18). As a consequence, the concatenation of all the ω_k allows to reach the objective marking m in N. □

Lemma 5. *Let N_I' be a P-complete and PI-complete subnet of a bounded Petri net with inhibitor arcs N_I. Let m' be a partial marking of N_I' and let m be a partial marking of N_I so that $supp(m) = supp(m')$ and for all $p \in supp(m)$, $m(p) = m'(p)$. If there exists a sequence of transitions ω such that $m_0' \xrightarrow{\omega} m'$ in N_I', then $m_0 \xrightarrow{\omega} m$ in N_I.*

Proof. The only difference with Lemma 2 is that some places in N_I may inhibit a transition of ω. This is prevented by the fact that N_I' is PI-complete: all these place must also exist in N_I' as well. A similar induction proof can thus be performed. □

5 Experimental Evaluation

We have implemented the algorithms in the tool ROMEO [16][2].

---

[2] 64bits Linux binaries for ROMEO and converters from pnml (MCC) to cts (ROMEO), and full results are at http://lara.rts-software.org/.

Note that since ROMEO deals with a very expressive formalism, encompassing inhibitor arcs, we have only implemented the concretisation function described in Sect. 4, even if it is less efficient than the one in Sect. 3 for the nets used in the experiments, which do not contain inhibitor arcs.

We implement line 4 in Algorithm 3 by always choosing the biggest possible T'_k. Actually to account for the added expressiveness of ROMEO, where transitions can modify the marking in an arbitrarily complex way, we add even a bit more: every transition that may modify the marking in P_k, i.e., ${}^{\bullet N_I} P_k \cup P_k{}^{N_I \bullet}$.

For reachability, we perform a simple explicit state exploration with no particular optimization.

ROMEO has then been run in a setting as close as we could of the Reachability Cardinality category of the 2020 edition of the model checking contest [10], and we compare it with the results of the other tools, which we do not recall here for the sake of space but which are fully available in [10]. For each model of the contest, we gave ROMEO one hour to solve the same 16 formulas that the contestant had to solve during the contest. The machine we used is not as powerful as the machines used for running the contest (it has four Intel Xeon E5-2620 processors and 128 GB of memory), but we did not run ROMEO on a virtual machine (during the MCC it would have been the case).

On a factual point of view, ROMEO, using only the algorithm presented here, has thus been faced to 1016 different models, among which it fully solved 120 (that is, it solved the 16 formulas corresponding to the model within one hour) and partially solved 288 (that is, solved at least 1 of the formulas within one hour but not the 16 of them). In total, ROMEO solved 2654 formulas among 16256. We have also run our algorithm on the 2018 version of the contest, thus on a subset of the models but with different formulas, with similar overall results.

While this overall result is not outstanding, the details are a lot more interesting. First, each result that ROMEO returned was the same as the result obtained by the majority of the tools during the actual contest, and hence we can assume with some confidence that all those results are correct.

Second, there is also a lot of room left to improve the algorithm itself, by choosing more cleverly which places to add, which transitions, how much of each partial net to compute, etc. And actually, since we require nothing in terms of exploration order, the technique we propose here should easily be combinable with, e.g., stubborn sets [14] for much better results.

Finally, and most significantly, ROMEO managed to completely solve (all 16 formulas) one model (GPPP-PT-C0010N1000000000) that no other tool could handle (they solved at most the first formula). We have had the same results for the 2018 formulas and at that time the other tools had solved none of them. This is very relevant because most of the best performing tools actually have a portfolio approach, in which several techniques are tried in parallel. It thus seems that this model is particularly difficult for current state-of-the-art tools, none of the currently implemented approaches is efficient to handle it, and hardly any progress has been made on it for the last four years (the model was introduced in the 2016 edition).

In contrast, the algorithm we have proposed here performs very well on that model and is thus likely to be a worthy addition to the portfolio approach.

6   Conclusion

In this paper, we have presented an algorithm for reachability analysis in possibly unbounded Petri nets and in bounded Petri Nets with inhibitor arcs. This algorithm heuristically performs reachability queries on subnets of the original net, in a lazy manner: it works on subnets of increasing size, trying to answer as soon as possible. We have proven that, in each case (unbounded Petri nets, bounded Petri nets with inhibitor arcs), the algorithm terminates, always answers, and always gives correct answers.

We have implemented the approach in the tool Romeo and performed a large scale experimental evaluation based on the models from the 2018 edition of the *model-checking contest*, showing that on all the models we indeed answered correctly. Moreover, it revealed that our implementation can solve problems that state-of-the-art model-checking tools cannot handle. While for many other problems, those tools outperform our implementation, we believe our algorithm is still a good candidate for inclusion in a portfolio approach.

Future work consists in incorporating the rest of the features of Roméo in the lazy framework: timing parameters, cost optimization, properties beyond reachability, control, etc.

References

1. Akshay, S., Chakraborty, S., Das, A., Jagannath, V., Sandeep, S.: On Petri nets with hierarchical special arcs. In: CONCUR, pp. 40:1–40:17 (2017)
2. Behrmann, G., David, A., Larsen, K.G.: A tutorial on UPPAAL. In: International School on Formal Methods for the Design of Computer, Communication and Software Systems, pp. 200–236 (2004)
3. Bonet, B., Haslum, P., Hickmott, S., Thiébaux, S.: Directed unfolding of Petri nets. ToPNOC **1**(1), 172–198 (2008)
4. Chatain, T., Paulevé, L.: Goal-driven unfolding of Petri nets. In: CONCUR, pp. 18:1–18:16 (2017)
5. Couvreur, J.-M., Thierry-Mieg, Y.: Hierarchical decision diagrams to exploit model structure. In: Wang, F., (ed.) FORTE, pp. 443–457 (2005)
6. Esparza, J., Römer, S., Vogler, W.: An improvement of McMillan's unfolding algorithm. In: TACAS, pp. 87–106 (1996)
7. Holzmann, G.J., Peled, D.: An improvement in formal verification. In: FORTE, pp. 197–211 (1994)
8. Jezequel, L., Lime, D.: Lazy reachability analysis in distributed systems. In: CONCUR, pp. 17:1–17:14 (2016)
9. Jezequel, L., Lime, D.: Let's be lazy, we have time - or, lazy reachability analysis for timed automata. In: FORMATS, pp. 247–263 (2017)
10. Kordon, F., et al.: Complete Results for the 2020 Edition of the Model Checking Contest, June 2020. http://mcc.lip6.fr/2020/results.php

11. Kordon, F., et al.: MCC'2015 - the fifth model checking contest. ToPNOC **11**, 262–273 (2016)

12. Rao Kosaraju, S.: Decidability of reachability in vector addition systems (preliminary version). In: Lewis, H.R., Simons, B.B., Burkhard, W.A., Landweber, L.H. (eds.) STOC, pp. 267–281. ACM (1982)

13. Lambert, J.-L.: A structure to decide reachability in Petri nets. TCS **99**(1), 79–104 (1992)

14. Lehmann, A., Lohmann, N., Wolf, K.: Stubborn sets for simple linear time properties. In: ICATPN, pp. 228–247 (2012)

15. Leroux, J., Schmitz, S.: Demystifying reachability in vector addition systems. In: LICS, pp. 56–67. IEEE Computer Society (2015)

16. Lime, D., Roux, O.H., Seidner, C., Traonouez, L.-M.: Romeo: a parametric model-checker for Petri nets with stopwatches. In: TACAS, pp. 54–57 (2009)

17. Mayr, E.W.: An algorithm for the general Petri net reachability problem. SIAM J. Comput. **13**(3), 441–460 (1984)

18. McMillan, K.: Using unfoldings to avoid the state explosion problem in the verification of asynchronous circuits. In: CAV, pp. 164–177 (1993)

19. Miner, A., Babar, J.: Meddly: multi-terminal and edge-valued decision diagram library. In: QEST, pp. 195–196 (2010)

20. Reinhardt, K.: Reachability in Petri nets with inhibitor arcs. ENTCS **223**, 239–264 (2008)

21. Weiser, M.: Program slicing. IEEE Trans. Softw. Eng. **SE-10**(4), 352–357 (1984)

22. Wolf, K.: Running LoLA 2.0 in a model checking competition. ToPNOC **11**, 274–285 (2016)

Abstraction-Based Incremental Inductive Coverability for Petri Nets

Jiawen Kang[1,2], Yunjun Bai[1,2], and Li Jiao[1,2(✉)]

[1] State Key Laboratory of Computer Science, Institute of Software,
Chinese Academy of Sciences, Beijing, China
ljiao@ios.ac.cn
[2] University of Chinese Academy of Sciences, Beijing, China

Abstract. We present a novel approach to check the coverability problem of Petri nets which is based on a tight integration of IC3 with place-merge abstraction. Place-merge abstraction can reduce the dimensionality of state spaces by trying to merge some places that may be not critical for proving the property. In this scenario, IC3 runs only on abstract Petri nets with lower dimensionality. When the current abstraction allows for a spurious counterexample, it is refined by splitting candidate abstract places. Furthermore, this can be done in a completely incremental way without discarding results found in previous abstractions. The experimental evaluation on the standard Petri net benchmarks shows the effectiveness and competitiveness of our approach.

Keywords: Petri nets · Inductive invariants · Coverability · IC3 · Place-merge abstraction

1 Introduction

IC3, proposed in [4], is an efficient algorithm for the verification of the safety property of hardware systems. Different from the bounded model checking and k-induction, IC3 does not require unrolling the transition relation. It tries to find an inductive invariant by maintaining over-approximations of the sets of forward-reachable states and strengthening them based on counterexamples obtained by searching backward from bad states. A reconstruction of IC3 together with an efficient implementation, presented in [10], demonstrates its extreme competitiveness for the verification of hardware models.

There have been several contributions to the extension of IC3 to software systems. One of them is generalizing IC3 from SAT to the case of SMT [5,6] where software programs are described by first-order logic formulas, and subsequently, it has been lifted to infinite-state transition systems combined with predicate abstraction [7]. There exists another extension of IC3 that represents software programs in the form of control-flow automata [19]. IC3 has been adapted to timed systems [17] and Markov decision processes [2], respectively. Furthermore,

© Springer Nature Switzerland AG 2021
D. Buchs and J. Carmona (Eds.): PETRI NETS 2021, LNCS 12734, pp. 379–398, 2021.
https://doi.org/10.1007/978-3-030-76983-3_19

IC3 algorithm has been generalized to solve the coverability problem of the class of downward-finite well-structured transition systems (WSTS) in [18].

We consider here the case of Petri nets, a subclass of WSTS [13], which provide a simple and natural automata-like method modeling concurrent systems. For all WSTS, a large class of safety properties can be solved by conversion to the coverability problem, which is decidable for WSTS [1], hence for Petri nets. Even though the coverability problem of Petri nets has an EXPSPACE-completeness complexity [15], a lot of research works have explored many efficient techniques to solve the coverability problem. The general and classical algorithm solves the coverability problem by backward exploration of state space [1,13]. The approach based on the marking equation and traps has achieved good results profiting from the SMT solver in [11]. Combining with properties of Petri nets, generalization of IC3 to WSTS has been implemented on Petri nets to solve the coverability problem without using SMT solvers [18]. However, when the number of places in Petri nets becomes larger, most approaches suffer from the explosion caused by the high dimensionality of state spaces. A method of forward and backward search based on abstraction has been developed in [14] trying to solve the coverability problem by manipulating lower dimensional sets, where the way of abstraction is called *place-merge abstraction*.

In this paper, we present a novel IC3-like method to solve the coverability problem of Petri nets based on place-merge abstraction. Place-merge abstraction is an efficient abstraction technique for Petri nets trying to gather places that may possibly be not important for satisfaction of properties, and it can reduce the dimensionality of the state space while preserving the satisfaction of properties. In our algorithm, we try to improve the outperformance of IC3 in computing inductive invariants by combining IC3 with dimensionality reduction benefited from place-merge abstraction. The main idea of our approach is to let IC3 proceed on abstract Petri nets generated by place-merge abstraction. We set our algorithm in the Counter-Example Guided Abstraction-Refinement (CEGAR) framework [8]. When an abstract counterexample is found by IC3, it is necessary to simulate it on the original Petri net. Note that if the abstract counterexample is spurious, the current abstract Petri net is not precise enough to get the conclusion, then the abstract Petri net can be refined by splitting abstract places that lead to the spurious counterexample path.

The proposed algorithm has several advantages as follows. First, the computation of inductive invariants now works on abstract Petri nets with lower dimensionality, which is often as effective and correct as the analysis on the original Petri net, but faster and more efficient. Second, the algorithm proceeds in a completely incremental way, allowing to keep all results found in previous abstractions, and the previous spurious counterexamples cannot exist after refinement.

We experimentally evaluated our approach on a set of benchmarks from several sources [14,16,20]. The results show that our approach is more effective than the original IC3 algorithm in terms of run times on a lot of the benchmarks, and benefiting from place-merged abstraction, the number of places of Petri nets that allows to conclude is reduced effectively.

The paper is organized as follows. In Sect. 2, we present some backgrounds about the coverability problem of Petri nets and IC3. In Sect. 3, we show detailed descriptions and prove the correctness of our algorithm. We experimentally evaluate our approach in Sect. 4. In Sect. 5, we discuss the related work. Conclusions and future work are given in Sect. 6.

2 Preliminaries

2.1 Notations

Well-Quasi-Orderings. For a set M and a relation $\preceq \subseteq M \times M$, $\preceq$ is a quasi-ordering iff $\preceq$ is a reflexive and transitive relation. A quasi-ordering $\preceq$ is a *well-quasi-ordering* (*wqo* for short) if for any infinite sequence $m_0, m_1, m_2, \cdots$ in M, there exists indices $i < j$ such that $m_i \preceq m_j$. For a wqo $\preceq$, we denote the converse ordering by $\succeq$, the equivalence relation $\preceq \cap \succeq$ by $=$, and strict ordering $\preceq \setminus = $ by $\prec$.

Upward-Closed Sets and *Downward-Closed Sets.* Given a wqo $\preceq \subseteq M \times M$, a set $U \subseteq M$ is *upward-closed* if for any $u \in U$, $u \preceq m$ implies $m \in U$. $x^\uparrow = \{m | x \preceq m \wedge m \in M\}$ is the *upward-closure* of $x \in M$. For any set $X \subseteq M$, its upward closure is the set $X^\uparrow = \bigcup_{x \in X} x^\uparrow$. Symmetrically, a set $D \subseteq M$ is *downward-closed* if for any $d \in D$ and $m \preceq d$ implies $m \in D$. $x^\downarrow = \{m | m \preceq x \wedge m \in M\}$ is the *downward closure* of $x \in M$. For any set $X \subseteq M$, its dowmward closure is the set $X^\downarrow = \bigcup_{x \in X} x^\downarrow$.

For an upward-closed set U, a basis of U is a finite subset $U_b \subseteq U$ if $U = U_b^\uparrow$. Clearly, any upward-closed set U has a finite basis consisting of the minimal elements since $\preceq$ is a wqo. Such a basis allows for a finite representation of an upward-closed set. Furthermore, the complement of an upward-closed set is downward-closed, and vice versa. Therefore, downward-closed sets can be represented by a basis of their complements.

2.2 Petri Nets

We briefly describe some basic concepts of Petri nets, more concepts and definitions are available in [21].

Definition 1 (Petri nets). *A Petri net is a tuple $N = \langle P, T, W, m_0 \rangle$ where:*

- *P is a finite set of places.*
- *T is a finite set of transitions such that $P \cap T = \emptyset$.*
- *W is an arc function: $(P \times T) \cup (T \times P) \to \mathbb{N}$ describing the relationship between places and transitions.*
- *m_0 is the initial marking. A marking $m \in \mathbb{N}^{|P|}$ is a vector specifying a number $m(p)$ of tokens for each place $p \in P$.*

We introduce a partial order $\preceq$ on the set of markings space $\mathbb{N}^{|P|}$ such that for all $m, m' \in \mathbb{N}^{|P|}$, $m \preceq m'$ iff for every $p \in P$: $m(p) \leq m'(p)$. It turns out that $\preceq$ is wqo on $\mathbb{N}^{|P|}$. A wqo $(X, \preceq)$ is *downward-finite* if for every $x \in X$, the downward closure $x^{\downarrow}$ is finite. It turns out that Petri nets are downward-finite systems, since markings can consist of only non-negative integers and $(\mathbb{N}^{|P|}, \preceq)$ is a downward-finite wqo.

Definition 2. *Let $N = \langle P, T, W, m_0 \rangle$ be a Petri net.*

- *A transition $t \in T$ is enabled at marking m iff $W(p, t) \leq m(p)$ for all $p \in P$.*
- *A transition t can fire at marking m iff t is enabled at m, yielding a new marking m' and $m'(p) = m(p) - W(p, t) + W(t, p)$ for all $p \in P$, which we write $m \xrightarrow{t} m'$.*
- *The set of predecessors of m is the set $pre(m) = \{m' | \exists t \in T : m' \xrightarrow{t} m\}$.*
- *The set of successors of m is the set $post(m) = \{m' | \exists t \in T : m \xrightarrow{t} m'\}$.*
- *A marking m' is reachable from m iff $m = m'$ or there exist a transition sequence $\sigma = t_1 t_2 \ldots t_k$ and a marking sequence $m_1, m_2 \ldots m_{k-1}$ such that $m \xrightarrow{t_1} m_1 \xrightarrow{t_2} m_2 \ldots m_{k-1} \xrightarrow{t_k} m'$ which we write $m \xrightarrow{\sigma} m'$.*
- *The set of successors in k-steps of m is the set $post^k(m) = \{m' | \exists \sigma : |\sigma| = k \wedge m \xrightarrow{\sigma} m'\}$, and specially $post^0(m) = \{m\}$ where σ is empty.*
- *The set of reachable markings from m within k-steps is the set $Reach_k(m) = \bigcup_{0 \leq i \leq k} post^i(m)$.*
- *The set of all reachable markings from m is $Reach(m) = \bigcup_{k \geq 0} Reach_k(m)$.*

Transitions of Petri nets and the wqo $\preceq \subseteq \mathbb{N}^{|P|} \times \mathbb{N}^{|P|}$ satisfy the monotonicity property: for all markings $m_1, m_2, m_3 \in \mathbb{N}^{|P|}$, if $m_1 \xrightarrow{t} m_2$ and $m_1 \preceq m_3$, then there exists m_4 such that $m_3 \xrightarrow{t} m_4$ and $m_2 \preceq m_4$.

2.3 The Coverability Problem

We recall some important notions about the coverability problem [12].

The Coverability Problem. Let $N = \langle P, T, W, m_0 \rangle$ be a Petri net. The marking m_t is a target marking. The coverability problem is to prove whether there exists a reachable marking $m_r \in Reach(m_0)$ such that $m_t \preceq m_r$.

If there does exist such a marking m_r, then the marking m_t is coverable. Moreover, it is equivalent to prove whether the set $Reach(m_0)$ has an intersection with $m_t^{\uparrow}$. The coverability problem of Petri nets can be converted to the safety property $\mathcal{P} = \mathbb{N}^{|P|} \setminus m_t^{\uparrow}$ by checking if $N \vDash \mathcal{P}$, i.e. if all reachable markings $Reach(m_0) \subseteq \mathcal{P}$. The safety property $\mathcal{P}$ is false iff the target marking m_t is coverable. Thus, we can say that the upward-closed set $m_t^{\uparrow}$ is the set of *bad* markings, while the complement $\mathbb{N}^{|P|} \setminus m_t^{\uparrow}$ represents the set of *good* markings.

To solve the coverability problem, we introduce coverable sets, defined as the downward-closure of reachable sets.

Definition 3. *Let* $N = \langle P, T, W, m_0 \rangle$ *be a Petri net.*

- *The coverable set of* N *within* k*-steps is the set* $Cover_k(N) = \{m | \exists m' \in Reach_k(m_0) : m \preceq m'\}$, *i.e.* $Cover_k(N) = Reach_k(m_0)^{\downarrow}$.
- *The coverable set of* N *is the set* $Cover(N) = Reach(m_0)^{\downarrow}$.

The coverability problem has an EXPSPACE-complete complexity and is decidable for Petri nets, resulting from a general decidability result [1]. The classical algorithm for solving this problem is a backward algorithm, it is based on the following formula:

$$U_0 = m_t, U_{i+1}^{\uparrow} = pre(U_i^{\uparrow}) \cup U_i^{\uparrow}$$

The $U_i^{\uparrow}$ increases gradually and must stabilize. When the sequence stabilizes, we denote stabilized $U_i^{\uparrow}$ by $U^{\uparrow}$. It is easy to see that if $m_0 \notin U^{\uparrow}$ then $Cover(N) \cap U^{\uparrow} = \emptyset$, i.e. the target marking m_t is not coverable in the Petri net.

2.4 IC3 for Petri Nets

IC3 [4] has drawn extensive concern as an efficient algorithm for the computation of inductive invariants. Combining with properties of Petri nets, generalization of IC3 to the class of downward-finite WSTS has been implemented on Petri nets to solve the coverability problem without using SMT solvers [18]. In the following, we present its main idea. More details can be found in [4,18].

Let $N = \langle P, T, W, m_0 \rangle$ be a Petri net and m_t be a target marking, the IC3 algorithm tries to find an inductive invariant F for the coverability problem such that $F \vDash \mathcal{P}$. F is an inductive invariant for the coverability problem iff (a) $m_0 \in F$, (b) F is a downward-closed set, and (c) $post(F) \subseteq F$. If the inductive invariant $F \subseteq \mathcal{P}$, we have $Cover(N) \subseteq F \subseteq \mathcal{P}$. Therefore, our goal is to build such an inductive invariant.

In order to find an inductive invariant F, the IC3 algorithm maintains a sequence $F_0, F_1 \ldots F_k$ such that for all $0 \leq i < k$:

$$m_0 \in F_i$$
$$post(F_i) \subseteq F_{i+1}$$
$$F_i \subseteq F_{i+1}$$
$$F_i \subseteq \mathcal{P}$$

where F_i is a downward-closed set called *frame* which over-approximates the set of coverable markings within i steps $Cover_i(N)$, F_0 plays a special role and always represents the downward closure of the initial marking, i.e. $F_0 = m_0^{\downarrow}$.

A description of IC3 is shown in Algorithm 1 as a pseudo-code which is summarized from [18]. The algorithm generally proceeds by alternating two phases: the blocking phase and the propagation phase.

In the blocking phase (lines 6–8), IC3 tries to block m_t in k-th frame or build a path from $m_0^{\downarrow}$ to $m_t^{\uparrow}$ by searching backward. In this phase, IC3 maintains a

Input: a Petri net $N = \langle P, T, W, m_0 \rangle$ and the target marking m_t
Output: TRUE or FALSE
1 **Function** IC3(N, m_t):
2 | **if** $m_0 \in m_t^\uparrow$ **then return** $TRUE$;
3 | $k = 1, F_0 = m_0^\downarrow, F_k = \mathbb{N}^{|P|}$
4 | $\pi = \emptyset$ // a list recording counterexample
5 | **while** $TRUE$ **do**
   | | // blocking phase
6 | | **if** $not\ recBlock(m_t, k)$ **then**
7 | | | **return** $TRUE$ // counterexample π found
8 | | **end**
   | | // propagation phase
9 | | $k = k + 1, F_k = \mathbb{N}^{|P|}, \pi = \emptyset$
10 | | **for** $i = 1$ to k **do**
11 | | | **foreach** $marking\ m_b\ blocked\ in\ F_i$ **do**
12 | | | | **if** $pre(m_b^\uparrow) \cap F_i = \emptyset$ **then** remove $m_b^\uparrow$ from F_{i+1};
13 | | | **end**
14 | | | **if** $F_i == F_{i+1}$ **then**
15 | | | | **return** $FALSE$// inductive invariant F_i found
16 | | | **end**
17 | | **end**
18 | **end**
19 **Function** recBlock(m, i):
20 | **if** $i = 0$ **then return** $False$;
21 | **while** $pre(m^\uparrow) \cap F_{i-1} \setminus m^\uparrow \neq \emptyset$ **do**
22 | | select a marking m_p in $pre(m^\uparrow) \cap F_{i-1} \setminus m^\uparrow$ along transition t
23 | | **if** $not\ recBlock(m_p, i-1)$ **then**
24 | | | π.append(t)
25 | | | **return** $False$
26 | | **end**
27 | **end**
28 | $m_g = generalize(m, i)$
29 | **for** $j = 1$ to i **do** remove $m_g^\uparrow$ from F_j;
30 | **return** $True$

Algorithm 1: IC3 for coverability problem of Petri net (summarized from [18])

set of pairs (m, i), where m is a bad marking or can lead to bad markings, and i is the position of m in the sequence of frames. For the pair (m, i), IC3 tries to block m in F_i by checking if $m^\uparrow$ is reachable from F_{i-1}. If unreachable, there is no path from $m_0^\downarrow$ to $m^\uparrow$ within i steps, and the marking m is not coverable within i steps. Then we can block m in F_i, i.e. remove $m^\uparrow$ from F_i safely. If reachable, it is not strong enough to show that $m^\uparrow$ is unreachable from $m_0^\downarrow$. In this scenario, let m_p be a predecessor of $m^\uparrow$ in F_{i-1} such that m_p can lead to $m^\uparrow$ in one step, then a new pair $(m_p, i-1)$ is generated and IC3 tries to block

m_p in F_{i-1}. The algorithm continues recursively, until either a pair $(m_a, 0)$ is generated, which means that a path from $m_0^{\downarrow}$ to $m_t^{\uparrow}$ is found and the target marking m_t is coverable, or the blocking operation succeeds somewhere meaning that the marking m in original pair (m, i) can be blocked in F_i.

In the propagation phase (lines 9–17), IC3 tries to extend the sequence with a new frame F_{k+1} that contains all markings. For every blocked marking m_b in $F_i(1 \leq i \leq k)$, if the set of predecessors of $m_b^{\uparrow}$ has no intersection with F_i, blocked marking m_b can be propagated to F_{i+1} and $m_b^{\uparrow}$ can be removed in F_{i+1}. During this process, if two consecutive frames become equivalent $F_i = F_{i+1}$, then we can conclude $post(F_i) \subseteq F_i$ and an inductive invariant F_i is found, i.e. m_t is not coverable.

More importantly, when the marking m can be blocked in F_i, we can remove even a bigger set $m_g^{\uparrow}$ instead of $m^{\uparrow}$, m_g is computed as follows [18]:

$$generalize(m, i) = \{m_g | m_g \preceq m \wedge m_0 \notin m_g^{\uparrow} \wedge pre(m_g^{\uparrow}) \cap F_{i-1} \setminus m_g^{\uparrow} = \emptyset\}$$

It is easy to see that m_g can also be blocked in F_i. This generalization can speed up search significantly for the construction of inductive invariants in a given search space.

In particular, we can get the predecessors of $m^{\uparrow}$ by the following Lemma 1.

Lemma 1. *Give a Petri net $N = \langle P, T, W, m_0 \rangle$. Let $t \in T$ be a transition. Then $m' \in pre(m^{\uparrow})$ is a predecessor of $m^{\uparrow}$ along t iff for all $p \in P$*

$$m'(p) \geq max\{m(p) + W(p, t) - W(t, p), W(p, t)\} \tag{1}$$

Proof. Suppose m' is a predecessor of $m^{\uparrow}$ along t. $m'(p) \geq W(p, t)$ for all $p \in P$, because the transition t is enabled in marking m'. Transition t fires yielding a new marking m'' such that $m \preceq m''$, i.e. $m'(p) - W(p, t) + W(t, p) \geq m(p)$ for all $p \in P$. Therefore, predecessors of $m^{\uparrow}$ make (1) satisfiable.

For the other direction, transition t is enabled at m' because $m'(p) \geq W(p, t)$ for all $p \in P$. After t firing, the resulting marking m'' satisfies $m \preceq m''$. because $m'(p) \geq m(p) - W(t, p) + W(p, t)$ for all $p \in P$. Thus, the marking m' making (1) satisfiable is a predecessor of $m^{\uparrow}$ along t.

Moreover, such a marking b satisfying $b(p) = max\{m(p) + W(p, t) - W(t, p), W(p, t)\}$ itself is a predecessor of $m^{\uparrow}$ along t, and b is the minimal marking of $pre(m^{\uparrow})$. $\qquad\square$

Clearly, all predecessors of $m^{\uparrow}$ form also an upward-closed set by Lemma 1 and the monotonicity property of Petri nets. It facilitates the computation of all predecessors of an upward-closed set and the process of backward search.

3   Combining IC3 with Place-Merge Abstraction

3.1   The Place-Merge Abstraction

Abstraction [9] in model checking is a very powerful approach to reduce the complexity of verification in search space while preserving the satisfaction of some properties. Therefore, the abstract model can be used to verify the satisfiability of some properties of the concrete model.

Place-merge abstraction [14] is an efficient abstraction method for Petri nets. With the help of place-merge abstraction, we get new abstract Petri nets with fewer places by merging unimportant places into one single place, and therefore reduce the dimensionality of marking spaces. In the following, we present some important definitions and propositions about place-merge abstraction.

Definition 4. *Given a Petri net $N = \langle P, T, W, m_0 \rangle$, where $P = \{p_1, p_2 \ldots p_k\}$. The abstraction function is a surjective function $\alpha : P \rightarrow \hat{P}$, where $\hat{P} = \{\hat{p}_1, \hat{p}_2 \ldots \hat{p}_{\hat{k}}\}$ and $\hat{k} \leq k$. The corresponding concretization function is an injection $\gamma_\alpha : \hat{P} \rightarrow 2^P$ such that $\gamma_\alpha(\hat{p}) = \{p \in P | \alpha(p) = \hat{p}\}$.*

Intuitively, the abstraction of Petri nets is a partition of the set of places. It divides k places into $\hat{k}$ classes. The partition will be used to compute abstract versions of markings with lower dimensionality. Given a marking m(denoted by a vector over $\mathbb{N}^k$), the abstraction converts m to an abstract marking $\hat{m}$ (denoted by a vector over $\mathbb{N}^{\hat{k}}$). Since transitions are related to places in Petri nets, we introduce the following definition of markings and transitions under place-merge abstraction. It is worth noting that the number of transitions is the same after abstraction.

Definition 5. *Given an abstraction function α and a marking m, $\hat{m}$ is the abstract version of m, and $\hat{m}(\hat{p}) = \sum_{\alpha(p)=\hat{p}} m(p)$. Given a transition t, the relationship between places and transitions of abstraction can be denoted by $\hat{W}(\hat{p}, t) = \sum_{\alpha(p)=\hat{p}} W(p, t)$ and $\hat{W}(t, \hat{p}) = \sum_{\alpha(p)=\hat{p}} W(t, p)$.*

From Definitions 4 and 5, we get the abstraction of the original Petri net $\hat{N} = \langle \hat{P}, T, \hat{W}, \hat{m}_0 \rangle$ which is also a Petri net.

The following propositions can be obtained to show that abstract Petri nets can admit behaviors in original Petri nets but the converse does not hold.

Proposition 1. *Given a Petri net $N = \langle P, T, W, m_0 \rangle$ and an abstraction function α, we get the abstract Petri net $\hat{N} = \langle \hat{P}, T, \hat{W}, \hat{m}_0 \rangle$ using Definitions 4 and 5. m and m' are markings in N. $\hat{m}$ and $\hat{m}'$ in $\hat{N}$ are the abstract version of m and m', respectively. $t \in T$ is a transition. If $m \xrightarrow{t} m'$ in the Petri net, then $\hat{m} \xrightarrow{t} \hat{m}'$ in the abstract Petri net.*

Proof. Following the fact that $m \xrightarrow{t} m'$, it implies that t is enabled at marking m in N where $m(p) \geq W(p, t)$ and $m'(p) = m(p) - W(p, t) + W(t, p)$ for all $p \in P$. According to Definition 5, $\hat{m}(\hat{p}) = \sum_{\alpha(p)=\hat{p}} m(p)$, $\hat{W}(\hat{p}, t) = \sum_{\alpha(p)=\hat{p}} W(p, t)$ and $\hat{W}(t, \hat{p}) = \sum_{\alpha(p)=\hat{p}} W(t, p)$ for all $\hat{p} \in \hat{P}$. Therefore, $\hat{m}(\hat{p}) \geq \hat{W}(\hat{p}, t)$ for all $\hat{p} \in \hat{P}$, i.e. t is enabled at the abstract marking $\hat{m}$. Transition t fires at $\hat{m}$ yielding $\hat{m}'(\hat{p}) = \hat{m}(\hat{p}) - \hat{W}(\hat{p}, t) + \hat{W}(t, \hat{p}) = \sum_{\alpha(p)=\hat{p}} m(p) - \sum_{\alpha(p)=\hat{p}} W(p, t) + \sum_{\alpha(p)=\hat{p}} W(t, p)$. Since $m'(p) = m(p) - W(p, t) + W(t, p)$, it allows to conclude that $\hat{m}'(\hat{p}) = \sum_{\alpha(p)=\hat{p}} m'(p)$ for all $\hat{p} \in \hat{P}$. $\qquad\square$

Proposition 2. *Given a Petri net $N = \langle P, T, W, m_0 \rangle$ and an abstraction function α, the abstract Petri net $\hat{N} = \langle \hat{P}, T, \hat{W}, \hat{m}_0 \rangle$ can be obtained by Definitions 4 and 5. m_r is a marking in N and $\hat{m}_r$ is its abstract version in $\hat{N}$. $\sigma = t_0, t_1 \ldots t_k$ is a transition sequence where $t_i \in T$ for all $0 \leq i \leq k$. If $m_0 \xrightarrow{\sigma} m_r$ in N, then $\hat{m}_0 \xrightarrow{\sigma} \hat{m}_r$ in $\hat{N}$.*

Proof. According to Proposition 1, if $m \xrightarrow{t} m'$ in N, then $\hat{m} \xrightarrow{t} \hat{m}$ in $\hat{N}$. Let the transitions in the sequence fire in order, it is easy to see that $\hat{m}_r$ is reachable from $\hat{m}_0$ in $\hat{N}$ along the same transition sequence σ.   □

Proposition 3. *Given a Petri net $N = \langle P, T, W, m_0 \rangle$ and an abstraction function α, the abstract Petri net $\hat{N} = \langle \hat{P}, T, \hat{W}, \hat{m}_0 \rangle$, $\hat{m}_t$ can be obtained by Definitions 4 and 5. m_t is a target marking in N and $\hat{m}_t$ is the abstract version of m_t in $\hat{N}$. If m_t is coverable in N, then $\hat{m}_t$ is coverable in $\hat{N}$.*

Proof. Supposed m_t is coverable in N, then there exists a reachable marking m_r in N such that $m_t \preceq m_r$. According to Proposition 2, the abstract version $\hat{m}_r$ is also reachable from $\hat{m}_0$ in $\hat{N}$. By Definition 5, $\hat{m}_t \preceq \hat{m}_r$ holds since $m_t \preceq m_r$. Thus, $\hat{m}_t$ is coverable in $\hat{N}$.   □

For an abstract Petri net, each abstract place $\hat{p}$ corresponds to a set of places of the original Petri net. By abuse of notation, we write $\hat{p} = \gamma_\alpha(\hat{p})$ directly and write a marking as a vector where the components are arranged by indices of places in Petri nets.

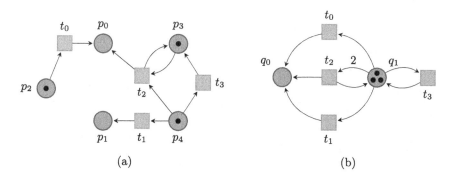

Fig. 1. A Petri net (a) and its abstraction (b). It assumes that all weights of arcs are equal to one, except for $W(q_1, t_2) = 2$.

Example 1. Consider the Petri nets shown in Fig. 1, the current distribution of tokens represents the initial marking. It is clear to see that the Petri net $\hat{N}$ in (b) is an abstraction of N in (a) with $q_0 = \{p_0, p_1\}, q_1 = \{p_2, p_3, p_4\}$. Our goal is to prove whether the target marking $m_t = (1, 1, 0, 0, 0)$ is coverable in N. In the abstraction, it is converted to prove whether the abstract marking $\hat{m}_t = (2,0)$ is coverable. If the abstract marking $\hat{m}_t$ is not coverable in $\hat{N}$, then m_t is also not coverable in N according to Proposition 3. Therefore, we can analyze the property of the original Petri net in its abstraction whose state space is $\mathbb{N}^2$ instead of $\mathbb{N}^5$.

Clearly, a marking of abstraction corresponds to a set of markings in the original Petri nets. It is safe to conclude that m_t is coverable in N implies $\hat{m}_t$ is coverable in $\hat{N}$ by Proposition 3. However, the converse does not hold, since there may exist counterexamples in $\hat{N}$ which have no counterpart in N, in this case, these counterexamples are spurious.

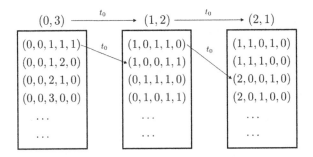

Fig. 2. A spurious counterexample path of the abstract Petri net in Fig. 1(b). Each vector in the rectangle represents a concrete marking and each solid arrow is a transition. Each rectangle represents an abstract marking and its value is on the top of the rectangle.

Example 2. Consider Fig. 2, the figure shows a spurious counterexample path of Example 1. Each vector is a marking and solid arrows are transitions. Each rectangle represents an abstract marking and its value is on the top of the rectangle, the concrete markings in the same rectangle correspond to the same abstract marking. It is clear that the marking $\hat{m}_t = (2,0)$ is coverable in $\hat{N}$ since there exists a reachable marking $\hat{m}_r = (2,1)$ along the counterexample path $\sigma = t_0 t_0$ such that $\hat{m}_t \preceq \hat{m}_r$. However, the path σ has no counterpart in the original Petri net N, since t_0 fires at the initial marking of N yielding $m_1 = (1,0,0,1,1)$, but the next transition t_0 is not enabled at m_1. Therefore, $\sigma = t_0 t_0$ is a spurious counterexample path.

It is possible there exist spurious counterexample paths if the abstraction is too coarse. The abstraction should be refined to rule out these spurious counterexample paths. CEGAR [8] is a fast-developing algorithm framework and is highly focused on software model checking. It can automatically refine an abstraction by extracting information from spurious counterexamples. In place-merge abstraction, we can refine the abstraction by splitting the large abstract places which contain too many original places.

3.2 The Algorithm

In this section, we propose a new algorithm (IC3+PMA for short) that combines IC3 and place-merge abstraction to solve the coverability problem of Petri nets more efficiently. The main idea of IC3+PMA is to let IC3 work on abstract Petri nets with lower dimensionality.

The IC3+PMA algorithm is shown in Algorithm 2 as pseudo-code and it has the same structure as IC3. The key differences between IC3+PMA and IC3 in Algorithm 1 arise from the fact that we are working on abstract marking spaces with lower dimensionality and the fact that counterexample paths may be spurious, and in this case, a new abstraction is obtained by refinement.

Just as IC3 shown in Algorithm 1, IC3+PMA also maintains a sequence of frames to find an inductive invariant, and consists of a loop, where each iteration is divided into the blocking phase and the propagation phase.

Initially, we use an initial abstraction function α that merges all places of the original Petri net N into one single abstract place, then we get the abstract Petri net $\hat{N}$ and the abstract version of the target $\hat{m}_t$ by Definitions 4 and 5 (line 2). Different from IC3, IC3+PMA initializes the sequence of frames with $\mathbb{N}^{|\hat{P}|}$ (line 7).

In the blocking phase (lines 9–17), IC3+PMA tries to block $\hat{m}_t$ or build a path from $\hat{m}_0^{\downarrow}$ to $\hat{m}_t^{\uparrow}$ in the current abstract Petri net by searching backward as IC3 does. IC3+PMA maintains a set of pairs $(\hat{m}, i)$, where $\hat{m}$ is a bad marking or can lead to bad markings in $\hat{N}$. IC3+PMA continues recursively (recBlock procedure at line 28) on abstract Petri net. If IC3+PMA generates a pair $(\hat{m}_a, 0)$, there exists such a counterexample path from $\hat{m}_0^{\downarrow}$ to $\hat{m}_t^{\uparrow}$, i.e. $\hat{m}_t$ is coverable in $\hat{N}$. In that case, the counterexample path should be simulated in the original Petri net N (line 11). If the simulation succeeds, it is sufficient to conclude m_t is coverable in the original Petri net along this path (line 15). Otherwise, the counterexample is spurious, thus the abstraction can be refined by splitting specific abstract places into two, and a new abstraction $\hat{N}$ is generated so that the spurious counterexample is ruled out in the new abstraction (line 12). Furthermore, it is necessary to update the sequence of frames to adapt it to the new abstraction after refinement. Then IC3+PMA continues to try to redo the search of counterexamples in the new abstraction without rebuilding the sequence of frames. If the blocking operation succeeds somewhere, the marking $\hat{m}$ in original pair $(\hat{m}, i)$ can be blocked in F_i. $\hat{m}_t$ is not coverable in $\hat{N}$ within k steps, it is sufficient to conclude that m_t is not coverable on N within k steps by Proposition 3.

In the propagation phase (lines 18-26), IC3 tries to extend the sequence with a new frame F_{k+1} with $\mathbb{N}^{|\hat{P}|}$ that contains all markings in $\hat{N}$. Blocked markings in F_i whose predecessors have no intersection with F_i are propagated to the frame F_{i+1} (line 21). During this process, same as IC3 in Algorithm 1, if two consecutive frames become equivalent, we can conclude $\hat{m}_t$ is not coverable in the abstract Petri net, and thus m_t is not coverable in the original one.

Input: a Petri net $N = \langle P, T, W, m_0 \rangle$, and the target marking m_t
Output: TRUE or FALSE

1 **Function** IC3+PMA(N, m_t):
       // α merges all places into a single abstract place initially
2      $\hat{N}, \hat{m}_t = abstraction(N, m_t, \alpha)$, $\pi = \emptyset$
3   **if** $\hat{m}_t \preceq \hat{m}_0$ **then**
4   **if** $m_t \preceq m_0$ **then return** *UNSAFE*;
5          $\hat{N}, \hat{m}_t = refinement(\pi)$// refinement at initial marking
6   **end**
7      $k = 1, F_0 = \hat{m}_0{}^{\downarrow}, F_k = \mathbb{N}^{|\hat{P}|}$
8   **while** *TRUE* **do**
           // blocking phase
9   **if** *not recBlock($\hat{m}_t$, k)* **then**
10   find an abstract counterexample path π
11   **if** *not simulation(N, m_t, π)* **then**
12                 $\hat{N}, \hat{m}_t = refinement(\pi)$, $\pi = \emptyset$
13   *continue*
14   **else**
15   **return** *TRUE* // counterexample π found
16   **end**
17   **end**
           // propagation phase
18         $k = k + 1, F_k = \mathbb{N}^{|\hat{P}|}, \pi = \emptyset$
19   **for** $i = 1$ *to* k **do**
20   **foreach** *marking m_b blocked in F_i* **do**
21   **if** *in $\hat{N}$. $pre(m_b^{\uparrow}) \cap F_i = \emptyset$* **then** remove $m_b^{\uparrow}$ from F_{i+1};
22   **end**
23   **if** $F_i == F_{i+1}$ **then**
24   **return** *FALSE*// inductive invariant F_i found
25   **end**
26   **end**
27   **end**
28 **Function** recBlock(m, i):
29   **if** $i = 0$ **then return** *False*;
30   **while** *in $\hat{N}$, $pre(m^{\uparrow}) \cap F_i \setminus m^{\uparrow} \neq \emptyset$* **do**
31   select a marking m_p in $pre(m^{\uparrow}) \cap F_{i-1} \setminus m^{\uparrow}$ along transition t
32   **if** *not recBlock($m_p, i - 1$)* **then**
33   π.append(t)
34   **return** *False*
35   **end**
36   **end**
37     $m_g = \text{generalize}(m_p, i)$
38   **for** $j = 1$ *to* i **do** remove $m_g^{\uparrow}$ from F_j;
39   **return** *True*

Algorithm 2: IC3 with place-merge abstraction (with changes w.r.t. Algorithm 1 in red)

3.3   Simulation and Refinement

When a counterexample, denoted by a transition sequence $\pi = t_0 t_1 \ldots t_{k-1}$, is found on current abstraction $\hat{N}$, we check if it can be simulated, i.e. it has a counterpart on N. The simulation procedure is called to check if such a counterexample π exists, and if not, refinement must be called to increase the precision of the abstraction by generating a new partition of places.

The abstract counterexample holds on the original Petri net iff

$$m_0 \xrightarrow{\pi} m_k \wedge m_t \preceq m_k \tag{2}$$

Note that if (2) is satisfiable, there exists a concrete counterexample on the original Petri net, i.e. m_t is coverable. Otherwise, if the formula is unsatisfiable, either there exists $i(0 \leq i \leq k-1)$ such that t_i is not enabled at marking m_i, or $m_t \npreceq m_k$, in these cases, π is spurious and the abstraction must be refined.

There are two refinement methods corresponding to the above two cases where (2) is unsatisfiable. In the first case, t_i is not enabled at marking m_i, there must exist $p_u \in P$ such that $m_i(p_u) < W(p_u, t_i)$. IC3+PMA finds all such p_u's denoted by a set $\mathcal{U}$, then splits all abstract places $\hat{p} \in \hat{P}$ satisfying $\gamma_\alpha(\hat{p}) \cap \mathcal{U} \neq \emptyset$ into two parts, $\gamma_\alpha(\hat{p}'_1) = \{p | p \in \gamma_\alpha(\hat{p}) \wedge p \in \mathcal{U}\}$ and $\gamma_\alpha(\hat{p}'_2) = \{p | p \in \gamma_\alpha(\hat{p}) \wedge p \notin \mathcal{U}\}$. In the second case, each transition $t_i \in \pi$ is enabled at marking m_i but finally $m_t \npreceq m_k$. The method of refinement is similar to the first case, IC3+PMA marks out places p_u satisfying $m_k(p_u) < m_t(p_u)$ by a set $\mathcal{U}$. For each abstract place $\hat{p} \in \hat{P}$ satisfying $\gamma_\alpha(\hat{p}) \cap \mathcal{U} \neq \emptyset$, IC3+PMA also splits it into two parts, $\gamma_\alpha(\hat{p}'_1) = \{p | p \in \gamma_\alpha(\hat{p}) \wedge p \in \mathcal{U}\}$ and $\gamma_\alpha(\hat{p}'_2) = \{p | p \in \gamma_\alpha(\hat{p}) \wedge p \notin \mathcal{U}\}$.

It is worth noting that each refinement may add more than one place in our approach. For example, the set of places of previous abstraction is $\hat{P} = \{q_0, q_1\}$, where $q_0 = \{p_0, p_1, p_2\}$ and $q_1 = \{p_3, p_4\}$. If there exists a spurious counterexample in the previous abstraction such that (2) is unsatisfiable and $\mathcal{U} = \{p_2, p_4\}$. According to the method of refinement explained above, all $\hat{p} \in \hat{P}$ satisfying are split $\gamma_\alpha(\hat{p}) \cap \mathcal{U} \neq \emptyset$ into two parts. i.e. $q_0 = \{p_0, p_1, p_2\}$ is split into to part $q_0 = \{p_0, p_1\}$ and $q_2 = \{p_2\}$, adn $q_1 = \{p_3, p_4\}$ is split into to part $q_1 = \{p_3\}$ and $q_3 = \{p_4\}$. Thus, the set of places of new abstraction is $\hat{P} = \{q_0, q_1, q_2, q_3\}$.

Clearly, the refinement method can generate a new partition of places of original Petri nets. After the refinement, we can get a new abstraction Petri net according to Definitions 4 and 5, and IC3+PMA continues to proceed in the new abstraction after processing the sequence of frames to adapt it to the new abstraction. The following theorem shows that the new abstract Petri net will rule out the counterexample π.

Theorem 1 (Refinement). *Given a Petri net $N = \langle P, T, W, m_0 \rangle$, an abstraction $\hat{N}$ and a counterexample path $\pi = t_0 t_1 \ldots t_{k-1}$ of $\hat{N}$ such that formula (2) is unsatisfiable on N. If $\hat{N}'$ is a new abstract Petri net obtained by the refinement explained above, (2) is unsatisfiable on $\hat{N}'$.*

Proof. According to the fact that π is a spurious counterexample and the two cases where (2) is unsatisfiable on N, we have the following analysis.

In the first case, t_i is not enabled at marking m_i for some $t_i (0 \le i \le k-1)$ in N. According to the refinement explained above, we have $\mathcal{U} = \{p \in P | m_i(p) < W(p, t_i)\}$. The abstract places q in $\hat{N}$ which did not allow for t_i to be enabled are split into two abstract places q_1 and q_2 where $\gamma_\alpha(q_1) = \gamma_\alpha(q) \cap \mathcal{U}$ and $\gamma_\alpha(q_2) = \gamma_\alpha(q) \setminus \gamma_\alpha(q_1)$. The new abstraction $\hat{N}'$ is obtained by the new partition after refinement by Definitions 4 and 5. Because m_i reachable along $t_0, t_1 \ldots t_{i-1}$ in N, the corresponding abstract marking $\hat{m}_i'$ is reachable from $\hat{m}_0'$ along $t_0, t_1 \ldots t_{i-1}$ in $\hat{N}'$ by Proposition 2. Clearly, for all $p \in \gamma_\alpha(q_1)$, $m_i(p) < W(p, t_i)$ since t_i is not enabled at marking m_i in the original Petri net. At the abstract marking $\hat{m}_i'$, there exist at least such a q_1 in $\hat{N}'$ satisfying $\hat{m}_i'(q_1) < \hat{W}(q_1, t_i)$ in the new abstract Petri net $\hat{N}'$ by Definition 5, and thus t_i is not enabled at $\hat{m}_i'$ in the new abstraction. Therefore, (2) is unsatisfiable on $\hat{N}'$ and the counterexample π does not hold after refinement.

In the second case, we can get that (2) is unsatisfiable on $\hat{N}'$ similar to the first one. Therefore, we can conclude that the method of refinement is sufficient to rule out the spurious counterexample. □

Example 3. Let us consider the Petri net $N = \langle P, T, W, m_0 \rangle$ of Example 1, where $P = \{p_0, p_1, p_2, p_3, p_4\}$, $T = \{t_0, t_1, t_2, t_3\}$, $m_0 = (0, 0, 1, 1, 1)$ and the arc function $W(p, t) = W(t, p) = 1$ for all $p \in P$ and $t \in T$ as shown in Fig. 1(a). To show the main ideas of IC3+PMA, we describe its key steps when checking the coverability of $m_t = (1, 1, 0, 0, 0)$ in N.

Initial Abstraction. We merge all places into a single abstract place q_0, i.e. $q_0 = \{p_0, p_1, p_2, p_3, p_4\}$, to get the abstract Petri net $\hat{N}$ and the abstract target marking $\hat{m}_t = (2)$. At line 3 in Algorithm 2, IC3+PMA checks that $\hat{m}_t \preceq \hat{m}_0$ and gets a new partition $\hat{P} = \{q_0, q_1\}$ where $q_0 = \{p_0, p_1\}$ and $q_1 = \{p_2, p_3, p_4\}$.

Second Abstraction. According to the new partition, IC3+PMA gets the new abstraction $\hat{N}$ with $\hat{P} = \{q_0, q_1\}$ shown in Fig. 1(b) and $\hat{m}_t = (2, 0)$. The status of frames is $F_0 = (0, 3)^\downarrow$ and $F_1 = \mathbb{N}^2$. In the blocking phase, IC3+PMA generates a pair $(\hat{m}_t, 1)$ and tries to blocks $\hat{m}_t$ in the frame F_1. In fact, $pre(\hat{m}_t^\uparrow) \cap F_1 \setminus \hat{m}_t^\uparrow = \emptyset$, therefore $\hat{m}_t$ can be blocked in F_1 and $\hat{m}_t$ has no possible generalization, then $F_1 = \mathbb{N}^2 \setminus (2, 0)^\uparrow$. The current sequence of frames is not sufficient to get the conclusion, and thus IC3+PMA adds a new frame $F_2 = \mathbb{N}^2$ and in the propagation phase no markings blocked in a frame that can be propagated to the successive frame. IC3+PMA generates a new pair $(\hat{m}_t, 2)$ and finds a chain of pairs $(\hat{m}_t, 2), ((1, 1), 1)$ and $((0, 2), 0)$ by recursively calling the function recBlock along the path $\pi = t_0 t_0$, i.e. $(0, 2) \xrightarrow{t_0} (1, 1) \xrightarrow{t_0} \hat{m}_t$ holds in the abstraction. Thus, an abstract counterexample is found but it cannot be simulated on the concrete system since t_0 fires at the initial marking of N yielding $m_1 = (1, 0, 0, 1, 1)$, but the next transition t_0 is not enabled at m_1. According to the result of simulation, IC3+PMA gets a new partition $\hat{P} = \{q_0, q_1, q_2\}$ where $q_0 = \{p_0, p_1\}$, $q_1 = \{p_3, p_4\}$ and $q_2 = \{p_2\}$ since $m_1(p_2) < W(p_2, t_0)$.

Third Abstraction. According to the new partition, IC3+PMA gets the new abstraction $\hat{N}$ with $\hat{P} = \{q_0, q_1, q_2\}$ and $\hat{m}_t = (2, 0, 0)$. Furthermore, IC3+PMA converts the sequence of frames to $F_0 = (0, 2, 1)^{\downarrow}$, $F_1 = \mathbb{N}^3 \setminus (2, 0, 0)^{\uparrow}$ and $F_2 = \mathbb{N}^3$, and resets π to empty set. After refinement, IC3+PMA still generates a chain of pairs $(\hat{m}_t, 2)$, $((1, 0, 1), 1)$ and $((0, 1, 1), 0)$ by recursively calling the function recBlock and an abstract counterexample path $\pi = t_1 t_0$ is found, i.e. $(0, 1, 1) \xrightarrow{t_1} (1, 0, 1) \xrightarrow{t_0} \hat{m}_t$ holds in the abstraction. Moreover, the simulation operation succeeds since $m_0 \xrightarrow{t_1} m_2 \xrightarrow{t_0} m_3$ in original Petri net N where $m_2 = (0, 1, 1, 1, 0)$ and $m_3 = (1, 1, 0, 1, 0)$ such that $m_t \preceq m_3$.

Finally, we can conclude that m_t is coverable in N.

3.4 Correctness

It is easy to see that IC3+PMA will work on the original Petri net in the worst case, which is equivalent to IC3 running on the original Petri net directly. Thus, the correctness of IC3 contributes to the correctness of IC3+PMA. The following two theorems briefly demonstrate the correctness of IC3+PMA.

Theorem 2 (Soundness). *Given a Petri net $N = \langle P, T, W, m_0 \rangle$ and a target marking m_t. When IC3+PMA terminates with "TRUE", there must exist a counterexample path from m_0 to $m_t^{\uparrow}$, i.e. m_t is coverable. And when it terminates with "FALSE", then there exists no path from m_0 to $m_t^{\uparrow}$, i.e. m_t is not coverable.*

Proof. If IC3+PMA returns TRUE, then the simulation of counterexample π of abstract Petri net on the original Petri net succeeded, and thus m_t is coverable on the original Petri net. If IC3+PMA returns FALSE, $F_{k-1} == F_k$ holds on the abstract Petri net. We have that F_k is an inductive invariant of the abstract Petri net, it means that $Cover(\hat{N}) \subseteq \hat{P}$ and $\hat{m}_t$ is not coverable on $\hat{N}$. Thus, m_t is uncoverable in the original Petri net by Proposition 3.   □

Theorem 3 (Termination). *Given a Petri net $N = \langle P, T, W, m_0 \rangle$ and a target marking m_t. If m_t is coverable, IC3+PMA will terminate with "TRUE". If m_t is not coverable, IC3+PMA will terminate with "FALSE".*

Proof. In the worst case, IC3+PMA finally works on the original Petri nets. If the target marking m_t is coverable on N, there must exists a finite k such that $Cover_k(N) \cap m_t^{\uparrow} \neq \emptyset$. Each frame F_i over-approximates coverable set $Cover_i(N)$. Thus, IC3+PMA will terminate with the sequence of k frames $F_0, F_1, F_2 \ldots F_k$ and find a counterexample while returning "TRUE".

Recall the backward coverability algorithm in Sect. 2.3 which IC3 based on, we define a set D_i which captures the basis of all new elements introduced into $U_i^{\uparrow}$. $D_i = \emptyset$ will hold since U_i must be stabilized with a finite number of steps. In the backward search phase of our algorithm, every marking we try to block in F_i is from D_i, there are only a finite number of different frames F_i. Thus the frame sequence must converge when the length of the sequence increases. If the target marking m_t is not coverable on N, m_0 has no intersection with U_i when U_i stabilizes and there must exist a finite number k such that $F_{k-1} = F_k$. By soundness, it must terminate with "FALSE".   □

From the above theorems, we know that our algorithm terminates and returns the right result. In terms of the worst case, more details of the proof can be obtained in [18].

4   Experimental Evaluation

We have implemented Algorithm 1 (IC3) and Algorithm 2(IC3+PMA) in Python 3.6 and both of them use the input format of MIST[1]. In our implementation, we refer to the implementation details of IC3 on Petri nets [18] as much as possible such as delta-encoding and the method of generalization. It is worth noting that the implementation of Algorithm 1 (IC3) and Algorithm 2 (IC3+PMA) is exactly the same except for the place-merge abstraction, and facilitates the comparisons of IC3+PMA with IC3.

Benchmarks. For the input of our algorithms, we have collected three benchmarks of coverability problem instances for Petri nets from several sources. The first part of benchmarks is a collection of Petri net examples from the MIST toolkit used in [14]. The second part consists of message passing benchmarks used in [20] which come from medical systems and bug-tracking systems. The third part consists of Petri net instances originating from the analysis of concurrent C programs [16].

Our experimental evaluation has two goals. First, we wanted to show that the number of places of the abstraction that allows to conclude is fewer than the original Petri net and the dimensionality of the coverability problem is reduced effectively. The second goal is to measure the performance of IC3+PMA and to compare it with the IC3 algorithm. All experiments were performed on identical machines, equipped with Intel(R) Core i5-10210U 1.60 GHz CPUs and 4 GB of memory, running Linux 3.10.0 in 64 bit mode. Execution time was limited to 1 h and memory to 5 GB.

Table 1 shows run times and the number of places (final in the case of IC3+PMA) on the set of benchmarks running IC3 and IC3+PMA. For each row, the results in bold show the optimal values (run times and number of places) for each benchmark. We can see that when IC3+PMA terminates, the number of places of the abstract Petri nets has decreased by 63.34% on average comparing with the original Petri nets. For example of the benchmark newrtp with 9 places, IC3+PMA abstracts it as an abstract Petri net with 1 place and tries to compute an inductive invariant in the marking space $\mathbb{N}$ instead of $\mathbb{N}^9$.

Table 2 shows comparisons of IC3+PMA with IC3 while IC3 performed much better. We have identified the main reasons for the worse performance of IC3+PMA on these benchmarks. The first reason is that the efficiency of refinement is not very high on these benchmarks. It is clear to see that the number of refinements on these benchmarks is more than half the number of places of original Petri nets. The second reason is that IC3+PMA may generate some unnecessary markings to explore when handling the sequence of frames. When a

---

[1] https://github.com/pierreganty/mist.

Table 1. Experimental results: comparison of running time and the number of places for IC3+PMA and IC3 on Petri net benchmarks. Running time is in seconds. "Places" is the number of the places of original Petri nets. "AbsPlaces" is the number of places of the abstraction which allows to get the conclusion. "Ref" is the number of refinements.

Benchmark	Places	IC3+PMA AbsPlaces	IC3+PMA Ref	IC3+PMA time(s)	IC3 time(s)
Uncoverable instances					
newrtp	9	**1**	0	**<0.01**	0.06
kanban (bounded)	16	**1**	0	**<0.01**	1.22
manufacturing	13	**1**	0	**<0.01**	0.16
fms	22	**4**	3	**<0.01**	**<0.01**
fms_attic	22	**4**	3	**0.01**	0.04
mesh2x2	32	**5**	4	**0.01**	0.03
mesh3x2	52	**5**	4	**0.02**	0.08
pingpong	6	**5**	4	**<0.01**	**<0.01**
RandCAS 2	110	**8**	7	**0.08**	0.44
Conditionals 2	214	**26**	25	**1.39**	5.79
Coverable instances					
leabasicapproach	16	**5**	4	**<0.01**	**<0.01**
Dekker 1	41	**27**	25	**2.08**	3.23
DoubleLock1 1	64	**35**	32	**11.26**	13.31
Pthread5 1	80	**47**	44	**97.28**	Timeout
RandLock0 2	110	**48**	46	**21.40**	24.89
Spin2003 2	56	**38**	35	**67.35**	Timeout
Szymanski 1	61	**46**	44	**19.62**	32.69
Constants 1	26	**14**	13	**0.03**	**0.03**
FuncPtr3 1	40	**16**	13	**0.19**	0.33

Table 2. Experimental results: the meaning of each column is as Table 1.

Benchmark	Places	IC3+PMA AbsPlaces	IC3+PMA Ref	IC3+PMA time(s)	IC3 time(s)
Uncoverable instances					
Peterson	14	**10**	8	0.35	**0.13**
Lamport	11	**7**	6	0.06	**0.02**
Ext. ReadWrite (small consts)	24	**14**	13	1.23	**0.28**
x0_AA_q1	312	#	#	Timeout	**70.28**
csm	14	**9**	8	0.19	**0.02**
Coverable instances					
RandCAS 1	48	**34**	33	0.85	**0.67**
StackCAS0 1	41	**30**	29	3.72	**2.14**
StackLock0 1	37	**26**	25	2.33	**1.06**
Lu-fig2 1	39	**20**	19	0.22	**0.12**
Lu-fig2 2	61	**35**	32	43.06	**9.05**

new abstract Petri net is generated, IC3+PMA has to convert the markings of the old abstraction in frames to markings of the new abstraction. This conversion may cause the number of markings in frames to increase dramatically, because a marking of the old abstraction maybe corresponds to multiple markings of the new abstraction. Thus, when IC3+PMA needs to propagate all its frames via counter examples to induction, it needs much more expense. For example, a frame $F_i = \mathbb{N} \setminus (1)^\uparrow$ in the old abstraction with one place can be converted to $F_i = \mathbb{N}^2 \setminus \{(1,0)^\uparrow, (0,1)^\uparrow\}$ in the new abstraction with two places, but $(0,1)$ may be a redundant and unnecessary marking to explore in the new abstraction. Thus, both reasons can cause IC3+PMA to be less efficient on some benchmarks than expected.

Based on Table 1 and Table 2, IC3+PMA performs better than IC3 on some coverable benchmarks though the number of refinements is more than half the number of places of original Petri nets. This is because in coverable instances, IC3+PMA just tries to build a true counterexample path, while in uncoverable instances, it is necessary to generate an inductive invariant, the former expenses less than the latter.

We can conclude that based on experimental results, IC3+PMA is a practical technique to solve the coverability problem on some benchmarks.

5   Related Work

In this paper, we focus on two effective techniques of verification: abstraction and IC3.

Abstraction [9] is a very powerful and efficient approach to reduce the complexity of model checking. Among the existing abstraction techniques, place-merge abstraction has been proposed and successfully applied to solve the coverability problem of Petri nets by computing fixpoints using forward analysis on abstract Petri nets [14]. In our work, we take advantage of place-merge abstraction in IC3 to attack the dimensionality problem of Petri nets during exploration in marking space.

The IC3 [4] has been widely generalized to software systems to prove safety property [5,19]. IC3 has been adapted to the class of downward-finite WSTS and implemented on Petri nets [18]. IC3 was first combined with implicit abstraction in a CEGAR loop in [6], and subsequently the approach was lifted to infinite-state systems in [7]. Furthermore, IC3 in counterexample to induction-guided abstraction-refinement (CTIGAR) focuses on single steps of the transition relation which reduce the expense of refinement and eliminating the need for full traces [3]. Inspired by the approach combining IC3 and predicate abstraction on symbolic transition systems [3,6,7], we try to develop an IC3-like approach combined with place-merge abstraction in the CEGAR framework to solve the coverability problem of Petri nets. The inspiring approach has a similar structure to our work, but benefiting from the original IC3 on Petri nets our approach does not depend on SMT solvers [18]. Different from the original IC3, our approach makes IC3 work on abstract Petri nets with fewer places that can sometimes speed up the exploration.

6   Conclusions

In this paper, we have proposed IC3+PMA, a new IC3-like algorithm to solve the coverability problem of Petri nets, based on an extension of IC3 with place-merge abstraction. The main feature of our algorithm is that IC3 proceeds on abstract Petri nets obtained by the place-merge abstraction. When current abstraction allows for a spurious counterexample, it is refined by splitting candidate abstract places. Moreover, the refinement can be done in a completely incremental way without discarding markings explored in previous abstractions.

The key advantage of our approach is that it can process some Petri nets with high dimensionality efficiently on some benchmarks, by converting them into abstract Petri nets with lower dimensionality, and taking advantage of the efficiency of the IC3 algorithm. The advantage is showed by experimental results on a set of benchmarks. In conclusion, IC3+PMA is a practical and efficient technique for the coverability problem of Petri nets on some benchmarks.

In the future, we plan to optimize the implementation to achieve better results and apply the approach to analyze more properties.

Acknowledgements. We thank Dr. Weifeng Wang for helpful suggestions on this paper, and we also thank the anonymous referees for their constructive comments. This work is partly funded by NSFC-62072443 and NSFC-61972385.

References

1. Abdulla, P.A., Cerans, K., Jonsson, B., Tsay, Y.: General decidability theorems for infinite-state systems. In: Proceedings, 11th Annual IEEE Symposium on Logic in Computer Science, 1996. pp. 313–321. IEEE Computer Society (1996). https://doi.org/10.1109/LICS.1996.561359
2. Batz, K., Junges, S., Kaminski, B.L., Katoen, J.-P., Matheja, C., Schröer, P.: PrIC3: property directed reachability for MDPs. In: Lahiri, S.K., Wang, C. (eds.) CAV 2020. LNCS, vol. 12225, pp. 512–538. Springer, Cham (2020). https://doi.org/10.1007/978-3-030-53291-8_27
3. Birgmeier, J., Bradley, A.R., Weissenbacher, G.: Counterexample to Induction-Guided Abstraction-Refinement (CTIGAR). In: Biere, A., Bloem, R. (eds.) CAV 2014. LNCS, vol. 8559, pp. 831–848. Springer, Cham (2014). https://doi.org/10.1007/978-3-319-08867-9_55
4. Bradley, A.R.: SAT-based model checking without unrolling. In: Jhala, R., Schmidt, D. (eds.) VMCAI 2011. LNCS, vol. 6538, pp. 70–87. Springer, Heidelberg (2011). https://doi.org/10.1007/978-3-642-18275-4_7
5. Cimatti, A., Griggio, A.: Software model checking via IC3. In: Madhusudan, P., Seshia, S.A. (eds.) CAV 2012. LNCS, vol. 7358, pp. 277–293. Springer, Heidelberg (2012). https://doi.org/10.1007/978-3-642-31424-7_23
6. Cimatti, A., Griggio, A., Mover, S., Tonetta, S.: IC3 modulo theories via implicit predicate abstraction. In: Ábrahám, E., Havelund, K. (eds.) TACAS 2014. LNCS, vol. 8413, pp. 46–61. Springer, Heidelberg (2014). https://doi.org/10.1007/978-3-642-54862-8_4

7. Cimatti, A., Griggio, A., Mover, S., Tonetta, S.: Infinite-state invariant checking with IC3 and predicate abstraction. Formal Methods Syst. Des. **49**(3), 190–218 (2016). https://doi.org/10.1007/s10703-016-0257-4

8. Clarke, E.M., Grumberg, O., Jha, S., Lu, Y., Veith, H.: Counterexample-guided abstraction refinement for symbolic model checking. J. ACM (JACM) **50**(5), 752–794 (2003). https://doi.org/10.1145/876638.876643

9. Clarke, E.M., Grumberg, O., Long, D.E.: Model checking and abstraction. ACM Trans. Programm. Lang. Syst. (TOPLAS) **16**(5), 1512–1542 (1994). https://doi.org/10.1145/186025.186051

10. Eén, N., Mishchenko, A., Brayton, R.K.: Efficient implementation of property directed reachability. In: Bjesse, P., Slobodová, A. (eds.) International Conference on Formal Methods in Computer-Aided Design, FMCAD 2011, pp. 125–134. FMCAD Inc. (2011)

11. Esparza, J., Ledesma-Garza, R., Majumdar, R., Meyer, P., Niksic, F.: An SMT-based approach to coverability analysis. In: Biere, A., Bloem, R. (eds.) CAV 2014. LNCS, vol. 8559, pp. 603–619. Springer, Cham (2014). https://doi.org/10.1007/978-3-319-08867-9_40

12. Finkel, A., Leroux, J.: Recent and simple algorithms for Petri nets. Softw. Syst. Model. **14**(2), 719–725 (2014). https://doi.org/10.1007/s10270-014-0426-0

13. Finkel, A., Schnoebelen, P.: Well-structured transition systems everywhere! Theor. Comput. Sci. **256**(1–2), 63–92 (2001)

14. Ganty, P., Raskin, J.-F., Van Begin, L.: From many places to few: automatic abstraction refinement for petri nets. In: Kleijn, J., Yakovlev, A. (eds.) ICATPN 2007. LNCS, vol. 4546, pp. 124–143. Springer, Heidelberg (2007). https://doi.org/10.1007/978-3-540-73094-1_10

15. Jones, N.D., Landweber, L.H., Lien, Y.E.: Complexity of some problems in Petri nets. Theor. Comput. Sci. **4**(3), 277–299 (1977)

16. Kaiser, A., Kroening, D., Wahl, T.: Efficient coverability analysis by proof minimization. In: Koutny, M., Ulidowski, I. (eds.) CONCUR 2012. LNCS, vol. 7454, pp. 500–515. Springer, Heidelberg (2012). https://doi.org/10.1007/978-3-642-32940-1_35

17. Kindermann, R., Junttila, T., Niemelä, I.: SMT-based induction methods for timed systems. In: Jurdziński, M., Ničković, D. (eds.) FORMATS 2012. LNCS, vol. 7595, pp. 171–187. Springer, Heidelberg (2012). https://doi.org/10.1007/978-3-642-33365-1_13

18. Kloos, J., Majumdar, R., Niksic, F., Piskac, R.: Incremental, inductive coverability. In: Sharygina, N., Veith, H. (eds.) CAV 2013. LNCS, vol. 8044, pp. 158–173. Springer, Heidelberg (2013). https://doi.org/10.1007/978-3-642-39799-8_10

19. Lange, T., Neuhäußer, M.R., Noll, T., Katoen, J.-P.: IC3 software model checking. Int. J. Softw. Tools Technol. Transf. **22**(2), 135–161 (2019). https://doi.org/10.1007/s10009-019-00547-x

20. Majumdar, R., Meyer, R., Wang, Z.: Static provenance verification for message passing programs. In: Logozzo, F., Fähndrich, M. (eds.) SAS 2013. LNCS, vol. 7935, pp. 366–387. Springer, Heidelberg (2013). https://doi.org/10.1007/978-3-642-38856-9_20

21. Reisig, W.: Petri Nets: An Introduction. EATCS Monographs on Theoretical Computer Science, vol. 4. Springer (1985)

Firing Partial Orders in a Petri Net

Robin Bergenthum[✉]

Faculty of Mathematics and Computer Science, FernUniversität in Hagen, Hagen,
Germany
`robin.bergenthum@fernuni-hagen.de`

Abstract. Petri nets have the simple firing rule that a transition is
enabled to fire if its preset of places is marked. The occurrence of a
transition is called an event. To check whether a sequence of events is
enabled, we simply try to fire the sequence from 'start' to 'end' in the
initial marking of the net. It is a bit of a stretch to call this an algorithm,
but its runtime complexity is in $O(|P| \cdot |V|)$, where P is the set of places
and V is the set of events.

Petri nets model distributed systems. An execution of a distributed
system is a partial order of events rather than a sequence. Compact
tokenflows are tailored to an efficient algorithm that decides if a partial
order of events is enabled in a Petri net. Yet, the runtime complexity of
this algorithm is in $O(|P| \cdot |V|^3)$.

In practical applications dealing with a huge amount of behavioral
data, the gap between *just firing* a sequence and *deciding* if a partial
order is enabled, makes a big difference.

In this paper, we present an approach to *just firing* a partial order
of events in a Petri net. By firing a partial order, we obtain a lot of
information about whether or not the partial order is enabled. We show
that *just firing* is often enough if done correctly.

1 Introduction

Petri nets model distributed systems. They have formal semantics, an intuitive
graphical representation, and are able to express concurrency among the occurrence of events [1,2,8,9,21,22]. Petri nets have the simple firing rule that a
transition can fire if its prefix is marked. This definition implies so-called firing
sequences, i.e., sequences of subsequently enabled transitions. In many practical
applications, we equate the set of firing sequences with the behavior of the net.
Thus, it is very easy to check if specified or recorded behavior is 'in' a Petri net
model. We simply fire the sequence from 'start' to 'end' and immediately obtain
a result.

Then again, the behavior of a concurrent system is often defined as a set
of scenarios [4,7,10–13] expressing causal dependencies and concurrency among
the events of the systems behavior. Obviously, such scenarios cannot be modeled
by sequences, only by partially ordered sets of events.

Even though partially ordered sets of events are a very intuitive approach to
modeling the behavior of a distributed system, checking if such order is 'in' a

© Springer Nature Switzerland AG 2021
D. Buchs and J. Carmona (Eds.): PETRI NETS 2021, LNCS 12734, pp. 399–419, 2021.
https://doi.org/10.1007/978-3-030-76983-3_20

Petri net model is not trivial. There are even different semantics which all define the same partial language of a Petri net. In this sense, the notion of executing a partially ordered set of events in a Petri net is ambiguous.

(i) Step semantics of Petri nets [15]: A partially ordered set of events is in a Petri net model if and only if each maximal set of unordered events of the partial order is enabled after the occurrence of its prefix.

(ii) Process net semantics of Petri net [16]: A partially ordered set of events is in a Petri net model if and only if there is a process net (occurrence net) of the Petri net, so that the run extends the order relation between events of this process.

(iii) Tokenflow semantics of Petri net [4,17]: A partially ordered set of events is in a Petri net model if and only if there is a valid distribution of tokens between events only using the relations specified by the partial order.

These three semantics are equivalent [17,19,23], i.e., they (fortunately) define the same partial language. However, each semantic implies a different algorithm deciding if a partially ordered set of events is enabled. Utilizing step or process net semantics, the number of process nets and the number of maximal sets of unordered events grow exponentially with the size of the partial order, producing slow algorithms. Only algorithms utilizing tokenflow semantics run in polynomial time [4].

A tokenflow is a distribution of tokens between events along the relations of a partial order. A tokenflow is valid if every event receives enough tokens to occur from its prefix, and no event has to produce more tokens than it is able to. If there is a valid tokenflow for every place of a Petri net, the partial order is enabled. We test if there is a valid tokenflow for a place by solving a related flow optimization problem [14]. There a many highly specialized flow optimization algorithms pushing the boundaries of their worst-case runtime [3]. For practical applications, however, the famous pre-flow-push algorithm [18] is easy to implement and has a very good runtime for most examples. The pre-flow-push is in $O(n^3)$, where n is the number of nodes of the flow network. Thus, deciding if a set of partially ordered events is enabled in a Petri net is in $O(|P| \cdot |V|^3)$, where P is the set of places and V is the set of events.

In the area of process mining, some of the recently developed algorithms exploit the idea that firing sequences of events is really cheap. For example, the eST-miner [20] records huge sets of data observing the behavior of a business process and tries to automatically generate a fitting process model. The eST-miner fires the observed sequences over and over again in some initial Petri net to generate more fitting places to complete the model. However, increasingly more process mining papers state that the observed log files are actually partial orders, not only sequences. Thus, there is a clear need to also *fire* partial orders fast.

In this paper, we revisit the problem of deciding if a partially ordered set of events is enabled in a Petri net and decompose the problem along the set of places. We *brute force* fire the partial order in the net once to build a first tokenflow for every place. If the size of the Petri net is fixed, this is possible in linear runtime. The constructed tokenflow will be valid for a subset of places. Experimental results will show that this set is actually quite large for most practical applications. During the firing of the partial order, we, furthermore, present how to easily check if there are alternative distributions of tokens. If there is only one possibility to distribute tokens and our first firing fails, the partial order is not enabled. Only if enabledness is not decided yet, we tackle the subset of *not yet*-decided places by a dedicated compact tokenflow algorithm. In the remainder of the paper, we present the new algorithm, discuss its runtime, and show experimental results using models taken from practical applications.

2   Preliminaries

Let f be a function and B be a subset of the domain of f. We write $f|_B$ to denote the restriction of f to B. As usual, we call a function $m : A \to \mathbb{N}$ a multiset and write $m = \sum_{a \in A} m(a) \cdot a$ to denote multiplicities of elements in m. Let $m' : A \to \mathbb{N}$ be another multiset. We write $m \geq m'$ if $\forall a \in A : m(a) \geq m'(a)$ holds. We denote the transitive closure of an acyclic and finite relation $<$ by $<^*$. We denote the skeleton of $<$ by $<^\circ$. The skeleton of $<$ is the smallest relation $\lhd$, so that $\lhd^* = <^*$ holds. Let $(V, <)$ be some acyclic and finite graph, $(V, <^\circ)$ is called its Hasse diagram. We model distributed systems by place/transition nets [9,21,22].

Definition 1. *A place/transition net (p/t-net) is a tuple (P, T, W) where P is a finite set of places, T is a finite set of transitions so that $P \cap T = \emptyset$ holds, and $W : (P \times T) \cup (T \times P) \to \mathbb{N}$ is a multiset of arcs. A marking of (P, T, W) is a multiset $m : P \to \mathbb{N}$. Let m_0 be a marking, we call the tuple $N = (P, T, W, m_0)$ a marked p/t-net and m_0 the initial marking of N.*

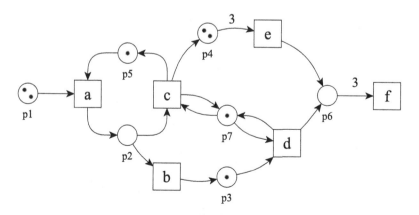

Fig. 1. A marked p/t-net.

Figure 1 depicts a marked p/t-net. Transitions are rectangles, places are circles, the multiset of arcs is represented by weighted arcs, and the initial marking is represented by black dots called tokens.

There is a simple firing rule for transitions of a p/t-net. Let t be a transition of a marked p/t-net (P, T, W, m_0). We denote $\circ t = \sum_{p \in P} W(p, t) \cdot p$ the weighted preset of t. We denote $t\circ = \sum_{p \in P} W(t, p) \cdot p$ the weighted postset of t. A transition t is enabled (can fire) at marking m if $m \geq \circ t$ holds. Once transition t fires, the marking of the p/t-net changes from m to $m' = m - \circ t + t\circ$.

In our exemplary marked p/t-net, transitions a and d can fire at the initial marking. If a fires, this removes one token from $p1$ and the token from $p5$. Additionally, firing a will produce a new token in $p2$. In this new marking, transitions c and d can fire. a is not enabled anymore because there is no more token in $p5$. Firing transition c will enable transition a again and enable transition e. Transition e needs three tokens in $p4$ to be enabled.

Repeatedly processing the firing rule produces so-called firing sequences. These firing sequences are the most basic behavioral model of Petri nets. For example, the sequence $a\ d\ c\ a\ b\ d\ e\ f$ is enabled in the marked p/t-net of Fig. 1. Let N be a marked p/t-net, the set of all enabled firing sequences of N is the (sequential) language of N.

Petri nets are able to express concurrency between events. For example, transitions a and d can fire independently from one another. Roughly speaking, they can fire in any order while not sharing tokens. If we fire transition a, transitions c and d can fire in any order but not concurrently because they share the token in $p7$.

To specify concurrency between events, we formalize executions of a p/t-net by means of labeled partial orders.

Definition 2. *Let T be a set of labels. A labeled partial order is a triple $(V, \ll, l)$ where V is a finite set of events, $\ll\ \subseteq V \times V$ is a transitive and irreflexive relation, and the labeling function $l : V \to T$ assigns a label to every event. A run is a triple $(V, <, l)$ iff $(V, <^*, l)$ is a labeled partial order. A run $(V, <, l)$ is also called a labeled Hasse diagram iff $<^\circ = <$ holds.*

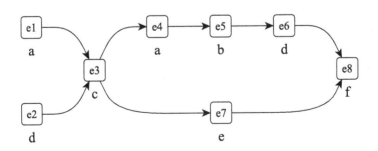

Fig. 2. A run.

Just like a firing sequence, a run can be enabled in a marked p/t-net. A run is enabled if we can replay the order by firing transitions where unordered parts of the partial order can fire concurrently. As stated in the introduction, there are different semantics to formally define whether a run is enabled, but the compact tokenflow semantic is the most efficient [4].

A compact tokenflow is a distribution of tokens along the relations and nodes of a run. A run is in the partial language of a p/t-net if there is a compact tokenflow distributing tokens so that three conditions hold: first, every event receives enough tokens, second, no event has to pass too many tokens, and third, the initial marking is not exceeded. Tokens must be received from the particular presets of events. Thus, we ensure that consumed tokens are available before the actual event occurs. If a transition produces tokens, the related events are allowed to produce tokenflow in the run and pass these tokens to their particular postsets. If an event receives tokens, it consumes the tokenflow needed and passes the redundant tokenflow to later events. Tokens of the initial marking are free for all, i.e., any event can consume or pass tokens from the initial marking.

Definition 3. *Let $N = (P, T, W, m_0)$ be a marked p/t-net and $run = (V, <, l)$ be a run so that $l(V) \subseteq T$ holds. A compact tokenflow is a function $x : (V \cup <) \to \mathbb{N}$. x is valid for $p \in P$ iff the following conditions hold:*
(i) $\forall v \in V: x(v) + \sum_{v' < v} x(v', v) \geq W(p, l(v))$,
(ii) $\forall v \in V: \sum_{v < v'} x(v, v') \leq x(v) + \sum_{v' < v} x(v', v) - W(p, l(v)) + W(l(v), p)$,
(iii) $\sum_{v \in V} x(v) \leq m_0(p)$.
run is valid for N iff there is a compact valid tokenflow for every $p \in P$.

Figure 3, Fig. 4, and Fig. 5 depict three compact tokenflows for three different places of the marked p/t-net of Fig. 1 and the run of Fig. 2 (integer 0 is not shown).

Figure 3 depicts a valid compact tokenflow for the place $p2$ of Fig. 1. The transitions related to the two events labeled b and c need to receive one token in $p2$. The transition related to the events labeled a can produce one token in $p2$. Initially, there are no tokens in this place but no event consumes tokens from the initial marking. Thus, this is a valid tokenflow for $p2$.

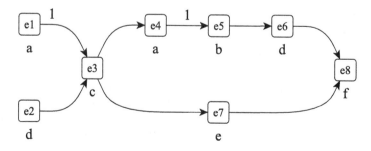

Fig. 3. Valid compact tokenflow for $p2$ of Fig. 1.

Figure 4 depicts a valid compact tokenflow for the place $p6$ of Fig. 1. The event labeled f needs to receive three tokens. All three events labeled d or e can produce one token. All other events just receive and push tokens to later events. Again, we do not need any token from the initial marking. Thus, this is a valid tokenflow for $p6$.

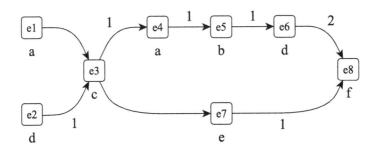

Fig. 4. Valid compact tokenflow for $p6$ of Fig. 1.

Figure 5 depicts a valid compact tokenflow for the place $p7$ of Fig. 1. The events labeled c or d need to receive one token. Because of the short loops at place $p7$, these events can also produce one token in $p6$. Obviously, an event is not allowed to consume the tokens it produces itself. This is why event $e2$ consumes a token from the initial marking before pushing its own token to $e3$. $e3$ consumes the token and pushes a new token to $e6$. This is a valid tokenflow for $p7$.

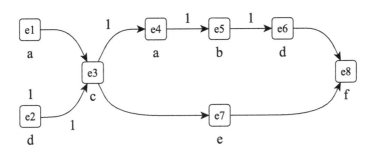

Fig. 5. Valid compact tokenflow for $p7$ of Fig. 1.

If there is a valid tokenflow for every place of a marked p/t-net, the run is valid. The set of valid runs coincides with the (partial) language of a p/t-net. Here, we refer the reader to [4,5] and state the following theorem.

Theorem 1. *The language of a marked p/t-net is well-defined by the set of valid runs [4].*

3   Deciding Enabledness and Firing Runs

In this section, we decide if a run is enabled in a marked p/t-net. In the first subsection, we recap the algorithm that decides if a run is enabled using compact tokenflows in polynomial runtime, originally introduced in [5]. In the second subsection, we present an approach to firing a run in a p/t-net to decide if the run is enabled for a subset of the places of the p/t-net in linear runtime. In the last subsection, we present the idea of firing backwards and combine all approaches to obtain a new and faster algorithm to decide enabledness for runs in p/t-nets.

3.1   Tokenflows and Flow Networks

We decide if a run is enabled in a marked p/t-net by constructing a flow network and a maximal flow for every place. A flow network (see for example [3]) is a directed graph with two specific nodes: A source, the only node having no ingoing arcs, and a sink, the only node having no outgoing arcs. Each arc has a capacity, and a flow is a function from the arcs to the non-negative integers assigning a value of flow to each arc. This flow function needs to respect the capacity of each arc and the so-called flow conservation. The flow conservation states that the sum of flow reaching a node is equal to the sum of flow leaving a node for every inner node of the flow network. Thus, flow is only generated at the source and flows along different paths till it reaches the sink. The value of a flow function in a flow network is the sum of flow reaching the sink. The maximal flow problem is to find a flow function that has a maximal value.

For a place of a p/t-net and a run, we construct the so-called associated flow network. The flow in the associated flow network directly coincides with a compact tokenflow in the related partial order. For each event, we create two nodes in the flow network: an *in-node* and an *out-node*. The flow at the in-node is the value of tokenflow received by the related event. This value has to be greater than the number of tokens needed by the related transition. We rout the number of tokens needed from the in-node to the sink representing tokens consumed by the occurrence of the transition. We distribute additional flow further through the network by adding an arc from every in-node to its out-node. The out-node will distribute flow to later events. The maximal amount of flow this node can push on is the amount of flow received from its in-node plus the number of tokens produced by the occurrence of the related transition. We rout the number of additionally produced tokens from the source to the out-node. In addition, all pairs of in-nodes and out-nodes are connected just like the partial order of events. Whenever there is an arc from one event to another, there is a related arc in the flow network connecting the out-node of the first event with the in-node of the second event. Finally, we add one additional node to the flow network to represent the initial marking.

Figure 6 depicts the associated flow network for Fig. 1, the place $p7$ of the same figure, and Fig. 2. We already saw a related compact tokenflow in Fig. 5. At the top of Fig. 6 is the source, at the bottom is the sink. The node furthest to the left relates to the initial marking. The initial marking of $p7$ is 1; thus, this node is connected to the source with capacity 1. Events $e2$, $e3$, and $e6$ can produce a token each. Thus, the related out-nodes are connected to the source as well. Roughly speaking, one piece of flow, i.e., one token in the run, can enter the flow network at the initial marking and at all events labeled by c or d. Events $e2$, $e3$, and $e6$ need to receive a token each. Thus, the related in-nodes are connected to the sink. Again, one piece of flow, i.e., one token in the run, can leave the flow network at these nodes. The capacity of all inner arcs is not limited. Only looking at place $p7$, the run is enabled if the nodes related to events $e2$, $e3$, and $e6$ can consume a token each. Due to the construction of the associated flow network, there is a valid compact tokenflow if there is a flow saturating all arcs leading to the sink (value 3 in this example).

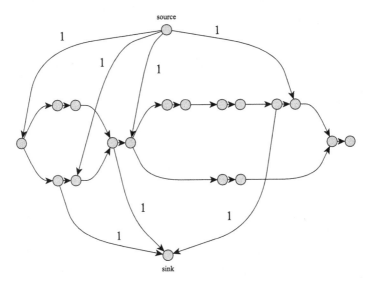

Fig. 6. Associated flow network for Fig. 1, the place $p7$ of the same figure, and Fig. 2.

Figure 7 depicts a maximal flow in the flow network of Fig. 6. This flow saturates all arcs going to the sink; thus, by construction, the flow directly relates to a valid compact tokenflow and the run is enabled. If the value of a maximal flow does not saturate all arcs leading to the sink, there is no valid compact tokenflow because for every distribution of tokens, at least one event cannot occur and the run is not enabled.

Constructing a maximal flow in a flow network is the well-known maximal flow problem (see, for example, [3]). There are various algorithms solving the maximal flow problem in polynomial time. For the application of calculating the value of a maximal flow in an associated flow network, we consider a pre-flow-push algorithm using a so-called gap heuristic. The worst-case time complexity of the pre-flow-push algorithm is in $O(n^3)$, where n is the number of nodes.

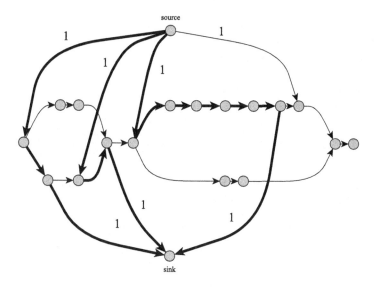

Fig. 7. A maximal flow in Fig. 6.

We recap the algorithm that decides if a run is enabled in a marked p/t-net using compact tokenflows. Additionally, Algorithm 1 computes the set of places that hinders the execution of the run. Obviously, the run is enabled if and only if this set of non-valid places is empty. We can simply stop the algorithm as soon as we find the first non-valid place, but in applications, it may be very helpful to know the set of all non-valid places to fix model or run.

Algorithm 1. *Calculates the set of non-valid places of a marked p/t-net for a run.*

1: **input:** marked p/t-net (P, T, W, m_0), run $(V, <, l)$
2: **for each** $p \in P$ **do**
3:    $G \leftarrow$ associated flow network of (P, T, W, m_0), $(V, <, l)$, p
4:    $x \leftarrow$ sum of capacities of arcs leading to the sink of G
5:    $w \leftarrow$ pre-flow-push of G
6:    **if** $(w < x)$ P_{nvalid} **add** p
7: **return** P_{nvalid}

The runtime of Algorithm 1 is in $O(|P| \cdot |V|^3)$.

3.2 Firing Runs

In this subsection, we introduce the concept of firing runs in a marked p/t-net. The initial marking is a multiset of places. We (randomly) distribute this marking to the set of minimal nodes of the run, creating a set of local marking at each event. We fire each event in its local marking and (randomly) push the resulting local markings to later events. The four conceptual differences to the construction of a valid tokenflow are: (a) Instead of building a network for every place, we handle all places at once. (b) We want to fire in linear time; thus, we cannot redistribute markings or search for paths. We just randomly push local markings to later events to fire every event exactly once. (c) Valid compact tokenflows are tailored to a fast flow network algorithm. Every event only has to receive enough tokens to occur. Thus, conditions (i) and (ii) of Definition 3 are formulated as inequalities to keep the number of tokens small. When firing an event in a local marking, every event will produce its maximal number of tokens. Thus, no token is lost, and we also construct a final marking. (d) A local marking can be negative.

We fire a run in a p/t-net as a first step to decide if the run is enabled or not. The main idea is that the enabledness problem can easily be decomposed along the set of places. For some of the places, we need the maximal flow algorithm to distribute and re-distribute tokens to decide if a valid tokenflow exist. For other places, obviously highly depending on the specific run, a valid compact token-flow may be very easy to construct. Thus, we tackle our problem in two steps: first *brute force* fire a run in a p/t-net, constructing a so-called multi-tokenflow describing a distribution of local markings. For some of the places, these markings will directly relate to valid compact tokenflows and we will not have to consider these places further. If the multi-tokenflow is not valid for some place, we will gain additional information about the existence of alternative token distributions to even see if a re-distribution is possible. If a re-distribution is possible, we redistribute using the algorithm presented in the previous subsection.

To fire a run in a marked p/t-net, constructing a multi-tokenflow as a distribution of (local) markings, we first extend the run by introducing two additional events, one initial and one final event.

Definition 4. *Let* $N = (P, T, W, m_0)$ *be a marked p/t-net and* $run = (V, <, l)$ *be a run so that* $l(V) \subseteq T$ *holds. We denote* $V_{min} \subseteq V$ *the set of events with an empty preset and* $V_{max} \subseteq V$ *the set of events with an empty postset. Let* $v_i, v_f \notin V$ *be two events and define an extended relation* $\prec$ *by* $\prec := < \cup (v_i \times V_{min}) \cup (V_{max} \times v_f)$. *We denote* $run^+ = (V, v_i, v_f, \prec, l)$ *the extended run of run. A function* $X : \prec \to \mathbb{Z}^P$ *is a multi-tokenflow for run iff the following conditions hold:*

(I) $\sum_{v_i \prec v'} X(v_i, v') = m_0$.

(II) $\forall v \in V: \sum_{v \prec v'} X(v, v') = \sum_{v' \prec v} X(v', v) - \circ l(v) + l(v) \circ$,

(III) $\forall v \in V \cup \{v_i\}: (\sum_{v \prec v'} X|_p(v, v') \geq 0 \Longrightarrow \forall v'' \in V: X|_p(v, v'') \geq 0)$.

We call $m_f := \sum_{v' \prec v_f} X(v', v_f)$ *the final marking of* X.

Note that local markings of a multi-tokenflow can be negative. Condition (I) distributes the initial marking to the minimal events of the run, condition (II) ensures that the local markings reflect the firing rule, and condition (III) ensures that tokens are distributed and not just appearing from nowhere by adding negative values to nearby arcs. Thus, a multi-tokenflow is a distribution of actually produced tokens whenever possible.

Figure 8 depicts a sequential run. Obviously, the concept of a multi-tokenflow for a sequential run is just the concept of markings of a firing sequence.

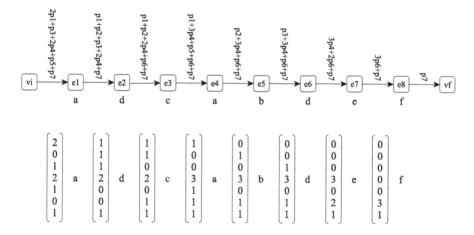

Fig. 8. A multi-tokenflow and markings of a firing sequence.

Figure 9 depicts a multi-tokenflow for the run depicted in Fig. 2. In comparison to Fig. 8, a multi-tokenflow implements the concept of local markings. Just like a tokenflow, the marking is distributed whenever the partial order branches. Every event receives a sum of local markings and fires to push the resulting local marking to later events. In contrast to tokenflows, every produced token has to be pushed until it is consumed or until it reaches the final marking. In contrast to markings, we allow negative values and distribute markings at every branch in the partial order.

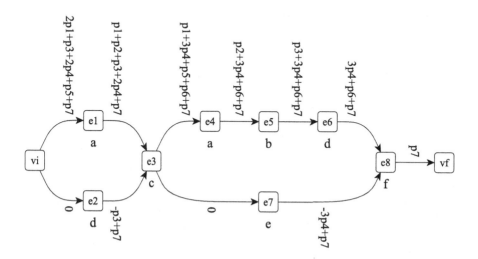

Fig. 9. A multi-tokenflow in the run of Fig. 2.

We formalize the relation between a tokenflow and a multi-tokenflow in the following theorem.

Theorem 2. *Let $N = (P, T, W, m_0)$ be a marked p/t-net, run $= (V, <, l)$ be a run so that $l(V) \subseteq T$ holds, $run^+ = (V, v_i, v_f, \prec, l)$ be the extended run. There is a valid compact tokenflow x in run for N and p, if and only if there is a multi-tokenflow X in run^+, so that $X|p$ enables every event in its sum of in-going local markings in N only considering p.*

Proof. If there is a valid compact tokenflow for a place p in run, we simply construct a multi-tokenflow enabling every event for place p. We can construct this p-component of a multi-tokenflow by copying the tokenflow to the extended run and moving the initial tokenflow to paths outgoing of the initial event. Whenever there is an event not producing its full number of tokens (i.e., (ii) holds, but not yet (II)), we find a path from this event to the final event. We can add tokenflow to this path without making conditions (i), (ii), and (iii) not valid until (II) holds. The same holds for the initial marking; we can go from (iii) to (I) by adding tokens on paths from the initial to the final event. Every valid tokenflow is non-negative, thus, (III) holds as well.

If there is a multi-tokenflow enabling every event for place p, there is a valid compact tokenflow for p. If all the events are enabled for p, we show that every local marking is positive for p. Assume there is some negative value. Without loss of generality, we choose an arc with a negative value so that there is no earlier arc with a negative value. This is always possible because the initial marking is non-negative. Because of (III), the start-event of a first negative arc has a negative sum of outgoing local markings for p but a non-negative sum of in-going local markings. We apply the firing rule to see that this event is not enabled for p in its in-going local markings. Thus, the multi-tokenflow is non-negative for component p. By copying component p of the multi-tokenflow from the extended run into the run and moving initial tokenflow to the minimal nodes of the run, we directly obtain a valid compact tokenflow, because if every event is enabled, (i) holds, (II) implies (ii), and (I) implies (iii).   □

Thus, if we fix a run, we can construct a multi-tokenflow for a marked p/t-net, and if this flow enables all the events of the run for a subset of places, the run is enabled according to this set.

Furthermore, if there is only one possibility to construct a p-component of a multi-tokenflow, if the set of local markings does not enable every event for p, there is no valid compact tokenflow for p.

Lemma 1. *Let $N = (P, T, W, m_0)$ be a marked p/t-net, run $= (V, <, l)$ be a run so that $l(V) \subseteq T$ holds, $run^+ = (V, v_i, v_f, \prec, l)$ be the extended run. Let X be a multi-tokenflow in run^+. For every $p \in P$, where at least one event is not-enabled for p in its sum of in-going local markings, if $X|_p \leq 0$ for every event with multiple out-going arcs, there is no valid compact tokenflow in run for p and N.*

Proof. With the preconditions of this lemma: assume there is a valid compact tokenflow for p. We can construct another multi-tokenflow enabling all events for p, but $X|_p$ is unique.   □

In the next lemma, we use the concept of a final marking of a multi-tokenflow. We show that this marking is actually unique and if it is negative for one p-component, the related run is not enabled.

Lemma 2. *Let $N = (P, T, W, m_0)$ be a marked p/t-net, $run = (V, <, l)$ be a run so that $l(V) \subseteq T$ holds, $run^+ = (V, v_i, v_f, \prec, l)$ be the extended run. Let X be a multi-tokenflow in run^+ and m_f be the final marking. If $m_f(p) < 0$ holds, there is no valid compact tokenflow for p in run.*

Proof. Due to the construction of the extended run, every event is on a path from the initial to the final event. For this reason and because of (II), a multi-tokenflow does not lose tokens and the final marking is $m_f = m_0 + (\sum_{v \in V} l(v) \circ - \circ l(v))$ for every multi-tokenflow. The final marking of a multi-tokenflow is independent from the distribution of tokens. If this final marking is negative for a component p, there is no valid compact tokenflow for p, because, if we assume run is enabled, then also every firing sequence respecting the order of run is enabled and leads to a negative local marking in p according to the (usual) firing rule for sequences. □

In Fig. 9, the depicted multi-tokenflow enables all events considering places $p1$, $p2$, $p5$, $p6$. Thus, with the help of Theorem 2 and only using one multi-tokenflow, we decide enabledness for four of the seven places. The multi-tokenflow branches for places $p3$, $p4$, and $p7$ at the initial event. Thus, although these three components do not enable all events, we cannot apply Lemma 1 because there are other possible distributions of tokens. We cannot apply Lemma 2, either, because the final marking is not negative. Thus, we have to decide places $p3$, $p4$, and $p7$ using Algorithm 1.

At the end of this subsection, we present the algorithm firing a run in a marked p/t-net, deciding enabledness and non-enabledness for a subset of places using Theorem 2, Lemma 1, and Lemma 2.

Algorithm 2. *Calculates a set of valid and a set of non-valid places of a marked p/t-net for a run.*

1: **input:** marked p/t-net (P, T, W, m_0), run $(V, <, l)$
2: $(V, v_i, v_f, \prec, l) \leftarrow$ extension of $(V, <, l)$.
3: (first successor of v_i).marking **add** m_0
4: **for each** $e \in V$ in $\prec$-order **do**
5:     $P_{fnvaild}$ **add** $\{p \in P | e.marking(p) < W(p, l(e))\}$
6:     $P_{fbranch}$ **add** $\{p \in P | e.marking(p) > 0, |e \bullet| > 1\}$
7:     (first successor of e).marking **add** $e.marking - \circ l(e) + l(e) \circ$
8: $P_{fnvaild}$ **add** $\{p \in P | v_f.marking(p) < 0\}$
9: $P_{valid} \leftarrow P \setminus P_{fnvaild}$
10: $P_{nvalid} \leftarrow P_{fnvaild} \cup (P \setminus P_{fbranch})$
11: **return** (P_{valid}, P_{nvalid})

To conclude this subsection, we take a look at the runtime of Algorithm 2. There is a problem in line 4, where we have to consider all events in some total order respecting $\prec$. We can very easily calculate such an order in a preprocessing step but not in linear time. If this order is part of the input, we only consider every event once. For every event, we only touch one outgoing arc. We store the sum of in-going arcs at events whenever we push local markings. Thus, we never have to iterate a set of arcs to calculate a sum of local markings. The runtime of Algorithm 2 is in $O(|P| \cdot |V|)$, or in $O(|V|^2 + |P| \cdot |V|)$ if we need to calculate a total order first.

3.3   Firing Backwards

The reason we have to tackle places $p3$, $p4$, and $p7$ from Fig. 1 deciding enabledness of 2 by Algorithm 2 is that they are marked at the two forward-branched events v_i and $e3$. The components of these three places do not enable all events, but we cannot apply Lemma 1 because there might be another valid distribution. The flow of $p2$ is unique because it is only positive at non-branching events. Yet, the multi-tokenflow for places $p1$ and $p5$ is positive at forward-branched events. However, since the multi-tokenflow components $p1$ and $p5$ already enable all events, we do not have to distribute further. In some sense, we were lucky to find these valid distributions for $p1$ and $p5$ at the first attempt. In this section, we will introduce the concept of firing backwards to offer a heuristic to find valid distributions more often.

We already mentioned that it is important not to re-distribute local markings or even look for paths to be able to fire in linear time. The most efficient strategy to construct a multi-tokenflow is to push the complete local marking of every event to its first subsequent event. Thus, the number of push operations is the number of events, not the number of arcs. We call this strategy the forward-strategy in the remainder of the paper. The multi-tokenflow depicted in Fig. 9 is produced by the forward-strategy, i.e., the complete initial marking is pushed to $e1$, the local marking of $e3$ is pushed to $e4$. Thus, the arcs $(v_i, e2)$ and $(e3, e7)$ are never touched. Obviously, the constructed multi-tokenflow randomly depends on the order of events.

Figure 10 depicts a typical structure of a run. In Fig. 10, all arcs leaving forward-branched events are depicted by dashed arcs. When firing this run in a p/t-net, we can actually decide enabledness in linear time using the forward-strategy of pushing local markings for every place which is not marked at the two highlighted events. If there is a local marking for a place at the branching events, we may be lucky to randomly construct a valid distribution. However, if we need to share the marking between different subsequent events, the forward-strategy will always fail. For this reason, it is not a good idea to fire the run forward again using some modified distribution strategy.

We *brute force* fire the run again but starting from the (unique) final marking and backwards. We already constructed this marking firing forward once. Starting from this final marking at the final event, we push this marking backwards to the first predecessor. We fire the event backwards to calculate the ingoing local

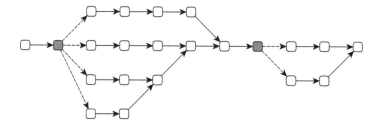

Fig. 10. The set of forward-branched events.

marking to this event. This local marking is pushed to the next first predecessor and so on. This will construct a multi-tokenflow as well.

Figure 11 depicts the run of Fig. 10 and highlights another set of arcs and events. When firing this run in a p/t-net, we can actually decide enabledness in linear time using the backward-strategy of pushing local markings from the final marking of a run for every place which is not marked at the four highlighted events.

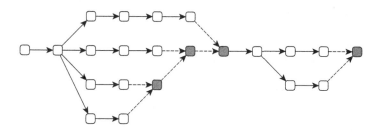

Fig. 11. The set of backward-branched events.

The main advantage of combining a forward-strategy with the backward-strategy in the example run of Fig. 10 is that the example does not contain any forward and backward-branched events. Thus, the set of difficult events is disjoint. If the forward-strategy is not able to decide enabledness, the backward-strategy only fails as well if the related place is also marked at some backward-branched event. If we think of typical workflow Petri nets, for example, that are relatively well-structured using workflow patterns like and/xor-splits and joins, these kinds of places are very rare. In the next section, we present experimental results on how many places can be decided in linear time using the combination of the forward-strategy and the backward-strategy for models taken from practical applications. Yet, before we move on, we combine the forward-, backward-, and flow network-strategies to present the new algorithm that decides if a run is enabled in a marked p/t-net.

Algorithm 3. *Calculates the set of non-valid places of a marked p/t-net for a run.*

1: **input:** marked p/t-net (P, T, W, m_0), run $(V, <, l)$
2: $(V, v_i, v_f, \prec, l) \leftarrow$ extension of $(V, <, l)$.
3: (first successor of v_i).marking **add** m_0
4: **for each** $e \in V$ **in** $\prec$-order **do**
5:    $P_{fnvaild}$ **add** $\{p \in P | e.marking(p) < W(p, l(e))\}$
6:    $P_{fbranch}$ **add** $\{p \in P | e.marking(p) > 0, |e \bullet| > 1\}$
7:    (first successor of e).marking **add** $e.marking - \circ l(e) + l(e) \circ$
8: $P_{fnvaild}$ **add** $\{p \in P | v_f.marking(p) < 0\}$
9: $P_{valid} \leftarrow P \setminus P_{fnvaild}$
10: $P_{nvalid} \leftarrow P_{fnvaild} \cup (P \setminus P_{fbranch})$
11: $P' \leftarrow P \setminus (P_{vaild} \cup P_{nvalid})$
12: $(P, T, W, m_0) \leftarrow (P', T, W|_{P' \times P'}, m_0|_{P'})$
13: (first predecessor of v_f).marking2 **add** $v_f.marking|_P$
14: **for each** $e \in V$ **in** reverse $\prec$-order **do**
15:    $P_{bnvaild}$ **add** $\{p \in P | e.marking2(p) < W(l(e), p)\}$
16:    $P_{bbranch}$ **add** $\{p \in P | e.marking2(p) + W(p, l(e)) - W(l(e), p) > 0, |\bullet e| > 1\}$
17:    (first predecessor of e).marking2 **add** $e.marking2 + \circ l(e) - l(e) \circ$
18: P_{valid} **add** $P \setminus P_{bnvaild}$
19: P_{nvalid} **add** $P_{bnvaild} \cup (P \setminus P_{bbranch})$
20: $P' \leftarrow P \setminus (P_{vaild} \cup P_{nvalid})$
21: $(P, T, W, m_0) \leftarrow (P', T, W|_{P' \times P'}, m_0|_{P'})$
22: P_{nvalid} **add Algorithm 1** $(P, T, W, m_0), (V, <, l)$
23: **return** P_{nvalid}

Algorithm 3 has three parts: lines 1 to 10 implement the forward-strategy. The set $P_{fbranch}$ keeps track of places marked at forward-branching events. Line 8 implements Lemma 2. In lines 11 and 12, we remove the set of places that we do not have to tackle anymore. Again, if a total order of events is part of the input, this part of the algorithm runs in linear run-time. In line 13, we start the backward-strategy reusing the final marking calculated in the first part of the algorithm. The set $P_{bbranch}$ keeps track of places marked at backward-branching events. Line 16 implements a backward version of Lemma 1. Note that there is no backwards version of Lemma 2 because firing backwards from the final marking will reconstruct the initial marking. In lines 20 and 21, we remove the set of places that we do not have to tackle further. Again, if a total order of events is part of the input, this second part of the algorithm runs in linear time as well. In line 20, we only keep places that have to be handled in cubic runtime by Algorithm 1.

4   Comparison and Experimental Results

In this section, we compare the runtimes of Algorithm 1 and Algorithm 3. To compare run-time, we denote (P, T, W, m_0) a marked p/t-net, $(V, <, l)$ a run, and assume a total order respecting $<$ is part of the input. In the remainder, we call all places we can check by brute force firing the run in a p/t-net *simple* places. We call the remaining places *complex* places. This is a bit misleading because whether a place is simple highly depends on the run as well. As stated above, dealing with sequential runs, every place is simple. Furthermore, whether a place is simple or complex does not only depend on the structure of a run, but also on the order of events. Figure 12 depicts a very simple p/t-net where a transition b can fire if transition a produces a token in $p1$ and a run with some kind of w-structure. Figure 12 depicts a distribution for tokens produced by events $e1$, $e2$, and $e3$ on solid arcs. The dashed arcs depict a redistribution of tokens redistributing all previous tokens adding the token from $e4$ to $e8$. This re-distribution is done by flow network algorithms looking for paths, and considering already produced flow as a possible step backwards. Although the net of Fig. 12 is very simple, it is a complex place.

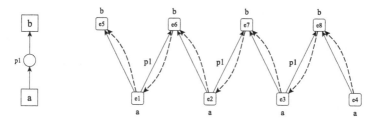

Fig. 12. Redistribution of tokens.

The worst-case runtimes of Algorithm 1 and Algorithm 3 is in $O(|P| \cdot |V|^3)$. If the set of simple-places is empty, the run-time of Algorithm 3 is, obviously, the runtime of Algorithm 1 plus the runtime of two times firing the run. If the set of complex places is empty, the runtime of Algorithm 3 is two times firing the run, i.e., in linear runtime.

In the remainder of this section, we will take a look at examples from practical applications to have a feeling for an average number of simple-places. In a first experiment, we take a look at the latest example net of the model checking contest (https://mcc.lip6.fr/2020/) to have a variety of different models. We calculate a set of 1000 runs of every net by simply randomly unfolding the net [6]. We implement Algorithm 3 and Algorithm 1 in Java to decide if the run is valid. Obviously, every run will be valid as a part of the unfolding. Yet, this will only increase the runtime of both algorithms. For a run of the language of a net, we have to calculate a distribution of tokens for every place. If a run is not valid, both algorithms can stop as soon as they find one place that is not valid. Roughly speaking, stopping early is an advantage for Algorithm 3, because the first two thirds of the algorithm are very fast. Using the examples, we compare the runtime of both algorithms and depict the number of simple places.

We perform the following two experiments on an Intel Core i5 3.30 GHz (4 CPUs) machine with 8 GB RAM running a Windows 10 operating system. The implementation of both algorithms is available at https://www.fernuni-hagen.de/ilovepetrinets/.

Experiment 1. *We consider the most recent example called SatelliteMemory from the Model Checking Contest 2020. SatelliteMemory has two parameters X and Y defining the maximal number of tokens per place. We refer the reader to https://mcc.lip6.fr/2020/pdf/SatelliteMemory-form.pdf for a detailed description of the example and the parameters. Figure 13 depicts the structure of the p/t-nets, i.e., markings with many tokens, arc weights, short loops, and cyclic behavior. For each example, we randomly compute 1000 runs for every number 100, 200, 300, and 400 of events and decide enabledness using Algorithm 1 and Algorithm 3. Figure 13 depicts: (1) percentage of all places decided by firing once with the forward strategy, (2) percentage of all places decided by firing twice, once with the forward and once with the backward-strategy, (3) overall average run-time of Algorithm 3, (4) overall average run-time of Algorithm 1. We set the parameters to (a) X=100 Y=3 (100 tokens), (b) X=1000 Y=32 (1000 tokens), (c) X=1500 Y=46 (1500 tokens), (d) X=3000 Y=94 (3000 tokens), (e) X=65535 Y=2048 (65535 tokens).*

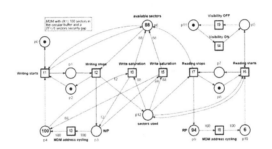

	100 events	200 events	300 events	400 events
a.	.34 .48 5ms 11ms	.31 .45 44ms 98ms	.28 .43 147ms 330ms	.26 .40 391ms 800ms
b.	.40 .55 4ms 15ms	.38 .54 34ms 115ms	.34 .50 115ms 310ms	.29 .41 287ms 608ms
c.	.39 .54 4ms 14ms	.38 .54 32ms 109ms	.36 .52 110ms 356ms	.35 .49 220ms 581ms
d.	.39 .54 4ms 14ms	.38 .53 32ms 105ms	.39 .54 108ms 352ms	.35 .53 267ms 936ms
e.	.39 .54 4ms 14ms	.38 .54 32ms 108ms	.38 .54 111ms 363ms	.36 .51 224ms 709ms

Fig. 13. Model and results of Experiment 1.

Experiment 1 considers a quite complex p/t-net model. The number of places in every combination of parameters is 13. We increase the number of tokens and the number of events. Experiment 1 shows that half of the places are simple-places in this example. The runtime of both algorithms grows quadratic with the size of the input, i.e., number of events, in this example. This fits perfectly with our considerations because, if tokens don't have to be redistributed often and the number of places is fixed, the runtime of Algorithm 1 is in $O(|V|^2)$. Furthermore, if half of the places are simple, the run-time of Algorithm 3 is twice as fast but still in $O(|V|^2))$.

Experiment 2. *We consider the data set from the Process Discovery Contest 2020. The model has parameters defining the control-flow, i.e., dependent tasks, loops, or-constructs, routing constructs, optional tasks, and duplicate tasks, of the model. We refer the reader to https://icpmconference.org/2020/process-discovery-contest/data-set/ for a detailed description of the example and the parameters. The structure of the p/t-nets are typical workflow Petri nets with an initial and a final marking. For each example, we randomly compute 1000 runs from start to end and decide enabledness using Algorithm 1 and Algorithm 3. Figure 14 depicts: (1) file-name (2) average number of events per run (3) percentage of all places decided by firing once with the forward-strategy, (4) percentage of all places decided by firing twice, once with the forward and once with the backward-strategy, (5) overall average runtime of Algorithm 3, (6) overall average runtime of Algorithm 1.*

pdc_2020_0010000.pnml	15	.93	.99	0.043ms	0.284ms
pdc_2020_1000000.pnml	16	.95	1.0	0.036ms	0.296ms
pdc_2020_0001000.pnml	21	.87	.99	0.060ms	0.389ms
pdc_2020_0000000.pnml	21	.87	.99	0.051ms	0.412ms
pdc_2020_0000100.pnml	21	.87	1.0	0.041ms	0.419ms
pdc_2020_0000010.pnml	22	.87	.99	0.041ms	0.387ms
pdc_2020_1111110.pnml	25	.99	1.0	0.014ms	0.845ms
pdc_2020_1211110.pnml	37	.99	1.0	0.019ms	1.733ms
pdc_2020_1210110.pnml	37	.98	1.0	0.023ms	2.412ms
pdc_2020_0100000.pnml	50	.86	.97	0.208ms	2.432ms
pdc_2020_0200000.pnml	86	.82	.97	0.579ms	7.540ms

Fig. 14. Model and results of Experiment 2.

Experiment 2 considers a workflow p/t-net model with standard workflow patterns. All examples in this experiment have an initial and a final marking; thus, we cannot scale the size of the input as we did in Experiment 1. Only if the model contains loops, i.e., the second parameter in this example, the number of events of the randomly generated runs increases. Almost every place of this example is a simple place. In workflow models, most of the places are empty most of the time and can, thus, be handled by firing very easily. The runtime of Algorithm 1 grows quadratic with the size of the input. In Experiment 2, the overall runtime of Algorithm 1 is almost exactly $|V|^2/1000$ ms. The overall runtime of Algorithm 3 is much smaller and highly depends on the number of simple places. The algorithm is very fast for the examples in lines 7, 8, and 9 where almost every place is decided by the forward-strategy of the algorithm. Obviously, for those examples, the algorithm runs in linear time.

5   Conclusion and Future Work

This paper presents an approach to firing a partially ordered set of events in a Petri net model. The new approach also introduces the concept of local markings of a run and a marked p/t-net. With the help of this definition, it is possible to define a set of simple places and to decide enabledness fast.

The paper presents two experiments deciding enabledness of a run in models taken from two very different but well-known contests in the area of Petri nets. In the latest example of the Model Checking Contest, we deal with nets having cyclic behavior, complex net structure, and many tokens. In the latest example of the Process Discovery Contest, we deal with workflow nets with initial and final markings, workflow-patterns, control-flow structure, and only few tokens.

In both experiments, the new algorithm clearly outperforms the algorithm using compact tokenflows only. In that sense, there is never a disadvantage in trying to fire first. The new algorithm is especially fast in workflow-net-like p/t-nets. Here, we open the door to further applications in the area of business process modelling.

In future work, we would like to check if the set of simple places is a good indicator for the complexity of a process model. Let us say we want to discover or synthesize a p/t-net model from behavioral data recorded in terms of partial orders of events; maybe it is sufficient to only generate simple places to obtain a readable, well-structured process model.

References

1. van der Aalst, W.M.P., van Dongen, B.F.: Discovering petri nets from event logs. In: Jensen, K., van der Aalst, W.M.P., Balbo, G., Koutny, M., Wolf, K. (eds.) Transactions on Petri Nets and Other Models of Concurrency VII. LNCS, vol. 7480, pp. 372–422. Springer, Heidelberg (2013). https://doi.org/10.1007/978-3-642-38143-0_10
2. van der Aalst, W.M.P.: The application of petri nets to workflow management. J. Circ. Syst. Comput. 8(1), 21–66 (1998)
3. Ahuja, R.K., Magnanti, T.L., Orlin, J.B.: Network Flows: Theory, Algorithms, and Applications. Prentice Hall, Englewood Cliffs (1993)
4. Bergenthum, R., Lorenz, R.: Verification of scenarios in petri nets using compact tokenflows. Fundamenta Informaticae 137, 117–142 (2015)
5. Bergenthum, R.: Faster verification of partially ordered runs in petri nets using compact tokenflows. In: Colom, J.-M., Desel, J. (eds.) PETRI NETS 2013. LNCS, vol. 7927, pp. 330–348. Springer, Heidelberg (2013). https://doi.org/10.1007/978-3-642-38697-8_18
6. Bergenthum, R., Lorenz, R., Mauser, S.: Faster unfolding of general petri nets based on token flows. In: van Hee, K.M., Valk, R. (eds.) PETRI NETS 2008. LNCS, vol. 5062, pp. 13–32. Springer, Heidelberg (2008). https://doi.org/10.1007/978-3-540-68746-7_6
7. Desel, J., Juhás, G., Lorenz, R., Neumair, C.: Modelling and validation with Vip-Tool. Bus. Process Manag. 2003, 380–389 (2003)

8. Desel, J., Juhás, G.: "What is a petri net?" informal answers for the informed reader. In: Ehrig, H., Padberg, J., Juhás, G., Rozenberg, G. (eds.) Unifying Petri Nets. LNCS, vol. 2128, pp. 1–25. Springer, Heidelberg (2001). https://doi.org/10.1007/3-540-45541-8_1

9. Desel, J., Reisig, W.: Place/transition petri nets. In: Reisig, W., Rozenberg, G. (eds.) ACPN 1996. LNCS, vol. 1491, pp. 122–173. Springer, Heidelberg (1998). https://doi.org/10.1007/3-540-65306-6_15

10. Dumas, M., García-Bañuelos, L.: Process mining reloaded: event structures as a unified representation of process models and event logs. In: Devillers, R., Valmari, A. (eds.) PETRI NETS 2015. LNCS, vol. 9115, pp. 33–48. Springer, Cham (2015). https://doi.org/10.1007/978-3-319-19488-2_2

11. Desel, J., Erwin, T.: Quantitative Engineering of Business Processes with *VIPbusiness*. In: Ehrig, H., Reisig, W., Rozenberg, G., Weber, H. (eds.) Petri Net Technology for Communication-Based Systems. LNCS, vol. 2472, pp. 219–242. Springer, Heidelberg (2003). https://doi.org/10.1007/978-3-540-40022-6_11

12. Fahland, D.: Scenario-based process modeling with Greta. BPM Demonstration Track 2010, CEUR 615 (2010)

13. Fahland, D.: Oclets – scenario-based modeling with petri nets. In: Franceschinis, G., Wolf, K. (eds.) PETRI NETS 2009. LNCS, vol. 5606, pp. 223–242. Springer, Heidelberg (2009). https://doi.org/10.1007/978-3-642-02424-5_14

14. Ford, L.R., Fulkerson, D.R.: Maximal flow through a network. Can. J. Math. **8**, 399–404 (1956)

15. Grabowski, J.: On partial languages. Fundamenta Informaticae **4**, 427–498 (1981)

16. Goltz, U., Reisig, W.: Processes of place/transition-nets. In: Diaz, J. (ed.) ICALP 1983. LNCS, vol. 154, pp. 264–277. Springer, Heidelberg (1983). https://doi.org/10.1007/BFb0036914

17. Juhás, G., Lorenz, R., Desel, J.: Can i execute my scenario in your net? In: Ciardo, G., Darondeau, P. (eds.) ICATPN 2005. LNCS, vol. 3536, pp. 289–308. Springer, Heidelberg (2005). https://doi.org/10.1007/11494744_17

18. Karzanov, A.: Determining the maximal flow in a network by the method of preflows. Doklady Math. **15**, 434–437 (1974)

19. Kiehn, A.: On the interrelation between synchronized and non-synchronized behaviour of petri nets. Elektronische Informationsverarbeitung und Kybernetik **2**(1/2), 3–18 (1988)

20. Mannel, L.L., van der Aalst, W.M.P.: Finding complex process-structures by exploiting the token-game. In: Donatelli, S., Haar, S. (eds.) PETRI NETS 2019. LNCS, vol. 11522, pp. 258–278. Springer, Cham (2019). https://doi.org/10.1007/978-3-030-21571-2_15

21. Peterson, J.L.: Petri Net Theory and the Modeling of Systems. Prentice-Hall, Englewood Cliffs (1981)

22. Reisig, W.: Understanding Petri Nets - Modeling Techniques, Analysis Methods, Case Studies. Springer, Heidelberg (2013). https://doi.org/10.1007/978-3-642-33278-4

23. Vogler, W. (ed.): Modular Construction and Partial Order Semantics of Petri Nets. LNCS, vol. 625. Springer, Heidelberg (1992). https://doi.org/10.1007/3-540-55767-9

Semantics

Deterministic Concurrent Systems

Samy Abbes$^{(\boxtimes)}$ (ID)

Université de Paris - IRIF (CNRS UMR 8243), Paris, France
abbes@irif.fr

Abstract. Deterministic concurrent system are "locally commutative" concurrent systems. We characterise these systems by means of their combinatorial properties.

Keywords: Trace monoid · Möbius transform · Concurrency · Lattice

1 Introduction

Trace monoids are well known models of concurrency. They represent systems able to perform several types of actions, represented by letters in a given alphabet, and with the feature that some actions may occur concurrently. If a and b are two concurrent actions, then the system does not distinguish between the two sequences of actions a-then-b and b-then-a. Instead, a unique compound action $a \cdot b = b \cdot a$ may be performed. This feature is typically used when one wishes to work on the logical order between actions rather than on the chronological order.

Mathematically, a trace monoid $\mathcal{M}$ is a monoid generated by an alphabet Σ, and with relations of the form $ab = ba$ for some fixed pairs of letters $(a, b) \in \Sigma \times \Sigma$. The identity $ab = ba$ in $\mathcal{M}$ renders the concurrency of the two actions a and b.

The use of trace monoids in concurrency theory goes back at least to the 1980s with survey works such as [6,7]. Trace monoids had also been studied in Combinatorics under different names, as free partially commutative monoids and heaps of pieces in the seminal works [4] and [13] respectively. Hence, trace monoids stand at a junction point between computer science and combinatorics.

Despite their successful use as models of concurrency for databases for instance, trace monoids lack an essential feature present in most real-life systems, namely they lack a notion of state. Indeed, any action can be performed at any time when considering a trace monoid model; whereas, in real-life systems, some actions may only be enabled when the system enters some specified state, and then one expects the system to enter a new state, determined by the former state and by the action performed.

A natural model combining both the "built-in" concurrency feature of trace monoids and the notion of state arises when considering a partially defined monoid action of a trace monoid $\mathcal{M}$ on a finite set of states X. Equivalently, instead of considering that the monoid action is only partially defined, it is more convenient to introduce a sink state $\perp$ and to consider a total monoid action

© Springer Nature Switzerland AG 2021
D. Buchs and J. Carmona (Eds.): PETRI NETS 2021, LNCS 12734, pp. 423–442, 2021.
https://doi.org/10.1007/978-3-030-76983-3_21

$(X \cup \{\bot\}) \times \mathcal{M} \to (X \cup \{\bot\})$. Hence, if the system is in state α, performing the letter $a \in \Sigma$ brings the system into the new state $\alpha \cdot a$, with the convention that a was actually not allowed if $\alpha \cdot a = \bot$. This notion of concurrent system, introduced in [1], encompasses in particular popular models of concurrency such as bounded Petri nets [10,11].

In the present paper, we use some results previously obtained in [1,3] in order to study a particular case of concurrent systems, namely the class of *deterministic concurrent systems*. Intuitively, a deterministic concurrent system (DCS) is a concurrent system where no conflict between different actions can ever arise. Hence the only non-determinism left results solely from the concurrency of the model, combined with the constraints imposed by the monoid action. Deterministic concurrent systems can be related, for instance, to causal nets and to elementary event structures found in 1980s papers [10]. We prove in particular that deterministic concurrent systems correspond to concurrent systems which are "locally commutative".

Compared to general concurrent systems, deterministic concurrent systems appear as limit cases. For instance, we prove that their space of maximal executions is at most countable—whereas it is uncountable in general; if the system is moreover irreducible, we prove that it carries a unique probabilistic dynamics—whereas there is a continuum of them in general. Yet, proving these properties is not trivial. The definition of DCS is formulated in elementary terms; their specific properties are formulated in elementary terms; but the proof of these properties relies on the combinatorics of partially ordered sets.

Beside the general properties of deterministic concurrent systems, our main contribution is to give several equivalent characterisations of concurrent systems which are both deterministic and irreducible: an algebraic characterisation; a probabilistic characterisation; a characterisation from the Analytic combinatorics viewpoint; and a characterisation through set-theoretic properties of the set of infinite executions. The multiplicity of these viewpoints suggests that the notion is worth exploring it.

Another contribution is a generalisation of the well known fact that commutative free monoids have a polynomial growth. The property that we obtain in Corollary 1 is general enough to be of interest *per se.*

Although quite specific, the class of deterministic concurrent systems has a non trivial modelisation power. We also believe that understanding deterministic concurrent systems is useful for the deeper understanding of general concurrent systems.

Organisation of the Paper. Section 2 is devoted to preliminaries, and is divided into three subsections. Section 2.1 and 2.2 survey respectively basic notions on trace monoids and on concurrent systems; Sect. 2.3 is devoted to an elementary, yet original result of trace theory, that we tried to formulate in a way not too specific so that it could be of general interest, and that will be used later in the paper. Deterministic concurrent systems are introduced in Sect. 3. Section 4 is devoted to the study of concurrent systems which are both deterministic and irreducible.

2 Preliminaries

2.1 Trace Monoids and Their Combinatorics

The background material introduced in this section is standard, see for instance [6,7], excepted for the probabilistic notions which are borrowed from [2].

Independence and Dependence Pairs. An *alphabet* is a finite set, which we usually denote by Σ, the elements of which are called *letters*. An *independence pair* is a pair (Σ, I), where I is a binary symmetric and irreflexive relation on Σ, called an *independence relation*. A *dependence pair* is a pair (Σ, D), where D is a binary symmetric and reflexive relation on Σ, called a *dependence relation*. With Σ fixed, dependence and independence relations correspond bijectively to each others, through the association $D = (\Sigma \times \Sigma) \setminus I$.

In the remaining of Sect. 2.1, we fix an independence pair (Σ, I), with corresponding dependence pair (Σ, D).

Traces. The *trace monoid*[1] $\mathcal{M}(\Sigma, I)$ is the presented monoid $\mathcal{M} = \langle \Sigma \mid ab = ba \text{ for } (a, b) \in I \rangle$. Elements of $\mathcal{M}$ are called *traces*. The unit element, also called *empty trace*, is denoted by ε, and the concatenation of $x, y \in \mathcal{M}$ is denoted by $x \cdot y$. We identify letters of the alphabet with their images in $\mathcal{M}$ through the canonical mappings $\Sigma \to \Sigma^* \to \mathcal{M}$.

The trace monoid $\mathcal{M}$ is *irreducible* if the dependence pair (Σ, D), seen as a graph, is connected.

Length. Occurrence of Letters. Every trace $x \in \mathcal{M}$ corresponds to the congruence class of some word $u \in \Sigma^*$. The *length* of x, denoted by $|x|$, is the length of u. For each letter $a \in \Sigma$, we write $a \in x$ whenever a has at least one occurrence in u, and we write $a \notin x$ otherwise.

Divisibility Order. The preorder $(\mathcal{M}, \leq)$ inherited from the left divisibility in $\mathcal{M}$ is defined by: $x \leq y \iff (\exists z \in \mathcal{M} \ \ y = x \cdot z)$. This preorder is actually a partial order. If $x \leq y$, the element $z \in \mathcal{M}$ such that $y = x \cdot z$ is unique since trace monoids are left cancelable. We denote this element by $z = x \backslash y$.

Cliques. A *clique* of $\mathcal{M}$ is a trace of the form $x = a_1 \cdot \ldots \cdot a_i$, where all a_is are letters such that $i \neq j \implies (a_i, a_j) \in I$. Since all a_is commute with each other, we identify the clique $x \in \mathcal{M}$ with the subset $\{a_1, \ldots, a_i\} \in \mathcal{P}(\Sigma)$. If $\mathscr{C}$ denotes the set of cliques of $\mathcal{M}$, the restricted partial order $(\mathscr{C}, \leq)$ corresponds to a sub-partial order of $(\mathcal{P}(\Sigma), \subseteq)$. We note that $\mathscr{C}$ is always downward closed in $(\mathcal{P}(\Sigma), \subseteq)$, and that $\mathscr{C}$ corresponds to the full powerset $\mathcal{P}(\Sigma)$ if and only if $\mathcal{M}$ is the free commutative monoid on Σ.

---

[1] In the literature, trace monoids are also called free partially commutative monoids, and they also correspond to right-angled Artin-Tits monoids.

A *non empty clique* is a clique $x \neq \varepsilon$. The set of non empty cliques of $\mathcal{M}$ is denoted by $\mathfrak{C}$. Minimal elements of $(\mathfrak{C}, \leq)$ correspond to the letters of Σ.

Parallel Cliques. Lower and Upper Bounds. Any two traces $x, y \in \mathcal{M}$ have a greatest lower bound (*glb*) in $(\mathcal{M}, \leq)$, which we denote by $x \wedge y$. They have a least upper bound (*lub*) in $(\mathcal{M}, \leq)$, denoted by $x \vee y$ if it exists, if and only if they have a common upper bound.

If x and y are cliques, then $x \wedge y$ is the clique corresponding to the subset $x \cap y \in \mathcal{P}(\Sigma)$. We say that x and y are *parallel*, denoted by $x \parallel y$, if $x \times y \subseteq I$, where x and y are seen as subsets of Σ. In this case, $x \vee y$ exists and is given by $x \vee y = x \cdot y = y \cdot x$.

Normal Sequences. A pair $(x, y) \in \mathscr{C} \times \mathscr{C}$ is a *normal pair* if: $\forall b \in y$ $\exists a \in x$ $(a, b) \in D$. This relation is denoted by $x \rightarrow y$. A sequence $(c_i)_i$ of cliques, the sequence being either finite or infinite, is a *normal sequence* if (c_i, c_{i+1}) is a normal pair for all pairs of indices $(i, i+1)$.

Note that the empty clique satisfies $x \rightarrow \varepsilon$ for all $x \in \mathscr{C}$, and $\varepsilon \rightarrow x$ if and only if $x = \varepsilon$.

Normal Form and Generalised Normal Form. [4] For any trace $x \neq \varepsilon$, there exists a unique integer $k \geq 1$ and a unique normal sequence $(c_1, \ldots, c_k)$ of non empty cliques such that $x = c_1 \cdot \ldots \cdot c_k$. The sequence $(c_1, \ldots, c_k)$ is the *Cartier-Foata normal form of x*, or the *normal form of x* for short. The integer k is the *height* of x, denoted by $k = \tau(x)$.

The *generalised normal form of x* is the infinite normal sequence $(c_i)_{i \geq 1}$ defined by $c_i = \varepsilon$ for $i > k$. By definition, the generalised normal form of ε is the normal sequence $(\varepsilon, \varepsilon, \ldots)$.

For every integer $i \geq 1$, we introduce the mapping $C_i : \mathcal{M} \rightarrow \mathscr{C}$ defined by $C_i(x) = c_i$, where $(c_i)_{i \geq 1}$ is the generalised normal form of x.

Generalised Traces and Infinite Traces. A *generalised trace* is any infinite normal sequence $\xi = (c_i)_{i \geq 1}$ of cliques. If $c_i = \varepsilon$ for some integer i, then $c_j = \varepsilon$ for all $j \geq i$, and then ξ is the generalised normal form of a unique element of $\mathcal{M}$. If $c_i \neq \varepsilon$ for all $i \geq 1$, then ξ is said to be an *infinite trace*.

We denote by $\overline{\mathcal{M}}$ the set of generalised traces, and by $\partial \mathcal{M}$ the set of infinite traces—the latter set is called the *boundary at infinity* of $\mathcal{M}$. We note that $\partial \mathcal{M}$ is non empty as soon as $\Sigma \neq \emptyset$.

We define a partial order on $(\overline{\mathcal{M}}, \leq)$ by putting, for $\xi = (c_i)_{i \geq 1}$ and $\zeta = (d_i)_{i \geq 1}$ two generalised traces:

$$\xi \leq \zeta \iff (\forall i \geq 1 \quad c_i \leq d_i).$$

The injection $\mathcal{M} \rightarrow \overline{\mathcal{M}}$ induces an embedding of partial orders $(\mathcal{M}, \leq) \rightarrow (\overline{\mathcal{M}}, \leq)$, so we simply identify $\mathcal{M}$ with its image in $\overline{\mathcal{M}}$. With this identification, we have $\overline{\mathcal{M}} = \mathcal{M} + \partial \mathcal{M}$, where '+' denotes the disjoint union.

The family of mappings $(C_i)_{i \geq 1}$ extends in the obvious way to the natural projections $C_i : \overline{\mathcal{M}} \to \mathscr{C}$, with restrictions $C_i : \partial \mathcal{M} \to \mathfrak{C}$.

The digraph $(\mathscr{C}, \to)$ is called the *digraph of cliques* of the monoid. Generalised traces correspond bijectively to infinite paths in $(\mathscr{C}, \to)$, with finite traces corresponding to paths hitting the empty clique ε, and infinite traces corresponding to paths never hitting the empty clique.

Möbius Transform. Let $f : \mathscr{C} \to A$ be a function where A is any commutative group. The *Möbius transform* [12] of f is the function $h : \mathscr{C} \to A$ defined by:

$$\forall c \in \mathscr{C} \quad h(c) = \sum_{c' \in \mathscr{C} \,:\, c \leq c'} (-1)^{|c'| - |c|} f(c'). \tag{1}$$

The function f can be retrieved from h thanks to the *Möbius inversion formula*, which is a kind of generalised inclusion-exclusion formula:

$$\forall c \in \mathscr{C} \quad f(c) = \sum_{c' \in \mathscr{C} \,:\, c \leq c'} h(c'). \tag{2}$$

In particular, one has:

$$f(\varepsilon) = \sum_{c \in \mathscr{C}} h(c). \tag{3}$$

Valuations and Probabilistic Valuations. [2] A *valuation* is a monoid homomorphism $f : (\mathcal{M}, \cdot) \to (\mathbb{R}_{\geq 0}, \times)$. One instance is the constant valuation $f = 1$. More generally, any assignation of non negative numbers λ_a to letters a of Σ yields a valuation f, obviously unique, such that $f(a) = \lambda_a$ for $a \in \Sigma$.

Let $h : \mathscr{C} \to \mathbb{R}$ be the Möbius transform of a valuation f, restricted to $\mathscr{C}$. Then f is a *probabilistic valuation* whenever:

$$\big(h(\varepsilon) = 0\big) \quad \wedge \quad \big(\forall c \in \mathfrak{C} \quad h(c) \geq 0\big). \tag{4}$$

In this case, the vector $\big(h(c)\big)_{c \in \mathfrak{C}}$ is a probability vector. Indeed, it is non negative and it sums up to 1 thanks to (3), since $f(\varepsilon) = 1$ and $h(\varepsilon) = 0$.

Markov Chain of Cliques. [2] If f is a probabilistic valuation, then there exists a unique probability measure ν on $\partial \mathcal{M}$ equipped with the natural Borel σ-algebra, such that $\nu(\uparrow x) = f(x)$ for all $x \in \mathcal{M}$, where $\uparrow x$ is the *visual cylinder* defined by $\uparrow x = \{\omega \in \partial \mathcal{M} \mid x \leq \omega\}$.

With respect to this probability measure, the sequence of mappings $C_i : \partial \mathcal{M} \to \mathfrak{C}$, seen as a sequence of random variables, is a homogeneous Markov chain. Its initial distribution is given by: $\forall c \in \mathfrak{C} \quad \nu(C_1 = c) = h(c)$, where h is the Möbius transform of f. The transition matrix of the chain can also be described, but we shall not need it in the sequel.

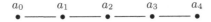

Fig. 1. Coxeter graph of the trace monoid $\mathcal{M}(\Sigma, I)$ with $\Sigma = \{a_0, \ldots, a_4\}$ and $(a_i, a_j) \in I \iff |i - j| \geq 2$. The set of cliques is $\mathscr{C} = \{\varepsilon, \ a_0, \ldots, a_4, \ a_0 \cdot a_2, a_0 \cdot a_3, \ a_0 \cdot a_4, \ a_1 \cdot a_3, \ a_1 \cdot a_4, \ a_2 \cdot a_4, \ \ a_0 \cdot a_2 \cdot a_4\}$.

Example. Let $\mathcal{M} = \langle a, b, c, d \,|\, ad = da, \ bd = db \rangle$. The set of cliques is $\mathscr{C} = \{\varepsilon, \ a, b, c, d, \ ad, bd\}$. Let us simply denote by a, b, etc., the values of $f(a)$, $f(b)$, etc., for some valuation f. The normalization conditions (4) for f to be a probabilistic valuation are:

$$1 - a - b - c - d + ad + bd = 0$$
$$a - ad \geq 0, \quad b - bd \geq 0, \quad c \geq 0, \quad d \geq 0, \quad ad \geq 0, \quad bd \geq 0.$$

A solution is to put $a = b = 1/3$ and $c = d = 1/4$. Another solution is to put $a = b = c = d = 1 - \sqrt{2}/2$. The later value is the root of smallest modulus of the polynomial $1 - 4p + 2p^2$, which we encounter below as the Möbius polynomial of the monoid.

Growth Series and Möbius Polynomials. The *growth series* $G(z)$ and the *Möbius polynomial* $\mu(z)$ of $\mathcal{M}$ are defined as follows:

$$G(z) = \sum_{x \in \mathcal{M}} z^{|x|}, \qquad \mu(z) = \sum_{c \in \mathscr{C}} (-1)^{|c|} z^{|c|}.$$

[4] The series $G(z)$ is rational, and it is the formal inverse of the Möbius polynomial: $G(z)\mu(z) = 1$.

[8,9] If $\Sigma \neq \emptyset$, the Möbius polynomial has a unique root of smallest modulus. This root, say r, is real and lies in $(0, 1]$. If $\Sigma = \emptyset$, we put $r = \infty$. In all cases, the radius of convergence of $G(z)$ is r.

We note that: $r \geq 1$ *if and only if* $\mathcal{M}$ *is commutative*—an elementary result to be generalised when dealing with deterministic concurrent systems in Sect. 3 and 4. Indeed, if $\mathcal{M}$ is not commutative, then $\mathcal{M}$ contains the free monoid on two generators as a submonoid, hence $r \leq 1/2$. Whereas, if $\mathcal{M}$ is commutative and Σ has $N \geq 0$ elements, then $\mu(z) = (1 - z)^N$ and therefore $r = 1$ or $r = \infty$. In this case, one recovers from the formula $G(z) = 1/(1 - z)^N$ the standard elementary result that commutative free monoids have a polynomial growth.

Representation of Traces. The alphabet Σ is usually represented by its *Coxeter graph* [5], which is the graph (Σ, D) with all self-loops omitted. Hence two distinct letters commute with each other if and only if they are not joined by an edge; see an example depicted on Fig. 1.

A convenient representation of traces is provided by the identification of traces with the *heaps of pieces* introduced in [13]. Picture each letter as a piece falling to the ground, in such a way that distinct letters which commute with each other fall along parallel lines; whereas non commutative letters fall in such a way that they block each other. The heaps of pieces thus obtained are combinatorial object corresponding bijectively to the elements of the trace monoid, by reading the letters labelling the pieces from bottom to top. The cliques of the normal form of a trace correspond to the horizontal layers that appear in the heap of pieces. See an illustration on Fig. 2.

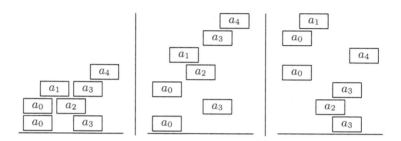

Fig. 2. In this example the commutation relations are those of the Coxeter graph depicted on Fig. 1. *Left:* representation as a heap of piece of the trace which normal form is $(a_0 a_3, a_0 a_2, a_1 a_3, a_4)$. *Middle and right:* representations of two words in the congruence class of the trace x: a_0-a_3-a_0-a_2-a_1-a_3-a_4 (middle) and a_3-a_2-a_3-a_0-a_4-a_0-a_1 (right).

2.2 Concurrent Systems and their Combinatorics

The background material presented in this section is borrowed from [1,3].

Concurrent Systems and Executions. A *concurrent system* is a triple $(\mathcal{M}, X, \perp)$ where $\mathcal{M}$ is a trace monoid, X is a finite set of *states* and $\perp$ is a special symbol not in X, together with a right monoid action of $\mathcal{M}$ on $X \cup \{\perp\}$, denoted by $(\alpha, x) \mapsto \alpha \cdot x$, and such that $\perp \cdot x = \perp$ for all $x \in \mathcal{M}$. By definition of a monoid action, one has thus $\alpha \cdot (x \cdot y) = (\alpha \cdot x) \cdot y$ for all $(\alpha, x, y) \in X \times \mathcal{M} \times \mathcal{M}$, and $\alpha \cdot \varepsilon = \alpha$ for all $\alpha \in X$.

The concurrent system $\mathcal{X}$ is *trivial* if $\alpha \cdot a = \perp$ for all $\alpha \in X$ and for all $a \in \Sigma$. It is *non trivial* otherwise.

The symbol $\perp$ represents a sink state. So we are interested, for every $\alpha, \beta \in X$, in the following subsets of $\mathcal{M}$:

$$\mathcal{M}_{\alpha,\beta} = \{x \in \mathcal{M} \mid \alpha \cdot x = \beta\}, \qquad \mathcal{M}_\alpha = \{x \in \mathcal{M} \mid \alpha \cdot x \neq \perp\}.$$

Traces of $\mathcal{M}_\alpha$ are called *executions starting from* α, or *executions* for short if the context is clear. Note that $\mathcal{M}_\alpha$ is always downward closed in $(\mathcal{M}, \leq)$.

We introduce the following useful notations, for $\alpha, \beta \in X$:

$$\Sigma_\alpha = \Sigma \cap \mathcal{M}_\alpha \qquad \mathscr{C}_\alpha = \mathscr{C} \cap \mathcal{M}_\alpha \qquad \mathfrak{C}_\alpha = \mathfrak{C} \cap \mathcal{M}_\alpha \qquad \mathscr{C}_{\alpha,\beta} = \mathscr{C} \cap \mathcal{M}_{\alpha,\beta}$$

A *generalised execution from* α is an element $\xi \in \overline{\mathcal{M}}$ such that:

$$\forall x \in \mathcal{M} \quad x \le \xi \implies x \in \mathcal{M}_\alpha.$$

Their set is denoted $\overline{\mathcal{M}}_\alpha$, and we also put $\partial \mathcal{M}_\alpha = \overline{\mathcal{M}}_\alpha \cap \partial \mathcal{M}$.

As a running example for a "general concurrent system", we use the 1-safe Petri net depicted in Fig. 3(a). The underlying trace monoid is generated by the transitions, with commutative transitions t and t' whenever ${}^\bullet t^\bullet \cap {}^\bullet t'^\bullet = \emptyset$, thus $\mathcal{M} = \langle a, b, c, d \mid ad = da,\ db = db \rangle$. The corresponding Coxeter graph is depicted on Fig. 3(b), and the graph of marking is depicted on Fig. 3(c).

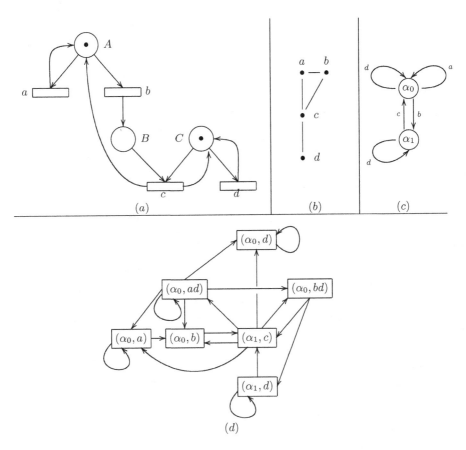

Fig. 3. (a)—A safe Petri net with its initial marking $\alpha_0 = \{A, C\}$ depicted. The two reachable markings are α_0 and $\alpha_1 = \{B, C\}$. (b)—The Coxeter graph of the associated trace monoid. (c)—Graph of markings of the net. (d)—Digraph of states-and-cliques of the associated concurrent system.

Digraph of States-and-Cliques. Generalised executions of a concurrent system $\mathcal{X} = (\mathcal{M}, X, \perp)$ are generalised traces of $\mathcal{M}$. As seen in Sect. 2.1, generalised traces correspond to paths in the digraph of cliques $(\mathscr{C}, \rightarrow)$. Not all paths of $(\mathscr{C}, \rightarrow)$ however correspond, in general, to executions of $\mathcal{X}$. In order to take into account the constraints induced by the monoid action, we introduce the *digraph of states-and-cliques* $(\mathscr{D}, \rightarrow)$, the vertices of which are pairs (α, c) with α ranging over X and c ranging over $\mathscr{C}_\alpha$. There is an arrow $(\alpha, c) \rightarrow (\beta, d)$ in $\mathscr{D}$ if $\beta = \alpha \cdot c$ and if (c, d) is a normal pair of cliques.

To every generalised execution $\xi = (c_i)_{i \geq 1}$ with $\xi \in \overline{\mathcal{M}}_\alpha$, is associated the path $(\alpha_{i-1}, c_i)_{i \geq 1}$ in $\mathscr{D}$, where α_i is defined by $\alpha_0 = \alpha$ and $\alpha_i = \alpha \cdot (c_1 \cdot \ldots \cdot c_i)$ for $i \geq 1$. We put $Y_i(\xi) = (\alpha_{i-1}, c_i)$ for every integer $i \geq 1$.

Conversely, every infinite path in $\mathscr{D}$ corresponds to a unique generalised execution. Consider the subgraph $\mathfrak{D}$ of $\mathscr{D}$ with all vertices of the form (α, c) with $c \neq \varepsilon$. Then infinite paths in $\mathfrak{D}$ correspond bijectively to infinite executions.

For our running example, the digraph of states-and-cliques is depicted on Fig. 3(d).

Characteristic Root. The combinatorics of a concurrent system $\mathcal{X} = (\mathcal{M}, X, \perp)$ involves not only the combinatorics of $\mathcal{M}$, but also of the monoid action $X \times \mathcal{M} \rightarrow X$. Consider the *Möbius matrix* $\mu(z) = (\mu_{\alpha,\beta}(z))_{(\alpha,\beta) \in X \times X}$, the polynomial $\theta(z)$, and the growth matrix $G(z) = (G_{\alpha,\beta}(z))_{(\alpha,\beta) \in X \times X}$ defined by:

$$\mu_{\alpha,\beta}(z) = \sum_{c \in \mathscr{C}_{\alpha,\beta}} (-1)^{|c|} z^{|c|} \qquad \theta(z) = \det \mu(z) \qquad G_{\alpha,\beta}(z) = \sum_{x \in \mathcal{M}_{\alpha,\beta}} z^{|x|}$$

Then $G(z)$ is a matrix of rational series, and it is the inverse of the Möbius matrix: $G(z)\mu(z) = \mathrm{Id}$. One of the roots of smallest modulus of the polynomial $\theta(z)$ is real and lies in $(0, 1] \cup \{\infty\}$, with the convention that it is ∞ if $\theta(z)$ is a non zero constant. This non negative real or ∞ is the *characteristic root* of the concurrent system $\mathcal{X}$. The characteristic root r is the minimum of all convergence radii of the generating series $G_{\alpha,\beta}(z)$, for (α, β) ranging over $X \times X$.

For our running example, the Möbius matrix is given by:

$$\mu(z) = \begin{matrix} \alpha_0 \\ \alpha_1 \end{matrix} \begin{pmatrix} 1 - 2z + z^2 & -z + z^2 \\ -z & 1 - z \end{pmatrix}$$

with determinant $\theta(z) = (1-z)^2(1-2z)$. The characteristic root is thus $r = 1/2$.

Irreducibility and the Spectral Property. A concurrent system $\mathcal{X} = (\mathcal{M}, X, \perp)$ is *irreducible* if: 1) The monoid $\mathcal{M}$ is irreducible; 2) $\mathcal{M}_{\alpha,\beta} \neq \emptyset$ for all $\alpha, \beta \in X$; 3) For every $\alpha \in X$ and for every letter $a \in \Sigma$ there exists $x \in \mathcal{M}_\alpha$ such that $a \in x$.

If Σ' is any subset of Σ, and if $\mathcal{M}' = \langle \Sigma' \rangle$ is the submonoid of $\mathcal{M}$ generated by Σ', then the restriction of the action $(X \cup \{\perp\}) \times \mathcal{M}' \rightarrow X \cup \{\perp\}$ defines clearly a new concurrent system $\mathcal{X}' = (\mathcal{M}', X, \perp)$, said to be *induced by restriction*. In particular, let $\mathcal{X}^a$ denote the concurrent system induced by restriction with $\Sigma' = \Sigma \setminus \{a\}$, and let r^a be the characteristic root of $\mathcal{X}^a$.

A key property, that we shall use later, is the *spectral property* [3] which states: *if $\mathcal{X}$ is irreducible, then $r^a > r$ for every $a \in \Sigma$.*

The concurrent system in our running example from Fig. 3 is irreducible.

Valuations and Probabilistic Valuations. Markov Chain of States-and-Cliques. A *valuation* on a concurrent system $\mathcal{X} = (\mathcal{M}, X, \perp)$ is a family $f = (f_\alpha)_{\alpha \in X}$ of mappings $f_\alpha : \mathcal{M} \to \mathbb{R}_{\geq 0}$ satisfying the three following properties:

$$\forall \alpha \in X \quad \forall x \in \mathcal{M} \quad \alpha \cdot x = \perp \implies f_\alpha(x) = 0 \tag{5}$$

$$\forall \alpha \in X \quad \forall x \in \mathcal{M}_\alpha \quad \forall y \in \mathcal{M}_{\alpha \cdot x} \quad f_\alpha(x \cdot y) = f_\alpha(x) f_{\alpha \cdot x}(y) \tag{6}$$

$$\forall \alpha \in X \quad f_\alpha(\varepsilon) = 1 \tag{7}$$

Let $f = (f_\alpha)_{\alpha \in X}$ be a valuation and for each $\alpha \in X$, let $h_\alpha : \mathscr{C} \to \mathbb{R}$ be the Möbius transform of the restriction $f_\alpha|_\mathscr{C} : \mathscr{C} \to \mathbb{R}_{\geq 0}$. Note first that $h_\alpha(x) = 0$ if $x \notin \mathcal{M}_\alpha$. We say that f is a *probabilistic valuation* if:

$$\forall \alpha \in X \quad \left(h_\alpha(\varepsilon) = 0 \quad \wedge \quad (\forall c \in \mathfrak{C}_\alpha \quad h_\alpha(c) \geq 0) \right). \tag{8}$$

In this case, there exists a unique family $\nu = (\nu_\alpha)_{\alpha \in X}$, where ν_α is a probability measure on $\partial \mathcal{M}_\alpha$, such that $\nu_\alpha(\uparrow x) = f_\alpha(x)$ for all $\alpha \in X$ and for all $x \in \mathcal{M}_\alpha$. Of course the existence of a probabilistic valuation implies in particular that $\partial \mathcal{M}_\alpha \neq \emptyset$, a property which might not be satisfied in general even if $\Sigma \neq \emptyset$.

If $\nu = (\nu_\alpha)_{\alpha \in X}$ is associated as above with a probabilistic valuation $f = (f_\alpha)_{\alpha \in X}$, then for each state $\alpha \in X$, and with respect to the probability measure ν_α, the family of mappings $Y_i : \partial \mathcal{M}_\alpha \to \mathfrak{D}$ defined earlier is a homogeneous Markov chain, called the *Markov chain of states-and-cliques*. Its initial distribution is given by $\mathbf{1}_\alpha \otimes h_\alpha$; hence in particular:

$$\forall \alpha \in X \quad \forall c \in \mathfrak{C}_\alpha \quad \nu_\alpha(C_1 = c) = h_\alpha(c). \tag{9}$$

Let us determine all the probabilistic valuations for the running example of Fig. 3. Any probabilistic valuation $f = (f_\alpha)_{\alpha \in X}$ is entirely determined by the *finite* family of values $f_\alpha(u)$ for (α, u) ranging over $\{\alpha_0, \alpha_1\} \times \Sigma$, since then the other values $f_\alpha(x)$ are obtained by the chain rule $f_\alpha(xy) = f_\alpha(x) f_{\alpha \cdot x}(y)$.

Since $f_{\alpha_0}(c) = f_{\alpha_1}(a) = f_{\alpha_2}(b) = 0$, the remaining parameters for f are $p = f_{\alpha_0}(a)$, $q = f_{\alpha_0}(b)$, $s = f_{\alpha_0}(d)$, $t = f_{\alpha_1}(c)$, $u = f_{\alpha_1}(d)$. The parameters are not independent; to cope with the commutativity relations induced by the trace monoid, one must have $f_{\alpha_0}(a) f_{\alpha_0 \cdot a}(d) = f_{\alpha_0}(d) f_{\alpha_0 \cdot d}(a)$, since $ad = da$, and $f_{\alpha_0}(b) f_{\alpha_1}(d) = f_{\alpha_0}(d) f_{\alpha_0 \cdot d}(b)$ since $bd = db$; yielding simply $u = s$ here.

The Möbius tranform of f_{α_0} evaluated for instance at b is $h_{\alpha_0}(b) = f_{\alpha_0}(b) - f_{\alpha_0}(bd) = f_{\alpha_0}(b) - f_{\alpha_0}(b) f_{\alpha_1}(d) = q - qs$. Other computations are done similarly, and we gather the results in Table 1. According to (8), the normalization contraints on the parameters for the valuation f to be probabilistic are thus:

$$1 - p - q - s + ps + qs = 0, \qquad\qquad 1 - t - s = 0, \tag{10}$$

Table 1. Möbius tranform of a generic valuation for the running example depicted in Fig. 3, with parameters $p = f_{\alpha_0}(a)$, $q = f_{\alpha_0}(b)$, $s = f_{\alpha_0}(d) = f_{\alpha_1}(d)$ and $t = f_{\alpha_1}(c)$.

state α	$h_\alpha(\varepsilon)$	$h_\alpha(a)$	$h_\alpha(b)$	$h_\alpha(c)$	$h_\alpha(d)$	$h_\alpha(ad)$	$h_\alpha(bd)$
α_0	$1 - p - q - s + ps + qs$	$p - ps$	$q - qs$	0	$s - ps - qs$	ps	qs
α_1	$1 - t - s$	0	0	t	s	0	0

plus all inequalities $h_{\alpha_0}(a) \geq 0$, etc., which in this case amount to specify that all parameters vary between 0 and 1. The second equality in (10) is standard: since there is no concurrenycy enabled at α_1, the events of firing c and d are disjoint, hence their probabilities sum up to 1. The first equality in (10) is less standard. It takes into account the existence of concurrency enabled at α_0 and shows a degree greater than 1, resulting form the existence of cliques of order 2.

Here, the equality $h_{\alpha_0}(\varepsilon) = 0$ rewrites as $(1 - p - q)(1 - s) = 0$. It follows that, if $s \neq 1$, then $1 - p - q = 0$ and therefore $h_{\alpha_0}(d) = s(1 - p - q) = 0$. Hence the node (α_0, d) is never reached, which meets well the intuition. We say that (α_0, d) is a *null node*. See [3] for more details about the notion of null node.

Representation of Concurrent Systems and of Executions. To represent a concurrent system $\mathcal{X} = (\mathcal{M}, X, \perp)$, we first use the Coxeter graph of $\mathcal{M}$, as in Fig. 1. We also depict the *labelled multigraph of states*, which vertices are the elements of X, and with an edge from α to β labelled by the letter $a \in \Sigma$ if $\alpha \cdot a = \beta$, as in Fig. 3(c). For representing executions, we stick to the representation by heaps of pieces introduced earlier for traces.

Remark 1. Any multigraph V with edges labelled by elements from a set Σ represents an action of the free monoid $(V \cup \{\perp\}) \times \Sigma^* \to (V \cup \{\perp\})$, provided that for any node $v \in V$, there is no two edges starting from v and labelled with the same letter. It requires an additional verification to check that it also represents an action of a trace monoid $\mathcal{M} = \mathcal{M}(\Sigma, I)$ on V; namely, one has to check that $\alpha \cdot (ab) = \alpha \cdot (ba)$ for any two letters $(a, b) \in I$.

2.3   A Comparison Result

In this subsection, we state an elementary lemma and its corollary, both belonging to trace theory, and given in a form slightly more general than precisely needed in the sequel.

Consider an alphabet Σ and two independence relations I and I' on Σ such that $I \subseteq I'$, and consider the two trace monoids $\mathcal{M} = \mathcal{M}(\Sigma, I)$ and $\mathcal{N} = \mathcal{M}(\Sigma, I')$. There is a natural surjection $\pi : \mathcal{M} \to \mathcal{N}$, which entails in particular that $\mathcal{M}$ is "not smaller" than $\mathcal{N}$. It seems to have been unnoticed so far that, when restricted to the set of sub-traces of a given trace of $\mathcal{M}$, or even of $\overline{\mathcal{M}}$, then π becomes injective. This is the topic of the following lemma.

The lemma generalises the following elementary fact. Let $\mathcal{M} = \Sigma^*$ be a free monoid and let $u \in \Sigma^*$. Then any prefix word $x \leq u$ is entirely determined by the collection $(n_a)_{a \in \Sigma}$ where n_a is the number of occurrences of the letter a in x. Hence x is entirely determined by its image in the free commutative monoid generated by Σ.

Lemma 1. *Let $I \subseteq I'$ be two independence relations on an alphabet Σ, let $\mathcal{M} = \mathcal{M}(\Sigma, I)$ and $\mathcal{N} = \mathcal{M}(\Sigma, I')$, and let $\pi : \mathcal{M} \to \mathcal{N}$ be the natural surjection. Then π extends naturally to a surjection on generalised traces, as a mapping still denoted by $\pi : \overline{\mathcal{M}} \to \overline{\mathcal{N}}$. Let $\omega \in \overline{\mathcal{M}}$, and define: $\overline{\mathcal{M}}_{\leq \omega} = \{x \in \overline{\mathcal{M}} \mid x \leq \omega\}$. Then the restriction of π to $\overline{\mathcal{M}}_{\leq \omega}$ is injective.*

Proof. The extension of π to a mapping $\overline{\mathcal{M}} \to \overline{\mathcal{N}}$ follows from the definitions, hence we focus on proving that the restriction of π to $\overline{\mathcal{M}}_{\leq \omega}$ is injective. Let $x \in \overline{\mathcal{M}}_{\leq \omega}$ and let $y = \pi(x)$. Let c_1 be the first clique in the normal form of x, and let d_1 be the first clique in the normal form of y. Let also C_1 be the first clique in the normal form of ω. We assume with loss of generality that $x \neq \varepsilon$ since $\pi^{-1}(\{\varepsilon\}) = \{\varepsilon\}$.

We claim that $c_1 = d_1 \cap C_1$. The inclusion $c_1 \subseteq d_1 \cap C_1$ is clear since both inclusions $c_1 \subseteq d_1$ and $c_1 \subseteq C_1$ are obvious. For proving the converse inclusion, seeking a contradiction, we assume that there is a letter $a \in d_1 \cap C_1$ such that $a \notin c_1$. Then, since $y = \pi(x)$, the letter a belongs to some higher clique in the normal form of x. But, since $x \leq \omega$, and since $a \in C_1$, that entails that $a \in c_1$, contradicting the assumption $a \notin c_1$. Hence $c_1 = d_1 \cap C_1$, as claimed.

Repeating inductively the same reasoning, with $x' = c_1 \backslash x$ and with $y' = \pi(x') = c_1 \backslash y$ and $\omega' = c_1 \backslash \omega$ in place of x and of y and of ω respectively[2], we see that all the cliques $(c_i)_{i \geq 1}$ of the generalised trace x can be reconstructed from y. This entails that π is injective. $\square$

Corollary 1. *Let $\mathcal{M}$ be a trace monoid, and let $\omega \in \partial \mathcal{M}$ be an infinite trace. For each integer $n \geq 0$, consider:*

$$\mathcal{M}_{\leq \omega}(n) = \{x \in \mathcal{M} \mid x \leq \omega \wedge |x| = n\}, \qquad p_n = \#\mathcal{M}_{\leq \omega}(n).$$

Then there is a polynomial $P \in \mathbb{Z}[X]$ such that $p_n \leq P(n)$ for all integers n. Furthermore, the set $\partial \mathcal{M}_{\leq \omega} = \{\xi \in \partial \mathcal{M} \mid \xi \leq \omega\}$ is at most countable. The polynomial P only depends on $\mathcal{M}$, and not on ω.

Proof. Let $\mathcal{M} = \mathcal{M}(\Sigma, I)$ and let $\mathcal{N}$ be the free commutative monoid generated by Σ, i.e., $\mathcal{N} = \mathcal{M}(\Sigma, I')$ with $I' = (\Sigma \times \Sigma) \backslash \Delta$ and $\Delta = \{(x, x) : x \in \Sigma\}$.

For each integer n, let $q_n = \#\mathcal{N}(n)$. Then it is well known that $q_n = P(n)$ for some polynomial $P \in \mathbb{Z}[X]$ (a short proof based on the Möbius inversion formula was given in Sect. 2.1). Since $I \subseteq I'$, it follows from Lemma 1 that $p(n) \leq q(n)$.

Furthermore, $\overline{\mathcal{N}}$ itself is at most countable since $\overline{\mathcal{N}}$ identifies with:

$$\overline{\mathcal{N}} \sim \{(x_i)_{i \in \Sigma} \mid x_i \in \mathbb{Z}_{\geq 0} \cup \{\infty\}, \quad \exists i \in \Sigma \ \ x_i = \infty\}.$$

Hence, the fact that $\partial \mathcal{M}_{\leq \omega}$ is at most countable also follows from Lemma 1. $\square$

---

[2] Recall that, if $c \leq u$ with $c, u \in \mathcal{M}$, we denote by $c \backslash u$ the left cancellation of u by c, which is the unique trace $v \in \mathcal{M}$ such that $c \cdot v = u$.

Remark 2. Of course, the direct argument:

$$\partial \mathcal{M}_{\leq \omega} \subseteq \{\xi \in \mathfrak{C}^{\mathbb{Z}_{\geq 1}} \mid \forall i \geq 1 \quad C_i(\xi) \subseteq C_i(\omega)\}$$

would not allow to conclude as in Corollary 1 that $\partial \mathcal{M}_{\leq \omega}$ is at most countable.

3   Deterministic Concurrent Systems

Definition 1. *A* deterministic concurrent system (DCS) *is a concurrent system* $\mathcal{X} = (\mathcal{M}, X, \perp)$ *such that for every state* $\alpha \in X$, *the partial order* $(\mathcal{M}_\alpha, \leq)$ *is a lattice.*

Remark 3. According to the background on *lub* and *glb* on trace monoids recalled in Sect. 2.1 on the one hand, and since $\mathcal{M}_\alpha$ is a downward closed subset of $\mathcal{M}$ on the other hand, we have for any two executions $x, y \in \mathcal{M}_\alpha$: 1) x and y have a *glb* in $\mathcal{M}_\alpha$, which coincides with their *glb* in $\mathcal{M}$; and 2) x and y have a *lub* in $\mathcal{M}_\alpha$ if and only they have a common upper bound in $\mathcal{M}_\alpha$, in which case their *lub* in $\mathcal{M}_\alpha$ coincides with their *lub* in $\mathcal{M}$. Note however that the existence of $x \vee y$ in $\mathcal{M}$ is not enough to insure that $x \vee y \in \mathcal{M}_\alpha$.

Henceforth, a concurrent system $(\mathcal{M}, X, \perp)$ is a DCS if and only if, for every state α, any two executions $x, y \in \mathcal{M}_\alpha$ have a common upper bound in $\mathcal{M}_\alpha$.

The following result says that DCS correspond to "locally commutative" concurrent systems.

Proposition 1. *Let* $\mathcal{X} = (\mathcal{M}, X, \perp)$ *be a concurrent system. Then the following properties are equivalent:*

(i) $\mathcal{X}$ *is deterministic.*
(ii) For every $\alpha \in X$, *the partial order* $(\mathscr{C}_\alpha, \leq)$ *is a lattice.*
(iii) For every $\alpha \in X$, *any two letters in* Σ_α *commute with each other.*

Proof. The equivalence (ii) $\iff$ (iii) and the implication (i) $\implies$ (iii) are clear. The interesting point is the implication (ii) $\implies$ (i).

Assume that $(\mathscr{C}_\alpha, \leq)$ is a lattice for every $\alpha \in X$. Fix $\alpha \in X$ and let $x, y \in \mathcal{M}_\alpha$. Assume first that $x \wedge y = \varepsilon$. Let $(c_1, \ldots, c_k)$ and $(d_1, \ldots, d_m)$ be the normal forms of x and of y. Maybe by adding the empty trace at the tail of one or the other normal form, we assume that $k = m$, at the cost of tolerating that some of the elements may be the empty trace.

On the one hand, since $c_1 \cdot c_2$ is an execution starting from α, one has $c_2 \in \mathscr{C}_{\alpha \cdot c_1}$. On the other hand, both c_1 and d_1 belong to $\mathscr{C}_\alpha$, which is a lattice by assumption. Hence $c_1 \vee d_1 \in \mathscr{C}_\alpha$. And since $c_1 \wedge d_1 = \varepsilon$ by assumption, one has $c_1 \vee d_1 = c_1 \cdot d_1 = d_1 \cdot c_1$. Therefore: $d_1 \in \mathscr{C}_{\alpha \cdot c_1}$. Since both cliques c_2 and d_1 belong to $\mathscr{C}_{\alpha \cdot c_1}$, which is a lattice, it follows that $c_2 \vee d_1 \in \mathscr{C}_{\alpha \cdot c_1}$.

Now we claim that $c_2 \wedge d_1 = \varepsilon$. Otherwise, there exists a letter a occurring in both c_2 and d_1. Since (c_1, c_2) is a normal pair of cliques, there exists $b \in c_1$ such that $(a, b) \in D$, the dependence pair of the monoid. Because of the assumption

$c_1 \wedge d_1 = \varepsilon$, the identity $a = b$ is impossible. But both a and b belong to Σ_α, and since $a \neq b$, the fact that $(a, b) \in D$ contradicts that $\mathscr{C}_\alpha$ is a lattice; our claim is proved.

We have obtained that $c_2 \vee d_1$ exists in $\mathscr{C}_{\alpha \cdot c_1}$ and that $c_2 \wedge d_1 = \varepsilon$. Hence $c_2 \vee d_1 = c_2 \cdot d_1 = d_1 \cdot c_2$. It implies that $c_2 \in \mathscr{C}_{\alpha \cdot (c_1 \vee d_1)}$. Symmetrically, we obtain that $d_2 \in \mathscr{C}_{\alpha \cdot (c_1 \vee d_1)}$. Since $\mathscr{C}_{\alpha \cdot (c_1 \vee d_1)}$ is a lattice, it follows that $d_2 \vee c_2 \in \mathscr{C}_{\alpha \cdot (c_1 \vee d_1)}$. But again, $d_2 \wedge c_2 = \varepsilon$ hence $d_2 \vee c_2 = d_2 \cdot c_2 = c_2 \cdot d_2$. Therefore we obtain that the following trace belongs to $\mathcal{M}_\alpha$:

$$(c_1 \vee d_1) \cdot (c_2 \vee d_2) = (c_1 \cdot c_2) \cdot (d_1 \cdot d_2) = (d_1 \cdot d_2) \cdot (c_1 \cdot c_2).$$

Repeating inductively the same reasoning, we finally obtain that $x \cdot y = y \cdot x \in \mathcal{M}_\alpha$, hence providing a common upper bound of x and of y in $\mathcal{M}_\alpha$. This proves the existence of $x \vee y$ in $\mathcal{M}_\alpha$ in the case where $x \wedge y = \varepsilon$.

The general case follows by considering $x' = (x \wedge y) \backslash x$ and $y' = (x \wedge y) \backslash y$ instead of x and y. □

Remark 4. In a DCS, for each state $\alpha \in X$, the partially ordered set of cliques $(\mathscr{C}_\alpha, \leq)$ identifies with the powerset $(\mathcal{P}(\Sigma_\alpha), \subseteq)$. In particular $\mathscr{C}_\alpha$ has a maximum $c_\alpha = \max(\mathscr{C}_\alpha) = \bigvee \Sigma_\alpha$, given by: $c_\alpha = \Sigma_\alpha$. We keep this notation in the statement of the following lemma.

Lemma 2. *Let $\mathcal{X} = (\mathcal{M}, X, \perp)$ be a deterministic concurrent system, and let $\alpha \in X$. Let $T_\alpha = (c_i)_{i \geq 1}$ be the sequence of cliques defined by $c_1 = c_\alpha$, and inductively by $c_{i+1} = c_{\alpha_i}$ where $\alpha_i = \alpha \cdot (c_1 \cdot \ldots \cdot c_i)$. Then T_α is a generalised execution which is the maximum of $(\overline{\mathcal{M}}_\alpha, \leq)$.*

Proof. We first observe that, for c_α the maximum of $\mathscr{C}_\alpha$, then $c_\alpha \to y$ holds[3] for every clique $y \in \mathscr{C}_{\alpha \cdot c_\alpha}$. Here in particular, $c_i \to c_{i+1}$ holds for all $i \geq 1$, hence T_α is indeed a generalised execution.

Let $x \in \overline{\mathcal{M}}_\alpha$, with $x = (d_i)_{i \geq 1}$. We prove that $x \leq T_\alpha$. Assume first that x is a finite trace, of height $k = \tau(x)$. Put $y = c_1 \cdot \ldots \cdot c_k$. Then x and y belong to $\mathcal{M}_\alpha$. Hence $z = x \vee y$ exists in $\mathcal{M}_\alpha$. Let $(e_1, \ldots, e_k)$ be the normal form of z (since x and y have the same height k, z also has height k). Then $c_j \leq e_j$ and thus $c_j = e_j$ for all j by maximality of c_j. Hence $d_j \leq c_j$ for all j, which was to be proved.

If $x = (c_i)_{i \geq 1}$ is now a generalised trace, we obtain the same result by applying the previous case to all sub-traces $(c_i)_{1 \leq i \leq k}$. □

Let us introduce a name for a valuation that will play a special role.

Definition 2. *Let $\mathcal{X} = (\mathcal{M}, X, \perp)$ be a concurrent system. The valuation $f = (f_\alpha)_{\alpha \in X}$ defined by:*

$$\forall \alpha \in X \quad \forall x \in \mathcal{M} \quad f_\alpha(x) = \begin{cases} 1, & \text{if } x \in \mathcal{M}_\alpha \\ 0, & \text{otherwise} \end{cases}$$

is called the dominant valuation *of $\mathcal{X}$.*

---

[3] This actually holds for any concurrent system, not necessarily deterministic, if c_α is taken to be any maximal element in $\mathscr{C}_\alpha$.

The family $f = (f_\alpha)_{\alpha \in X}$ given in Def. 2 is indeed a valuation. Indeed, using the axioms of the monoid action and the additional assumption $\bot \cdot z = \bot$ for all $z \in \mathcal{M}$, one sees that the following equivalence is true for every $\alpha \in X$ and for every traces $x, y \in \mathcal{M}$:

$$\alpha \cdot (x \cdot y) \neq \bot \iff (\alpha \cdot x \neq \bot \wedge (\alpha \cdot x) \cdot y \neq \bot),$$

which translates at once as the identity $f_\alpha(x \cdot y) = f_\alpha(x) f_{\alpha \cdot x}(y)$.

Theorem 1. *Let $\mathcal{X} = (\mathcal{M}, X, \bot)$ be a non trivial concurrent system.*

1. *If $\Sigma_\alpha \neq \emptyset$ for all $\alpha \in X$, then the two following statements are equivalent:*
   (i) $\mathcal{X}$ is deterministic.
   (ii) The dominant valuation of $\mathcal{X}$ is probabilistic.
2. *If $\mathcal{X}$ is deterministic, then all sets $\partial \mathcal{M}_\alpha$, for $\alpha \in X$, are at most countable and the characteristic root of $\mathcal{X}$ is $r = 1$ or $r = \infty$.*

Proof. Point 1. To prove the stated equivalence, assume (i), and let $f = (f_\alpha)_{\alpha \in X}$ be the dominant valuation. Let $\alpha \in X$, and let $c \in \mathscr{C}_\alpha$. Since $\mathscr{C}_\alpha$ identifies with $\mathcal{P}(\Sigma_\alpha)$, the Möbius transform of f_α evaluated at c is given by:

$$h_\alpha(c) = \sum_{c' \in \mathscr{C}_\alpha \,:\, c' \geq c} (-1)^{|c'| - |c|} = \begin{cases} 1, & \text{if } c = c_\alpha \text{ (the maximum of } \mathscr{C}_\alpha \text{)} \\ 0, & \text{otherwise.} \end{cases}$$

Since $\varepsilon \neq c_\alpha$ for all $\alpha \in X$, this shows that f is a probabilistic valuation.

Conversely, assume as in (ii) that f is probabilistic. Let $\alpha \in X$ be a state, and let c_α be a maximal element of $(\mathscr{C}_\alpha, \leq)$. Then, on the one hand, and since c_α is a maximal clique, one has $h_\alpha(c_\alpha) = f_\alpha(c_\alpha) = 1$. But on the other hand, h_α is nonnegative on $\mathscr{C}_\alpha$ and sums up to 1 on $\mathscr{C}_\alpha$. Hence h_α vanishes on all other cliques of $\mathscr{C}_\alpha$. Since this is true for every maximal element of $\mathscr{C}_\alpha$, it entails that $\mathscr{C}_\alpha$ has actually a unique maximal element, which is thus its maximum Σ_α. Hence $(\mathscr{C}_\alpha, \leq)$ is a lattice for every $\alpha \in X$, which proves (i) according to Proposition 1.

Point 2. We assume that $\mathcal{X}$ is a DCS. According to Lemma 2, the partial order $(\mathcal{M}_\alpha, \leq)$ has a maximum T_α for every $\alpha \in X$, hence $\overline{\mathcal{M}}_\alpha \subseteq \overline{\mathcal{M}}_{\leq T_\alpha}$. It follows at once from Corollary 1 that $\partial \mathcal{M}_\alpha$ is at most countable, and that $\# \mathcal{M}_\alpha(n) \leq P(n)$ for all integers n and for some polynomial P. All generating series $G_{\alpha, \beta}(z)$ are rational with non zero coefficients at least 1, and they have their coefficients dominated by some polynomial. They have therefore a radius of convergence either 1 or ∞. Hence $r \in \{1, \infty\}$. $\qquad \square$

Remark 5. In general, there might exist other probabilistic valuations than the dominant valuation, even for a DCS. See an example at the end of next section.

Since the dominant valuation f is probabilistic, there corresponds a family of probability measures as described in Sect. 2.2. The behaviour of the associated Markov chain of states-and-cliques is trivial, as shown by the following result.

Proposition 2. *Let $\mathcal{X} = (\mathcal{M}, X, \perp)$ be a non trivial* DCS *such that $\Sigma_\alpha \neq \emptyset$ for all $\alpha \in X$, and let $\nu = (\nu_\alpha)_{\alpha \in X}$ be the family of probability measures associated with the dominant valuation. Then for each initial state $\alpha \in X$, the probability measure ν_α is the Dirac distribution $\delta_{\{T_\alpha\}}$, where $T_\alpha = \max \overline{\mathcal{M}}_\alpha$.*

Proof. Assuming that $\mathcal{X}$ is a DCS, we keep using the notation $c_\alpha = \max \mathscr{C}_\alpha = \Sigma_\alpha$ for all $\alpha \in X$.

A direct proof is as follows. Fix $\alpha \in X$, and let $(\alpha_i, z_i)_{i \geq 0}$ be defined inductively by $\alpha_0 = \alpha$, $z_0 = \varepsilon$ and $z_{i+1} = z_i \cdot c_{\alpha_i}$, $\alpha_{i+1} = \alpha \cdot z_i$. On the one hand, we have $\bigvee_{i \geq 0} z_i = T_\alpha$ by the construction used in the proof of Lemma 2. But on the other hand, the characterisation of the probability measure ν_α yields $\nu_\alpha(\uparrow z_i) = f(z_i) = 1$ for all $i \geq 0$. Since $\uparrow z_{i+1} \subseteq \uparrow z_i$ for all $i \geq 0$, we have thus:

$$\nu_\alpha(\omega \geq T_\alpha) = \nu_\alpha\left(\bigcap_{i \geq 0} \uparrow z_i\right) = \lim_{i \to \infty} \nu_\alpha(\uparrow z_i) = 1.$$

Since $T_\alpha = \max \overline{\mathcal{M}}_\alpha$, it implies $\nu_\alpha(\omega = T_\alpha) = 1$.

An alternative proof is as follows. Let $(Y_i)_{i \geq 1}$ be the Markov chain of states-and-cliques associated to the dominant valuation, and let $\alpha \in X$. One has $\nu_\alpha(C_1 = c) = h_\alpha(c)$ for all $c \in \mathfrak{C}_\alpha$, by (9). The values of h_α computed in the proof of Th. 1 show that the initial distribution of the chain is $\delta_{\{(\alpha, c_\alpha)\}}$. It is shown in [1] that the (α, c)-row of the transition matrix of the chain is proportional to $h_{\alpha \cdot c}(\cdot)$. Hence all entries of the (α, c)-row are 0, except for the $((\alpha, c), (\beta, c_\beta))$ entry with $\beta = \alpha \cdot c$, where the entry is 1. Hence the execution T_α is given ν_α-probability 1. □

4   Irreducible Deterministic Concurrent Systems

Before stating the main result of this section, we need to prove two lemmas.

Lemma 3. *Let $\mathcal{X} = (\mathcal{M}, X, \perp)$ be a* DCS*. Let $\alpha \in X$ and let $c \in \mathscr{C}_\alpha$ be a clique such that $a \notin c$ for some letter $a \in \Sigma_\alpha$. Then:*

$$\forall x \in \overline{\mathcal{M}}_\alpha \quad C_1(x) = c \implies a \notin x.$$

Proof. Let α, a and c be as in the statement. Clearly, the implication stated in the lemma is true if we prove it to be true for x ranging over $\mathcal{M}_\alpha$ instead of $\overline{\mathcal{M}}_\alpha$. Hence, let $x \in \mathcal{M}_\alpha$ be such that $C_1(x) = c$. Let $(c_i)_{i \geq 1}$ be the generalised normal form of x, and define by induction $x_0 = \varepsilon$, $x_{i+1} = x_i \cdot c_{i+1}$ for all $i \geq 0$ and $\alpha_i = \alpha \cdot x_i$ for all $i \geq 0$. We prove by induction on $i \geq 1$ that: 1) $a \in \Sigma_{\alpha_{i-1}}$; and 2) $a \notin c_i$.

For $i = 1$, both properties derive from the assumptions of the lemma. Assume that both properties hold for some $i \geq 1$. By construction, $c_i \in \mathscr{C}_{\alpha_{i-1}}$, and $a \in \Sigma_{\alpha_{i-1}}$ by the induction hypothesis. Since the concurrent system is deterministic, it follows that $a \vee c_i \in \mathscr{C}_{\alpha_{i-1}}$. Since $a \notin c_i$ by the assumption hypothesis, this *lub* is given by $c_i \cdot a \in \mathscr{C}_{\alpha_{i-1}}$. This entails first that $a \in \mathscr{C}_{\alpha_{i-1} \cdot c_i}$, but $\alpha_{i-1} \cdot c_i = \alpha_i$ hence $a \in \Sigma_{\alpha_i}$. But it also entails that $a \notin c_{i+1}$, completing the induction step. The result of the lemma follows. □

Lemma 4. *Let $\mathcal{X} = (\mathcal{M}, X, \perp)$ be a concurrent system. Let $\alpha \in X$, and let r_α be the radius of convergence of the generating series $G_\alpha(z) = \sum_{x \in \mathcal{M}_\alpha} z^{|x|}$. Then the following properties are equivalent: (i) $\mathcal{M}_\alpha$ is finite; (ii) $\partial \mathcal{M}_\alpha = \emptyset$; (iii) $r_\alpha = \infty$.*

Proof. The implications (i) $\implies$ (ii) and (i) $\implies$ (iii) are clear.

Assume that $\mathcal{M}_\alpha$ is infinite. Then there exists executions in $\mathcal{M}_\alpha$ of length arbitrary large. Therefore there exists $x \in \mathcal{M}_\alpha$ and $y \neq \varepsilon$ such that $\alpha \cdot x = \alpha \cdot (x \cdot y)$. Then all traces $x_n = x \cdot y^n$ belong to $\mathcal{M}_\alpha$ for $n \geq 0$. This proves two things. First, if $k = |y|$, the coefficient of $z^{|x|+kn}$ in the series $G_\alpha(z)$ is ≥ 1 for all integers n, hence $r_\alpha < \infty$. Second, the execution $\xi = \bigvee_{n \geq 0} x_n$ is an element of $\partial \mathcal{M}_\alpha$, showing that $\partial \mathcal{M}_\alpha \neq \emptyset$. Hence we have proved both (ii) $\implies$ (i) and (iii) $\implies$ (i) by contraposition, completing the proof. $\qquad\square$

Theorem 2. *Let $\mathcal{X} = (\mathcal{M}, X, \perp)$ be an irreducible and non trivial concurrent system, of characteristic root r, and let f be the dominant valuation of $\mathcal{X}$. Then the following statements are equivalent:*

(i) $\mathcal{X}$ is deterministic.
(ii) f is a probabilistic valuation.
(iii) f is the only probabilistic valuation of $\mathcal{X}$.
(iv) $r = 1$.
(v) One set $\partial \mathcal{M}_\alpha$ is at most countable.
(vi) Every set $\partial \mathcal{M}_\alpha$ is at most countable.

Proof. Since $\mathcal{X}$ is both irreducible and non trivial, it satisfies in particular $\Sigma_\alpha \neq \emptyset$ for all $\alpha \in X$. Hence the equivalence (i) $\iff$ (ii) and the implications (i) $\implies$ (iv) and (i) $\implies$ (vi) derive already from Theorem 1. The implications (iii) $\implies$ (ii) and (vi) $\implies$ (v) are trivial.

(i) $\implies$ (iii). Let $f = (f_\alpha)_{\alpha \in X}$ be a probabilistic valuation, and let $\widetilde{f} = (\widetilde{f}_\alpha)_{\alpha \in X}$ be the dominant valuation. Let $\alpha \in X$ and let $c \in \mathfrak{C}_\alpha$ with $c \neq c_\alpha$, where $c_\alpha = \Sigma_\alpha$ is the maximum of $\mathscr{C}_\alpha$. There is thus a letter $a \in \Sigma_\alpha$ such that $a \notin c$. Let $\mathcal{M}^a$ be the submonoid of $\mathcal{M}$ generated by $\Sigma \setminus \{a\}$. It follows from Lemma 3 that $\{w \in \partial \mathcal{M}_\alpha \mid C_1(w) = c\} \subseteq \partial \mathcal{M}_\alpha^a$.

According to the spectral property recalled in Sect. 2.2, the characteristic root r^a of $\mathcal{X}^a = (\mathcal{M}^a, X, \perp)$ satisfies $r^a > r$ since $\mathcal{X}$ is assumed to be irreducible. But $r = 1$ since $\mathcal{X}$ is deterministic, and therefore $r^a = \infty$, which implies that $\partial \mathcal{M}_\alpha^a = \emptyset$ according to Lemma 4. Let $\nu = (\nu_\alpha)_{\alpha \in X}$ be the family of probability measures associated with the probabilistic valuation f, as explained in Sect. 2.2. Then $\nu_\alpha(\partial \mathcal{M}_\alpha^a) = 0$ and thus $\nu_\alpha(C_1 = c) = 0$. But one also has $h_\alpha(c) = \nu_\alpha(C_1 = c)$ according to (9), where h_α is the Möbius transform of f_α. Hence $h_\alpha(c) = 0$. We have proved that h_α vanishes on all cliques $c \in \mathscr{C}_\alpha$ such that $c \neq c_\alpha$. Since $(h_\alpha(c))_{c \in \mathfrak{C}_\alpha}$ is a probability vector, it entails that $h_\alpha(c_\alpha) = 1$. Thus h_α coincides with the Möbius transform of $\widetilde{f}_\alpha$, and $f = \widetilde{f}$.

(iv) $\implies$ (i) and (v) $\implies$ (i) . By contraposition, assume that $\mathcal{X}$ is not deterministic. Prop. 1 implies the existence of a state α and of two distinct letters $a, b \in \Sigma_\alpha$ such that $a \cdot b \neq b \cdot a$. Since $\mathcal{X}$ is assumed to be irreducible, there exists

$x \in \mathcal{M}_{\alpha \cdot a, \alpha}$ and $y \in \mathcal{M}_{\alpha \cdot b, \alpha}$. Put $x_a = a \cdot x$ and $x_b = b \cdot y$, and we can also assume without loss of generality that $|x_a| = |x_b|$. Then $\mathcal{M}_\alpha$ contains the submonoid generated by $\{x_a, x_b\}$, which is free. This implies two things: first, the generating series $G_\alpha(z) = \sum_{x \in \mathcal{M}_\alpha} z^{|x|}$ has radius of convergence smaller than 1, and thus $r < 1$; second, $\partial \mathcal{M}_\alpha$ is uncountable. The proof is complete.   □

For an irreducible DCS, the behaviour of the Markov chain of states-and-cliques associated to the unique probabilistic dynamics is the trivial dynamics described by Prop. 2. This is illustrated in the following example.

Example 1. Fig. 4 depicts an example of irreducible DCS. The digraph of states-and-cliques of the system is depicted on Fig. 5. Compare with the situation depicted next for a DCS which is not irreducible.

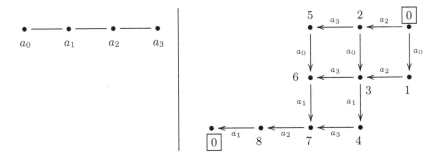

Fig. 4. Example of an irreducible and deterministic concurrent system $\mathcal{X} = (\mathcal{M}, X, \perp)$ with $\Sigma = \{a_0, \ldots, a_3\}$, $X = \{0, 1, \ldots, 8\}$. *Left:* Coxeter graph of the monoid $\mathcal{M}$. *Right:* multigraph of states of $\mathcal{X}$. The two framed labels $\boxed{0}$ are identified and correspond to the same state.

Example 2. Without the irreducibility assumption, the equivalence stated in Th. 2 may fail. We give below an example of a deterministic concurrent systems not irreducible, and not satisfying point (iii).

Let $\mathcal{X} = (\mathcal{M}, X, \perp)$ be the DCS depicted in Fig. 6. The system is not irreducible for several reasons: none of the three conditions for irreducibility is met. The probabilistic valuations of $\mathcal{X}$ are all of the following form, for some real $p \in [0, 1]$:

$$f_{\alpha_0}(a) = 1 \quad f_{\alpha_0}(c) = p \quad f_{\alpha_1}(b) = 1 \quad f_{\alpha_1}(c) = p \quad f_{\beta_0}(a) = 1 \quad f_{\beta_1}(b) = 1$$

Hence the dominant valuation is not the unique probabilistic valuation, contrary to irreducible systems as stated by point (iii) of Th. 2. The parameter p is to be interpreted as the "probability of playing c" in the course of the execution. But this decision—playing c or not—is made once, hence allowing all values between 0 or 1 for the probability. Whereas, in a sequential model of concurrency, that would typically be a decision repeated infinitely often, hence yielding the only two possible values 0 or 1 for this probability. The formula $\nu_\alpha(C_1 = \gamma) = h_\alpha(\gamma)$ for $\gamma \in \mathscr{C}_\alpha$ yields the following initial distribution of the Markov chain of states-and-cliques if, for instance, the initial state of the system is α_0:

$$\nu_{\alpha_0}(C_1 = a) = 1 - p \qquad \nu_{\alpha_0}(C_1 = c) = 0 \qquad \nu_{\alpha_0}(C_1 = ac) = p$$

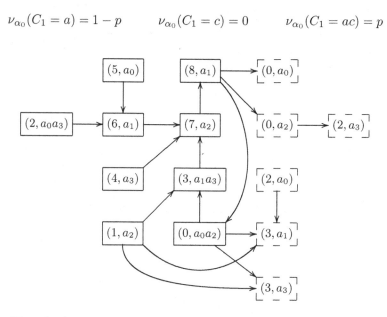

Fig. 5. Digraph of states-and-cliques for the DCS depicted on Fig. 4. Nodes with solid frames are nodes of the form (α, c_α) with $c_\alpha = \max \mathscr{C}_\alpha$. The probability for the Markov chain of states-and-cliques to jump from a solid frame node to a dashed frame node is 0; the probability of starting in a dashed node in 0.

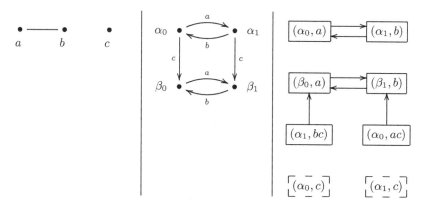

Fig. 6. A non irreducible DCS not satisfying property (iii) of Th. 2. *Left:* the Coxeter graph of the monoid. *Middle:* the multigraph of states of the DCS. *Right:* the digraph of states-and-cliques. The parameter p is only involved in the initial distribution of the Markov chain of states-and-cliques. The dashed nodes are isolated in the digraph of states-and-cliques and are immaterial to the Markov chain of states-and-cliques.

References

1. Abbes, S.: Markovian dynamics of concurrent systems. Discrete Event Dyn. Syst. **29**(4), 527–566 (2019). https://doi.org/10.1007/s10626-019-00291-z
2. Abbes, S., Mairesse, J.: Uniform and Bernoulli measures on the boundary of trace monoids. J. Comb. Theor. Ser. A **135**, 201–236 (2015)
3. Abbes, S., Mairesse, J., Chen, Y.-T.: A spectral property for concurrent systems and some probabilistic applications. Submitted for publication. Available at https://arxiv.org/abs/2003.03762 (2020)
4. Cartier, P., Foata, D.: Problèmes combinatoires de commutation et réarrangements. LNM, vol. 85. Springer, Heidelberg (1969). https://doi.org/10.1007/BFb0079468
5. Dehornoy, P., Digne, F., Godelle, E., Krammer, D., Michel, J.: Foundations of Garside Theory. EMS (2015)
6. Diekert, V.: Combinatorics on Traces. LNCS, vol. 454. Springer, Heidelberg (1990). https://doi.org/10.1007/3-540-53031-2
7. Diekert, V., Rozenberg, G. (eds.): The Book of Traces. World Scientific (1995)
8. Goldwurm, M., Santini, M.: Clique polynomials have a unique root of smallest modulus. Inform. Process. Lett. **75**(3), 127–132 (2000)
9. Krob, D., Mairesse, J., Michos, I.: Computing the average parallelism in trace monoids. Discrete Math. **273**, 131–162 (2003)
10. Nielsen, M., Plotkin, G., Winskel, G.: Petri nets, event structures and domains, part I. Theor. Comput. Sci. **13**, 85–108 (1981)
11. Reisig, W.: Petri Nets- An Introduction. Springer, Heidelberg (1985). https://doi.org/10.1007/978-3-642-69968-9
12. Rota, G.-C.: On the foundations of combinatorial theory I. Theory of Möbius functions. Z. Wahrscheinlichkeitstheorie **2**, 340–368 (1964). https://doi.org/10.1007/BF00531932
13. Viennot, G.X.: Heaps of pieces, I: Basic definitions and combinatorial lemmas. In: Labelle, G., Leroux, P. (eds.) Combinatoire énumérative. LNM, vol. 1234, pp. 321–350. Springer, Heidelberg (1986). https://doi.org/10.1007/BFb0072524

Deciphering the Co-Car Anomaly of Circular Traffic Queues Using Petri Nets

Rüdiger Valk[(✉)]

Department of Informatics, University of Hamburg, Hamburg, Germany
valk@informatik.uni-hamburg.de

Abstract. The co-car anomaly appears in the study of circular traffic queues. An unfolding of the corresponding coloured net is proved to be isomorphic to a particular cycloid. Then the anomaly is reduced to a combinatorial property of path lengths in the cycloid. Methods of the cycloid algebra are used to derive iterations of cycloids. Different such iterations correspond to different models of traffic queues, but only those with observable co-traffic items show the anomaly.

Keywords: Circular traffic queues · Coloured petri nets · Unfoldings · Structure of petri nets · Cycloids · Cycloid algebra · Iteration of cycloids

1 Introduction

During the study of circular traffic queues an interesting anomaly came to our attention. In the model under investigation, traffic items like cars, trains, aircrafts, production goods, computer tasks or electronic particles, are divided into two sorts, namely those moving from left to right and those moving to the opposite direction. While the former are called *cars* in this introduction the latter are denoted as *co-cars*. In each step of the system, when in face of a co-car a car may interchange its position with it. The circular traffic queue of Fig. 1 contains a number of $c = 3$ cars a_0, a_1 and a_2 and $g = 5$ co-cars. In the current state of this system car a_0 in position 2 can swap with co-car u_2 in position 3, as well as a_2 in position 7 with co-car u_0 in position 0. For modelling and programming the size of such systems is of importance. The size is strongly connected with the minimal length $\Xi(c, g)$ of a *recurrent transition sequence*, which is defined as a sequence that reproduces a given initial state. Experiments have shown that for the system in Fig. 1 we obtain $\Xi(3, 5) = 120$, but by slightly increasing one of the parameters to $g = 6$ this number is not increased, but reduced to $\Xi(3, 6) = 54$. This effect is called the *co-car anomaly of circular traffic queues*. In this paper we give an explanation of this anomaly using the graphical representation of T-nets. To this end we first model circular traffic queues by a standard representation of coloured nets and then unfold this net (also by a standard method) into a

© Springer Nature Switzerland AG 2021
D. Buchs and J. Carmona (Eds.): PETRI NETS 2021, LNCS 12734, pp. 443–462, 2021.
https://doi.org/10.1007/978-3-030-76983-3_22

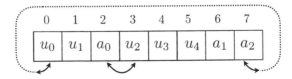

Fig. 1. Circular traffic queue with $c = 3$ cars and $g = 5$ co-cars.

T-net, which obviously shows more graphical structure than the coloured net. Instead of directly investigating such T-nets we will use the elaborated theory of cycloids. Cycloids have been introduced by C.A. Petri in [4]. Cycloids define a subclass of partial cyclic orders and hence generalize the well known token ring structure (a total cyclic order) that is at the core of many solutions to the distributed mutual exclusion problem. This also includes virtual token rings that have been employed in group communication middleware (e.g. the Spread system). We conjecture that cycloids could more generally play a role as coordination models in new middleware architectures. The methods and results of this article lead to understand and construct coordination mechanisms of cooperating processes. For instance the problem to make such a system *wait-free* [1] is solved in this context (not in this paper).

Petri used cycloids to model very different phenomena among which were special forms of circular traffic queues. When replacing the co-cars by gaps or anonymous items, a simpler system is obtained which was called $tq\text{-}g(c, g)$ in [7], where a proof can be found that $tq\text{-}g(c, g)$ is *behavioural* equivalent to a particular cycloid. In this paper we prove a much stronger relation, namely that the mentioned unfolding of a coloured net is *syntactically* isomorphic to a different cycloid.

The results of this article are summarized as follows (some notions will be introduced later in the article): In Sect. 2 circular traffic queues are formally defined and the theorem on the number of recurrent transition sequences from [7] is cited. The model is represented as a coloured net in Sect. 3 together with an unfolding which follows the standard construction of occurrence nets (the Petri net processes). Section 4 recalls the needed results on cycloids [5] and of regular cycloids [7]. A new theorem is presented which gives for an arbitrary element of the Petri space its equivalent element in the fundamental parallelogram. Using the Chinese remainder theorem, in Sect. 5 the unfolding of the coloured net from Sect. 3 is proved to be isomorphic to the cycloid $\mathcal{C}(g, c, \frac{g \cdot c}{\Delta}, \frac{g \cdot c}{\Delta})$. This cycloid plays an important role in Sect. 6, where the co-car anomaly is reduced to the f-factor, which is the quotient of the lengths of certain paths in the cycloid $\mathcal{C}(g, c, c, c)$. This cycloid models circular traffic queues with gaps $tq\text{-}g(c, g)$, where co-cars are reduced to anonymously named items. The relation of both kinds of cycloids is studied in Sect. 7, where the iteration of cycloids is defined using the cycloid algebra.

The author is grateful to the anonymous referees for proposing numerous improvements.

We recall some standard notations for set theoretical relations. If $R \subseteq A \times B$ is a relation and $U \subseteq A$ then $R[U] := \{b \mid \exists u \in U : (u, b) \in R\}$ is the *image* of U and $R[a]$ stands for $R[\{a\}]$. R^{-1} is the *inverse relation* and R^+ is the *transitive closure* of R if $A = B$. Also, if $R \subseteq A \times A$ is an equivalence relation then $[\![a]\!]_R$ is the *equivalence class* of the quotient A/R containing a. Furthermore $\mathbb{N}_+$, $\mathbb{Z}$ and $\mathbb{R}$ denote the sets of positive integer, integer and real numbers, respectively. For integers: $a|b$ if a is a factor of b. The *modulo*-function is used in the form $a \bmod b = a - b \cdot \lfloor \frac{a}{b} \rfloor$, which also holds for negative integers $a \in \mathbb{Z}$. In particular, $-a \bmod b = b - a$ for $0 < a \leq b$. Furthermore we will use $(a \otimes b) \bmod n = (a \bmod n \otimes b \bmod n) \bmod n$ for $\otimes \in \{+, -\}$ and $(z + k \cdot p) \bmod p = z \bmod p$ for all $k \in \mathbb{N}, z \in \mathbb{Z}$. As a short notation we write $x \oplus_r y$ for $(x + y) \bmod r$ and $x \ominus_r y$ for $(x - y) \bmod r$. Different to the article [7] we use indices in sets of size n in the form $B = \{b_0, \cdots, b_{n-1}\}$ rather than $B = \{b_1, \cdots, b_n\}$.

2   Circular Traffic Queues

Circular traffic queues are composed by a number of sequential and interacting processes of traffic items $a \in C$ and co-items $u \in G$. An intuitive notation would be to consider a state as a word of length n over the alphabet $C \cup G$ with distinct letters only, and the rewrite rule $au \to ua$ with $a \in C$, $u \in G$ when inside the word and $u \cdots a \to a \cdots u$ at the borders. An example of two such transitions with $C = \{a, b, c\}, G = \{u, v, w, x\}$ is $u\,a\,b\,v\,w\,x\,c \to u\,a\,v\,b\,w\,x\,c \to c\,a\,v\,b\,w\,x\,u$. When defining the elements of G to be indistinguishable, they can be interpreted as gaps interchanging with the traffic items from C.

Definition 1. *A circular traffic queue $tq(c, g)$ is defined by two positive integers c and g. Implicitly with these integers we consider two finite and disjoint sets of traffic items $C = \{a_0, \cdots, a_{c-1}\}$ and $G = \{u_0, \cdots, u_{g-1}\}$ with cardinalities c and g, respectively. A* state *is a bijective index function* ind : $\{0, \cdots, n-1\} \to C \cup G$, *hence $c + g = n$. The labelled transition system $LTS(c, g) = (States, T, tr, ind_0)$ of $tq(c, g)$ is defined by a set States of states, a set of transitions $T = \{\langle\!\langle t_i, a_j \rangle\!\rangle | 0 \leq i < n, 0 \leq j < c\}$, a transition relation tr and a regular initial state ind_0. The regular initial state is given by $ind_0(i) = a_i$ for $0 \leq i < c$ and $ind_0(i) = u_{i-c}$ for $c \leq i < n$. The transition relation $tr \subseteq States \times T \times States$ is defined by $(ind_1, \langle\!\langle t_i, a_j \rangle\!\rangle, ind_2) \in tr \Leftrightarrow$*
$$ind_1(i \oplus_n 1) = ind_2(i) \in G \ \wedge \ ind_2(i \oplus_n 1) = ind_1(i) = a_j \ \wedge$$
$$ind_2(m) = ind_1(m) \ \text{for all} \ m \notin \{i, i \oplus_n 1\}.$$
This is written as $ind_1 \xrightarrow{\langle\!\langle t_i, a_j \rangle\!\rangle} ind_2$ or $ind_1 \to ind_2$. A transition sequence $ind_0 \to ind_1 \to \cdots \to ind_0$ of minimal length, leading from the initial state ind_0 back to ind_0 is called a recurrent sequence. *As usual $ind_1 \xrightarrow{*} ind_2$ denotes the reflexive and transitive closure of tr. We restrict the set of states to the states reachable from the initial state: $States := \mathcal{R}(LTS(c, g), ind_0) := \{ind | ind_0 \xrightarrow{*} ind\}$.*

Theorem 2. ([7]). *Let $\Delta = gcd(c, g)$ be the greatest common divisor of c and g. The length of each recurrent sequence of $tq(c, g)$ is $\Xi(c, g) := \frac{g}{\Delta} \cdot (c + g) \cdot c$.*

For the example in the introduction we obtain $\varXi(3,5) = \frac{5}{1} \cdot (3+5) \cdot 3 = 120$ and $\varXi(3,6) = \frac{6}{3} \cdot (3+6) \cdot 3 = 54$.

While the regular initial state is natural in the sense that the traffic items start without gaps in between, in a different context it is useful that the gaps are equally distributed, as in the following definition of a *standard initial state*. If for instance the numbers c and g are even, the queue in its initial state is composed of two equal subsystems with the parameters $\frac{c}{2}$ and $\frac{g}{2}$. An analogous situation holds for larger divisors. The *spacial iteration* of a cycloid, as defined in Sect. 7, formalizes such a composition.

Definition 3. *A standard initial state ind_0 of $tq(c,g)$ is defined by the state $ind_0(0)ind(1)_0 \cdots ind_0(n-1) = a_0 w_0 a_1 w_1 \cdots a_{c-1} w_{c-1}$ with $a_j \in C, w_j \in G^{r_j}$ (set of words of length r_j over G), $r_j = |\{ x \in \mathbb{Z} \mid j-1 \leq \frac{c}{g} \cdot x < j \}|$ for $0 \leq j < c$ and $w_0 w_1 \cdots w_{c-1} = u_0 u_1 \cdots u_{g-1}$.*

To give an example, we consider the case $c = 6, g = 4$. We obtain $(r_0, r_1, r_2, r_3, r_4, r_5) = (0, 1, 1, 0, 1, 1)$ and $ind_0 = \mathbf{a_0 a_1} u_0 \mathbf{a_2} u_1 \mathbf{a_2 a_4} u_2 \mathbf{a_5} u_3$. Contrary to the regular initial state the standard initial state is invariant to taking integer multiples of c and g. For instance the standard initial state of the circular traffic queue $tq(3,2)$ is $ind'_0 = \mathbf{a_0 a_1} u_0 \mathbf{a_2} u_1$ which is the first half of ind_0.

3   A Coloured Net and Its T-Equivalent

We define nets as they will be used in this article. In this section we use a simple form of a coloured net, for the definition of which we refer to the literature [2].

Definition 4. *As usual, a net $\mathcal{N} = (S, T, F)$ is defined by non-empty, disjoint sets S of places and T of transitions, connected by a flow relation $F \subseteq (S \times T) \cup (T \times S)$ and $X := S \cup T$. $\mathcal{N} \simeq \mathcal{N}'$ denote isomorphic nets. A transition $t \in T$ is active or enabled in a marking $M \subseteq S$ if ${}^\bullet t \subseteq M \wedge t^\bullet \cap M = \emptyset$ and in this case $M \xrightarrow{t} M'$ if $M' = M \backslash {}^\bullet t \cup t^\bullet$, where ${}^\bullet x := F^{-1}[x]$, $x^\bullet := F[x]$ denote the input and output elements of an element $x \in X$, respectively. $\xrightarrow{*}$ is the reflexive and transitive closure of $\rightarrow$. A net together with an initial marking $M_0 \subseteq S$ is called a net-system $(\mathcal{N}, M_0)$ with its reachability set $\mathcal{R}(\mathcal{N}, M_0) := \{M | M_0 \xrightarrow{*} M\}$.*

We start with modelling the circular traffic queue from Definition 1 by a coloured net. This is shown in Fig. 2 by a standard construction of the net $\mathcal{N}_{sym}(c,g)$ using places and their complementary counter-parts (with the restriction $c \geq 2$ due to the layout style).

Definition 5. *The coloured net $\mathcal{N}_{sym}(c,g) = (\bar{S}, T, F, var, M_0)$ is defined by a set of places $\bar{S} := S \cup S'$ with $S = \{s_0, \cdots, s_{n-1}\}$, $S' = \{s'_0, \cdots, s'_{n-1}\}$ and $n = c + g$, a set of transitions $T = \{t_0, \cdots, t_{n-1}\}$, the set of arrows $F := F^1 \cup F^2 \cup F^3 \cup F^4$ with $F^1 := \{(s_{i \ominus_n 1}, t_i) | 0 \leq i < n\}$, $F^2 := \{(t_i, s_i) | 0 \leq i < n\}$, $F^3 := \{(s'_{i \oplus_n 1}, t_i) | 0 \leq i < n\}$, $F^4 := \{(t_i, s'_i) | 0 \leq i < n\}$. The arrow labelling by variables $var : F \rightarrow \{\mathsf{x}, \mathsf{y}\}$ is defined by $var((x_1, x_2))$*
$$= \begin{cases} \mathsf{x} & \text{if } (x_1, x_2) \in F_1 \cup F_2 \\ \mathsf{y} & \text{if } (x_1, x_2) \in F_3 \cup F_4 \end{cases}.$$

The initial marking is given by $M_0(s_{i \ominus_n 1}) = a_i$ for $0 \leq i < c$ and $M_0(s'_i) = u_{i-c}$ for $c \leq i < n$. All different places are unmarked.

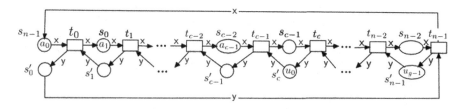

Fig. 2. Coloured net $\mathcal{N}_{sym}(c, g)$ of a circular traffic queue ($c \geq 2$).

The problem to be studied in this paper becomes more apparent in the T-net equivalent $\mathcal{N}_T(c, g)$ of this net. This net is built by using a well-known method of creating a net-process of $\mathcal{N}_{sym}(c, g)$. To begin with, we first define the occurrence modes[1] or bindings of a transition t_i, which is obtained by the fact that such a transition is interchanging a traffic item a_j in position i with a co-item u_h in position $i \oplus_n 1$ (see Fig. 3 a). From the representation of a transition t_i of $\mathcal{N}_{sym}(c, g)$ in Fig. 3 b) we construct the process transition $[t_i, a_j, u_h]$ in part c) of the figure. The semantics of the net elements from the last figure are as follows:

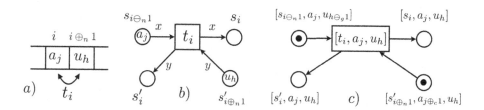

Fig. 3. Constructing a T-net equivalent from $\mathcal{N}_{sym}(c, g)$ in Fig. 2.

a) $[t_i, a_j, u_h]$: t_i swaps a_j at position i with u_h at position $i \oplus_n 1$.
b) $[s_{i \ominus_n 1}, a_j, u_{h \ominus_g 1}]$: a_j is at position i and the next item to swap with is u_h by transition t_i.
c) $[s'_{i \oplus_n 1}, a_{j \oplus_c 1}, u_h]$: u_h is at position $i \oplus_n 1$ and the next item to swap with is a_j by transition t_i.

---

[1] For the notion of occurrence mode of a coloured net see [2], page 35.

In the following (pseudo-)process construction we start from the given initial marking and then in an iterated way add some transition from T_{base} having its input places in the marking obtained so far. But different to a real process construction we do not create copies of places or transitions, giving a finite net containing cycles. To keep the following definition simpler we begin with all transitions of T_{base} and then restrict to the reachable places and transitions.

Definition 6. *For the coloured tq-net $\mathcal{N}_{sym}(c,g) = (S \cup S', T, F, var, M_0)$ a net $\mathcal{N}_T(c,g) = (S_2, T_2, F_2, M_2^0)$, called T-net-equivalent of $\mathcal{N}_{sym}(c,g)$, is defined as follows with $n=c+g$:*

$S_{base} := \{[s,a,u]|s \in S, a \in C, u \in G\}$, $S'_{base} := \{[s',a,u]|s' \in S', a \in C, u \in G\}$,
$T_{base} := \{[t,a,u]|t \in T, a \in C, u \in G\}$, $F_{base} := F_2^1 \cup F_2^2 \cup F_2^3 \cup F_2^4$,
$F_2^1 := \{([s_{i \ominus_n 1}, a_j, u_{h \ominus_g 1}], [t_i, a_j, u_h])|0 \leq i < n, 0 \leq j < c, 0 \leq h < g\}$,
$F_2^2 := \{([t_i, a_j, u_h], [s_i, a_j, u_h])|0 \leq i < n, 0 \leq j < c, 0 \leq h < g\}$,
$F_2^3 := \{([s'_{i \oplus_n 1}, a_{j \oplus_c 1}, u_h], [t_i, a_j, u_h])|0 \leq i < n, 0 \leq j < c, 0 \leq h < g\}$,
$F_2^4 := \{([t_i, a_j, u_h]), [s'_i, a_j, u_h]|0 \leq i < n, 0 \leq j < c, 0 \leq h < g\}$,
$M_0^2 := \{[s_{i \ominus_n 1}, a_i, u_{g-1}]|0 \leq i < c\} \cup \{[s'_i, a_0, u_{i-c}]|c \leq i < n\}$.
T_2 is defined as the set of transitions which are obtained by starting from M_0^2 and inductively generating transitions of T_{base} with their output places using the arrows from F_{base}. Then $S_2 := T_2^\bullet \cup {}^\bullet T_2$ and $F_2 := F_{base} \cap ((S_2 \times T_2) \cup (T_2 \times S_2))$.

As in the case of occurrence nets or net processes $\mathcal{N}_{sym}(c,g)$ is a homomorphic image of $\mathcal{N}_T(c,g)$. But contrary to this case $\mathcal{N}_T(c,g)$ is necessarily finite. The number of transitions is given in the following lemma. This result also follows from the isomorphism proved in Sect. 5.

Lemma 7. $\mathcal{N}_{sym}(c,g) = (\bar{S}, T, F, var, M_0)$ *is a homomorphic image of $\mathcal{N}_T(c,g)$ and $\mathcal{N}_T(c,g)$ contains $\frac{g}{\Delta} \cdot n \cdot c$ transitions and twice as much places.*

Proof. The homomorphism is defined by $\rho_1 : S_2 \to \bar{S}$, $\rho_1([s,a,u]) := s$ and $\rho_2 : T_2 \to T_1$, $\rho_1([t,a,u]) := t$. It has to be proved $(s,t) \in F_2^1 \cup F_2^3 \Rightarrow (\rho_1(s), \rho_2(t)) \in F^1 \cup F^3$ as well as $(t,s) \in F_2^2 \cup F_2^4 \Rightarrow (\rho_2(t), \rho_1(s)) \in F^2 \cup F^4$. This follows easily by comparing Definitions 5 and 6. The morphism is also compatible with the initial markings: $[s_{i \ominus_n 1}, a_i, u_{g-1}] \in M_0^2 \Rightarrow M_0(\rho_1([s_{i \ominus_n 1}, a_i, u_{g-1}])) = M_0(s_{i \ominus_n 1}) = a_i$ for $0 \leq i < c$ and $[s'_i, a_0, u_{i-c}] \in M_0^2 \Rightarrow M_0(\rho_1([s'_i, a_0, u_{i-c}])) = M_0(s'_i) = u_{i-c}$ for $c \leq i < n$. To compute the number of transitions we consider the process cycle T_j of an item a_j: $T_j = [t_0, a_j, u_{h_0}], [t_1, a_j, u_{h_1}], [t_2, a_j, u_{h_2}], \cdots$, where the indices of t_i repeat the sequence $0, \cdots, n-1$ a number of times and, similarly, the indices of u_{h_k} repeat $0, \cdots, g-1$. The cycle is closed at the least common multiple $lcm(n,g)$ of n and g. Therefore we obtain for the length of the cycle (the *process length*) $p = lcm(n,g) = \frac{n \cdot g}{gcd(n,g)} = \frac{n \cdot g}{gcd(c+g,g)} = \frac{n \cdot g}{gcd(c,g)} = \frac{n \cdot g}{\Delta}$. As there are c such disjoint cycles of transition the total number of transitions is at most $\frac{g}{\Delta} \cdot n \cdot c$. In the calculation the equations $lcm(a,b) = \frac{|a \cdot b|}{gcd(a,b)}$ and $gcd(b+a,a) = gcd(a,b)$, holding for all integers a and b are used. As for each transition there are exactly two places it is sufficient to prove that all transitions are different[2]. The proof

---

[2] This also follows from the Chinese remainder theorem as discussed in Sect. 5.

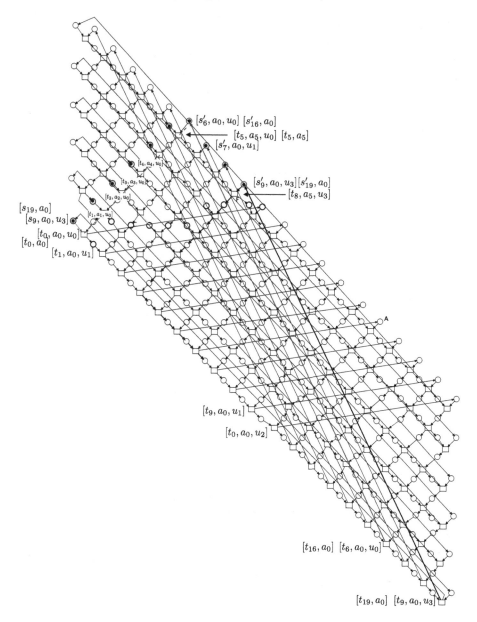

Fig. 4. The net $\mathcal{N}_T(c, g)$ for $c = 6, g = 4$. Pairs like $[s_{19}, a_0]$ are regular coordinates.

becomes easier to read if we restrict it to $j = 0$, while the other cases follow by symmetry. Since the second component of each element of T_0 is a_0 we consider the sequence of the pairs composed of the first and third component $\Omega = (t_0, u_{h_0}), (t_1, u_{h_1}), (t_2, u_{h_2}), \cdots$ of length $p = \frac{g}{\Delta} \cdot n$. The first components form a sequence of classes modulo n: $0, 1, \cdots, n-1, 0, 1, \cdots, n-1, \cdots$ repeated $\frac{g}{\Delta}$

times. The distance of a swap of a_0 with some u_h to the next time of such a swap is g, since there are $g-1$ swaps to pass in between. Hence for the second components we have a sequence of classes modulo g: $0, 1, \cdots, g-1, 0, 1, \cdots, g-1, \cdots$ repeated $\frac{n}{\Delta}$ since g divides p. Next we prove by contradiction that all pairs in the sequence Ω are different: suppose that the same element (x, y) is at positions $r \in \mathbb{Z}$ and $s \in \mathbb{Z}$. Their distance is smaller than the length of the whole sequence: $r - s = q \cdot n$ with $q < \frac{g}{\Delta}$. Furthermore, since the elements are equal, we have $s - r = q \cdot n = q' \cdot g$ for some integers q and q'. We conclude that $q' = \frac{1}{g} \cdot q \cdot n = \frac{q}{g} \cdot (c + g) = \frac{q \cdot c}{g} + q \in \mathbb{Z}$. With $\Delta = gcd(g, c)$ we obtain $\frac{q \cdot c}{g} = \frac{q \cdot c_1 \cdot \Delta}{g_1 \cdot \Delta} = \frac{q \cdot c_1}{g_1} \in \mathbb{Z}$ with $gcd(g_1, c_1) = 1$. Therefore g_1 has to be a divisor of q, which is a contradiction to $q < \frac{g}{\Delta} = g_1$.   □

An example of the construction for the case $g = 4, c = 6$ is given in Fig. 4. The process length is $p = \frac{g}{\Delta} \cdot n = \frac{4}{2} \cdot 10 = 20$. For the transition $[t_9, a_0, u_3]$ we obtain $[t_9, a_0, u_3]^\bullet = \{[s_9, a_0, u_3], [s_9', a_0, u_3]\}$. Furthermore $[s_9', a_0, u_3]^\bullet = \{[t_8, a_5, u_3]\}$. Pairs like $[s_{19}, a_0]$ are for later references. The reader might wonder a little why not smaller parameters are chosen for this example. The reason is that, with respect to later considerations in this paper, the parameters should have the property $gcd(g, c) > 1$ but one of them should not be a divisor of the other.

4   Cycloids

In this section cycloids are defined. Some results are cited from [5,6] and [7], whereas Theorem 18 is new.

Definition 8. *A Petri space is defined by the net* $\mathcal{PS}_1 := (S_1, T_1, F_1)$ *where* $S_1 = S_1^\rightarrow \cup S_1^\leftarrow$, $S_1^\rightarrow = \{s_{\xi,\eta}^\rightarrow \mid \xi, \eta \in \mathbb{Z}\}$, $S_1^\leftarrow = \{s_{\xi,\eta}^\leftarrow \mid \xi, \eta \in \mathbb{Z}\}$, $S_1^\rightarrow \cap S_1^\leftarrow = \emptyset$, $T_1 = \{t_{\xi,\eta} \mid \xi, \eta \in \mathbb{Z}\}$, $F_1 = \{(t_{\xi,\eta}, s_{\xi,\eta}^\rightarrow) \mid \xi, \eta \in \mathbb{Z}\} \cup \{(s_{\xi,\eta}^\rightarrow, t_{\xi+1,\eta}) \mid \xi, \eta \in \mathbb{Z}\} \cup \{(t_{\xi,\eta}, s_{\xi,\eta}^\leftarrow) \mid \xi, \eta \in \mathbb{Z}\} \cup \{(s_{\xi,\eta}^\leftarrow, t_{\xi,\eta+1}) \mid \xi, \eta \in \mathbb{Z}\}$ *(cutout in Fig. 5 a).* $S_1^\rightarrow$ *is the set of* forward places *and* $S_1^\leftarrow$ *the set of* backward places. $^\bullet t_{\xi,\eta} := s_{\xi-1,\eta}^\rightarrow$ *is the forward input place of* $t_{\xi,\eta}$ *and in the same way* $^\bullet t_{\xi,\eta} := s_{\xi,\eta-1}^\leftarrow$, $t_{\xi,\eta}^\bullet := s_{\xi,\eta}^\rightarrow$ *and* $t_{\xi,\eta}^\bullet := s_{\xi,\eta}^\leftarrow$ *(see Fig. 5 a).*

By a twofold folding with respect to time and space we obtain the cyclic structure of a cycloid. See [5,6] for motivation and Fig. 6 a) for an example of a cycloid.

Definition 9. *([5,6]).   A cycloid is a net* $\mathcal{C}(\alpha, \beta, \gamma, \delta) = (S, T, F)$, *defined by parameters* $\alpha, \beta, \gamma, \delta \in \mathbb{N}_+$, *by a quotient of the Petri space* $\mathcal{PS}_1 := (S_1, T_1, F_1)$ *with respect to the equivalence relation* $\equiv \subseteq X_1 \times X_1$ *with* $X_1 = S_1 \cup T_1$, $\equiv [S_1^\rightarrow] \subseteq S_1^\rightarrow$, $\equiv [S_1^\leftarrow] \subseteq S_1^\leftarrow$, $\equiv [T_1] \subseteq T_1$, $x_{\xi,\eta} \equiv x_{\xi+m\alpha+n\gamma, \eta-m\beta+n\delta}$ *for all* $\xi, \eta, m, n \in \mathbb{Z}$, $X = X_1/_\equiv$, $[x]_\equiv F [y]_\equiv \Leftrightarrow \exists x' \in [x]_\equiv \exists y' \in [y]_\equiv : x' F_1 y'$ *for all* $x, y \in X_1$. *The matrix* $\mathbf{A} = \begin{pmatrix} \alpha & \gamma \\ -\beta & \delta \end{pmatrix}$ *is called the matrix of the cycloid. Petri denoted the number* $|T|$ *of transitions as the area A of the cycloid and proved in [4] its value to* $|T| = A = \alpha\delta + \beta\gamma$ *which equals the determinant* $A = det(\mathbf{A})$. *Cycloids are*

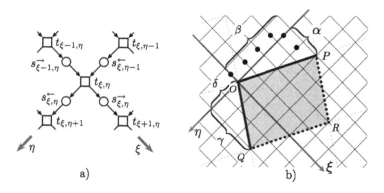

a)   b)

Fig. 5. a) Petri space, b) Fundamental parallelogram of $\mathcal{C}(\alpha, \beta, \gamma, \delta) = \mathcal{C}(2, 4, 3, 2)$ with regular initial marking.

safe T-nets with $|{}^\bullet s| = |s^\bullet| = 1$ for all places $s \in S$. The embedding of a cycloid in the Petri space is called **fundamental parallelogram** (see Fig. 5 b), but ignore the tokens for the moment). If the cycloid is represented as a net $\mathcal{N}$ without explicitly giving the parameters $\alpha, \beta, \gamma, \delta$, we call it a cycloid in net form $C(\mathcal{N})$.

For proving the equivalence of two points in the Petri space the following procedure is useful.

Theorem 10. ([7]). *Two points* $x_1, x_2 \in X_1$ *are equivalent* $x_1 \equiv x_2$ *if and only if* $\pi(v) = \pi(x_2 - x_1)$ *has integer values, where* $\pi(v) = \frac{1}{A} \cdot \mathbf{B} \cdot v$ *with area A and*
$\mathbf{B} = \begin{pmatrix} \delta & -\gamma \\ \beta & \alpha \end{pmatrix}$. *With* m, n *from Definition 9 we obtain* $v = \mathbf{A} \begin{pmatrix} m \\ n \end{pmatrix}$.

Definition 11. *For a cycloid* $\mathcal{C}(\alpha, \beta, \gamma, \delta)$ *we define a cycloid-system* $\mathcal{C}(\alpha, \beta, \gamma, \delta, M_0)$ *or* $C(\mathcal{N}, M_0)$ *by adding the standard initial marking:*
$$M_0 = \{s_{\xi,\eta}^{\rightarrow} \in S_1^{\rightarrow} \mid \beta\xi + \alpha\eta \leq 0 \wedge \beta(\xi+1) + \alpha\eta > 0\}/_\equiv \cup$$
$$\{s_{\xi,\eta}^{\leftarrow} \in S_1^{\leftarrow} \mid \beta\xi + \alpha\eta \leq 0 \wedge \beta\xi + \alpha(\eta+1) > 0\}/_\equiv$$

The motivation of this definition is given in [5] and [6]. See Fig. 6 a) for an example of a cycloid with standard initial marking. We define a regular initial marking for cycloids, but not necessarily within the fundamental parallelogram (the black tokens in Fig. 5 b). It is characterized by the absence of gaps between the traffic items whereby only a single transition on the top of the queue is enabled. behavioural

Definition 12. ([7]). *For a cycloid* $\mathcal{C}(\alpha, \beta, \gamma, \delta)$ *a regular initial marking is defined by a number of* β *forward places* $\{s_{-1,i}^{\rightarrow} \mid 0 \geq i \geq 1 - \beta\}$ *and a number of* α *backward places* $\{s_{i,-\beta}^{\leftarrow} \mid 0 \leq i \leq \alpha - 1\}$.

By the construction of a behavioural equivalent cycloid from a circular traffic queue $tq(c, g)$ in [7] the regular and standard initial marking are preserved. Circular traffic queues are composed of c many sequential and interacting processes

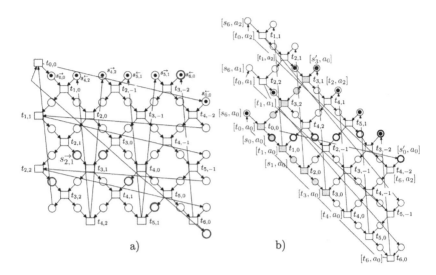

Fig. 6. Cycloid $\mathcal{C}(4,3,3,3)$ in a) and with regular coordinates in b).

of equal length. In the formalism of cycloids this corresponds to a number of β disjoint processes of equal length p. Cycloids with such a property are called regular.

Definition 13. ([7]) *A cycloid $\mathcal{C}(\alpha,\beta,\gamma,\delta)$ with area A is called* regular *if for each $\eta \in \{0, \cdots, 1-\beta\}$ the set $\{t_{\xi,\eta} | 0 \le \xi < p\}$ with $p = \frac{A}{\beta}$ of transitions forms an elementary cycle[3] and all these sets are disjoint. $p \in \mathbb{N}_+$ is called the process length of the regular cycloid. A regular cycloid together with its regular initial marking M_0 is called a* regular cycloid system $\mathcal{C}(\alpha,\beta,\gamma,\delta,M_0)$.

Similar to the processes of cars in Definition 13, processes for co-cars can be defined. They start with the output transitions of tokens in backward places of the regular initial marking and follow through backward output places until a cycle is closed. A cycloid is called *co-regular* if there are a number of α many such disjoint cycles of length $p' = \frac{A}{\alpha}$.

Theorem 14. ([7]). *A cycloid $\mathcal{C}(\alpha,\beta,\gamma,\delta)$ is regular if and only if $\beta|\delta$.*

Similar to the proof of Theorem 14 in [7], it can be shown that a cycloid $\mathcal{C}(\alpha,\beta,\gamma,\delta)$ is co-regular if and only if $\alpha|\gamma$.   A regular cycloid can be seen as a system of β disjoint sequential and cooperating processes. To exploit this structure we define specific coordinates, called *regular coordinates*. The process of a traffic item a_0 starts with transition $t_{0,0}$ which is denoted $[t_0, a_0]$, having the input place $[s_{p-1}, a_0]$. The next transitions are $[t_1, a_0]$ up to $[t_{p-1}, a_0]$ and then returning to $[t_0, a_0]$. The other processes for a_1 to a_{c-1} (with $\beta = c$) are denoted in the same way (see Fig. 6 b). As the process of a_j starts in position j of the queue, its initial token is in $[s_{j\ominus_p 1}, a_j]$.

---

[3] An elementary cycle is a cycle where all nodes are different.

Definition 15. ([7]). *Given a regular cycloid $\mathcal{C}(\alpha, \beta, \gamma, \delta)$, regular coordinates are defined as follows: transitions of process $j \in \{1, \cdots, \beta\}$, each with length p, are denoted by $\{[t_0, a_j], \cdots, [t_{p-1}, a_j]\}$. For each transition we define $[t_i, a_j]^{\overset{\bullet}{\rightarrow}} := [s_i, a_j]$ and $[t_i, a_j]^{\overset{\bullet}{\leftarrow}} := [s_i', a_j]$ and $[s_i, a_j]^\bullet := [t_{i\oplus_p 1}, a_j]$ for $0 \le i < p, 0 \le j < c$. Regular coordinates are related to standard coordinates of the Petri space by defining the following initial condition $[t_0, a_j] := t_{-j,-j}$ for $0 \le j < c$ (taking the equivalent transition of $t_{-j,-j}$ in the fundamental parallelogram).*

For instance, in Fig. 6 b) we obtain for the last formula in Definition 15: $[t_0, a_2] := t_{-2,-2} \equiv t_{1,1}$. While the output place $[s_i', a_j]$ in regular coordinates takes its name from the input transition, it remains to determine its output transition according to the corresponding standard coordinates.

Lemma 16. ([7]). *In a regular cycloid the injective mapping* stand *from regular to standard coordinates is given by $stand([t_i, a_j]) = t_{i-j,-j}$ for $0 \le i < p$ and $0 \le j < c$ (modulo equivalent transitions). The output transition is $[s_i', a_0]^\bullet = [t_{(i+\beta+\alpha-1) \bmod p}, a_{\beta-1}]$ (Case 1), while for $0 < j < c$ we have $[s_i', a_j]^\bullet = [t_{i\ominus_p 1}, a_{j\ominus_\beta 1}]$ (Case 2).*

Corollary 17. ([7]). *The regular initial marking of a regular cycloid system $\mathcal{C}(\alpha, \beta, \gamma, \delta, M_0)$ with process length p in regular coordinates is*
$$M_0 = \{[s_{p-1}, a_0]\} \cup \{[s_i, a_{i+1}] | 0 \le i < \beta - 1\} \cup \{[s_i', a_0] | p - \alpha \le i < p\}.$$
For later reference, we note $^{\overset{\bullet}{\leftarrow}}[t_{\beta-1}, a_{\beta-1}] = [s_{p-\alpha}', a_0]$.

In the regular cycloid system $\mathcal{C}(4, 3, 3, 3, M_0)$ in Fig. 6 b) we obtain $^{\overset{\bullet}{\leftarrow}}[t_2, a_2] = [s_{p-\alpha}', a_0] = [s_{7-4}', a_0]$. The given regular initial marking is $\{[s_6, a_0], [s_0, a_1], [s_1, a_2], [s_3', a_0], [s_4', a_0], [s_5', a_0], [s_6', a_0]\}$. The standard initial marking is given by bold circles. When working with cycloids it is sometimes important to find for a transition outside the fundamental parallelogram the equivalent element inside. For instance the first set in Definition 12 of a regular initial marking contains the element $s_{-1,1-\beta}^{\rightarrow}$. It is named by its input transition $t_{-1,1-\beta}$, which is outside the fundamental parallelogram. To obtain the corresponding place of the cycloid net the equivalent transition of $t_{-1,1-\beta}$ inside the fundamental parallelogram has be computed. In general, by enumerating all elements of the fundamental diagram (using Theorem 7 in [5]) and applying the equivalence test from Theorem 10 a runtime is obtained, which already fails for small cycloids. The following theorem allows for a better algorithm, which is linear with respect to the cycloid parameters.

Theorem 18. *For any element $\boldsymbol{u} = (u, v)$ of the Petri space the (unique) equivalent element of the fundamental parallelogram is $\boldsymbol{x} = \boldsymbol{u} - \mathbf{A}\begin{pmatrix} m \\ n \end{pmatrix}$ where $m = \lfloor \frac{1}{A}(u\delta - v\gamma) \rfloor$ and $n = \lfloor \frac{1}{A}(v\alpha + u\beta) \rfloor$.*

Proof. As *m-sector* we denote the band between the lines $\overline{OQ}$ and $\overline{PR}$ including the points of $\overline{OQ}$, but not those of $\overline{PR}$ (see Fig. 7). The idea of the proof is to reach the *m-sector* on a line parallel to $\overline{QR}$ starting in u until a point c is reached[4]. The point c is the intersection of this line and the line through x parallel to $\overline{OQ}$. Then m, n are the integer multiples of the vectors $\begin{pmatrix} -\alpha \\ \beta \end{pmatrix}$, $\begin{pmatrix} \gamma \\ \delta \end{pmatrix}$, resulting in the line segments u to c and c to x, respectively.

More formally we derive: $\boldsymbol{u} \equiv \boldsymbol{x} \Leftrightarrow \boldsymbol{u} - \boldsymbol{x} = \mathbf{A} \begin{pmatrix} m \\ n \end{pmatrix}$ (by Theorem 10). Hence

we obtain: $\boldsymbol{x} = \boldsymbol{u} - \mathbf{A} \begin{pmatrix} m \\ n \end{pmatrix} = \boldsymbol{u} - \begin{pmatrix} \alpha & \gamma \\ -\beta & \delta \end{pmatrix} \begin{pmatrix} m \\ n \end{pmatrix} = \boldsymbol{u} - \begin{pmatrix} m \cdot \alpha + n \cdot \gamma \\ -m \cdot \beta + n \cdot \delta \end{pmatrix} =$

$\boldsymbol{u} + m \begin{pmatrix} -\alpha \\ \beta \end{pmatrix} - n \begin{pmatrix} \gamma \\ \delta \end{pmatrix}$ or

$$\boldsymbol{x} + n \begin{pmatrix} \gamma \\ \delta \end{pmatrix} = \boldsymbol{u} + m \begin{pmatrix} -\alpha \\ \beta \end{pmatrix}. \tag{1}$$

This vector equation results in two linear equations for four unknown n, m and $\boldsymbol{x} = (x, y)$. With $\lambda \in \mathbb{R}$ and $\boldsymbol{u} + \lambda \begin{pmatrix} -\alpha \\ \beta \end{pmatrix}$ the right hand side of equation (1) defines the line through u parallel to the line $\overline{QR}$. It is intersecting the line $\overline{OQ}$ in point d and the line $\overline{PR}$ in point b. In the same way equation (1) defines with $\mu \in \mathbb{R}$ and $\boldsymbol{x} + \mu \begin{pmatrix} \gamma \\ \delta \end{pmatrix}$ the line through x parallel to the line containing O und Q. Both lines intersect in d, which is defined by the equation

$$\begin{pmatrix} 0 \\ 0 \end{pmatrix} + \mu \begin{pmatrix} \gamma \\ \delta \end{pmatrix} = \begin{pmatrix} u \\ v \end{pmatrix} + \lambda \begin{pmatrix} -\alpha \\ \beta \end{pmatrix} \tag{2}$$

This vector equation provides two linear equations with the solutions $\lambda = \frac{1}{A}(u\delta - v\gamma)$ und $\mu = \frac{1}{A}(v\alpha + u\beta)$. If λ is not negative, then u is on the half-line starting with (and including) the point d in the direction of c and b, hence $\boldsymbol{d} = \boldsymbol{u} + \lambda \begin{pmatrix} -\alpha \\ \beta \end{pmatrix}$ and $\boldsymbol{b} = \boldsymbol{d} - \begin{pmatrix} -\alpha \\ \beta \end{pmatrix} = \boldsymbol{u} + (\lambda - 1) \begin{pmatrix} -\alpha \\ \beta \end{pmatrix}$. Since c is between d and b on the same line we obtain for λ_1 satisfying $\boldsymbol{c} = \boldsymbol{u} + \lambda_1 \begin{pmatrix} -\alpha \\ \beta \end{pmatrix}$ the inequality $\lambda \geq \lambda_1 > \lambda - 1$ and since $\lambda_1 \in \mathbb{Z}$ the result $m = \lfloor \lambda \rfloor$. If λ is negative, then u, now denoted as u', is on the complementary half-line until (but not containing) d and not containing b. The distance between u' and c now is **greater** (or equal) than the distance between u' and d. Instead of the inequality $\lambda \geq \lambda_1 > \lambda - 1$ in the case before, we now have $\lambda \leq \lambda_1 < \lambda - 1$ and the number $\lambda_1 \in \mathbb{Z}$ is again obtained by $m = \lambda_1 = \lfloor \lambda \rfloor$. The formula for $n = \lfloor \mu \rfloor$ is derived analogously: it remains to determine the vector $\overrightarrow{cx} := \boldsymbol{x} - \boldsymbol{c}$. To this end we introduce $\boldsymbol{a} := \boldsymbol{d} + \overrightarrow{cx}$. As in the case where we concluded $m = \lambda_1 = \lfloor \lambda \rfloor$ from λ we compare $\begin{pmatrix} 0 \\ 0 \end{pmatrix} = \boldsymbol{d} - \mu \begin{pmatrix} \gamma \\ \delta \end{pmatrix}$

with $\boldsymbol{a} = \boldsymbol{d} - \mu_1 \begin{pmatrix} \gamma \\ \delta \end{pmatrix}$ to obtain $n = \mu_1 = \lfloor \mu \rfloor$ and $\overrightarrow{cx} = \boldsymbol{a} - \boldsymbol{d} = \boldsymbol{d} - \mu_1 \begin{pmatrix} \gamma \\ \delta \end{pmatrix} - \boldsymbol{d} =$

---

[4] $u, a, b, c, \cdots$ denote the points in the Petri space, while the vectors $\boldsymbol{u}, \boldsymbol{a}, \boldsymbol{b}, \boldsymbol{c}, \cdots$ are pointing to them when originated in the origin $(0, 0)$, respectively.

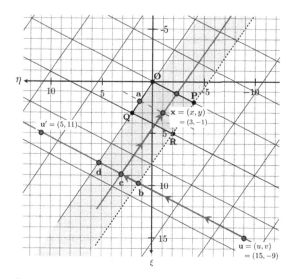

Fig. 7. Cycloid $\mathcal{C}(2,4,3,2)$ in the Petri space to illustrate the proof of Theorem 18.

$-\mu_1 \begin{pmatrix} \gamma \\ \delta \end{pmatrix}$. Combining this result with the vector $\boldsymbol{c}$ we obtain for the wanted point

$$\boldsymbol{x} = \boldsymbol{c} + \overrightarrow{c\,x} = \boldsymbol{u} + \lfloor\lambda\rfloor \begin{pmatrix} -\alpha \\ \beta \end{pmatrix} + \lfloor\mu\rfloor \begin{pmatrix} -\gamma \\ -\delta \end{pmatrix} = \boldsymbol{u} + m \begin{pmatrix} -\alpha \\ \beta \end{pmatrix} + n \begin{pmatrix} -\gamma \\ -\delta \end{pmatrix} = \boldsymbol{u} - \mathbf{A} \begin{pmatrix} m \\ n \end{pmatrix}$$

the formula just before equation (1). □

An example for the preceding proof is shown in Fig. 7 for the cycloid $\mathcal{C}(2,4,3,2)$, which is the same as in Fig. 5 b): for $\boldsymbol{u} = (u,v) = (15,-9)$ we obtain $(\lambda,\mu) = (\frac{57}{16}, \frac{21}{8})$, $(m,n) = (3,2)$, $\boldsymbol{c} = (9,3)$, $(x,y) = (3,-1)$. For the point $(u',v') = (5,11)$ on the same line $\overline{(d,b)}$ we obtain $(\lambda,\mu) = (-\frac{23}{16}, \frac{21}{8})$, $(m,n) = (-2,2)$, $\boldsymbol{c} = (9,3)$, $(x,y) = (3,-1)$. Next we apply the theorem for $\mathcal{C}(4,3,3,3)$ (Fig. 6) to the element $t_{-1,1-\beta}$, which was mentioned just before the theorem. In this case we obtain: $(u,v) = (-1,-2)$, $m = \lfloor\frac{3}{21}\rfloor = 0$, $n = \lfloor-\frac{11}{21}\rfloor = -1$ and $\boldsymbol{x} = (2,1)$. The corresponding place $s_{2,1}^{\rightarrow}$ is highlighted in Fig. 6 a) as an element of the regular initial marking.

5   The Net Isomorphism

In this section it will be proved that the T-net-equivalent $\mathcal{N}_T(c,g)$ of the coloured net $\mathcal{N}_{sym}(c,g)$ is isomorphic to the cycloid-system $(\mathcal{C}(g,c), M_3^0)$. This is of particular interest as the nets arise from different contexts.

Theorem 19. *The T-net-equivalent $\mathcal{N}_T(c,g) = (S_2, T_2, F_2, M_2^0)$ (Definition 6) of $\mathcal{N}_{sym}(c,g)$ is isomorphic to the cycloid-system $(\mathcal{C}(g,c), M_3^0) := \mathcal{C}(g,c, \frac{g\cdot c}{\Delta}, \frac{g\cdot c}{\Delta}, M_3^0)$ with regular initial marking M_3^0.*

Proof. The sequences of the first and second components of the sequence Ω in the proof of Lemma 7 are instances of the Chinese remainder theorem. In this theorem a system of two (or more) congruences are considered which are in our case:

$$x = a \bmod n \qquad\qquad x = b \bmod g \qquad\qquad (3)$$

Different to the classical case n and g are not coprime in our cases. In a generalized version of the theorem however [3], page 60, there is a unique solution modulo $\frac{g \cdot n}{d}$ if and only if $d|(b-a)$ or, equivalently $a = b \bmod d$ where $d = gcd(n, g) = gcd(g + c, g) = \Delta$. This condition holds in our case since we will take a and b from a common position x of Ω, starting the count with 0. For some integers i, h, n', g' from this follows $x = i \cdot n + a = i \cdot (n' \cdot \Delta) + a = (i \cdot n') \cdot \Delta + a$ and $x = h \cdot g + b = h \cdot (g' \cdot \Delta) + b = (h \cdot g') \cdot \Delta + b$ with $gcd(n', g') = 1$. Therefore we obtain $(i \cdot n') \cdot \Delta + a = (h \cdot g') \cdot \Delta + b$ and the wanted condition $a = b \bmod \Delta$. Now consider $x = \frac{1}{\Delta} \cdot (a \cdot v \cdot g + b \cdot u \cdot n)$, where u, v solve Bézout's identity $u \cdot n + v \cdot g = \Delta$, [3], page 7. If by this identity the term $v \cdot g = \Delta - u \cdot n$ is introduced into the term of x we obtain $x = (\frac{b-a}{\Delta} \cdot u) \cdot n + a$, which proves that it solves the lefthand part of (3). In a similar way the same is proved for the right hand part of (3). Also, by [3], page 60, the general solution is a single congruence class modulo $lcm(g, n) = \frac{g \cdot n}{\Delta}$. We will use $\chi(a, b, n, g) := (\frac{1}{\Delta} \cdot (a \cdot v \cdot g + b \cdot u \cdot n)) \bmod p$ with $p = \frac{g \cdot n}{\Delta}$ and the condition $u \cdot n + v \cdot g = \Delta$ to denote the unique solution of (3).

In the following we assume n and g to be given and define $\chi_0(a, b) := \chi(a, b, n, g)$. As will be seen in the definition of the isomorphism, $\chi_0(a, b)$ is defined with respect to the process of a_0. The corresponding function for a_j is $\chi_j(a, b) := \chi_0(a, b + j)^5$ for $0 \le j < c$. Before defining and proving the isomorphism we derive some equations of the function χ_0:

$$\chi_0(a, a) = a \quad \text{for} \quad 0 \le a < p \qquad\qquad (4)$$

$$\chi_0(a + k, b + k) = \chi_0(a, b) \oplus_p k \quad \text{for} \quad k \in \mathbb{Z} \qquad\qquad (5)$$

$$\chi_0(n - k, g - k) = p - k \quad \text{for} \quad k \in \mathbb{N}, \ 0 \le k < p \qquad\qquad (6)$$

$$\chi_0(a, b + c) = \chi_0(a, b) \oplus_p n \quad (c \text{ is the constant from } \mathcal{N}_T(c, g)) \qquad\qquad (7)$$

To prove (4) consider $\chi_0(a, a) = \frac{1}{\Delta} \cdot (a \cdot v \cdot g + a \cdot u \cdot n) \bmod p = \frac{1}{\Delta} \cdot (a \cdot \Delta) \bmod p = a$ since $u \cdot n + v \cdot g = \Delta$. To prove (5) consider $\chi_0(a + k, b + k) \bmod p = \frac{1}{\Delta} \cdot ((a + k) \cdot v \cdot g + (b + k) \cdot u \cdot n) \bmod p = (\frac{1}{\Delta} \cdot (a \cdot v \cdot g + b \cdot u \cdot n) + \frac{1}{\Delta} (v \cdot g + u \cdot n) \cdot k) \bmod p = (\chi_0(a, b) + k) \bmod p$. To prove (6) consider $\chi_0(n - k, b - k) = (\frac{1}{\Delta} \cdot ((n - k) \cdot v \cdot g + (g - k) \cdot u \cdot n) \bmod p = (\frac{1}{\Delta} \cdot g \cdot n \cdot (v + u) - k) \bmod p = (p \cdot (v + u) - k) \bmod p = (0 + (-k)) \bmod p = (0 \bmod p + (-k) \bmod p) \bmod p = (0 + p - k) \bmod p = p - k$. To prove (7) we calculate $\chi_0(a, b + c) - \chi_0(a, b) = \frac{1}{\Delta} \cdot c \cdot u \cdot n$. In this difference u is unknown, but the value does not depend on the arguments a and b. We compute it using the particular values $a = n$ and $b = g$ using Eqs. (4) and (6): $\chi_0(n, g + c) - \chi_0(n, g) = \chi_0(n, n) - \chi_0(n, g) = n - p$. Finally: $(n - p) \bmod p = n$.

For the cycloid-system $(\mathcal{C}(g, c), M_0) := \mathcal{C}(g, c, \frac{g \cdot c}{\Delta}, \frac{g \cdot c}{\Delta}, M_0)$ we consider the regular coordinates from Definition 15 and denote its net elements as

---

^{5} $\chi_0(a, b \oplus_g j)$ would be insufficient in case of $c > g$.

$(\mathcal{C}(g,c), M_0) = (S_3, T_3, F_3, M_3^0)$. To prove that the net systems $\mathcal{N}_T(c,g)$ and $(\mathcal{C}(g,c), M_0)$ are isomorphic we have to define bijections $\psi_1 : S_2 \to S_3$ and $\psi_2 : T_2 \to T_3$ such that $(s,t) \in F_2 \Leftrightarrow (\psi_1(s), \psi_2(t)) \in F_3$ and $(t,s) \in F_2 \Leftrightarrow (\psi_2(t), \psi_1(s)) \in F_3$. Furthermore the initial markings must be consistent: $\{\psi_1(m)|m \in M_2^0\} = M_3^0$. Using the functions χ_j the bijections are defined by $\psi_1([s_i, a_j, u_h]) := [s_{\chi_j(i,h)}, a_j]$, $\psi_1([s_i', a_j, u_h]) := [s_{\chi_j(i,h)}', a_j]$ and $\psi_2([t_i, a_j, u_h]) := [t_{\chi_j(i,h)}, a_j]$. They are bijective since for a given a_j the χ_j are bijective on their domain. The required properties of ψ_1 and ψ_2 are verified for the parts of $F_2 := F_2^1 \cup F_2^2 \cup F_2^3 \cup F_2^4$. Since the cases for F_2^2 and F_2^4 are obvious we concentrate on the cases for F_2^1 and F_2^3.

ad F_2^1: $(\psi_1([s_{i\ominus_n 1}, a_j, u_{h\ominus_g 1}]), \psi_2([t_i, a_j, u_h])) =$

$$([s_{\chi_j(i\ominus_n 1, h\ominus_g 1)}, a_j], [t_{\chi_j(i,h)}, a_j]) \tag{8}$$

Assume first $i \neq 0, h \neq g$, hence $i \ominus_n 1 = i - 1, h \ominus_g 1 = h - 1$. Then we continue with (8) $= ([s_{\chi_j(i-1,h-1)}, a_j], [t_{\chi_j(i,h)}, a_j]) \underset{(5)}{=} ([s_{\chi_j(i,h)\ominus_p 1}, a_j], [t_{\chi_j(i,h)}, a_j]) \in F_3$ by equation (5) and the case $[s_i, a_j]^\bullet := [t_{i\oplus_p 1}, a_j]$ of Definition 15.

If $i = 0, h = g$, hence $i \ominus_n 1 = n-1, h \ominus_g 1 = g-1$, then (8) $= ([s_{\chi_0(n-1,g-1+j)}, a_j], [t_{\chi_0(0,g+j)}, a_j]) \underset{(5)}{=} [s_{\chi_0(n,g+j)\ominus_p 1}, a_j], [t_{\chi_0(0,g+j)}, a_j]) \in F_3$. This holds since

$\chi_0(n,b) = \chi_0(0,b)$ for all $b \in \mathbb{N}$. The latter property of χ_0 is proved by $\chi_0(n,b) - \chi_0(0,b) = (\frac{1}{\Delta} \cdot n \cdot v \cdot g) \bmod \frac{g \cdot n}{\Delta} = 0$. The remainig cases $i = 0, h \neq g$ and $i \neq 0, h = g$ are similarly proved.

ad F_2^3: $(\psi_1([s_{i\oplus_n 1}', a_{j\ominus_c 1}, u_h]), \psi_2([t_i, a_j, u_h])) =$

$$([s_{\chi_{j\ominus_c 1}(i\oplus_n 1, h)}', a_{j\ominus_c 1}], [t_{\chi_j(i,h)}, a_j]) \tag{9}$$

In Case 1 of Lemma 16 we have $j = c - 1$ and $j \oplus_c 1 = 0$. For easier computation we continue with an index shift $i \oplus_n 1$ to i and i to $i \ominus_n 1$. In a first subcase we assume $i \neq 0$, hence $i \ominus_n 1 = i - 1$ and continue the calculation from (9) by: $([s_{\chi_0(i,h)}', a_0], [t_{\chi_{c-1}(i-1,h)}, a_{c-1}])$. The second component of this pair is evaluated to $[t_{\chi_0(i-1,h+c-1)}, a_{c-1}] \underset{(5)}{=} [t_{\chi_0(i,h+c)-1}, a_{c-1}] \underset{(7)}{=} [t_{\chi_0(i,h)\oplus_p(n-1)}, a_{c-1}]$. Together with the first component this result is matching the case $[s_i', a_0]^\bullet = [t_{(i+\beta+\alpha-1) \bmod p}, a_{\beta-1}] = [t_{(i\oplus_p(n-1))}, a_{c-1}]$ in Lemma 16. In the subcase $i = 0$ we evaluate the second component to $[t_{\chi_0(0\ominus_n 1, h+c-1)}, a_{c-1}] = [t_{\chi_0(n-1,h+c-1)}, a_{c-1}] \underset{(5)}{=} [t_{\chi_0(n,h+c)-1}, a_{c-1}] \underset{(7)}{=} [t_{\chi_0(n,h)\oplus_p(n-1)}, a_{c-1}]$. Since $\chi_0(n,b) = \chi_0(0,b)$ for all $b \in \mathbb{N}$, as proved before, this reduces to the first subcase. In case 2 of Lemma 16 we have $0 < j < c$. For the subcase $i \neq n - 1, j \neq c - 1$, resulting in $i \oplus_n 1 = i+1, j \oplus_c 1 = j+1$, we obtain (9) $= ([s_{\chi_0(i+1,h+j+1)}', a_{j+1}], [t_{\chi_0(i,h+j)}, a_j]) \underset{(5)}{=} ([s_{\chi_0(i,h+j)+1}', a_{j+1}], [t_{\chi_0(i,h+j)}, a_j]) \in F_3$ matching Case 2 of Lemma 16. The remaining three subcases are similar to prove: in the cases where $j = c - 1$ the proof is similar to the preceding proof for $j \oplus_c 1 = 0$, while in the remaining case $i = n - 1, j \neq c - 1$ it is similar to the first subcase.

In order to make the proof for the initial markings easier we use the initial marking of $\mathcal{N}_T(c,g)$ (Definition 6) in the form $M_0^2 = A \cup B \cup C$

where $A = \{[s_{n-1}, a_0, u_{g-1}]\}$, $B = \{[s_i, a_{i \oplus_c 1}, u_{g-1}] | 0 \le i < c-1\}$ and $C = \{[s'_i, a_0, u_{i-c}] | c \le i < c+g\}$. With respect to the regular initial marking of $(\mathcal{C}(g, c), M_3^0)$ (in regular coordinates: Corollary 17) with $\alpha = g$ and $\beta = c$ we have to prove: $\psi_1(A) = \{[s_{p-1}, a_0]\} =: U$, $\psi_1(B) = \{[s_i, a_{i \oplus_c 1}] | 0 \le i < c-1\} =: V$ and $\psi_1(C) = \{[s'_i, a_0] | p - g \le i < p\} =: W$. For the first part we obtain $\psi_1(A) = \{[s_{\chi_0(n-1, g-1)}, a_0]\} \underset{(6)}{=} \{[s_{p-1}, a_0]\} = U$. With respect to $\psi_1(B)$, since $i \ne c-1$ (hence $i \oplus_c 1 = i+1$) we have $\psi_1([s_i, a_{i \oplus_c 1}, u_{g-1}]) = [s_{\chi_{i \oplus_c 1}(i, g-1)}, a_{i \oplus_c 1}] = [s_{\chi_0(i, (g-1)) + (i+1))}, a_{i \oplus_c 1}] \underset{(5)}{=} [s_{\chi_0(0, g) + i}, a_{i \oplus_c 1}] = [s_i, a_{i \oplus_c 1}]$. Here $\chi_0(0, g) = (\frac{g \cdot n}{\Delta} \cdot u) \, mod(\frac{g \cdot n}{\Delta}) = 0$ is used and $\psi_1(B) = \{[s_i, a_{i \oplus_c 1}] | 0 \le i < c-1\} = V$. $\psi_1(C) = \{[s'_{\chi_0(i, i-c)}, a_0] | c \le i < c+g\} = W$. This holds since the bounds of the inequality match when introduced as indices. For the lower bound $i = c$ we obtain $\chi_0(c, c-c) = \chi_0(c+g-g, g-g) = \chi_0(n-g, g-g) \underset{(6)}{=} p-g$, while the upper bound $i = c+g-1$ gives $\chi_0(c+g-1, c+g-1-c) = \chi_0(n-1, g-1) \underset{(6)}{=} p-1$. $\square$

In the cycloid $\mathcal{C}(4, 6, 12, 12)$ of Fig. 4 we obtain as an example for the isomorphism $\psi_2([t_9, a_0, u_3]) = [t_{19}, a_0]$, $\psi_1([s_9, a_0, u_3]) = [s_{19}, a_0]$ and $\psi_1([s'_9, a_0, u_3]) = [s'_{19}, a_0]$ since $\chi_0(9, 3) = 19$. By $\chi(6, 0) = 16$ the transition $[t_6, a_0, u_0]$ transforms to $[t_{16}, a_0]$ in regular coordinates. In the computation of χ_0 the values $u = 1, v = -2$ are used to fulfil the condition $n \cdot u + g \cdot v = \Delta = 2$. To illustrate the first part of case F_2^3 we compute $(\psi_1([s'_6, a_0, u_0]), \psi_2([t_5, a_5, u_0])) = ([s'_{\chi_0(6, 0)}, a_0], [t_{\chi_5(5, 0)}, a_5]) = ([s'_{16}, a_0], [t_{\chi_0(5, 0+5)}, a_5]) \underset{(4)}{=} ([s'_{16}, a_0], [t_5, a_5]) \in F_3$.

6 The f-factor of Regular Cycloids

A *circular traffic queue with gaps* is obtained from the model given by Definition 1 by replacing the co-cars by indistinguishable items. In [7] it has been proved that for c cars and g gaps this model is behaviour equivalent to the (regular) cycloid $\mathcal{C}(g, c, c, c)$, which is the topic of this section. Definition 1 is modified as follows to define circular traffic queue with gaps.

Definition 20. *A circular traffic queue with gaps tq-g(c, g) is defined as in Definition 1, with the difference that $|G| = 1$ and the index function ind is not bijective in general, but only on the co-image ind$^{-1}(C)$. In addition we require that there is at least one gap: ind$^{-1}(G) \ne \emptyset$. As the number g from Definition 1 is not longer needed, we use it here to define the number of gaps: $g := n - c \ge 1$.*

With $G = \{\times\}$ for tq-g$(3, 4)$ the example given before Definition 1 modifies to $\times a\, b \times \times \times c \rightarrow \times a \times b \times \times c \rightarrow c\, a \times b \times \times \times$. When replacing the co-items $u \in G$ by black tokens and removing the variables y in the coloured net $\mathcal{N}_{sym}(c, g)$ of Fig. 2 we obtain the coloured net $\mathcal{N}_{coul}(c, g)$. In [7] is has been shown that it is behaviour equivalent to the cycloid $\mathcal{C}_1(g, c) := \mathcal{C}(g, c, c, c)$. In Theorem 2 the *co-car anomaly* was proved outside the theory of cycloids. For a given circular traffic queue with gaps tq-g(c, g) the length of a recurrent transition

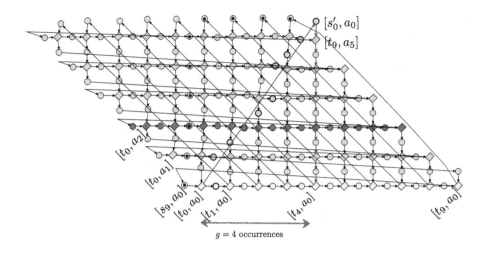

Fig. 8. Cycloid $\mathcal{C}(4, 6, 6, 6)$ with $\Delta > 1$.

sequence is $\Gamma(c, g) := c \cdot (c + g)$. When switching to the case where the gaps are replaced by individual cars this value is not increased in general by the factor g, which is the number of different co-cars. Instead, the factor is $\frac{g}{\Delta}$ where $\Delta = gcd(c, g)$. In the following, we deduce this result by a proof, which is using the results of the cycloid theory. Furthermore, the effect which intuitively describes a form of resonance between the streams of cars and co-cars, becomes more intuitive by the graphical representation of cycloids. By the *f-factor*, as defined in the next definition, the regular cycloid $\mathcal{C}(g, c, c)$ is extended to the cycloid $\mathcal{C}(g, c, \frac{g \cdot c}{\Delta}, \frac{g \cdot c}{\Delta})$, which is regular *and* co-regular (see the remarks after Definition 13 and Theorem 14). In a regular cycloid the run of a co-car is modelled by a so-called *co-process*.

Definition 21. *A* co-process *is an elementary cycle in a regular cycloid, containing only backward places of $S_1^{\leftarrow}$ (Definition 8) starting in a marked place of the regular initial marking. For given integers g, c, f consider the regular cycloid $\mathcal{C}_f := \mathcal{C}(g, c, f \cdot c, f \cdot c)$, where all tokens are individual (modelling cars on processes and co-cars on co-processes). The smallest positive integer f, such that $\mathcal{C}_f$ represents a cycloid, where each transition sequence containing all transitions once, reproduces the initial state of cars as well of co-cars is called f-factor.*

Theorem 22. *Consider the cycloid $\mathcal{C}(g, c, c, c)$ and $n = c + g$.*

a) The f-factor of the cycloid is $f = \frac{g}{\Delta}$.

b) The length p' of a co-process is $p' = \frac{c \cdot n}{\Delta}$.

c) The number of (disjoint) co-processes is $\Delta = gcd(g, c)$.

d) The number of tokens in a co-process is $\frac{g}{\Delta}$.

Proof. a) The f-factor increases the process length $n = g + c$ of $\mathcal{C}(g, c, c, c)$ to the process length $p_1 = \frac{A}{c} = \frac{1}{c} \cdot (g \cdot f \cdot c + c \cdot f \cdot c) = f \cdot n$. Without loss of generality by the symmetry of regular cycloids it is sufficient to consider the process of any car a_i. Let us take the first one a_0 and any of its transitions, say $[t_i, a_0]$ for $0 \leq i < n$. When $[t_i, a_0]$ is occurring a token from its forward input place $\overset{\bullet}{\rightarrow}[t_i, a_0]$ is moved as well as a token from its backward input place $\overset{\bullet}{\leftarrow}[t_i, a_0]$. While the former token is describing the state of the car a_0 we can interpret the latter as giving state of a co-car u_h. We now determine the next transition of the process where a_0 is interacting again with u_h. After the occurrence of $[t_i, a_0]$ the position of u_h is $[t_i, a_0]^{\overset{\bullet}{\leftarrow}} = [s_i', a_0]$. By Lemma 16 the output transition of this place is $[s_i', a_0]^{\bullet} = [t_{(i+g+c-1) \bmod p}, a_{c-1}]$ with $p = n$. With this transition u_h runs through a *release message chain* (see [7]) (see Fig. 8): $[t_{(i+g+c-1) \bmod n}, a_{c-1}], [t_{((i+g+c-1)-1) \bmod n}, a_{c-2}], [t_{((i+g+c-1)-2) \bmod n}, a_{c-3}], \cdots,$ $[t_{((i+g+c-1)-(c-1)) \bmod n}, a_{(c-1)-(c-1)}] = [t_{(i+g) \bmod n}, a_0]$. We conclude that the process of a_0 is meeting u_h exactly after g transition occurrences. To determine the f-factor, the process of a_0 of length p has to be repeated until it reaches the length $p_1 = f \cdot n$, which is the least common multiple $lcm(n, g)$ of n and g. Therefore we obtain $p_1 = x \cdot g = lcm(n, g) = \frac{n \cdot g}{gcd(n,g)} = \frac{n \cdot g}{gcd(c+g,g)} = \frac{n \cdot g}{gcd(c,g)} = \frac{n \cdot g}{\Delta}$ and $f = \frac{p_1}{n} = \frac{g}{\Delta}$.

b) Consider again the co-process starting in $[s_i', a_0]$. It reaches the process of a_0 after c transitions in a transition $[t_q, a_0]$. This is repeated x-times and the co-process length is $p' = x \cdot c$. After $[t_q, a_0]$ the co-process passes the transitions $[t_{q \oplus_n g}, a_0], [t_{q \oplus_n (2 \cdot g)}, a_0], \cdots [t_{q \oplus_n (x \cdot g)}, a_0]$ until the cycle is closed with $q \oplus_n (x \cdot g) = q$. The factor $x \cdot g$ is computed in the same way as p_1 in part a), hence $x \cdot g = p_1$ and $x = \frac{n}{\Delta}$. Finally we obtain $p' = c \cdot x = c \cdot \frac{n}{\Delta}$.

c) and d) By the area $A = c \cdot n$ of $\mathcal{C}(g, c, c, c)$ there are $\frac{A}{p'} = \frac{c \cdot n}{p'} = \Delta$ co-processes. The number of g marked places in S_1^{-} are equally distributed over the number Δ of co-processes.   □

In the cycloid $\mathcal{C}(4, 6, 6, 6)$ of Fig. 8 we take $[t_0, a_0]$ as an instance of $[t_i, a_0]$ in the preceding proof. Then $[t_0, a_0]^{\overset{\bullet}{\leftarrow}} = [s_0', a_0]$ and $[s_0', a_0]^{\bullet} = [t_{(0+g+c-1) \bmod p}, a_{c-1}] = [t_{(0+4+6-1) \bmod 10}, a_5] = [t_9, a_5]$. Furthermore, the here starting release message chain is $[t_9, a_5], [t_8, a_4], [t_7, a_3], [t_6, a_2], [t_5, a_1], [t_4, a_0]$ and the new meeting arises after $g = 4$ occurrences. Therefore $n = 10$, $p_1 = lcm(n, g) = \frac{n \cdot g}{gcd(n,g)} = \frac{10 \cdot 4}{gcd(10,4)} = 20$, $f = \frac{p_1}{n} = 2$ and $p' = 30$. The augmented cycloid $\mathcal{C}(4, 6, f \cdot 6, f \cdot 6)$ as shown in Fig. 4 is regular and co-regular. Theorem 22 gives an alternative explanation of the co-car anomaly: in a circular traffic queue $tq(c, g)$ there is a number of g disjoint co-processes of length $p' = \frac{c \cdot n}{\Delta}$. Therefore the length of a recurrent transition sequence is $g \cdot \frac{c \cdot n}{\Delta}$. The cycloid $\mathcal{C}(g, c, f \cdot c, f \cdot c)$ with $f = \frac{g}{\Delta}$ is the smallest co-regular extension of $\mathcal{C}(g, c, c, c)$.

7   Compositions of Cycloids

By the f-factor, introduced in the section before, a regular cycloid is enlarged in the temporal dimension and the process cycles become larger. Using cycloid algebra such extensions can be formulated for the general case of cycloids. This is also a form of composing cycloids.

Theorem 23. *Let $C_1(\alpha, \beta, \gamma, \delta)$ be a cycloid and $a, b \in \mathbb{N}_+$ such that a is a divisor of α and β as well as b is a divisor of γ and δ. Then the equivalence relation $\equiv_1$ of $C_1(\alpha, \beta, \gamma, \delta)$ is included in the equivalence relation $\equiv_2$ of $C_2(\frac{\alpha}{a}, \frac{\beta}{a}, \frac{\gamma}{b}, \frac{\delta}{b})$, more precisely $\equiv_1 \subseteq \equiv_2$.*

Proof. Let be $v = x_2 - x_1$ and π_1 and π_2 the parameter vector function of C_1 and C_2, respectively. Then we have to prove that $\pi_2(v) \in \mathbb{Z}^2$ if $\pi_1(v) \in \mathbb{Z}^2$. This is done by the following deduction, where A_1 is the area of C_1 and $A_2 = \frac{1}{a \cdot b} A_1$ is the area of C_2.

$$\pi_2(v) = \frac{1}{A_2} \begin{pmatrix} \frac{\delta}{b} & \frac{-\gamma}{b} \\ \frac{\beta}{a} & \frac{\alpha}{a} \end{pmatrix} v = \frac{a \cdot b}{A_1} \begin{pmatrix} \frac{\delta}{b} & \frac{-\gamma}{b} \\ \frac{\beta}{a} & \frac{\alpha}{a} \end{pmatrix} v = \frac{1}{A_1} \begin{pmatrix} a \cdot \delta & -a \cdot \gamma \\ b \cdot \beta & b \cdot \alpha \end{pmatrix} v =$$

$$\frac{1}{A_1} \begin{pmatrix} a & 0 \\ 0 & b \end{pmatrix} \begin{pmatrix} \delta & -\gamma \\ \beta & \alpha \end{pmatrix} v = \begin{pmatrix} a & 0 \\ 0 & b \end{pmatrix} \frac{1}{A_1} \begin{pmatrix} \delta & -\gamma \\ \beta & \alpha \end{pmatrix} v = \begin{pmatrix} a & 0 \\ 0 & b \end{pmatrix} \pi_1(v).$$

Since $\pi_1(v) \in \mathbb{Z}^2$ by assumption also $\pi_2(v) \in \mathbb{Z}^2$.   □

Theorem 23 allows to define iterations of cycloids, both with respect to time and space. In a cycloid transitions that are occurring only sequentially, are ordered in time, while those that may occur concurrently model a space-oriented order.

Definition 24. *For a cycloid $C(\alpha, \beta, \gamma, \delta)$ and integers $n, m \in \mathbb{N}_+$ the spacio-temporal iteration is defined by $C(\alpha, \beta, \gamma, \delta)^{[n]}_{[m]} := C(m \cdot \alpha, m \cdot \beta, n \cdot \gamma, n \cdot \delta)$.*

In particular, $C(\alpha, \beta, \gamma, \delta)^{[n]} := C(\alpha, \beta, \gamma, \delta)^{[n]}_{[1]}$ is the temporal iteration and $C(\alpha, \beta, \gamma, \delta)_{[m]} := C(\alpha, \beta, \gamma, \delta)^{[1]}_{[m]}$ is the spacial iteration.

Lemma 25. *The iteration $C^{[n]}(\alpha, \beta, \gamma, \delta)$ of a regular cycloid $C(\alpha, \beta, \gamma, \delta)$ is regular with process length $p^{[n]} = n \cdot p$ if p is the process length of $C(\alpha, \beta, \gamma, \delta)$.*

Proof. If β is a divisor of δ then also of $n \cdot \delta$. If $A^{[n]}$ and A are the areas of the two cycloids then $A^{[n]} = \alpha \cdot n \cdot \delta + \beta \cdot n \cdot \delta = n \cdot A$ and $p^{[n]} = \frac{A^{[n]}}{\beta} = \frac{n \cdot A}{\beta} = n \cdot p$ □

When replacing all items $a \in C$ and co-items $u \in G$ by black tokens and removing the variables x and y in the coloured net $\mathcal{N}_{sym}(c, g)$ of Fig. 2 we obtain the net $\mathcal{N}_{basic}(c, g)$. In [7] is has been shown that it is isomorphic to the cycloid $C_0(g, c) := C(g, c, 1, 1)$. This allows to characterize the two most important cycloids of this article by iterations of this cycloid.

Corollary 26. a) *The cycloid $C_1(g, c) = C(g, c, c, c)$, which is behaviour equivalent to the circular traffic queue $tq\text{-}g(c, g)$ (Definition 20), is isomorphic to the c-fold temporal iteration: $C_0(g, c)^{[c]} = C(g, c, 1, 1)^{[c]} \simeq C(g, c, c, c)$.*

b) *The cycloid $C_2(g, c) = C(g, c, \frac{g \cdot c}{\Delta}, \frac{g \cdot c}{\Delta})$, which is behaviour equivalent to the circular traffic queue $tq(c, g)$, is isomorphic to the $\frac{g \cdot c}{\Delta}$-fold temporal iteration $C_0(g, c)^{[\frac{g \cdot c}{\Delta}]} = C(g, c, 1, 1)^{[\frac{g \cdot c}{\Delta}]} \simeq C(g, c, \frac{g \cdot c}{\Delta}, \frac{g \cdot c}{\Delta})$.*

c) *The cycloid $C_2(g, c)$ is also isomorphic to the $\frac{g}{\Delta}$-fold temporal iteration of the cycloid $C_1(g, c)^{[\frac{g}{\Delta}]} \simeq C_2(g, c)$.*

8   Conclusion

The co-car anomaly arises if the co-traffic items are observable. This step can be compared with the step from classical Petri nets to coloured nets. Using cycloids an alternative and more graphical explanation of the anomaly has been given. The unfolding of the coloured net, modelling circular traffic queues, giving a representation of individual tokens by indistinguishable black tokens, is shown to be isomorphic to a certain class of cycloids. This establishes a connection between two different research areas in the field of Petri nets. On the base of these results elementary synchronization mechanisms will be developed, such as making cooperating processes *wait-free* [1].

References

1. Attiya, H., Welch, J.: Distributed Computing. McGraw-Hill, New York (1998)
2. Girault, C., Valk, R. (eds.): Petri Nets for System Engineering - A Guide to Modelling, Verification and Applications. Springer, Berlin (2003) https://doi.org/10.1007/978-3-662-05324-9
3. Jones, G.A., Jones, J.M.: Elementary Number Theory. SUMS, pp. 163–189. Springer, London (1998). https://doi.org/10.1007/978-1-4471-0613-5_9
4. Petri, C.A.: Nets, Time and Space. Theor. Comput. Sci. **153**, 3–48 (1996)
5. Valk, R.: On the Structure of cycloids introduced by Carl Adam Petri. In: Khomenko, V., Roux, O.H. (eds.) PETRI NETS 2018. LNCS, vol. 10877, pp. 294–314. Springer, Cham (2018). https://doi.org/10.1007/978-3-319-91268-4_15
6. Valk, R.: Formal Properties of Petri's Cycloid Systems. Fundamenta Informaticae **169**, 85–121 (2019)
7. Valk, R.: Circular traffic queues and Petri's cycloids. In: Janicki, R., Sidorova, N., Chatain, T. (eds.) PETRI NETS 2020. LNCS, vol. 12152, pp. 176–195. Springer, Cham (2020). https://doi.org/10.1007/978-3-030-51831-8_9

Tools

Cortado—An Interactive Tool
for Data-Driven Process Discovery
and Modeling

Daniel Schuster[1]([✉])[iD], Sebastiaan J. van Zelst[1,2][iD],
and Wil M. P. van der Aalst[1,2][iD]

[1] Fraunhofer Institute for Applied Information Technology FIT,
Sankt Augustin, Germany
[2] RWTH Aachen University, Aachen, Germany
{daniel.schuster,sebastiaan.van.zelst}@fit.fraunhofer.de,
wvdaalst@pads.rwth-aachen.de

Abstract. Process mining aims to diagnose and improve operational processes. Process mining techniques allow analyzing the event data generated and recorded during the execution of (business) processes to gain valuable insights. Process discovery is a key discipline in process mining that comprises the discovery of process models on the basis of the recorded event data. Most process discovery algorithms work in a fully automated fashion. Apart from adjusting their configuration parameters, conventional process discovery algorithms offer limited to no user interaction, i.e., we either edit the discovered process model by hand or change the algorithm's input by, for instance, filtering the event data. However, recent work indicates that the integration of domain knowledge in (semi-)automated process discovery algorithms often enhances the quality of the process models discovered. Therefore, this paper introduces Cortado, a novel process discovery tool that leverages domain knowledge while incrementally discovering a process model from given event data. Starting from an initial process model, Cortado enables the user to incrementally add new process behavior to the process model under construction in a visual and intuitive manner. As such, Cortado unifies the world of manual process modeling with that of automated process discovery.

Keywords: Process mining · Interactive process discovery · Process trees · Block-structured workflow nets · Process modeling

1 Introduction

Process mining techniques allow analyzing the execution of (business) processes on the basis of event data collected by any type of information system, e.g., SAP, Oracle, and Salesforce. Next to *conformance checking* and *process enhancement*, *process discovery* is one of the three main sub-disciplines in process mining [3]. Process discovery aims to learn a process model from observed process behavior,

© Springer Nature Switzerland AG 2021
D. Buchs and J. Carmona (Eds.): PETRI NETS 2021, LNCS 12734, pp. 465–475, 2021.
https://doi.org/10.1007/978-3-030-76983-3_23

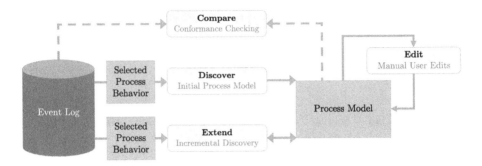

Fig. 1. Overview of Cortado's core functionality. The user discovers an initial model from user-selected process behavior. Next, the obtained process model can be incrementally extended by new process behavior from the event log. In addition, the user can edit the process model anytime and compare it with the event log

i.e., *event data*. Most process discovery algorithms are fully automated. Apart from adjusting configuration parameters of a discovery algorithm, which, for instance, can influence the complexity and quality of the resulting models, the user has no direct option to steer or interact with the algorithm. Further (indirect) user interaction is limited to either changing the input, i.e., the event data fed into the discovery algorithm, or manipulating the output, i.e., the discovered process model. Thus, conventional process discovery algorithms work like a *black box* from the user's perspective.

Several studies indicate that exploiting domain knowledge within (semi-) automated process discovery leads to better process models [4,6]. Recent work has proposed the tool ProDiGy [5], allowing the user to interact with automated process discovery. However, the tool approaches the user-interaction from a *modeling-perspective*, i.e., a human modeler supported by the underlying algorithms (including an *auto-complete* option) is central to the tool and makes the design decisions for the model. Thus, model creation is still a largely manual endeavor.

This paper introduces Cortado, an interactive tool for data-driven process discovery and modeling. Cortado exploits automated process discovery to construct process models from event data in an incremental fashion. Main functionalities of our tool are visualized in Fig. 1. The central idea of Cortado is the incremental discovery of a process model, which is considered to be "under construction". Cortado thereby utilizes the user's domain knowledge by delegating the decision to the user, which is about selecting the observed process behavior that gets added to the process model.

Cortado allows for discovering an initial process model from a user-selected subset of observed process behavior with a conventional process discovery algorithm (see **Discover** in Fig. 1). Alternatively, one can also import a process model into Cortado. Cortado allows incrementally extending an initially given process model, which is either imported or discovered, by adding process behavior that is not yet described by the process model "under construction". Thus, the user is required to incrementally select process behavior from the event log and to perform incremental process discovery. Our incremental discovery algorithm [10] takes the current process model and the selected process behavior and alters the process model such that the selected process behavior is described by the resulting model (see **Extend** in Fig. 1). By incrementally selecting process behavior, the user *guides the incremental process discovery algorithm* by providing feedback on the correctness of the observed event data. The user therefore actively selects the process behavior to be added. Since the incremental process discovery approach allows users to undo/redo steps at any time, they have more control over the process discovery phase of the model compared to conventional approaches. To improve the flexibility of Cortado, a process model editor is also embedded, allowing the user to alter the process model at any time (see **Edit** in Fig. 1). Furthermore, feedback mechanisms are implemented that notify the user of the quality of the discovered process models (see **Compare** in Fig. 1).

The remainder of this paper is structured as follows. In Sect. 2, we briefly introduce background knowledge. In Sect. 3, we explain the algorithmic foundation of Cortado, i.e., the incremental process discovery approach. In Sect. 4, we present our tool and explain its main functionality and usage. In Sect. 5, we briefly describe the underlying implementation. Section 6 concludes the paper.

2 Background

In this section, we briefly explain the concept of event data and present *process trees*, which is the process modeling formalism used by Cortado.

2.1 Event Data

The information systems used in companies, e.g., Customer Relationship Management (CRM) and Enterprise Resource Planning (ERP) systems, track the performed activities during the executions of a process in great detail.

Table 1 presents a simplified example of such event data, i.e., referred to as an *event log*. Each row represents an *event*, a recording related to some *activity instance* of the process. For example, the first row indicates that a fine with identifier A1 was created on July 24, 2006. The next line/event records that the same fine was sent. Note that the corresponding expense for sending the fine was €11.0, the Article of this violation is 157, the vehicle class is A, etc. Multiple rows have the same value for the *Fine*-column, i.e., often referred to as the *case identifier*; all these events are executed for the same instance of the process, e.g., for the same customer, the same patient, the same insurance claim, or, in the given case, for the same fine. We refer to the digital recording of a process instance as a *case*. As such, an *event log*, describes a *collection of cases*. In Cortado, we focus on *trace variants*, i.e., unique sequences of executed activities. For instance, for the fine A1 we observe the trace ⟨*Create Fine, Send Fine*⟩ and for the fine A100 ⟨*Create Fine, Send Fine, Insert Fine Notification, Add penalty, Send for Credit Collection*⟩. Note that, in general, there may be several cases for which the same sequence of activities has been performed.

Table 1. Example (simplified) event data, originating from the *Road Traffic Fine Management Process Event Log* [8]. Each row records an activity executed in the context of the process. The columns record various data related to the corresponding fine and the activity executed.

Fine	Event	Start	Complete	Amount	Notification	Expense	Payment	Article	Vehicle Class	Total Payment
A1	Create Fine	2006/07/24	2006/07/24	35.0				157	A	0.0
A1	Send Fine	2006/12/05	2006/12/05	35.0		11.0		157	A	0.0
A100	Create Fine	2006/08/02	2006/08/02	35.0				157	A	0.0
A100	Send Fine	2006/12/12	2006/12/12	35.0		11.0		157	A	0.0
A100	Insert Fine Notification	2007/01/15	2007/01/15	35.0	P	11.0		157	A	0.0
A100	Add penalty	2007/03/16	2007/03/16	71.5	P	11.0		157	A	0.0
A100	Send for Credit Collection	2009/03/30	2009/03/30	71.5	P	11.0		157	A	0.0
A10000	Create Fine	2007/03/09	2007/03/09	36.0				157	A	0.0
A10000	Send Fine	2007/07/17	2007/07/17	36.0		13.0		157	A	0.0
A10000	Add penalty	2007/10/01	2007/10/01	74.0	P	13.0		157	A	0.0
A10000	Payment	2008/09/09	2008/09/09	74.0	P	13.0	87.0	157	A	87.0
...	...	...	...	...	...	...	...	...	...	...

2.2 Process Trees

We use *process models* to describe the control-flow execution of a process. Some *process modeling formalisms* additionally allow specifying, for instance, what resources execute an activity and what data attributes in the information system might be read or written during the activity execution. In Cortado, we use *process trees* as a process modeling formalism. Process trees are a hierarchical process modeling notation that can be expressed as *sound Workflow nets* (sound WF-nets), i.e., a subclass of Petri nets, often used to model business processes. Process trees are annotated rooted trees and correspond to the class of *block-structured WF-nets*, a subclass of sound WF-nets. Process trees are used in various process discovery algorithms, e.g., the inductive miner [7].

In Fig. 2, we show two simplified models of the road fine management process, which is partially shown in Table 1. Figure 2a shows a sound WF-net. Figure 2b shows a process tree describing the same behavior as the model in Fig. 2a. Both models describe that the *Create Fine* activity is executed first. Secondly, the *Send Fine* activity is optionally executed. Then, the *Insert Fine Notification* activity is performed, followed by a block of concurrent behavior including *Add penalty* and potentially multiple executions of *Payment*.

The semantics of process trees are fairly simple, and, arguably, their hierarchical nature allows one to intuitively reason about the general process behavior. Reconsider Fig. 2b. We refer to the internal vertices as *operators* and use them to specify control-flow relations among their children. The leaves of the tree refer to *activities*. The *unobservable activity* is denoted by τ. In terms of operators, we distinguish four different types: the *sequence operator* $(\rightarrow)$, the *exclusive choice operator* $(\times)$, the *parallel operator* $(\wedge)$, and the *loop operator* $(\circlearrowright)$. The sequence operator $(\rightarrow)$ specifies the execution of its subtrees in the given order from left to right. The exclusive choice operator $(\times)$ specifies that *exactly one* of its subtrees gets executed. The parallel operator $(\wedge)$ specifies that all subtrees get executed in any order and possibly interleaved. The loop operator $(\circlearrowright)$ has

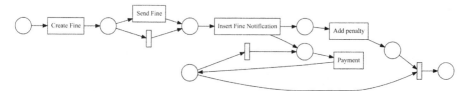

(a) Simple example Petri net (sound WF-net) modeling a road fine management process.

(b) A process tree modeling the same behavior as the Petri net in Figure 2a.

Fig. 2. Two process models, a Petri net (Fig. 2a) and a process tree (Fig. 2b), describing the same process behavior, i.e., a simplified fine management process

exactly two subtrees. The first subtree is called the "do-part", which has to be executed at least once. The second subtree is called the "redo-part", which is optionally executed. If the redo-part gets executed, the do-part is required to be executed again.

3 Algorithmic Foundation

In this section, we briefly describe the algorithmic foundation of Cortado's incremental process discovery approach [10]. Consider Fig. 3, in which we present a schematic overview on said algorithmic foundation.

As an input, we assume a process model M, which is either given initially or the result of a previous iteration of the incremental discovery algorithm. Additionally, a *trace* $\sigma' = \langle a_1, \ldots, a_n \rangle$, i.e., a sequence of executed activities $a_1, \ldots, a_n$, is given. We assume that the trace σ' is not yet part of the language of model M (visualized as $\sigma' \notin L(M)$). Note that σ' is selected by the user. If the incremental procedure has already been executed before, i.e., traces have been already added to the process model in previous iterations, we use those traces as input as well (visualized as $\{\sigma_1, \sigma_2, \sigma_3, \ldots\}$ in Fig. 3). The incremental process discovery algorithm transforms the three input artifacts into a new process model M' that describes the input trace σ' and the previously added traces $\{\sigma_1, \sigma_2, \sigma_3, \ldots\}$. In the next iteration, the user selects a new trace σ'' to be added and the set of previously added traces gets extended, i.e., $\{\sigma_1, \sigma_2, \sigma_3, \ldots\} \cup \{\sigma'\}$.

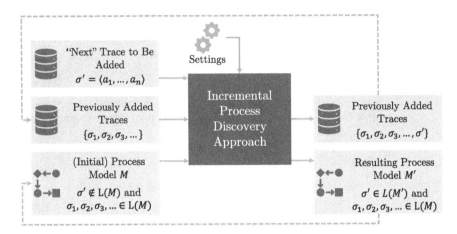

Fig. 3. Schematic overview of incremental process discovery (presented in our earlier work [10]), i.e., the algorithmic foundation of Cortado. Starting with an initial process model M and observed process behavior (a trace σ' capturing a sequence of executed process activities: $a_1, \ldots, a_n$) that is not yet captured by the model, the incremental discovery approach alters the given process model M into a new model M' that additionally accepts the given trace σ'

As mentioned before, Cortado uses process trees as a process model formalism. The incremental discovery approach [10] exploits the hierarchical structure of the input process tree M and pinpoints the subtrees where the given trace σ' deviates from the language described from the model. To identify the subtrees, the process tree is converted into a Petri net and alignments [2] are calculated. Subsequently, the identified subtrees get locally replaced, i.e., M' is a locally modified version of M.

4   Functionalities and User Interface

In this section, we present the main functionalities of Cortado. We do so along the lines of the user interface of Cortado as visualized in Fig. 4.

4.1   I/O Functionalities

Cortado supports various importing and exporting functionalities, which can be triggered by the user by clicking the import/export buttons visualized in the left sidebar, see Fig. 4. Cortado supports importing event data stored in the IEEE eXtensible Event Stream (XES) format [1]. Furthermore, Cortado supports importing process tree models stored as a .ptml-file, for instance, if an initial (manual) process model is available. Process tree model files (.ptml-files) can be generated, e.g., by process mining tools such as ProM[1] and PM4Py[2].

---
[1] https://www.promtools.org.
[2] https://pm4py.fit.fraunhofer.de/.

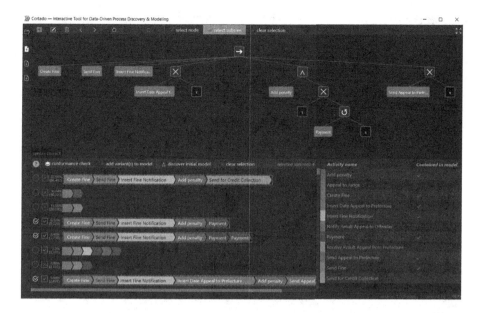

Fig. 4. Screenshot of the graphical user interface of Cortado. In the screenshot, we have loaded the *Road Traffic Fine Management Process Event Log* [8].

Next to importing, Cortado supports exporting the discovered process model both as a Petri net (.pnml-file) and as a process tree (.ptml-file). In short, Cortado offers a variety of I/0 functionalities and, hence, can be easily combined with other process mining tools.

4.2   Visualizing and Editing Process Trees

Cortado supports the visualization and editing of process trees. The "process tree under construction" – either loaded or iteratively discovered – is visualized in the upper half of the tool (Fig. 4). The user can interactively select subtrees or individual vertices of the process tree by clicking an operator or a leaf node. Various edit options, e.g., removing or shifting the selected subtree left or right, are available from the top bar of the application (Fig. 4). Apart from removing and shifting subtrees, the user can also add new subtrees to the process tree. Figure 5 shows a screenshot of the tree editor in detail. In the given screenshot, an inner node, a parallel operator ($\wedge$), is selected. Based on the selected inner node, the user can specify the position where to add a new node in the dropdown-menu by clicking on either **insert left**, **insert right** or **insert below**. In the given screenshot, **insert right** is selected. Next, the user can choose between an activity (a leaf node) or an operator (an inner node). By clicking on one of the options, the new node is added directly to the right of the selected node. In summary, the process tree editor in Cortado allows the user to alter the process tree at any time.

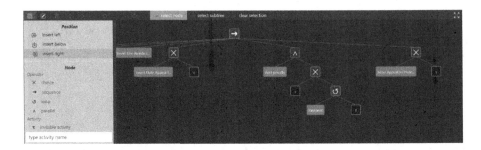

Fig. 5. Screenshot of the process tree editor in Cortado

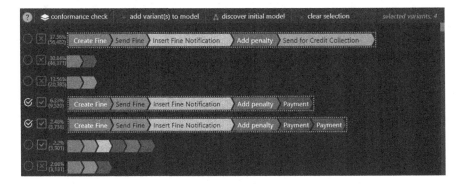

Fig. 6. Screenshot of the trace variants visualization in Cortado. There are two icons to the left of each trace variant. The left icon, a circle or a circle with a check mark, indicates whether a trace variant has been explicitly added to the model by the user. The right icon, a red cross or a green check mark, indicates if the trace variant is accepted by the current process model

4.3 Event Data Interaction

To visualize the loaded event data, Cortado uses trace variants. Clearly, multiple instances of a process can describe the exact same behavior regarding the sequence of observed activities. For example, the most frequently observed sequence of behavior in the *Road Traffic Fine Management Process Event Log* [8] (i.e., used in Fig. 4), describes the sequence: ⟨*Create Fine, Send Fine, Insert Fine Notification, Add Penalty, Send for Credit Collection*⟩. In total, the process behavior of 56,482 fines (37.56% of the total number of recorded fines) follows this sequence of activities.

Trace variants are visualized in Cortado as a sequence of *colored chevrons*. Each activity gets a unique color assigned. For instance, the activity *Create Fine* is assigned a blue color in Fig. 6. Cortado sorts the trace variants based on their frequency of occurrence, descending from top to bottom. By clicking a trace variant, the user "selects a variant". Selection of multiple variants is also supported. In case an initial model does not exist, clicking the `discover initial model` button discovers one from the selected trace variants using the Inductive

Miner [7], a process discovery algorithm that guarantees replay fitness on the given traces and returns a process tree. In case an initial model is present, the selected variants can be "added to the model" by clicking the add variant(s) to model button. In this case, Cortado performs incremental process discovery as described in Sect. 3.

Left to each trace variant, we see statistics about its occurrence in the event log and two icons. The left-most icon, an empty circle or a white check mark, indicates whether or not the trace variant has been explicitly added to the model by the user (Fig. 6). A variant has been explicitly added by the user if either the variant was used to discover an initial model or the variant has been added to an existing model by applying incremental discovery, i.e., the variant was selected and the user pressed the button add variant(s) to model. Note that it is possible that a particular trace variant which was not explicitly selected by the user is described by the process model; however, after incrementally adding further variants to the model, the variant is potentially no longer described. In contrast, Cortado guarantees that explicitly added trace variants are always described by any future model incrementally discovered. However, since Cortado allows for manual tree manipulation at any time, it might be the case that an explicitly added variant is not described anymore by the tree due to manual changes to the process tree.

The right-most icon is either a red cross or a green check mark (Fig. 6). These icons indicate whether a trace variant is described/accepted by the process model, i.e., if a trace variant is in the language of the process model. For the computation of these conformance statistics, we use *alignments* [2]. Therefore, we internally translate the process tree into a Petri net and execute the alignment calculation. For instance, the first three variants in Fig. 6 are not accepted by the current process model, but the last two variants are accepted. Similar to triggering incremental discovery, manipulations of the process tree potentially result in variants that are no longer described by the process model. To assess the conformity of traces after editing the process tree manually, the user can trigger a new conformity check by clicking the conformance check button.

Lastly, Cortado shows an overview list of all activities from the loaded event log. This overview is located in the lower right part of Cortado's user interface (Fig. 4). Besides listing the activity names, Cortado indicates – by using a check mark icon – which activities from the event log are already present in the process model under construction. Thereby, the user gets a quick overview of the status of the incremental discovery.

5   Implementation and Installation

The algorithmic core of Cortado is implemented in Python. For the core process mining functionality, we use the PM4Py[3] library, a python library that contains, for instance, event log handling and conformance checking functionality.

---

[3] https://pm4py.fit.fraunhofer.de/.

The GUI is implemented using web technologies, e.g., we chose the Electron[4] and Angular[5] framework to realize a cross-platform desktop application. For the graphical representation of the process tree and the trace variants we use the JavaScript library d3.js[6].

The tool is available as a desktop application and can be freely downloaded at https://cortado.fit.fraunhofer.de/. The provided archive, a ZIP-file, contains an executable file that will start the tool. Upon starting, the data used within this paper, i.e., Road Traffic Fine Management Process [8], gets automatically loaded. Moreover, the archive contains examples of other event logs available as XES-files in the directory example_event_logs.

6   Conclusion and Future Work

This paper presented Cortado, a novel tool for interactive process discovery and modeling. The tool enables the user to incrementally discover a process model based on observed process behavior. Therefore, Cortado allows to load an event log and visualizes the trace variants in an intuitive manner. Starting from an initial model, which can be either imported or discovered, the user can incrementally add observed process behavior to the process model under construction. Various feedback functionalities, e.g., conformance checking statistics and the activity overview, give the user an overview of the process model under construction anytime. Supporting common file formats such as XES and PNML, Cortado can be easily used with other process mining tools.

In future work, we plan to extend Cortado's functionality in various ways. First, we aim to offer more options for the user to interact with the underlying incremental discovery approach. For example, we plan to allow the user to *lock* specific subtrees during incremental discovery to prevent these from being modified further. We also plan, in case the user changes the tree in the editor, to provide improved and instant feedback on the conformance impact the changes have w.r.t. the loaded event log and the already explicitly added trace variants. However, since the calculation of conformance checking statistics – a crucial part for instant user feedback – is computationally complex, we plan to evaluate the extent to which approximation algorithms [9] can be integrated.

Next to further functionality, we plan to conduct case studies with industry partners. Thereby, we aim to focus on the practical usability of Cortado. The goal is to investigate which interaction options are meaningful and understandable for the user interacting with Cortado.

---

[4] https://www.electronjs.org/.
[5] https://angular.io/.
[6] https://d3js.org/.

References

1. IEEE standard for extensible event stream (XES) for achieving interoperability in event logs and event streams. IEEE Std 1849–2016 pp. 1–50 (2016). https://doi.org/10.1109/IEEESTD.2016.7740858
2. van der Aalst, W., Adriansyah, A., van Dongen, B.: Replaying history on process models for conformance checking and performance analysis. WIREs Data Min. Knowl. Disc. **2**(2), 182–192 (2012). https://doi.org/10.1002/widm.1045
3. van der Aalst, W.M.P.: Process Mining - Data Science in Action, Second Edition. Springer (2016). https://doi.org/10.1007/978-3-662-49851-4
4. Benevento, E., Dixit, P.M., Sani, M.F., Aloini, D., van der Aalst, W.M.P.: Evaluating the effectiveness of interactive process discovery in healthcare: a case study. In: Di Francescomarino, C., Dijkman, R., Zdun, U. (eds.) BPM 2019. LNBIP, vol. 362, pp. 508–519. Springer, Cham (2019). https://doi.org/10.1007/978-3-030-37453-2_41
5. Dixit, P.M., Buijs, J.C.A.M., van der Aalst, W.M.P.: ProDiGy : Human-in-the-loop process discovery. In: 12th International Conference on Research Challenges in Information Science, RCIS 2018, Nantes, France, 29–31 May 2018, pp. 1–12. IEEE (2018). https://doi.org/10.1109/RCIS.2018.8406657
6. Dixit, P.M., Verbeek, H.M.W., Buijs, J.C.A.M., van der Aalst, W.M.P.: Interactive data-driven process model construction. In: Trujillo, J.C. (ed.) ER 2018. LNCS, vol. 11157, pp. 251–265. Springer, Cham (2018). https://doi.org/10.1007/978-3-030-00847-5_19
7. Leemans, S.J.J., Fahland, D., van der Aalst, W.M.P.: Discovering block-structured process models from event logs - a constructive approach. In: Colom, J.-M., Desel, J. (eds.) PETRI NETS 2013. LNCS, vol. 7927, pp. 311–329. Springer, Heidelberg (2013). https://doi.org/10.1007/978-3-642-38697-8_17
8. de Leoni, M., Mannhardt, F.: Road traffic fine management process (Feb 2015). https://doi.org/10.4121/uuid:270fd440-1057-4fb9-89a9-b699b47990f5
9. Schuster, D., van Zelst, S., van der Aalst, W.M.P.: Alignment approximation for process trees. In: Leemans, S., Leopold, H. (eds.) Process Mining Workshops. ICPM 2020. Lecture Notes in Business Information Processing, vol. 406, pp. 247–259. Springer, Cham (2021). https://doi.org/10.1007/978-3-030-72693-5_19
10. Schuster, D., van Zelst, S.J., van der Aalst, W.M.P.: Incremental discovery of hierarchical process models. In: Dalpiaz, F., Zdravkovic, J., Loucopoulos, P. (eds.) RCIS 2020. LNBIP, vol. 385, pp. 417–433. Springer, Cham (2020). https://doi.org/10.1007/978-3-030-50316-1_25

PROVED: A Tool for Graph Representation and Analysis of Uncertain Event Data

Marco Pegoraro$^{(\boxtimes)}$ (ID), Merih Seran Uysal (ID), and Wil M.P. van der Aalst (ID)

Chair of Process and Data Science (PADS) Department of Computer Science,
RWTH Aachen University, Aachen, Germany
{pegoraro,uysal,wvdaalst}@pads.rwth-aachen.de
http://www.pads.rwth-aachen.de/

Abstract. The discipline of process mining aims to study processes in a data-driven manner by analyzing historical process executions, often employing Petri nets. Event data, extracted from information systems (e.g. SAP), serve as the starting point for process mining. Recently, novel types of event data have gathered interest among the process mining community, including uncertain event data. Uncertain events, process traces and logs contain attributes that are characterized by quantified imprecisions, e.g., a set of possible attribute values. The PROVED tool helps to explore, navigate and analyze such uncertain event data by abstracting the uncertain information using behavior graphs and nets, which have Petri nets semantics. Based on these constructs, the tool enables discovery and conformance checking.

Keywords: Process mining · Uncertain data · Partial order · Petri net tool

1 Introduction

Process mining is a branch of process sciences that performs analysis on processes focusing on a log of execution data [4]. From an event log of the process, it is possible to automatically discover a model that describes the flow of a case in the process, or measure the deviations between a normative model and the log.

The primary enabler of process mining analyses is the control-flow perspective of event data, which has been extensively investigated and utilized by researchers in this domain.

Modern information systems supporting processes can enable the extraction of more data perspectives: for instance, it is often possible to retrieve (and thus analyze) additional event attributes, such as the agent (resource) associated with the event, or the cost of a specific activity instance.

We thank the Alexander von Humboldt (AvH) Stiftung for supporting our research interactions.

© Springer Nature Switzerland AG 2021
D. Buchs and J. Carmona (Eds.): PETRI NETS 2021, LNCS 12734, pp. 476–486, 2021.
https://doi.org/10.1007/978-3-030-76983-3_24

Collected event data can be subjected to errors, imprecisions and anomalies; as a consequence, they can be affected by uncertainty. Uncertainty can be caused by many factors, such as sensitivity of sensors, human error, limitations of information systems, or failure of recording systems. The type of uncertainty we consider here is quantified: the event log includes some meta-attributes that describe the uncertainty affecting the event. For instance, the activity label of an event can be unknown, but we might have access to a set of possible activity labels for the event. In this case, in addition to the usual attributes constituting the event in the log, we have a meta-attribute containing a set of activity labels associated with the event. In principle, such meta-attributes can be natively supported by the information system; however, they are usually inferred after the extraction of the event log, in a pre-processing step to be undertaken before the analysis. Often, this pre-processing step necessitates domain knowledge to define, identify, and quantify different types of uncertainty in the event log.

In an event log, regular traces provide a static description of the events that occurred during the completion of a case in the process. Conversely, uncertain process traces contain behavior, and describe a number of possible scenarios that might have occurred in reality. Only one of these scenarios actually took place. It is possible to represent this inherent behavior of uncertain traces with graphical constructs, which are built from the data available in the event log. Some applications of process mining to uncertain data require a model with execution semantics, so to be able to execute all and only the possible real-life scenarios described by the uncertain attributes in the log. To this end, Petri nets are the model of choice to accomplish this, thanks to their ability to compactly represent complex constructs like exclusive choice, possibility of skipping activities, and most importantly, concurrency.

Process mining using uncertain event data is an emerging topic with only a few recent papers. The topic was first introduced in [12] and successively extended in [14]: here, the authors provide a taxonomy and a classification of the possible types of uncertainty that can appear in event data. Furthermore, they propose an approach to obtain measures for conformance score (upper and lower bounds) between uncertain process traces and a normative process model represented by a Petri net.

An additional application of process mining algorithms for uncertain event logs relates to the domain of process discovery. Here, the uncertain log is mined for possible directly-follows relationships between activities: the result, an Uncertain Directly-Follows Graph (UDFG), expresses the minimum and maximum possible strength of the relationship between pair of activities. In turn, this can be exploited to perform process discovery with established discovery techniques. For instance, the inductive miner algorithm can, given the UDFG and some filtering parameters, automatically discover a process model of the process which also embeds information about the uncertain behavior [13].

While the technological sector of process mining software has been flourishing in recent years, no existing tool – to the best of our knowledge – can analyze or handle event data with uncertainty. In this paper, we present a novel tool based on Petri nets, which is capable of performing process mining analyses on uncertain event logs. The PROVED (PRocess mining OVer uncErtain Data) software [2] is able to leverage uncertain mining techniques to deliver insights on the process without the need of discarding the information affected by uncertainty; on the contrary, uncertainty is exploited to obtain a more precise picture of all the possible behavior of the process. PROVED utilizes Petri nets as means to model uncertain behavior in a trace, associating every possible scenario with a complete firing sequence. This enables the analysis of uncertain event data.

The remainder of the paper is structured as follows: Sect. 2 provides an overview of the relevant literature on process mining over uncertainty. Section 3 presents the concept of uncertain event data with examples. Section 4 illustrates the architectural structure of the PROVED tool. Section 5 demonstrates some uses of the tool. Lastly, Sect. 6 concludes the paper.

2   Related Work

The problem of modeling systems containing or representing uncertain behavior is well-investigated and has many established research results. Systems where specific components are associated with time intervals can, for instance, be modeled with time Petri nets [6]. Large systems with more complex timed interoperations between components can be represented by interval-timed coloured Petri nets [3]. Probabilistic effects can be modeled and simulated in a system by formalisms such as generalized stochastic Petri nets [11]. It is important to notice, however, that the focus of process mining over uncertain event data is different: the aim is not to simulate the uncertain behavior in a model, but rather to perform data-driven analyses, some results of which can be represented by (regular) Petri nets.

The PROVED tool contains the implementation of existing techniques for process mining over uncertain event data. In this paper, we will show the capabilities of PROVED in performing the analysis presented in the literature mentioned above. In terms of tool functionalities, constructing a Petri net based on the description of specific behavior – known as synthesis in Petri net research – has some precedents: for instance, from transition systems [8] in the context of process discovery. More relevantly for this paper, the VipTool [5] allows to synthesize Petri nets based on partially ordered objects. While partial order between events is in itself a kind of uncertainty and a consequence of the presence of uncertain timestamps, in this tool paper we extend Petri net synthesis to additional types of uncertainty, and we add process mining functionalities.

3   Preliminary Concepts

The motivating problem behind the PROVED tool is the analysis of uncertain event data. Let us give an example of a process instance generating uncertain data.

An elderly patient enrolls in a clinical trial for an experimental treatment against myeloproliferative neoplasms, a class of blood cancers. The enrollment in this trial includes a lab exam and a visit with a specialist; then, the treatment can begin. The lab exam, performed on the 8th of July, finds a low level of platelets in the blood of the patient, a condition known as thrombocytopenia (TP). At the visit, on the 10th of May, the patient self-reports an episode of night sweats on the night of the 5th of July, prior to the lab exam: the medic notes this, but also hypothesized that it might not be a symptom, since it can be caused not by the condition but by external factors (such as very warm weather). The medic also reads the medical records of the patient and sees that, shortly prior to the lab exam, the patient was undergoing a heparine treatment (a blood-thinning medication) to prevent blood clots. The thrombocytopenia found with the lab exam can then be primary (caused by the blood cancer) or secondary (caused by other factors, such as a drug). Finally, the medic finds an enlargement of the spleen in the patient (splenomegaly). It is unclear when this condition has developed: it might have appeared at any moment prior to that point. The medic decides to admit the patient to the clinical trial, starting 12th of July.

These events are collected and recorded in the trace shown in Table 1 in the information system of the hospital. Uncertain activities are indicated as a set of possibilities. Uncertain timestamps are denoted as intervals. Some event are indicated with a "?" in the rightmost column; these so-called *indeterminate events* have been recorded, but it is unclear if they actually happened in reality. Regular (i.e., non-indeterminate) events are marked with "!". For the sake of readability, the timestamp field only indicates the day of the month.

Table 1. The uncertain trace of an instance of healthcare process used as a running example. For the sake of clarity, we have further simplified the notation in the timestamps column, by showing only the day of the month.

Case ID	Event ID	Timestamp	Activity	Indet. event
ID192	e_1	5	*NightSweats*	?
ID192	e_2	8	$\{PrTP, SecTP\}$	!
ID192	e_3	[4, 10]	*Splenomeg*	!
ID192	e_4	12	*Adm*	!

Throughout the paper, we will utilize the trace of Table 1 as a running example to showcase the functionalities of the PROVED tool.

4   Architecture

This section provides an overview of the architecture of the PROVED tool, as well as a presentation of the libraries and existing software that are used in the tool as dependencies.

Our tool has two distinct parts, a library (implemented in the PROVED Python package) and a user interface allowing to operate the functions in the library in a graphical, non-programmatic way.

The library is written in the Python programming language (compatible with versions 3.6.x through 3.8.x), and is distributed through the Python package manager `pip` [1]. Notable software dependencies include:

- PM4Py [7]: a process mining library for Python. PM4Py is able to provide many classical process mining functionalities needed for PROVED, including importing/exporting of logs and models, management of log objects, and conformance checking through alignments. Notice that PM4Py also provides functions to represent and manage Petri nets.
- NetworkX [10]: this library provides a set of graph algorithms for Python. It is used for the management of graph objects in PROVED.
- Graphviz [9]: this library adds visualization functionalities for graphs to PROVED, and is used to visualize directed graphs and Petri nets.

The aforementioned libraries enable the management, analysis and visualization of uncertain event data, and support the mining techniques of the PROVED toolset here illustrated. An uncertain log in PROVED is a log object of the PM4Py library; here, we will list only the novel functionalities introduced in PROVED, while omitting existing features inherited from PM4PY – such as importing/exporting and attribute manipulation.

4.1 Artifacts

As mentioned earlier, uncertain data contain behavior and, thus, dedicated constructs are necessary to enable process mining analysis. In the PROVED tool, the subpackage `proved.artifacts` contain the models and construction methods of such constructs. Two fundamental artifacts for uncertain data representation are available:

- `proved.artifacts.behavior_graph`: here are collected the PROVED functionalities related to the *behavior graph* of an uncertain trace. Behavior graphs are directed acyclic graphs that capture the variability caused by uncertain timestamps in the trace, and represent the partial order relationships between events. The behavior graph of the trace in Table 1 is shown in Fig. 1. The PROVED library can build behavior graphs efficiently (in quadratic time with respect to the number of events) by using an algorithm described in [15].
- `proved.artifacts.behavior_net`: this subpackage includes all the functionalities necessary to create and utilize *behavior nets*, which are acyclic Petri nets that can replay all possible sequences of activities (called *realizations*) contained in the uncertain trace. Behavior nets allow to simulate all "possible worlds" described by an uncertain trace, and are crucial for tasks such as computing conformance scores between uncertain traces and a normative model. The construction technique for behavior nets is detailed in [14].

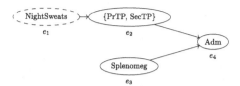

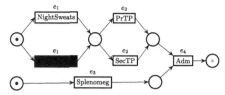

Fig. 1. The behavior graph of the trace in Table 1. All the nodes in the graph are connected based on precedence relationships. Pairs of nodes for which the order is certain are connected by a path in the graph; pairs of nodes for which the order is unknown are pairwise unreachable.

Fig. 2. The behavior net corresponding to the uncertain trace in Table 1. The labels above the transitions show the corresponding uncertain event. The initial marking is displayed; the gray "token slot" represents the final marking. This net is able to replay all and only the sequences of activities that might have happened in reality.

4.2 Algorithms

The algorithms contained in the PROVED tool are categorized in the three subpackages:

- `proved.algorithms.conformance`: this subpackage contains all the functionalities related to measuring conformance between uncertain data and a normative Petri net employing the alignment technique [12, 14]. It includes functions to compute upper and lower bounds for conformance score through exhaustive alignment of the realizations of an uncertain trace, and an optimized technique to efficiently compute the lower bound.
- `proved.algorithms.discovery`: this subpackage contains the functionalities needed to perform process discovery over uncertain event logs. It offers functionalities to compute a UDFG, a graph representing an extension of the concept of directly-follows relationship on uncertain data; this construct can be utilized to perform inductive mining [13].
- `proved.algorithms.simulation`: this subpackage contains some utility functions to simulate uncertainty within an existing event log. It is possible to add separately the different kinds of uncertainty described in the taxonomy of [14], while fine-tuning the dictionary of activity labels to sample and the amplitude of time intervals for timestamps.

4.3 Interface

Some of the functionalities of the PROVED tool are also supported by a graphical user interface. The PROVED interface is web-based, utilizing the Django framework in Python for the back-end, and the Bootstrap framework in Javascript and HTML for the front end. The user interface includes the PROVED library as a dependency, and is, thus, completely decoupled from the logic and algorithms in it. We will illustrate some parts of the user interface in the next section.

5   Usage

In this section, we will outline how to install and use our tool. Firstly, let us focus on the programmatic usage of the Python library.

The full source code for PROVED can be found on the GitHub project page[1]. Once installed Python on the system, PROVED is available through the `pip` package manager for Python, and can be installed with the terminal command `pip install proved`, which will also install all the necessary dependencies.

Thanks to the import and export functionalities inherited from PM4Py, which has full XES [16] certification, it is possible to start uncertain logs analysis easily and compactly. Let us examine the following example:

```
1  from pm4py.objects.log.importer.xes import importer as x_importer
2  from proved.artifacts import behavior_graph, behavior_net
3
4  uncertain_log = x_importer.apply('uncertain_event_log.xes')
5  uncertain_trace = uncertain_log[0]
6  beh_graph = behavior_graph.BehaviorGraph(uncertain_trace)
7  beh_net = behavior_net.BehaviorNet(beh_graph)
```

In this code snippet, an uncertain event log is imported, then the first trace of the log is selected, and the behavior graph and behavior net of the trace are obtained. Nodes and connections of behavior graphs and nets can be explored using the igraph functionalities and the PM4Py functionalities. We can also visualize both objects with Graphviz, obtaining graphics akin to the ones in Figs. 1 and 2.

```
1  from pm4py.objects.petri.importer import importer as p_importer
2  from proved.algorithms.conformance.alignments import alignment_bounds_su
3
4  net, i_mark, f_mark = p_importer.apply('model.pnml')
5
6  alignments = alignment_bounds_su_log(uncertain_log, net, i_mark, f_mark)
```

In the snippet given above, we can see the code that allows to compute upper and lower bounds for conformance score of all the traces in the uncertain log against a reference model that we import, utilizing the technique of alignments [14]. For each trace in the log, a pair of alignment objects is computed: the first one corresponds to an alignment with a cost equal to the lower bound for conformance cost, while the second object is an alignment with the maximum possible conformance cost. The object **alignments** is a list with one of such pairs for each trace in the log.

Let us now see some visual examples of the usages of the PROVED tool user interface[2]. The graphical tool can be executed in a local environment by starting the Django server in a terminal with the command `python manage.py runserver`.

Upon opening the tool and loading an uncertain event log, we are presented with a dashboard that summarizes the main information regarding the event log, as shown in Fig. 3.

---

[1] Available at https://github.com/proved-py/proved-core/.
[2] Available at https://github.com/proved-py/proved-app/.

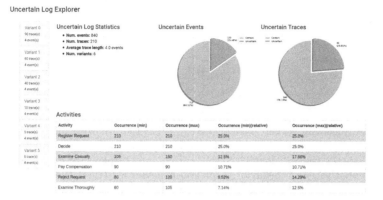

Fig. 3. The dashboard of the PROVED user interface. This screen contains general information regarding an uncertain event log, including the list of uncertain variants, the number of instances of each activity label (minimum and maximum), and statistics regarding the frequency of uncertain events and uncertain traces in the log.

In the center panel of the dashboard, we can see statistics regarding the uncertain log. On the top left, we find basic statistics such as the size of the log in the number of events and traces, the average trace length, and the number of uncertain variants. Note that the classical definition of *variant* is inconsistent in uncertain event logs; rather, uncertain variants group together traces which have mutually isomorphic behavior graphs [14]. We can also find pie charts indicating the percentage of uncertain events in the log (events with at least one uncertain attribute) and the percentage of uncertain traces in the log (traces with at least one uncertain event).

On the bottom, a table reports the counts of the number of occurrences for each activity label in the event log. Because of uncertainty on activity labels and indeterminate events, there is a minimum and maximum amount of occurrences of a specific activity label. The table reports both figures. There are two other tables in the dashboard, the Start Activities table and the End Activities table. Both are akin to the activity table depicted, but separately list activity labels appearing in the first or last event in a trace.

Upon clicking on one of the uncertain variants listed on the left, the user can access the graphical representation of the variant. It is possible to visualize both the behavior graph and the behavior net: the former is depicted in Fig. 4. The figure specifically shows information related to the trace depicted in Table 1.

Next to the variant menu on the left, we now have a trace menu, listing all the traces belonging to that uncertain variant. Clicking on a specific trace, the user is presented with data related to it, including a tabular view of the trace similar to that of Table 1, and a Gantt diagram representation of the trace. Similarly to the behavior graph, the Gantt diagram shows time information in a graphical manner; but, instead of showing the precedence relationship between

Fig. 4. The uncertain variant page of the PROVED tool, showing information regarding the variant obtained from the trace in Table 1. For a variant in an uncertain log, this page lists the traces belonging to that variant, and displays the graphical representations for that variant – behavior graph and behavior net (the latter is not displayed, but can be accessed through the tab on the top).

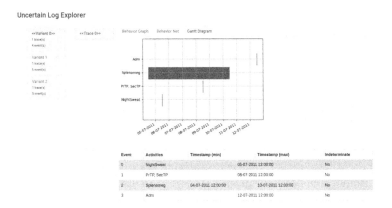

Fig. 5. Visualization dedicated to a specific trace in the PROVED tool, showing information related to the trace in Table 1. It is possible to see details on each event and on the uncertainty that might affect them, as well as a visualization showing the time relationship between uncertain event in scale.

events, it shows the time information in scale, representing the time intervals on an absolute scale. This visualization is presented in Fig. 5.

The interface allows the user to explore the features of an uncertain log, to "drill down" to variants, traces, event and single attributes, and visualize the uncertain data in a graphical manner without the need to resort to coding in Python.

Lastly, the menu on the left also allows for loading a Petri net, and obtaining alignments on uncertain event data.

As shown above, every uncertain trace can be represented by a behavior net. A conformance score can be computed between such behavior nets and a normative process model also represented by a Petri net: Fig. 6 illustrate the results of such alignment. For a given behavior net, two alignments are provided, together with the respective cost: one, showing a best-case scenario, and the other showing a worst-case scenario. This enables diagnostics on uncertain event data.

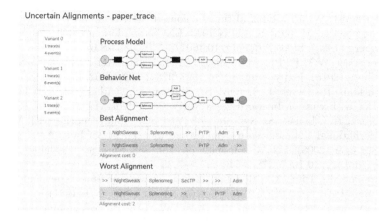

Fig. 6. Visualization of alignments of the uncertain trace in Table 1 and a normative process model. In this case, the optimal alignment in the best case scenario perfectly fits the model, while in the worst case scenario we have an alignment cost of 2, caused by one move on model and one move on log.

6   Conclusions

In many real-world scenarios, the applicability of process mining techniques is severely limited by data quality problems. In some situations, these anomalies causing an erroneous recording of data in an information system can be translated in uncertainty, which is described through meta-attributes included in the log itself. Such uncertain event log can still be analyzed and mined, thanks to specialized process mining techniques. The PROVED tool is a Python-based software that enables such analysis. It provides capabilities for importing and exporting uncertain event data in the XES format, for obtaining graphical representations of data that can capture the behavior generated by uncertain attributes, and for computing upper and lower bounds for conformance between uncertain process traces and a normative model in the form of a Petri net.

Future work on the tool includes the definition of a formal XES language extension with dedicated tags for uncertainty meta-attributes, the further development of front-end functionalities to include more process mining capabilities, and more interactive objects in the user interface. Moreover, the research effort on uncertainty affecting the data perspective of processes can be integrated with the model perspective, blending uncertainty research with formalisms such as stochastic Petri nets.

References

1. Pip - PyPi. https://pypi.org/project/pip/. Accessed 03 Feb 2020
2. The PROVED project on GitHub. https://github.com/proved-py/. Accessed 03 Feb 2021

3. van der Aalst, W.M.P.: Interval timed coloured Petri nets and their analysis. In: Ajmone Marsan, M. (ed.) ICATPN 1993. LNCS, vol. 691, pp. 453–472. Springer, Heidelberg (1993). https://doi.org/10.1007/3-540-56863-8_61
4. van der Aalst, W.M.P.: Process Mining: Data Science in Action, pp. 3–23. Springer, Heidelberg (2016). https://doi.org/10.1007/978-3-662-49851-4_1
5. Bergenthum, R., Desel, J., Lorenz, R., Mauser, S.: Synthesis of petri nets from scenarios with viptool. In: van Hee, K.M., Valk, R. (eds.) Bergenthum, R., Desel, J., Lorenz, R., Mauser, S.: Synthesis of Petri nets from scenarios with VipTool. In: International Conference on Applications and Theory of Petri Nets. pp. 388–398. Springer (2008). LNCS, vol. 5062, pp. 388–398. Springer, Heidelberg (2008). https://doi.org/10.1007/978-3-540-68746-7_25
6. Berthomieu, B., Diaz, M.: Modeling and verification of time dependent systems using time Petri nets. IEEE Trans. Software Eng. **17**(3), 259 (1991)
7. Berti, A., van Zelst, S.J., van der Aalst, W.M.P.: Process mining for python (PM4Py): bridging the gap between process and data science. In: ICPM Demo Track (CEUR 2374), pp. 13–16 (2019)
8. Carmona, J., Cortadella, J., Kishinevsky, M.: Genet: a tool for the synthesis and mining of Petri nets. In: 2009 Ninth International Conference on Application of Concurrency to System Design, pp. 181–185. IEEE (2009)
9. Ellson, J., Gansner, E., Koutsofios, L., North, S.C., Woodhull, G.: Graphviz: open source graph drawing tool. In: Mutzel, P., Jünger, M., Leipert, S. (eds.) GD 2001. LNCS, vol. 2265, pp. 483–484. Springer, Heidelberg (2002). https://doi.org/10.1007/3-540-45848-4_57
10. Hagberg, A., Swart, P., S Chult, D.: Exploring network structure, dynamics, and function using NetworkX. Technical report, Los Alamos National Lab. (LANL), Los Alamos, NM (United States) (2008)
11. Marsan, M.A., Balbo, G., Conte, G., Donatelli, S., Franceschinis, G.: Modelling with generalized stochastic Petri nets. ACM SIGMETRICS Perform. Eval. Rev. **26**(2), 2 (1998)
12. Pegoraro, M., van der Aalst, W.M.P.: Mining uncertain event data in process mining. In: 2019 International Conference on Process Mining (ICPM), pp. 89–96. IEEE (2019)
13. Pegoraro, M., Uysal, M.S., van der Aalst, W.M.P.: Discovering process models from uncertain event data. In: Di Francescomarino, C., Dijkman, R., Zdun, U. (eds.) BPM 2019. LNBIP, vol. 362, pp. 238–249. Springer, Cham (2019). https://doi.org/10.1007/978-3-030-37453-2_20
14. Pegoraro, M., Uysal, M.S., van der Aalst, W.M.P.: Conformance checking over uncertain event data. arXiv preprint - arXiv:2009.14452 (2020)
15. Pegoraro, M., Uysal, M.S., van der Aalst, W.M.P.: Efficient time and space representation of uncertain event data. Algorithms **13**(11), 285–312 (2020)
16. Verbeek, H.M.W., Buijs, J.C.A.M., van Dongen, B.F., van der Aalst, W.M.P.: XES, XESame, and ProM 6. In: Soffer, P., Proper, E. (eds.) CAiSE Forum 2010. LNBIP, vol. 72, pp. 60–75. Springer, Heidelberg (2011). https://doi.org/10.1007/978-3-642-17722-4_5

Author Index

Abbes, Samy   423
Amat, Nicolas   164

Bai, Yunjun   379
Bergenthum, Robin   399
Bernardinello, Luca   3
Berthomieu, Bernard   164
Bouvier, Pierre   339
Briday, Mikaël   55
Burke, Adam   312

Dal Zilio, Silvano   164
Devillers, Raymond   274
Didriksen, Martin   118

Esparza, Javier   141

Garavel, Hubert   339
Gieseking, Manuel   95

Haddad, Serge   76
Haustermann, Michael   230
Hélouët, Loïc   33

Jensen, Peter G.   118
Jezequel, Loïg   360
Jiao, Li   379
Jønler, Jonathan F.   118

Kang, Jiawen   379
Katona, Andrei-Ioan   118

Lama, Sangey D. L.   118
Leemans, Sander J. J.   312

Leroux, Jérôme   17
Lime, Didier   76, 360
Lottrup, Frederik B.   118

Miklos, Zoltan   33
Moldt, Daniel   230
Mosteller, David   230

Parrot, Rémi   55
Pegoraro, Marco   476

Raskin, Mikhail   141
Roux, Olivier H.   55, 76

Schuster, Daniel   465
Sérée, Bastien   360
Shajarat, Shahab   118
Singh, Rituraj   33
Srba, Jiří   118

Tredup, Ronny   253, 292

Uysal, Merih Seran   476

Valk, Rüdiger   443
van der Aalst, Wil M. P.   208, 465, 476
van Zelst, Sebastiaan J.   465

Wallner, Sophie   186
Welzel, Christoph   141
Wolf, Karsten   186
Würdemann, Nick   95
Wynn, Moe Thandar   312

Printed in the United States
by Baker & Taylor Publisher Services